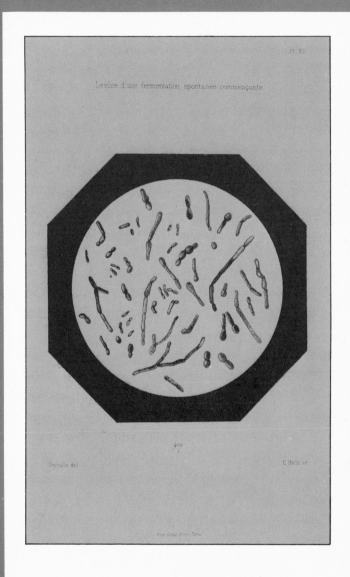

Plate xii from "Études sur la Bière"
depicting yeast grown on wort but identified as
responsible for spontaneous fermentation of wine

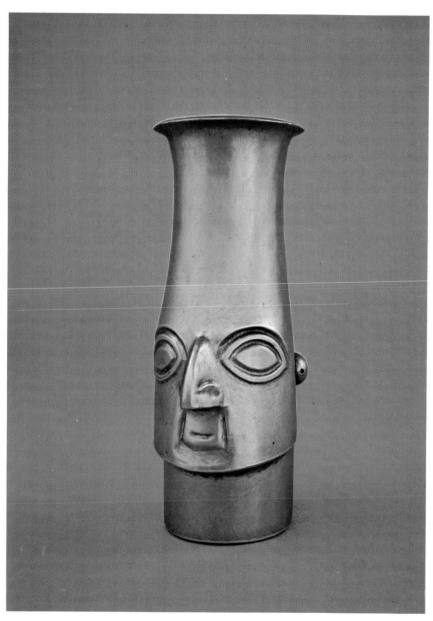

Inca Period (A.D. 1300–1532) gold beaker in the form of human head. Such beakers were used for serving maize beer. From the Ica Valley, south coast of Peru. Photo courtesy of the Art Institute of Chicago.

Fermented Food Beverages In Nutrition

EDITED BY

CLIFFORD F. GASTINEAU

Endocrinology, Nutrition, and Internal Medicine
The Mayo Clinic
Rochester, Minnesota

WILLIAM J. DARBY

The Nutrition Foundation, Inc.
New York, New York

THOMAS B. TURNER

The Johns Hopkins University
Baltimore, Maryland

ACADEMIC PRESS

New York   San Francisco   London   1979
A Subsidiary of Harcourt Brace Jovanovich, Publishers

ACADEMIC PRESS, INC.
111 Fifth Avenue, New York, New York 10003

United Kingdom Edition published by
ACADEMIC PRESS, INC. (LONDON) LTD.
24/28 Oval Road, London NW1 7DX

Library of Congress Cataloging in Publication Data
Main entry under title:

Fermented food beverages in nutrition.

(The Nutrition Foundation series)
Based in part on edited and revised papers prepared
for the International Symposium on Fermented Food
Beverages in Nutrition held in Rochester, Minnesota,
June 15–17, 1977.
Includes bibliographies and index.
1. Alcohol––Physiological effect––Congresses.
2. Alcoholic beverages––Therapeutic use––Congresses.
3. Nutrition––Congresses. 4. Beer––Physiological
effect––Congresses. 5. Wine––Physiological effect––
Congresses. 6. Alcoholism––Congresses. 7. Fermen-
tation––Congresses. 8. Drinking habits––Congresses.
I. Gastineau, Clifford Felix, Date II. Darby,
William Jefferson, Date III. Turner, Thomas
Bourne. IV. Series: Nutrition Foundation. [DNLM:
1. Alcoholic beverages––Congresses. 2. Alcoholism––
Congresses. WB444 S989f 1977]
QP801.A3F47   613.8'1   78–22526
ISBN 0–12–277050–1

CONTENTS

LIST OF CONTRIBUTORS   xv

PREFACE   xix

I HISTORICAL OVERVIEW

1 FERMENTED BEVERAGES IN ANTIQUITY
P. Ghalioungui

I. The Beginnings	4
II. Early Egypt	5
III. Recognition of Excess	12
IV. Pre-Islamic Near East	13
V. Changing Attitudes with Islam	14
References	18

2 MAIZE BEER IN THE ECONOMICS, POLITICS, AND RELIGION OF THE INCA EMPIRE
C. Morris

I. Native American Fermented Food Beverages	21
II. The Preparation of Maize Beer	22
III. Exchange in the Andes: Reciprocity and Redistribution	25
IV. Maize Beer (Chicha) in Pre-Spanish Times	26
V. State Management of the Brewing	27
VI. Social and Cultural Importance of Fermented Beverages	32
References	34

3 NUTRITIONALLY SIGNIFICANT INDIGENOUS FOODS INVOLVING AN ALCOHOLIC FERMENTATION
Keith H. Steinkraus

I. Introduction	36
II. Alcoholic Foods in Which Sugars Are the Major Fermentable Substrate	37
III. Alcoholic Foods in Which Saliva Is the Amylolytic Agent	41
IV. Alcoholic Food Fermentations Involving an Amylolytic Mold and Yeast	43

V. Alcoholic Foods in Which Starch Hydrolysis Involves a
   Malting (Germination) Step   46
VI. Alcoholic Fermentations Involving Use of a *Koji* for Starch Hydrolysis   52
VII. Alcoholic Milk Beverages   54
VIII. Significance of Indigenous Alcoholic Foods in Human Nutrition   55
IX. Summary   56
   References   57

4 THE NUTRIENT CONTRIBUTIONS OF FERMENTED BEVERAGES
W. J. Darby

I. Nature of Nutritional Contribution of Beverages   61
II. Time-Honored Precepts of Use   62
III. Excesses and Nutritional Deficiencies   66
IV. Direct Nutrient Contributions of Fermented Beverages   67
V. Clarification, Filtration, and Distillation   69
VI. Trace Elements   71
VII. Organic Compounds in Fermented Beverages   71
VIII. Summary   72
   References   73
   Editorial Comment   75

II FERMENTATION

5 A HISTORICAL PERSPECTIVE ON FERMENTATION BIOCHEMISTRY AND NUTRITION
E. Neige Todhunter

I. Introduction   83
II. Early Fermented Beverages   85
III. Health Value of Fermented Beverages   86
IV. Early Chemistry of Fermentation   90
V. Yeast: A Living Organism   91
VI. Discovery of Zymase   93
VII. Chemistry of Yeast-Juice Action   94
VIII. Yeast: Its Nutritional Aspects   94
   References   96
   Editorial Comment   98

6 BIOCHEMISTRY OF FERMENTATION
L. Krampitz

I. Introduction   99
II. The Alcoholic Fermentation: A Look into the Future   101
III. Glycerol Fermentation   103
IV. The Citric Acid Fermentation   104
   Reference   106

7 NUTRIENT SYNTHESIS BY YEAST
Harry P. Broquist

    I. Introduction 107
    II. Production of Nutrients by Yeast 110
    III. Conclusions 118
    References 119

8 THE BIOCHEMICAL PROCESSES INVOLVED IN WINE
FERMENTATION AND AGING
Maynard A. Amerine

    I. Introduction 121
    II. Products 122
    III. By-Products of Fermentation 123
    IV. Indirect By-Products 126
    V. Higher Alcohols 126
    VI. Heat 129
    VII. By-Products of Processing 129
    VIII. The Malo-Lactic Fermentation 130
    IX. Research Needs 131
    References 132

9 THE BREWING OF BEER
W. A. Hardwick

    I. Brewing Constituents 133
    II. Steps in Brewing 134
    III. Terminology 138
    Editorial Comment 140

III CONSUMPTION OF BEER AND WINE

10 NATIONAL PATTERNS OF CONSUMPTION AND
PRODUCTION OF BEER
Philip C. Katz

    I. Introduction 143
    II. History of Beer 144
    III. Production Trends 146
    IV. Consumption Trends 148
    V. World Production and Consumption 152
    VI. Conclusions 153
    References 154
    Supplementary Reading 154

11 PRODUCTION AND CONSUMPTION OF WINE: FACTS,
    OPINIONS, TENDENCIES
    Werner Becker

    I. Introduction   157
    II. Production and Distribution   163
    III. Consumption   170
    IV. Conclusions   183
    References   186

12 THE WINE INDUSTRY AND THE CHANGING ATTITUDES OF
    AMERICANS: AN OVERVIEW
    J. A. De Luca

    I. In the Beginning   188
    II. Prohibition Attitudes   189
    III. Recovery following Repeal   190
    IV. The Present   191
    V. The American Wine Consumer   191
    References   193

            IV METABOLISM AND THERAPEUTIC USE OF
                    ALCOHOLIC BEVERAGES

13 ABSORPTION, METABOLISM, AND EXCRETION OF ETHANOL
    INCLUDING EFFECTS ON WATER BALANCE AND
    NUTRITIONAL STATUS
    Robert E. Olson

    I. Introduction   197
    II. Absorption of Ethanol   198
    III. Distribution of Ethanol in the Body   199
    IV. Metabolism of Ethanol   199
    V. Excretion of Ethanol   202
    VI. Endogenous Biosynthesis of Ethanol in Animals   202
    VII. Effect of Ethanol Intake on Fluid Balance   205
    VIII. Effect of Ethanol Intake on Nutritional Status   206
    IX. Research Needs   209
    X. Summary   209
    References   210

14 THE ENERGY VALUE OF ALCOHOL
    R. Passmore

    I. Introduction   213
    II. Experiments of Atwater and Benedict   214   III. Utilization of Alcohol by Normal Man   220

IV. Utilization of Alcohol by Alcoholics   222
References   223

15 ROLE OF ALCOHOLIC BEVERAGES IN GERONTOLOGY
Donald M. Watkin

I. Introduction   226
II. Alcohol Abuse and Gerontology   226
III. Prevention of Alcohol Abuse   228
IV. Results of Carefully Controlled Alcoholic Beverage Consumption
Studies among Institutionalized Elderly   230
V. Discussion   235
VI. Research Needs   239
VII. Conclusion   240
References   240
Editorial Comment   243

16 MEDICATIONS, DRUGS, AND ALCOHOL
Frank A. Seixas

I. Absorption Effects   246
II. Congeners   246
III. Metabolism   247
IV. Microsomal Ethanol Oxidizing System   247
V. Disulfiram (Antabuse)   248
VI. Interaction of Alcohol with Other Psychoactive Drugs   249
VII. Depressants   250
VIII. Conclusions   251
References   251
Editorial Comment   255

17 INCORPORATING ALCOHOLIC BEVERAGES INTO THERAPEUTIC DIETS: SOME POTENTIALITIES AND PROBLEMS
K. M. West

I. Some Problems   257
II. Should Alcohol Be Allowed in the Diets of Fat People?   260
III. Does Alcohol Stimulate Weight Gain in the Undernourished?   261
References   262

18 ALCOHOL AND KETOGENESIS
F. John Service

I. Introduction   263
II. Alcoholic Ketoacidosis   264
III. Ethanol and Ketogenesis   264
References   270
Editorial Comment Concerning Alcohol Use in Diabetes   273

V EFFECTS OF MISUSE AND EXCESS

19 EFFECTS OF ALCOHOL ON GASTROINTESTINAL AND
   PANCREATIC FUNCTION IN ALCOHOLICS
   Esteban Mezey and Charles H. Halsted

    I. Introduction   277
    II. Absorption and Metabolism of Ethanol by the Gastrointestinal Tract   278
    III. Alcohol and the Stomach   281
    IV. Alcoholism and the Small Intestine   284
    V. Alcoholism and the Pancreas   291
    VI. Summary of Research Needs   296
    References   297

20 LIVER ABNORMALITIES IN ALCOHOLISM: ALCOHOL
   CONSUMPTION AND NUTRITION
   Esteban Mezey and Patricia B. Santora

    I. Introduction   303
    II. Alcohol Consumption and Beverage Choice   304
    III. Malnutrition and Liver Disease   311
    IV. Summary of Research Needs   313
    References   314
    Editorial Comment   316

21 EFFECTS OF ALCOHOL ON THE CARDIOVASCULAR SYSTEM
   Arthur L. Klatsky

    I. Introduction and Historical Review   317
    II. Effects of Alcohol on Cardiovascular Physiology, Biochemistry,
       and Structure   319
    III. Alcohol and Cardiovascular Disease   323
    IV. Summary and Research Needs   335
    References   336

22 EFFECTS OF ALCOHOL ON THE NERVOUS SYSTEM
   Pierre M. Dreyfus

    I. General Considerations   342
    II. Disorders Associated with High Blood Alcohol Levels   343
    III. Disorders Associated with Zero or Diminishing Blood Alcohol Levels   344
    IV. Neurological Disorders of Nutritional Cause Associated with Chronic
       Alcoholism   347
    V. Neurological Disorders of Undetermined Cause Associated with
       Chronic Alcoholism   351
    References   356

23 ALCOHOLISM: HOW DO YOU GET IT?
    R. M. Morse

    I. Common Concepts in Etiology of Alcoholism — 359
    II. Is Alcoholism Associated with Underlying Psychopathology? — 361
    III. Role of the Pleasurable Experience from Alcohol — 364
    IV. Habituation, Dependence, Problems — 365
    V. Tolerance and Need — 366
    VI. Is Everyone Vulnerable? — 367
    References — 367
    Editorial Comment — 369

24 NATURAL HISTORY OF ALCOHOL DEPENDENCE
    Jack H. Mendelson and Nancy K. Mello

    I. Introduction — 371
    II. Procedures — 373
    III. Results — 376
    IV. Discussion — 386
    V. Research Needs — 390
    VI. Summary — 391
    Appendix I: Social and Drinking History — 392
    Appendix II: Neurological Rating Form — 393
    References — 394

25 NUTRITIONAL STATUS OF ALCOHOLICS BEFORE AND AFTER
    ADMISSION TO AN ALCOHOLISM TREATMENT UNIT
    Richard D. Hurt, Ralph A. Nelson, E. Rolland Dickson, John A.
    Higgins, and Robert M. Morse

    I. Introduction: Skid Row Alcoholic versus the More Common Alcoholic — 397
    II. The Present Study — 399
    III. Methods for Nutritional Data — 400
    IV. Results — 401
    V. Conclusions — 407
    References — 407
    Editorial Comment — 408

26 THIAMINE STATUS OF AUSTRALIAN ALCOHOLICS
    Beverley Wood and Kerry J. Breen

    I. Introduction — 409
    II. Clinical Studies — 411
    III. Biochemical Studies — 412
    IV. Etiology of Thiamine Deficiency in Alcoholism — 419
    V. Thiamine Status of the Australian Population — 421
    VI. Summary — 422
    References — 423
    Editorial Comment — 426

27 CANCER AND ALCOHOLIC BEVERAGES
A. J. Tuyns

    I. Introduction 427
    II. Alcohol and Carcinogenesis 428
    III. Human Cancer in Relation to Alcohol Consumption:
        The Epidemiological Evidence 428
    IV. Summary 434
    References 435
    Editorial Comment 437

28 ALCOHOL IN PREGNANCY AND ITS EFFECTS ON OFFSPRING
Eileen M. Ouellette

    I. Introduction 439
    II. Historical Review 440
    III. Modern Studies 441
    IV. Discussion 445
    V. Research Needs 448
    References 449
    Editorial Comment 454

VI AN EXPERIMENTAL MODEL

29 MINIATURE SWINE AS A MODEL FOR THE STUDY OF HUMAN
ALCOHOLISM: THE WITHDRAWAL SYNDROME
*M. E. Tumbleson, D. P. Hutcheson, J. D. Dexter, and
C. C. Middleton*

    I. Introduction 457
    II. Experiments on Physical Dependence 460
    III. Discussion 471
    IV. Summary and Conclusions 472
    References 473

VII SOCIOECONOMIC CONSIDERATIONS

30 THE MEDICAL COSTS OF EXCESSIVE USE OF ALCOHOL
Milton Terris

    I. Cirrhosis of the Liver 481
    II. Accidents 482
    III. Suicide, Homicide, and Assault 483
    IV. Cancer 484
    V. Mental Disorders 485
    VI. Other Disorders 485

VII. Conclusions 485
References 486
Editorial Comment 488

31 SOCIOECONOMIC CONSIDERATIONS AND CULTURAL ATTITUDES: AN ANALYSIS OF THE INTERNATIONAL LITERATURE
M. H. Brenner

I. Introduction 489
II. The Supply-Demand-Stress Model 491
III. Indicators of Key Economic, Sociocultural, and Stress Variables 492
IV. Overview and Conceptual Organization of the Effects of Government Policies 494
V. Discussion 498
References 499
Editorial Comment 504

APPENDIX: Compounds Identified in Whiskey, Wine, and Beer 505

INDEX 519

LIST OF CONTRIBUTORS

Numbers in parentheses indicate the pages on which the authors' contributions begin.

Maynard A. Amerine (121), University of California, Davis, California 95616

Werner Becker (157), Deutscher Weinbauverband, Generalsekretaer Heussallee 26, D-5300 Bonn 1, West Germany

Kerry J. Breen (409), University of Melbourne, Department of Medicine, St. Vincent's Hospital, Melbourne, Australia

M. H. Brenner (489), Division of Operations Research, The Johns Hopkins University School of Hygiene and Public Health, Baltimore, Maryland 21218

Harry P. Broquist (107), Department of Biochemistry, Vanderbilt University School of Medicine, Nashville, Tennessee 37232

W. J. Darby (61), The Nutrition Foundation, Inc., New York, New York 10017

J. A. De Luca (187), Wine Institute, San Francisco, California 94108

J. D. Dexter (457), Department of Neurology, School of Medicine, University of Missouri—Columbia, Columbia, Missouri 65201

E. Rolland Dickson (397), Department of Medicine, Mayo Clinic and Mayo Foundation, Mayo Medical School, Rochester, Minnesota 55901

Pierre M. Dreyfus (341), Department of Neurology, School of Medicine, University of California, Sacramento, California 95817

P. Ghalioungui (3), 18 Gezireh Street, Zamalek, Cairo, Egypt

Charles H. Halsted (277), Section of Gastroenterology, School of Medicine, University of California, Davis, California 95616

W. A. Hardwick (133), Anheuser-Busch, Inc., St. Louis, Missouri 63118

John A. Higgins (397), Department of Medicine, Mayo Clinic and Mayo Foundation, Mayo Medical School, Rochester, Minnesota 55901

Richard D. Hurt (397), Department of Medicine, Mayo Clinic and Mayo Foundation, Mayo Medical School, Rochester, Minnesota 55901

D. P. Hutcheson* (457), Sinclair Comparative Medicine Research Farm, College of Veterinary Medicine and School of Medicine, University of Missouri, Columbia, Missouri 65201

Philip C. Katz (143), Research Services, United States Brewers Association, Inc., Washington, D.C. 20006

Arthur L. Klatsky (317), Division of Cardiology, Kaiser–Permanente Medical Center, Oakland, California 94611

L. Krampitz (99), Department of Microbiology, Case Western Reserve University, Cleveland, Ohio 44106

Nancy K. Mello (371), Department of Psychiatry, Alcohol and Drug Abuse Research Center, Harvard Medical School—McLean Hospital, Belmont, Massachusetts 02178

Jack H. Mendelson (371), Harvard Medical School—McLean Hospital, Belmont, Massachusetts 02178

Esteban Mezey (277, 303), The Johns Hopkins University, Baltimore City Hospitals, Baltimore, Maryland 21224

C. C. Middleton (457), Sinclair Comparative Medicine Research Farm, College of Veterinary Medicine and School of Medicine, University of Missouri, Columbia, Missouri 65201

C. Morris† (21), The American Museum of Natural History, c/o Comision Fulbright, Maxime Abril 599, Lima, Peru

Robert M. Morse (359, 397) Department of Psychiatry, Mayo Clinic and Mayo Foundation, Mayo Medical School, Rochester, Minnesota 55901

Ralph A. Nelson (397), Department of Internal Medicine, University of South Dakota, Sioux Falls, South Dakota 57105

Robert E. Olson (197), Department of Biochemistry, St. Louis University School of Medicine, St. Louis, Missouri 63104

Eileen M. Ouellette (439), Pediatric Neurology, Massachusetts General Hospital, Harvard Medical School, Boston, Massachusetts 02114

---

*Present address: Texas A & M Research and Extension Center, Amarillo, Texas 79106.
†Present address: Anthropology Department, The American Museum of Natural History, New York, New York 10024.

R. Passmore (213), Department of Physiology, University Medical School, Teviot Place, Edinburgh EH8 9AG, Scotland

Patricia B. Santora (303), The Johns Hopkins University School of Medicine, Baltimore, Maryland 21205

Frank A. Seixas* (245), National Council on Alcoholism, New York, New York 10017

F. John Service (263), Department of Medicine, Mayo Clinic and Mayo Foundation, Mayo Medical School, Rochester, Minnesota 55901

Keith H. Steinkraus (35), Cornell University—New York State Agriculture Experiment Station, Cornell University, Geneva, New York 14456

Milton Terris (481), Department of Community and Preventive Medicine, New York Medical College, New York, New York 10029

E. Neige Todhunter (83), Department of Biochemistry, Vanderbilt University School of Medicine, Division of Nutrition, Nashville, Tennessee 37232

M. E. Tumbleson (457), Sinclair Comparative Medicine Research Farm, College of Veterinary Medicine, University of Missouri, Columbia, Missouri 65201

A. J. Tuyns (427), International Agency for Research on Cancer, 69372 Lyon Cedex 2, France

Donald M. Watkin† (225), Office of State and Community Programs, Administration on Aging, Office of Human Development Services, Department of Health, Education, and Welfare, Washington, D.C. 20201

K. M. West (257), University of Oklahoma, Health Sciences Center, Oklahoma City, Oklahoma 73104

Beverley Wood (409), Department of Medicine, University of Melbourne, St. Vincent's Hospital, Melbourne, Australia

*Present address: 2 Summit Drive, Hastings-on-Hudson, New York 10706.
†Present address: Lipid Research Clinic, Department of Medicine, George Washington University Medical Center, Washington, D.C. 20037.

PREFACE

The contributions of alcoholic beverages in nutrition have not generally been examined. National statistics on per capita caloric supply, for example, fail to include either those beverages so commonly a part of the meal, beers and wines, or distilled spirits. Physicians and health workers often address the effects of obviously excessive intakes of alcoholic beverages, and the focus of health research has been on intake levels that clearly are detrimental. From such studies extrapolation downward from excessive intake levels frequently is made to draw conclusions concerning the effects of small intakes. Estimates so derived of the effects of moderate amounts of these beverages may thereby be distorted. At the same instant, claims based upon folklore concerning healthful effects of some fermented beverages—wines, beer, and certain distilled spirits—are often quite unfounded but widely stated and believed.

Recognizing the common consumption at mealtime of wines and beers by a major segment of the adult population and the caloric density of these beverages, due largely to alcohol, it was felt that an objective examination of the role of fermented food beverages in nutrition was desirable. This examination was planned to bring together existing factual knowledge concerning these beverages and their contributions to nutrition and health.

This edited comprehensive treatise is the product of a broad and critical examination of current scientific information placed in perspective with the evidence of man's long cultural and technologic concerns with these unique food/drug beverages. The concepts contained in the volume were discussed at an International Symposium on Fermented Food Beverages in Nutrition convened at the Mayo Clinic in Rochester, Minnesota, following which the respective chapters were revised and edited. The Symposium was one of the series supported by The Nutrition Foundation and made possible, in part, by grants to The Nutrition Foundation from the Medical Advisory Committee of the United States Brewers Association and from the Wine Institute. Other investigators were then invited to prepare reviews of additional areas in which they are actively working, in order to ensure comprehensive coverage.

The choice of the term "symposium" to describe the initial scientific session was an appropriate one. The word "symposium" was used in ancient Greece to mean a drinking party or feast, particularly a social gathering, in which there was free interchange of ideas. While the participants were constrained with the modern practice of abstinence during the proceedings, there was a lively exchange of ideas. We believe this volume offers the reader an enjoyable and informative distillate of the most active ferment in this field compounded by knowledgeable authorities from diverse disciplines.

In many discussions of the use of alcoholic beverages it is immediately assumed that the only motive for use is pharmacologic, i.e., one drinks for the sedative or mood-changing effects. Such an assumption is generally made in comparisons between the use of alcohol and of various licit and illicit drugs. A case can be made that motivation for the use of wine and beer in moderation is usually less pharmacologic than gustatory. The enjoyment of a good meal can be greatly enhanced by beer or wine. The taste of foods may be perceived with greater enjoyment when wine or beer accompanies a meal, perhaps from a solvent action which prevents persistent contact of the food with the taste organs, or perhaps only from the contrasting taste of the beer or wine and the food. Thirst can be effectively and pleasantly satisfied by a cold beer or wine, and the ritual of these beverages taken with meals seems to be a part of their enjoyment. Beer and wine in moderation often are enjoyed for these properties in amounts so small that the pharmacologic effects of the alcohol are scarcely felt, and the mild sedative effects of somewhat larger amounts can be pleasant and impose little risk. It is much more difficult to measure and describe these taste pleasures of wine and beer than it is to study the undesirable effects of immoderate and excessive use of alcohol. In planning this volume, an attempt was made to keep these matters in balance and in perspective, but perhaps unavoidably more attention and time were given to the more easily documented effects of excessive alcohol use.

The making and use of fermented food beverages are of widespread interest. In primitive times fermentation was one of the early means of food preservation, and mystical and religious qualities commonly have been assigned to wine and beer. Consider the conventions and the rituals with which we often surround the use of these beverages: the religious symbolism of wine, the giving of a toast, the loyalties we often give to a particular form of drink, our habits of taking beer, wine, or champagne only under certain circumstances—all of these are parts of the rituals of one's particular life style. Health-giving properties have been attributed to wine and beer in general, and frequently particular virtues have been credited, without substantiation, to specific types of wine or beer. Perhaps the fallacious latter claims have inhibited, until most recently, serious inquiry into the beneficial effects of moderate enjoyment of these beverages. On the other hand, the potentially harmful effects of the immoderate use of alcohol have long been recognized. The Greeks wore amulets mounted with amethyst, a wine-colored, semiprecious stone, to ward off drunkeness, hence the term "amethystic," meaning the property of preventing intoxication. The Italian term "fiasco" means a wicker-covered wine bottle, but it is not clear how this word has come to mean total confusion and failure.

The well-to-do Englishman of the eighteenth and nineteenth centuries was often caricatured as suffering from gout, and this disorder was blamed on his use of port. Only recently has the mechanism of this particular form of gout been unraveled. It appears that the port of that time was produced using stills made with lead solder. Specimens of wine from that era have been found to contain significant amounts of lead, and a similar epidemic of gout occurring in consumers of lead-contaminated home-distilled whiskey was identified in Alabama in 1969. The source of the lead in the American version apparently was the automobile radiators used as condensers in homemade stills. While lead intoxication appears to interfere more strongly with urate excretion than other indicators of renal function impairment, the provocative agent within the afflicted joint may be guanine rather than urate crystals. This form of gout is referred to as saturnine gout because of the alchemist's use of saturn as the symbol for lead. Lead has at times in the past been added to wines intentionally to alter the taste, and the fermentation of homemade wine in lead-glazed earthenware has caused lead poisoning (Ball, G. V., Two epidemics of gout. *Bull. Hist. Med.* **45**, 401–408, Sept.–Oct., 1971; Farkas, W. R., Stanawitz, T., and Schneider, M., Saturnine gout: Lead-induced formation of guanine crystals. *Science* **199**, 786–787, February, 1978). Harmful effects of contaminants or adulterants in all sorts of foods and beverages have been occasionally recorded throughout the centuries. Such episodes should not be interpreted as examples of inherent injurious properties of wholesome beverages, just as the injurious effects of immoderate or excessive uses of any beverage or food should not lead to labeling a desirable commodity as inherently bad or harmful.

Mankind's long and diverse utilization of fermented foods is illustrated, but not exhaustively detailed, in this work. Some aspects of this are revealed by the etymology of the word "alcohol" itself (p. 4). However, nutritionists, physicians, and planners widely ignore the implications of moderate uses of alcoholic beverages. Indeed, rarely, if ever, does one find their contributions to nutrient intake considered in calculations of individual diets (except those of recognized alcoholics) or of national food supplies. This volume was planned to place these matters in scientific and historical perspective. It provides modern objective information concerning fermented food beverages of interest to the physician, nutritionist, other health advisor, researcher, producer, planner, and regulator.

We are most grateful to all of the scientists and other scholars who have contributed to this volume and to the granting institutions for support. We are grateful to the staff of Academic Press for the quality of production maintained in this and other volumes in the Nutrition Foundation Series and for the pleasant associations that characterize our relationships.

Clifford F. Gastineau
William J. Darby
Thomas B. Turner

I

Historical Overview

1

FERMENTED BEVERAGES IN ANTIQUITY

P. Ghalioungui

I.	The Beginnings	4
II.	Early Egypt	5
	A. Beer	6
	B. Wine	9
III.	Recognition of Excess	12
IV.	Pre-Islamic Near East	13
V.	Changing Attitudes with Islam	14
	References	18

The aristocratic Greek poet, Theognis of Magara (1) wrote in the fifth–sixth century B.C., "If it is a disgrace to be drunk among the sober, it is disgraceful too to be sober among the drunk." In some fashionable restaurants, patrons asking for water are quietly led to the bathroom to avoid scandal. Maitre d's are thus applying Pliny's (2) motto: "There are two liquids agreeable to the body: wine inside and water outside," in the true French tradition of Baudelaire who said, "A man who drinks water has a secret to hide."

In ancient writings, chapter after chapter discuss wine and beer; very little is written about water, except as a warning. Whereas wine has been reported to inactivate some viruses, to lower blood cholesterol, and to resemble gastric juices in pH (3,4), naturally occurring water was not always wholesome near human settlements in antiquity. Ancient authors were very particular about recommending only specific springs and rapidly flowing waters. Theophrastus is quoted as saying that in a period of drought, Nile water had become poisonous (5) and a king of Persia is known to have drunk only boiled water (6).

In North America, as Dr. Todhunter points out (see Chapter 5), early settlers ascribed their sickness to water and insistently asked for brewers from Europe (7).

FERMENTED FOOD BEVERAGES IN NUTRITION

3

Before distilled liquor was available, popular beverages were beer and wines, and they were common to all the ancients, with the exception of Eskimos and Australian aborigines.

I. THE BEGINNINGS

Which came first, wine or beer, is debatable. The biblical story of Noah suggests it was wine. Greek authors are of the same opinion. The view of the Greeks, however, who were winelovers and considered beer an unmanly drink (8), might have reflected the perennial conflict between vine-growing and cereal-growing communities that, independently and possibly at the same time, discovered the exhilarating effects of fermented grapes and gruels.

This discovery marked one of the great moments in civilization, for drinking with no thirst, like eating with no hunger, distinguishes man from animals (although I am not sure that the distinction is always to our advantage).

Primitive man must have been awed by the phenomenon of fermentation. Here was decent grain and fruit suddenly foaming, bubbling, and deflagrating jars, and impregnating unfermented material with the same kind of life that turned decent ladies into riotous bacchantes. How could people who believed at that time that Achilles acquired the strength of a lion by eating its innards (9) escape the animistic belief that fermenting material was possessed by a spirit that penetrated drinkers (10)—the "spirit of wine," as alcohol was first called?

The phenomenon of distillation was recognized by Alexandrians around the fourth century B.C. and was utilized by sailors and alchemists. We do not know for certain, however, who first distilled the "spirits of wine" (11) that were held to be the living embodiment of the inebriating factor, until Büchner in 1896 extracted zymase from yeasts and reduced alcohol to the status of a lifeless chemical compound.

The word alcohol comes through a devious derivation from Arabic *alkohol,* the finely powdered antimony oxide used as black eye paint. Hence came Spanish, *alcoholar,* to turn black, and *alcoholado,* a beast with black circles around the eyes. In Europe, the word *alcohol* was first used, e.g., by Ambroise Paré, in connection with ocular applications. But in the sixteenth century, spirits of wine were called by Paracelsus, *alcohol vini,* i.e., the quintessence of vine, later abridged to *alcohol.*

Spirits of wine were at first used in the preparation of inks, lacquers, and the like. A manuscript doubtfully ascribed to Arnaud de Villeneuve (1235–1312 or 1248–1314) recommended it under the name of *aqua vitae* in preparations to preserve youth *(de conservatio iuventute).* The word was a great success: it gave Scandinavian *aquavit,* and after translation, French *eau-de-vie,* and Celtic *uisquebaugh,* hence whiskey.

Fig. 1. Hieroglyph of Shemw, god of the wine press.

II. EARLY EGYPT

Egyptians believed that beer was invented by Osiris (11a).* According to a legend, the god Rê, to stop the bloodthirsty lioness Sekhmet from utterly destroying mankind, cheated her by offering her red-dyed beer that she took for blood. Appeased, she became Hathor, a "Lady of Drunkenness, Music and Dance" (12), who adopted as an emblem the "sistrum" a musical rattle that was reproduced on the capitals of her temples. Herodotus (13) said that people feasted her by dancing and drinking, women unveiling themselves and loading watchers with abuse. It is a shame that by Ptolemaic times, possibly because of chronic beer drinking, this lovely goddess had turned into a flabby Greek matron.

Drunkenness was thus regarded as an inspired state: beer and wine deified, or offered to gods. It is no wonder that Egyptians, without giving offense, could call their children by such otherwise sacrilegious names as "How drunk is Cheops, or Mykerinus" (14).

Wine, too, was very old in Egypt.† The earliest evidence of cultivated grapes in history comes from el-Omari in Egypt and Hama in Syria (15). Wine jars have been found in predynastic tombs (16). Cellars and wine presses had an old god *Shemw* whose hieroglyph was a wine press (Fig. 1). The miraculous beverage in First Dynasty records is called "The Beverage of the Body of Horus." What

---

*For a fuller discussion the reader is referred to Chapter 13 of Darby *et al.* (11a).

†The reader is referred to Chapters 14 and 15 of Darby *et al.* (11a).

more could prove the sanctity of wine that, like the red beer offered to Sekhmet, was identified in many texts with blood (10,17).

The belief in the transsubstantiation of wine into blood is so fundamental in nonreformed churches, that an attempt to replace it with milk by persecuted Christians who were afraid of being betrayed by the smell of wine when they took early morning Communion, was emphatically condemned in 248 A.D. (18).

The influence of such beliefs in shaping civilization cannot be overemphasized. Deference to wine in nonreformed countries led to the development of the glorious French, Spanish, and Greek wines and, through the Spaniards and Jesuits in the Americas to the introduction of European varieties of wine in the Americas.

On the other hand, in Islamic countries, the ban on wine favored the breeding of sweet table grapes like Muscat and raisins, while wine countries preferred the less sweet and more acidic grapes. We shall see that in countries like Egypt, where a mixed population lives, the results varied with the rulers.

A. Beer

To make beer, starch has first to be split into fermentable sugars. In the Far East molds were used. Egypt relied on malting. One can follow the whole

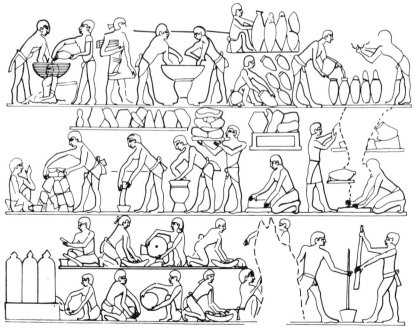

Fig. 2. Brewing. Ti's Tomb, Old Kingdom, Saqqara. Reproduced by permission of Institut Français d'Archaeologie Orientale.

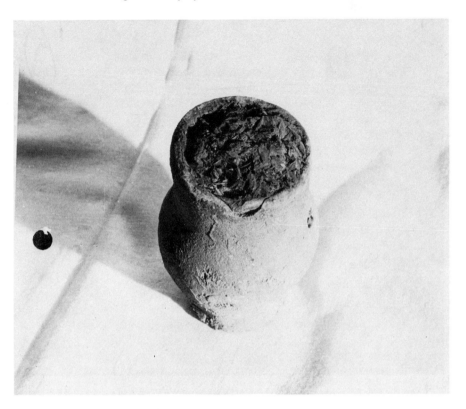

Fig. 3. Jar with beer residues. Cairo, Agricultural Museum.

brewing process in the scenes in many tombs, of which the most detailed is the tomb of Ti (Fig. 2).

In Ken-Amun's tomb is depicted what is probably malt being withdrawn from a jar and stored in two jars, one of which is capped with a measuring pot.

There is also evidence that leavening was used. This use is depicted in Ti's tomb, where two kinds of dough were mixed. A previously unknown variety of *Saccharomyces*, mixed with starch, molds, and bacteria, was identified in beer jars (19). In Eighteenth Dynasty remains, yeasts were found that were sufficiently pure to suggest that purified preparations were available. It is doubtful, however, whether pure yeast was known before Ptolemaic times, and it is likely that brewers, like bakers, relied on some kind of sour dough.

Finally, the insides of the jars were smeared, not with bitumen as elsewhere, but with clay. These interpretations of the drawings are supported by actual finds of vessels containing remains of barley strainings (Fig. 3) and of models of squeezing mills and beer strainers (Fig. 4).

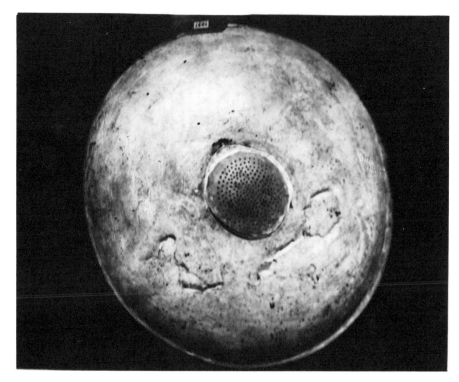

Fig. 4. Beer strainer. Tomb of Tut-ankh-Amon, Cairo Museum.

Medical papyri mention at least seventeen varieties of beer, but do not give any details. The only indications on variety we possess are the different proportions of wheat, barley, and additional ingredients in different papyri (20).

Greco-Roman authors seldom mention flavoring. When they do, the beers they describe are more medicated than flavored beers. However, we do know that flavoring, or priming, with dates was practiced, both from accounts in papyri and from scenes such as the one in Antefoker's tomb in which a date cake, for some of which a boy is begging, is sieved into a jar.

The result was different from beer as we know it today. It resembled popular drinks called today in Egypt *bouza* (21) (which has nothing to do with "booze") and in the Sudan, *merissa*. These are both made of lightly baked wheat or millet dough, mixed with germinated wheat, and primed with some old *bouza*.

Sieved clear *bouza* examined at the National Research Center in Cairo contains per 100 ml, 3.8 gm of alcohol, 1.2 gm of protein, 2.5 gm of fat, and 9.1 gm of carbohydrates, and, therefore, yields about 100 calories per 100 ml; unsieved, as it is often served, yields much more. More important is the fact that the amino acids and vitamins have been found to be increased during fermentation (22,23), and this enrichment, as pointed out by Darby (see Chapter 4), helps to explain the frequent freedom from nutritional deficiencies of some populations whose diets are apparently poor.

In fact, this kind of beer is as much a food as it is a beverage. Thus, in ancient Egypt, the phrase "Bread and Beer" was a common greeting formula. Like "bread and salt" today, it stood for all food; and written with the hieroglyphics for a loaf of bread and a jug, it meant invocation offerings to the gods.

B. Wine

Whereas beer was a commoner's beverage, wine was an aristocrat's drink. This was commented upon by Athenaeus (24) who, as a typical wine-loving Greek, stated that beer was invented for those who could not afford wine.

Even gods were equally snobbish. In Sumer, in southern Iraq, King Gudea is known to have offered beer to the god Ningursu, but wine to the more important Bel Marduk (25).

In Rome around 89 B.C., Greek wines were so expensive that only one cup of Greek wine was served at banquets (26). The cost difference is explainable by the more complex technology required by wine making, that could never be, like brewing, a housewife's occupation. Indeed, while wine making was a man's occupation, brewing is shown as a woman's job, both in artifacts and in the title "Controller of Women Brewers" given to Chief Brewers (27).

In the absence of texts describing wine making, we again have to rely on pictures. In some of these, grapes are pressed in a bag, either fixed at one end and twisted at the other (Fig. 5) or twisted at both ends by poles held apart by a third man. In a unique scene, the third party is a baboon. This may seem funny, but to Egyptologists, who are very serious people, it does not. They claim that it is a pun on $i^c ny$, the name of the baboon, and $^c ny$, to twist, the verb being there to indicate the action.

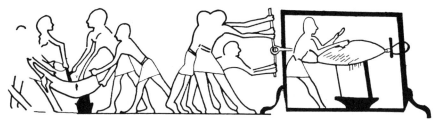

Fig. 5. (*Right*) Wine press. (*Left*) Straining the must through a cloth. Beni-Hassan, Middle Kingdom.

In other tombs, grapes are shown being trodden in large vats. What is the relationship between pressing and treading? It is known that the juices of both red and white grapes are white, and that red wine obtains its color by being left to extract color from the skins of red grapes. This distinction is quite clear in Champagne. Ordinarily, white Champagne is made with red grapes. Only the more refined *Blanc de blanc* (white of white) is made with white grapes. *Rosé* was born in the beginning of the nineteenth century when a cavist in Moët's cellars mistakenly left the must too long with the skins. This *Oeil de perdrix* (partridge eye), as it was called, was an immediate success. Rosé has been made ever since, some wineries like Pommery or Mumm sticking to the old "error"; others, like Moët or Veuve Clicquot mixing red with white wines to get the right color.

One explanation of the relation between treading and pressing was that trodden grapes were left in the vat to make red wine, whereas pressed grapes were immediately transferred from the vats to make white wine.

Against this explanation is the intriguing observation that the outflowing juice from both vats and presses is colored. Moreover, if the vat was only the preliminary step before the press, one would expect treading and squeezing shown in sequence in the same scenes, which nowhere is the case.

Lutz noted that the press bag is seen only in the tombs north of Thebes and treading only in and south of that city. He concluded that the difference might be geographical. I suppose, however, that the difference indicates a chronological evolution of techniques. Squeezing is seen only in Old and Middle Kingdom tombs, while treading is exclusively a New Kingdom feature.

It is true that members of the French Expedition saw in the last years of the eighteenth century the press bag still being used in the Fayum (28). However, the Fayum was at the time the only province where wine was made, probably for the large Coptic Congregation who lived there, and whose religion forbade treading grapes intended for the Mass.

I believe, therefore, that wringing in bags might have been the original technique, and that it was gradually superseded by treading which, incidentally, persisted during the Middle Ages right to the present day in Greece and many European countries. Regarding the taste of wine, treading may be preferable to mechanical presses because it extracts less astringent tannins from the skins and seeds, it therefore also extracts less of the "antiviral principles."

The dark color of the outflowing juices from both vats and bags could therefore have been an artistic device of showing the juices better in illustrations, rather than indicating the real color.

Many wine jars, e.g., in Tut-ankh-Amon's tomb, had a hole drilled in the lids (Fig. 6), which suggests that wine was bottled before the total arrest of fermentation. When this process was complete, the hole was sealed with clay and the jar was stamped and inscribed with the names of the owner and gardener, the date, vintage, quality, and degree of excellence of the wine.

Fig. 6. Wine jar. Tomb of Tut-ankh-Amon, Cairo Museum.

At some stage, wines were siphoned off or decanted and blended. This was probably done in what was the banquet hall, judging by the elaborate decoration of the furniture (Fig. 7) and the accompanying dancing and music.

For several reasons, Egyptian wines, of which at least 24 varieties were known, were different from Greek and modern wines. Because of the scarcity of wood, jars were not smeared inside with pitch, and the wine could not acquire the subtle qualities it gains from contact with wood. In addition, Athenaeus says that the wines were so sweet and rich that they did not mix with water easily, much as honey does and much as the modern Cypriot Commanderie and Coptic Church wines do. One wonders, however, whether the drinker in Antef's tomb, who is saying, "How sweet is this wine" is commenting on the sweetness of the wine or on the racy serving girl.

Fig. 7. Siphoning and mixing wine.

Texts also mention medicated and fruit wines, especially *shedeh,* the pomegranate wine that was praised in Solomon's song:

> I would lead you to my mother's house
> I would give you to drink spiced wine of the juice
> of my pomegranates.

It is believed that Egyptian wines did not keep as well as the heavily resinated Greek wines. But the notes of an Alexandrian merchant of the second or third century (29) that mention export to Arabia would prove the contrary. Tomb illustrations, too, indicate the possibility of storage. Jars filled Antef's cellar in which a drunk guardian answers insistent knocks at the door by saying, "No, I am not drunk."

Nevertheless, export from Egypt could not have been very extensive since many texts and illustrations show envoys bringing wine into the country.

III. RECOGNITION OF EXCESS

For the banquets of the rich, the flow of wine was free in the sumptuous silver, gold, and alabaster jars, cups, and other drinking vessels now exhibited in all museums. Humorists make good use of this vein. One painted a maiden vomiting in a banquet. Another drew a member of the New Kingdom "beautiful" set relieving herself in the middle of the festivities. Such could never have happened in Rome if it is true that no Roman lady was allowed to drink. A third satirized politics that recall modern, diplomatic mores, by drawing a retinue of cats humbly serving wine to an enthroned mouse.

To avoid inebriety, drinkers were encouraged by Greek authors to eat almonds (30) or cabbage. The custom of eating cabbage was adopted by the Greeks (31) and the Arabs (32) but it did not appear very effective if we judge by an artifact of the Greco-Roman period, a time when Eubulus (33) discussing ten cups of wine, said the first was to health, the second to pleasure, the third to sleep, the fourth to violence, the fifth to uproar, the sixth to revel, the seventh to black eyes, the eighth to policemen, the ninth to biliousness, and the tenth to hurling furniture.

The ancients were fond of such literary exercises. Much later, a fourteenth century Arab (34) restricted himself to four stages. The devil, he said, watered the vine successively with the blood of a peacock, an ape, a lion, and a pig. Hence drinkers are successively as garish as peacocks, as garrulous as apes, as violent as lions, and as slothful as pigs.

This last stage was depicted by an artist in Khety's tomb in Beni Hassan. A Persian miniaturist, however, possibly mindful of the classical descriptions, started at the top of his painting depicting angels, sipping their cup with befitting poise and dignity, and then he graded his effect with unsteadiness, increasing until it ended in rioting and total dejection.

No description from ancient times is available of chronic alcoholism. Ruffer (35) discovered fibrosis in the liver of a mummy. Since there was no evidence of bilharziasis, alcoholism was a possible cause.

On the other hand, some illustrations in a Saqqara tomb of abdominal distension with umbilical and inguinal herniae, and of gynecomastia, are more readily explained by schistosomiasis than by alcoholic cirrhosis, because of their association with genital hyperplasia in the same tomb.

IV. PRE-ISLAMIC NEAR EAST

In Mesopotamia, beer was probably the original drink of the cereal-growing south, and wine, the drink of the north, where the vine grew wild. Some of the oldest Sumerian deities were wine gods and goddesses. Literature and art from 3000 years B.C. show them receiving offerings and libations of wine.

In Arabia proper, previous to the advent of Islam, barley, wheat, sorghum, and other grain, mixed with herbs, served to prepare beer; dates, sometimes mixed with raisins, were made into *skr*. Mead (*bata^c*) was locally made from honey, but was also imported from Jordan (*al-mukdi,* from *al-Mukda* in Jordan). Fermented milk, especially camel milk, was popular among nomads. Fermentation was carried out in jars or gourds, with dates added in hollowed-out palm trunks (36).

Wine was made from grapes locally grown in the South, where the vine was a frequent decorative architectural motif and where grapes are now so abundant that merchants have difficulty in disposing of them.

Preference was given to Syrian and Iraqi wines that, too, were so syrupy that the same word *sharab* was used for syrup and wine. Three centuries later Abu Nawas, the Court poet, gave lyrical evidence of their sweetness and storage capacity (37).

> Come, pour it out, gentle boys
> A vintage ten years old
> That seems as though it were in the cup
> A lake of liquid gold
> And where the water mingles there
> To fancy's eye are set
> Pearls over shining pearls
> As in a carcanet.

Before Islam, drinking in houses, taverns, and dancing houses and in town and along caravan routes, was a serious problem; it was even used as a means of committing suicide, much as would be done today with barbiturates. It is, therefore, possible that in reaction to excess the countries that originally fostered wine, adopted its most powerful enemy, Islam. In fact, even before Islam, some Nabataeans never drank and, in the otherwise bibulous and wealthy Palmyra of Queen Zenobia, and the mysteriously pink city of Petra ("Rose red city half as old as Time"), worshiped a god called "the Abstemious" (38). In Arabia proper, the same restrictive reaction took place before the appearance of Islam, when some pagan Arabs (*the Hunafah*), though neither Jews nor yet Moslems, were monotheists and preached asceticism and abstemience (39).

V. CHANGING ATTITUDES WITH ISLAM

The attitude of the new religion was at first lenient, but it rapidly grew in severity in successive Koranic revelations.

Initially, paradise was compared to "rivers of wine, a joy to drinkers" (*Mohammed*, 15) and God, enumerating his boons, said:

> And from the fruit of the date palm and wine ye get wholesome drink and food (*The Bees*, 67)

But when drunken converts went to pray there came the verse

> O ye who believe! Draw not near unto prayer when ye are drunk, till ye know that which ye utter (*Women*, 43)

A similar nuance is expressed in the chapter of *The Cow,* 219, which says that the sin of wine is greater than the profit. However, when repeated warnings on the necessity of moderation were not heeded, drinkers were equated with idolators (40) and alcohol was described as an invention of the devil meant to turn people away from worship and to create dissensions (*The Table,* 90,91).

The paradox between early descriptions of women and wine in paradise and later revelations was the object of gentle fun from wine-loving Omar Khayyam (41):

> Heaven teems with beauty as a flowing well
> And fountains flow with wine, or so they tell
> If to love wine and woman there's no sin
> Is it a sin to love them here as well?

The final prohibition was, however, traumatizing. Yemenite converts anxiously asked the Prophet about their *mezr* (beer). The Prophet, on learning that it caused elation, prohibited it (42).

Abu Suffian, one of the Moslem proselytes, on hearing that the poet Al-A^csha wished to be converted, explained to him what he had to forsake, gave him a hundred camel, and told him to wait for a year before making a decision. The poet took the camels and went on drinking until he died (43).

Following the initial impact of Islam, attitudes varied. Abstemience was, and still is, the rule in the Arabian peninsula, especially among the Wahhabites of Nedjd. Opulent Damascus, Bagdad, Iran, and Cairo were more permissive, for although very few Muslims would touch pork, having been conditioned in childhood to regard it as impure, many do not shirk a merry cup.

These unorthodox drinkers rationalize their attitude with two arguments: One is the praise given by early revelations to intoxicants and the progressiveness of the ban that makes it a matter of social propriety or expediency rather than one of creed. The other is a very specious linguistic nicety. Since the word *khamr,* used for intoxicants, is derived from *khamira,* to ferment, they hold that it should only apply to fermented beverages, like beer and wine, not to distilled liqueurs, whisky, or brandy that, in point of fact, did not exist in the time of the Prophet.

In Egypt, Islam was tolerant, except under some despotic rulers. In 1005 A.D., the insane Al-Hâkim Bi-amrillah ordered that all wine jars be broken and their contents spilled in the streets. Earlier, in 852 A.D., an equally fanatic ruler forbade even the culture of the vine. At times, especially during the sacred month of Ramadan, grapes fetching prohibitive prices had to be smuggled by Christians from abroad for use in the Mass. Since these grapes reached Egypt dry, priests used the pressed juice of raisins soaked in water for three days (provided the feet were not used in pressing). This juice was not willfully fermented, but, of course, alcohol was formed in the process. Initially the raisin wine was just tolerated but later it became the usual medium of Holy Communion. It is called

Abarkah from Greek *Oparki,* ''primeur,'' because Christians were in the habit of offering to the Church the first products of their wine. In times of extreme necessity, date wine was used instead (44).

The State drew heavy revenue from wine taxes; and foreign travelers, who complained from the scarcity of wine and its outrageous prices, as well as Cairene historians, described daily scenes of drunkenness even among the authorities (45).

The poorer classes drank *bouza,* albeit in moderation, and date or palm wine; later, date brandy was used.

In the thirteenth century, court dignitaries proclaimed their ranks on the heraldic crests that decorated their cutlery. Like the *jukundars,* the Gentlemen of the Polo who claimed two polo sticks, the Sultan's cup bearers were not embarrassed in exhibiting the cups of their function (Fig. 8). In Muslim Spain, ivory casks and other objects were frequently decorated with drinking scenes. From twelfth century Bagdad, the city of the Arabian Nights, the book on ''Mechanical Devices'' of Court Engineer Al-Jazri illustrates a musical box from which a robot girl dispenses a cup to the Sovereign (46).

Physicians recognized the damages of excesses but did not lag in the praise of

Fig. 8.   Plate of the Sultan's cup bearer. Fourteenth-century. Cairo Museum of Islamic Art.

Fig. 9. "Willowy damsels...."

wine. Both Razes (47) and Avicenna (32a), the Princes of Medicine, described its deleterious effects, of which they mentioned anorexia, black bile vomiting (by which they possibly meaned hematemesis), palsies, vertigo, and loss of memory. They discussed the impact of different kinds of wine upon different individuals and temperaments. Nevertheless, they both extolled its therapeutic indications, and Avicenna ended his discourse by praising God who made of it a medicine that sustains the innate forces of man.

It was in Persia that poets and painters expressed in immortal masterpieces their love of wine, painting willowy damsels still able to aim wine correctly at a glass on the floor (Fig. 9), private and royal carousing, noblemen and gracious dancers. The nostalgic yearning of the country is expressed by the bard, Omar Khayyam in a final toast to the divine beverage:

> Bring me red wine, musk scented, to set free
> My soul from sound of words that weary me
> Give me a flask of wine ere envious fate
> Make jars of clay that was me and thee
>
> So should I drink that fragrance of the wine
> Even in the silence of the tomb shall still be mine
> That when a kindred spirit passes by
> May grateful odours round his senses twine.

REFERENCES

1. Theognis of Megara. "Poèmes Elégiaques" (transl. into Fr. by J. Carrière). Belles Lettres, Paris, 1962.
2. Pliny. "Natural History," XIV; XXIX, 150.
3. "Uses of Wine in Medical Practice," Wine Advis. Board, San Francisco, California.
4. Konowalchuk, J., and Speirs, J. I. Virus inactivation by grapes and wines. *Appl. Environ. Microbiol.* **32**(6), 757–763 (1976).
5. Athenaeus. "The Deipnosophists," II, 42.
6. Athenaeus. "The Deipnosophists," II, 45.
7. Baron, S. "Brewed in America: The History of Beer and Ale in the United States." Little, Brown, Boston, 1962.
8. Aechyles. "The Supplices" (transl. by T. G. Tucker), p. 953. Macmillan, London, 1889.
9. Apollodorus. "The Library" (transl. by J. G. Frazer), pp. 3, 6, 13. Putnam, New York, 1921.
10. Plutarch. Isis and Osiris. 353, 6. *In* "Moralia" (transl. by F. C. Babbitt), Vol. 5. Harvard Univ. Press, Cambridge, Massachusetts, 1936.
11. See Sarton, G. "Introduction to the History of Science," Vol. 1, pp. 534 and 723; Vol. 2, p. 29. Williams & Wilkins, Baltimore, Maryland, 1927.
11a. Darby, W. J., Ghalioungui, P., and Grivetti, L. "Food: The Gift of Osiris," Vol. 2. Academic Press, New York, 1977.
12. Junker, H., and Winter, E. "Das Geburthaus des Tempels der Isis im Phila." H. Bohlhaus, Vienna, 1965.
13. Herodotus. II, 60.
14. Reisner, G. A. "Mycerinus." Cambridge Univ. Press, London, 1931.
15. Helbaek, quoted in Renfrew, J. M. "Palaeoethnobotany," p. 127. Columbia Univ. Press, New York, 1973.

16. Emery, W. B. "A Funerary Repast in an Egyptian Tomb of the Archaic Period." Ned. Inst. Het. Nabije Oosten, Leiden, 1962.
17. Achilles Tatius. "Adventures of Leucippe and Clitophon," II; II, 4, 5. Putnam, New York, 1918.
18. Rohrbacher. "Histoire Universelle de l'Eglise Catholique," Vol. II, p. 544. Soc. Gen. Libr. Cathol., Paris, 1878.
19. Gruss, J. Untersuch. v. Broten aus der Aegypt. Samml. d. staatlich. Mus. z. Berlin. *Z. Aegypt. Sprache Altetumskunde* **68**, 79, 80 (1932).
20. Helck, W. "Das Bier im Alten Aegypten." Ges. Geschichte Bibliogr. Brauwesens, Berlin, 1971.
21. Lucas, A., and Harris, J. R. "Ancient Egyptian Materials and Industries," p. 11. Arnold, London, 1962.
22. Morcos, S. R., Hegazi, S. M., and El-Damhougy, S. T. Fermented foods in common use in Egypt, II. *J. Sci. Food Agric.* **24**, 1157–1161 (1973).
23. Nour Eldin, S. Biological Studies on Bouza. M. S. Thesis, Cairo Fuad Univ., Cairo, 1947.
24. Athenaeus. "Deipnosophists" (transl. by C. B. Gulick), I, 34 B. Putnam, New York, 1927–1941.
25. Lutz, H. F. "Viticulture and Brewing in the Ancient Orient." Hinrichs, Leipzig, 1922.
26. Pliny. Ref. 2, XIV; XVI, 96.
27. Murray, M. A., and Sethe, K. "Mastabas," Vol. II, p. 11. Bernard Quaritch, London, 1937.
28. Girard, M. P. S. *In* "Description de l'Egypte, Etat Moderne," Vol. II, pp. 491–714. Imprimerie Impériale, Paris, 1813.
29. Schoff, W. H. "The Periplus of the Erythraean Sea." Longmans, Green, New York, 1912.
30. Pliny. Ref. 2, XXIII; LXXV, 144, 145.
31. Deipnosophists. Ref. 5, I, 34 C.
32. Avicenna. "The Canon," Vol. I, p. 346. Dar el Sadr, Beirut, n.d.
32a. Avicenna. "The Canon," Vol. I, p. 442. Dar el Sadr, Beirut, n.d.
33. Deipnosophists. Ref. 5, 36 B-C.
34. Al-Demiry, K. M. "Hayat al-Hayawan al-Kubra," 3rd Ed., Vol. 1, p. 651. Halaby Press, Cairo, 1965 (in Arabic).
35. Ruffer, M. A. "Studies in the Palaeopathology of Egypt," p. 78. Univ. of Chicago Press, Chicago, Illinois, 1921.
36. Gawad, A. "Al-Mufassal fi tarikh el-Arab kabl al Islam," Vol. 4, pp. 666–669. Al-Nahda, Bagdad, 1968.
37. Alevi, S. M. B. "Arabian Poetry and Poets," p. 169. Jami Millia, Aligarh, India, 1924.
38. Gawad. Ref. 36, Vol. 6, p. 324.
39. Ronart, S., and Ronart, N. "Concise Encyclopedia of Arabic Civilization," p. 204. Praeger, New York, 1960.
40. Al-Lissan. Vol. 13, p. 195, quoted in Gawad, Ref. 36, Vol. 4, p. 665, under *dmn*.
41. Johnson Pasha. "The Rubaiyat of Omar el-Khayyam," p. 284. Kegan Paul, Trench, Trubner, London, 1913.
42. Gawad. Ref. 36, Vol. 4, p. 668.
43. Al-Aghání. Vol. 8, p. 77, quoted in Gawad, Ref. 36, Vol. 8, p. 77.
44. Wassef, C. W. "Pratiques Rituelles et Alimentaires des Coptes," pp. 108, 109. Imprimerie Inst. Fr. Archeol. Orient., Cairo, 1971.
45. Clerget, M. "Le Caire," Vol. II, p. 75. Schindler, Cairo, 1934.
46. Winder, R. B. "Badi^c^az-Zamān Isma^c^il bin ar-Razaz al Jazari." *In* "The Genius of Arab Civilization" (J. R. Hayes, ed.), p. 188. Nesterham Press, New York, 1975.
47. Said, H. M. Razi and treatment through nutritive correction. *In* "Hamdard Medicus," Vol. XIX, nos. 7–12, pp. 113–120. Hamdard Natl. Found., Karachi, Pakistan, 1976.

2

MAIZE BEER IN THE ECONOMICS, POLITICS, AND RELIGION OF THE INCA EMPIRE

C. Morris

I. Native American Fermented Food Beverages 21
II. The Preparation of Maize Beer 22
III. Exchange in the Andes: Reciprocity and Redistribution 25
IV. Maize Beer (*Chicha*) in Pre-Spanish Times 26
V. State Management of the Brewing 27
VI. Social and Cultural Importance of Fermented Beverages 32
    References 34

I. NATIVE AMERICAN FERMENTED FOOD BEVERAGES

It is hardly news that fermented beverages play an important role in economic and sociopolitical affairs. In fact, one wonders how history would ever have progressed without fermented beverages to toast successes, dull the pain of failure, and in general ease the establishment of interpersonal relations. Whatever the varying nutritional qualities and clinical effects may have been, their special socioeconomic significance is widespread in ancient and modern human societies. It is thus not unusual that maize beer played a critical role in Andean society at the time of the Inca empire, and a commentary similar to that which I offer here might be made for almost any fermented beverage in any society. But the degree to which the high level of the state administrative apparatus was involved with beer production and consumption, I believe, was truly extraordinary. Beer was not necessarily the secret ingredient that enabled the Inca to establish the largest native empire of the New World, and one of the largest in history, but it certainly was one of a series of interrelated key features that

21

enabled the Cuzco rulers to extend and maintain their power over a vast region.

Most of the principal native New World food crops such as potatoes, corn, and many varieties of beans and squash are well known to us, since they have been adopted into the European diet and now constitute a basic part of the world food supply. The fermented beverages prepared from these food plants, however, are still very unfamiliar outside native American communities. To my knowledge they have never been produced and distributed on an industrial scale.

Native beers in Andean South America are known collectively as *chicha* in Spanish, or *aqa* in the native Quecha language. *Chicha* is made of several products, including peanuts and the manioc root, which is very important in the tropical forest regions of Peru and elsewhere. But in the regions conquered by the Inca, *chicha* of maize was by far the most important both in terms of the quantity produced and served and in terms of its prestige. The beverage is prepared in a great variety of ways, the differences being both regional and related to the personal specialties of brewers. Several by-products also result from the process. One of these is a sweet paste or gel which is now usually reincorporated into the beverage, but in pre-Columbian times before the introduction of sugar it was probably an important and generally used sweetener.

II. THE PREPARATION OF MAIZE BEER

Two major variants in maize beer preparation involve the source of the diastase, which is used to increase alcoholic content and to improve flavor. In much of the Americas a common source of diastase is saliva. Dried ground corn is put into the mouth in slightly moist balls and worked with the tongue until it has absorbed saliva. This "salivated" maize flour may then be dried to form the basic raw material for *chicha*. At present, in most of Peru *chicha* is made from malted maize, called *jora*. The *jora* is made by soaking the maize in water overnight in a pottery jar. The following day the grains are placed in rather thick layers on leaves or straw and left for germination until the shoots are about the length of the grains. The sprouted grain is then dried in the sun. The resulting *jora* is then ground or, more correctly, cracked to make the base for the drink (Figs. 2–4).

The actual brewing of *chicha* is typically begun by filling a large wide-mouth jar to about one-third of capacity with the malted maize or *jora*; hot but not boiling water is added to just below the jar's rim. The mixture is thoroughly stirred, then allowed to cool and settle for about an hour. The liquid on top is separated into another wide-mouth jar. A semi-congealed layer resting on top of the sediment remaining in the jar is also usually removed and further condensed by slow simmering into a rather caramel-like sweet paste (the sweet substance mentioned above). The liquid is allowed to stand for about 2 days; it is then strained, boiled for 3 hours or so, and allowed to cool in a wide-mouth jar. When

Fig. 1. Inca terraces near Pisac in the Urubamba Valley, Southern Peru. Most of these terraces were probably used to grow maize eventually converted to *chicha*. (Photo courtesy of Edward Ranney.)

Fig. 2. Maize being dried after harvest, prior to *chicha* making. (Photo courtesy of Idilio Santillana.)

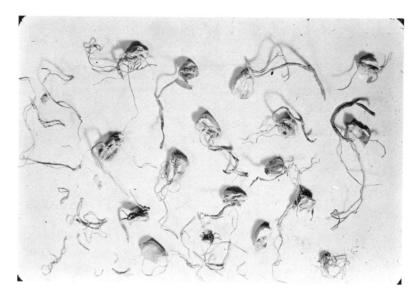

Fig. 3. Malted maize or *jora.* (Photo courtesy of Wilhelm Booz.)

Fig. 4. Modern brewer of *chicha,* showing some of the ceramic vessels used in the process. (Photo courtesy of Idilio Santillana.)

it cools some of the sweet by-product previously prepared is added. Fermentation then begins, and the brew is transferred to narrow-mouth jars, where fermentation will continue and from which the beverage will eventually be served. The speed of fermentation varies with the temperature of different regions. There are also various preferences as to the degree of fermentation; some people like a *chicha* that has only recently begun to ferment, whereas others prefer that it be kept for 5 or 6 days before drinking. The resulting beverage is usually rather cloudy, although the most prized *chichas* are a sparkling amber, with a taste probably closest to apple cider of the fermented beverages familiar to European palates. (For a more complete description, see Cutler and Cardenas, 1947; Nicholson, 1960.)

III. EXCHANGE IN THE ANDES: RECIPROCITY AND REDISTRIBUTION

In order to understand the role of *chicha* in Tawantinsnuyu, as the Inca state was called, it is necessary to understand certain aspects of the nonmarket Andean economy. Exchange in the Andes was based on two related forms of exchange which anthropologists refer to as reciprocity and redistribution. Basically, reciprocity and redistribution are forms of exchange similar to gift-giving. Both contrast with the market-exchange system which is the basis of most modern economies in that the value of commodities is not determined by supply and demand, in a more or less free or open exchange in which the focus of attention is on the commodity itself. Instead, social and economic aspects of the exchange are linked, and a critical feature of reciprocity and redistribution is that the actual "value" of the goods exchanged is very closely tied to the sociopolitical status of the people involved in the transaction. The value of a gift depends not just on what the gift is, but on who is giving it. A small token from a king is worth many baskets of food from a commoner—so long as the king's rank and identity are clearly identified with the gift. A second important aspect of reciprocity is that the exchange itself is not usually completed in a single act. A gift creates an obligation which may be fulfilled at a later date. It not only may be fulfilled, but should be, and this felt obligation to return a favor and the resulting prolongation of the exchange makes it conducive to supporting personal relationships between exchanging partners. In our own society we are most familiar with reciprocal exchange in terms of dinners and parties—events which are conceived of as purely social but at which an economic transaction, in a sense, still takes place. In fact, throughout history, it appears that food and drink have always been products that were particularly well suited to this type of exchange; it is just that in quantitative terms the amount of products exchanged in this manner is no longer very important in modern societies. This was not always so.

Redistributive exchange is really just a special kind of reciprocity, while reciprocity is the exchange of goods as gifts between relatively equal partners, redistribution brings a new social level into play. The giftlike exchanges are carried out between peoples of different sociopolitical status, typically a leader or chief and the people who are subject to him. The leader acts as a central receiving point for different kinds of goods from different people, reciprocating a gift with something different—perhaps something given him by a different kind of specialist or a resident of a different ecology. Goods are thereby redistributed so that the products of several different regions and specialists are available to all—even though there are no or few markets.

The anthropology lesson above is used to show that both the archaeological and historical evidence suggests that markets and trade as we usually understand them played only a peripheral role in the Andean economy. As we will see, I think that it is because of this that maize beer came to play such a special role. Given this evidence we have to explain in other ways how goods moved through such a vast and varied landscape. Since reciprocal and redistributive exchange systems tend to mesh the economic and the political aspects of a society into the same system and the same events, we will be looking at both as two sides of the same coin.

IV. MAIZE BEER (*CHICHA*) IN PRE-SPANISH TIMES

Evidence for the role of maize beer comes from two sources: the observation set down by early Spanish travelers, soldiers, and bureaucrats and the archaeological remains of Inca settlements we are now studying. Both kinds of evidence show clearly that maize beer was consumed in very large quantities in pre-Spanish times and that its consumption was an integral part of the religious and political ceremonies of the day. More than that, it was such a critical element in the whole process of reciprocity that the authority structure was weakened when the Spanish attempted to limit its use to curb what they viewed as drunkenness, disorder, and paganism associated with native drinking customs (Rostworoski de Diez Canseco, 1977). The Spanish, of course, failed to appreciate the subtleties of native politics and economics even though they depended on local leaders to carry out their orders.

The order restricting the use of *chicha* by the judge Gregorio Gonzales de Cuenca in 1556 revealed much of the special political and economic context of the drink (Rostworowski de Diez Canseco, 1977, p. 241):

Because said chiefs and leaders used to have "taverns" and places where all who came to them were given *chicha* to drink, and it is the cause of the drunkenness of the people, and it occupies many men and women in making the *chicha,* and it is a thing of bad example, and the said chiefs and leaders spend excessively on it; it is ordered that from now on they (the chiefs) not have such "taverns" nor public or secret places for drinking...

Maria Rostworowski de Ciez Canseco, director of the National Museum of History in Lima, has observed that it was not just in their residences that the local chiefs gave their followers *chicha* in liberal amounts. In addition, when a chief traveled, each time his litter rested, the common men came to him to drink at his expense. This required numerous bearers to carry the beverages in generous quantities and produced a great retinue that accompanied each chief. Of course, it was not just political loyalty that accrued to the chiefs in return for their generosity, it was also labor to till their fields and to perform numerous other services. Much of the economy worked on these exchanges of favors. There was always both an economic and a political aspect to the reciprocity.

In quantitative terms, potatoes and other tubers were certainly more important to the native highland Andean diet than was maize. But curiously the religious rituals of the state aimed at securing a good and bountiful harvest were almost all directed toward maize. This rather surprising evidence from historical sources has been analyzed by John V. Murra (1960), who concludes that it is the result of the Inca state's emphasis on maize as a prestige crop. In fact, he sees two separate agricultural systems: one at the local level which emphasized the root crops adapted to the highlands, with local rituals reflecting that reality; the other, at the state level, emphasized maize, essentially an imported crop which in most of the highland region grew only with difficulty. Whatever the relative quantities of the two crops produced on state lands, religious ceremony and agricultural ritual were geared to maize, as was state astronomy. The technological emphasis was also on maize. Incredible irrigated terraces were built in the Urumbamba valley near Cuzco to increase production (Fig. 1). The main reason for this emphasis on the part of the state was almost certainly the special prestige of the crop. And prestige came from its use in *chicha* and the particular economic and political value the brew represented.

Archaeological evidence for the brewing and drinking of maize beer is, of course, less direct. We deal with the broken containers and abandoned equipment used by the Inca brewers, and when preservation is good, we may find some of the maize which was its raw material. Evidence for *chicha* consumption is even more of a problem since the large wooden beakers in which the drink was most customarily served do not survive in the humid highland Andean soil.

V. STATE MANAGEMENT OF THE BREWING

In spite of these difficulties, our research in the Inca administrative city of Huánuico Pampa (Fig. 5) since 1971 has yielded evidence that not only confirms the overall importance of the beverage in diet and ritual, but also goes beyond that to show that it was brewed on a very large scale, apparently under direct state management.

Before we can appreciate the significance of a single urban activity such as brewing, we must emphasize that the city where it took place was very different from most cities with which we are familiar, either past or present. There is not time to consider the whole range of these differences here. For our purposes, it is most important to note that Huánuco Pampa, and some other Inca cities, were installations built on previously unoccupied sites by the state and linked together by a legendary system of roads. Their architecture and their pottery imitate those from the Inca homeland in Cuzco. They were constructed and populated quite rapidly between about 1470 and the Spanish invasion 60 years later. Most of the centers were depopulated even more rapidly than they were built once Inca power was broken in 1532. All of these things suggest that the cities were created and managed by the state. Even though they were large and housed some of the activities of cities in other parts of the world, which grew up more spontaneously as trading or production centers, they functioned mainly as links between the ruling elite in Cuzco and the hinterland populations and were also the major nodes in a vast supply and communications network which supported military and consolidation activities.

The evidence for *chicha* at Huánuco Pampa takes the form of concentrations of sherds of the large jars which were used in various stages of maize beer production: soaking the corn to produce malt or *jora*; boiling the *jora*; and fermenting, storing, and dispensing the results. We have also found the large rocker flattening stones used to crack the *jora* open.

The evidence was concentrated in two sectors of the city—two sections which constitute preindustrial breweries (Fig. 6). One was a walled compound of 50 buildings with only one tightly controlled entrance. Along with the large jars, and other evidence suggesting brewing, were numerous spinning and weaving tools. The occurrence of evidence for these two kinds of activities in a compound with such a tightly controlled access has led to the suggestion that the residents of the compound were a group of *mamakuna,* the famous "chosen women" of the Inca, sometimes referred to as "virgins of the sun." While their role in Inca society is usually interpreted as a honored one, and the state did apparently go to considerable lengths to protect them and give them high status, the lives of these women were hardly spent entirely in blissful leisure. Written sources indicate that they spun, wove, and made beer for the emperor. If our interpretation of the compound north of Huánuco Pampa's central plaza is correct, the jars and other tools present indicate that they were very busy indeed. The evidence also tells us something about the Inca state's labor practices. The *mamakuna* were certainly a special group, but we have to balance their image—apparently carefully cultivated by Inca officialdom—as a group of protected, honored, and supposedly beautiful women with their position as suppliers of two of the state's principal needs: beer and cloth. As state brewers and weavers they were one of the most important economic units in the Empire.

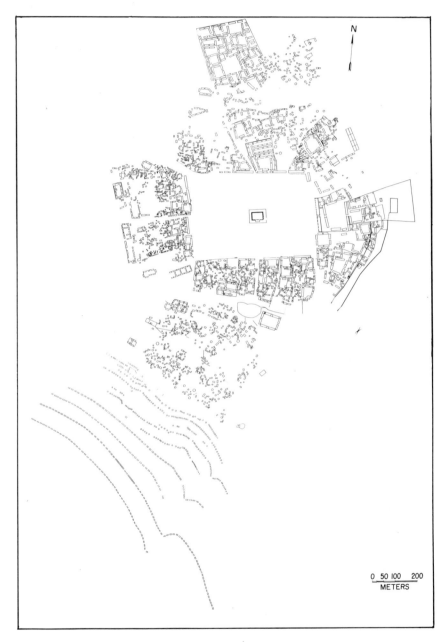

Fig. 5. Preliminary architectural base map of the Inca administrative city of Huánuco Pampa. Major brewing areas are located to the north and east of the central plaza.

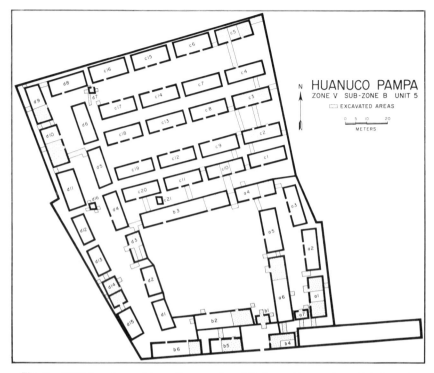

Fig. 6. Walled compound in northern sector of Huánuco Pampa thought to have been occupied only by women. The major activities in the compound appear to have been weaving and *chicha* brewing.

The other place where beer played an unusually important part in activities at Huánuco Pampa was in the zone of fine architecture east of the plaza. This part of the city is composed of large buildings and two spacious plazas joined by compounds containing small dressed-stone buildings, a bath, a pool, and a large terrace overlooking a small artificial lake inside a large trapezoidal enclosure (Fig. 7). The dressed-stone buildings near the baths are traditionally interpreted as the residence reserved for the Inca himself when he was passing through the city, which he could not actually have done very frequently given the size of his empire and the fact that he traveled by litter on mens' backs. One of the most surprising results of all our work came from excavations in the buildings surrounding the two plazas near this presumed lodging place of the Inca ruler. Excavations by Pat H. Stein from 1972 to 1974 revealed a whole complex of culinary pottery, food remains, cooking areas, and literally tons of the large jars thought to be primarily associated with *chicha* (Stein, 1975). These were no solemn halls devoted to decision-making and the slow movements of state

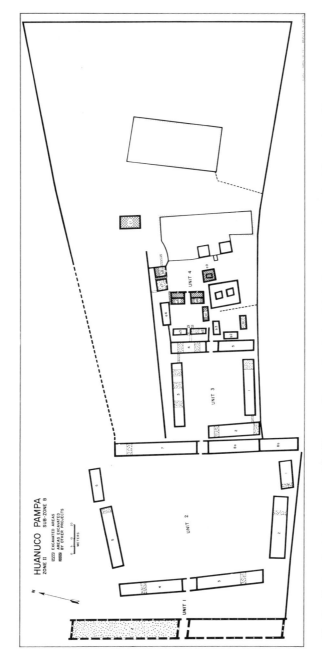

Fig. 7. Series of related compounds at Huanuco Pampa devoted mainly to the preparation and public serving of food and beer. Unit four contained the building probably reserved for the use of the emperor.

bureaucracy as we once thought. We obviously need to recall the records of the "taverns" mentioned by north coast chiefs to interpret this complex of buildings and plazas covering more than 6 acres.

Huánuco Pampa was a provincial capital of the Inca empire. The Inca had a residence here not because he frequently spent the night, but to maintain and increase his power. He had to entertain; he had to have a place where "all who came to him were given *chicha* to drink." It was central to keeping his armies on the move, preventing revolt, and maintaining the storehouse filled. Just as *chicha* making and other aspects of state hospitality were at the heart of the most elegant zone of the provincial capital, they were also at the heart of provincial administration.

What I have tried to show is that maize beer was not just another commodity to be bought and sold, drunk and enjoyed. It was one of a handful of products, also including cloth, that through the centuries had acquired a quite special significance. Its association with political and religious ceremony was basic to the maintenance of the whole political and economic system. It is not just that millions of gallons of *chicha* were brewed and consumed annually, it is that the way in which they were distributed made them basic to giving the leaders their authority. The state's ability to increase beer production was essential to its political and economic expansion. As in all societies authority can be abused. The Inca may have been guilty of that, but at least we must admire the accomplishments of the workers rewarded with the Inca's maize beer. They built more than 5000 miles of roads; they constructed the great fortress of Sacayhuaman above Cuzco, cutting and fitting blocks of stone weighing several tons; they built warehouses throughout the empire and stocked them with food and other goods (the warehousing facility at Huánuco Pampa alone had a capacity of more than one million bushels); they built the vast system of irrigated terraces of the Uramba valley which greatly increased the arable land of that warm and fertile region. Political relationships which *chicha* helped maintain at many levels of Andean society also facilitated the production and movement of many other kinds of goods.

VI. SOCIAL AND CULTURAL IMPORTANCE OF FERMENTED BEVERAGES

It would, of course, be an exaggeration to say that Andean civilization was built on a foundation of corn beer. Many other factors were also critical, but beer was certainly not just another beverage that people took with their meals. Don Cristóbal Payco, a chief from Jequetepeque on the north coast, quoted by Rostworowski de Diez Canseco (1977, p. 242) perhaps best sums up the situation in his lament and complaint about Spanish restrictions on the use of *chicha*.

It is a great inconvenience since the main reason that the people obey their leaders here, is through the custom that they [the leaders] have to give the people drink . . . and if they do not oblige by giving the people drink neither will the people plant their crops for them.

The conflicting attitudes and understandings of the Andean natives and the Spanish invaders regarding maize beer are in a sense symptomatic of the different interpretations of the value and positive and negative effects of fermented beverages. Their effects and values are different in different situations and subject to differing perceptions.

The final point I hope the Inca evidence has emphasized is that the social and cultural context within which fermented beverages are consumed should be considered along with their role in nutrition, illness, and behavior. In our own society, where most reciprocity has been replaced by markets and the relationship between economics and politics is presumably quite different from Inca times, fermented beverages are not so critical to keep the economy and polity functioning.

We are thus able to concentrate more on nutritional characteristics of various beverages and on their physiological and behavioral effects on the people who consume them. In societies which anthropologists often study (societies which are now changing from one cultural and economic condition to another) the situation is more complex. For example, in the Andes, much of the rich, warm irrigated land which was once cultivated to provide corn for beer as well as direct consumption, has now been converted to the production of sugar cane to provide a badly needed export crop. But in many areas part of this sugar cane is converted into *aguardiente*, a low-grade distilled alcoholic drink. This is specifically the case in much of the Huallaga Valley which once provided the maize for the *chicha* of Huánuco Pampa. *Aguardiente* is the cheapest way to produce alcohol and accounts for much of the alcoholic beverage consumption of the lower socioeconomic levels in places where *chicha* is now rare. Industrial beer is priced beyond the reach of much of the population except for very special occasions.

For whatever reason, the tendency to consume fermented beverages seems to be very nearly a universal in human societies, at least since the beginnings of agriculture. Given that, it appears to me that one of the important tasks facing us, particularly in terms of nutrition, is that of examining the positive and negative aspects of the *various* alcoholic beverages. I know of no comparative nutritional studies of maize beer and cane alcohol, but I would guess that the substitution of the latter for the former represents a significant nutritional shift, which if not compensated for would lead to deterioration in the overall diet.

Besides the possible dietary effects of changes in the kinds of alcohol consumed, there is evidence that the character of drinking is also altered by the contact of a non-Western society by Europeans. In the case of the Inca, and in the case of many other societies for that matter, most drinking is done in a social or ritual context. These contexts control the situations in which drinking is permitted or encouraged and greatly influence both the frequency and amount of drinking. The destruction of the ritual and social framework for drinking, coinciding with the introduction of distilled liquor (which frequently formed an important part of European trade with the natives), brought about unfortunate consequences. Most of us are familiar with the stories about "fire-water" among native North Americans shortly after the arrival of Europeans. I suspect the "drunkeness and disorder" that the Spanish complained about may have been amplified by the disorientation in the economic religious and political systems resulting from the imposition of foreign rule. The role of fermented beverages in a society is extraordinarily complex. We must orient our studies in such a way as to understand that great complexity in all its dimensions.

REFERENCES

Cuter, H. C., and Cardenas, M. (1947). Chicha, a native South American beer. *Bot. Mus. Leafl., Harv. Univ.* **13**(6), 33–60.

Murra, J. V. (1959). Rite and crop in the Inca state. *In* "Culture in History" (S. Diamond, ed.), pp. 390–407. Columbia Univ. Press, New York.

Nicholson, G. E. (1960). Chicha maize types and chicha manufacture in Peru. *Econ. Bot.* **14,** No. 4.

Rostworowski de Diez Canseco, M. (1977). "Etnía y Sociedad." Inst. Estud. Peru., Lima.

Stein, P. H. (1975). The Inca's hospitality: Food processing and distribution at Huánuco Viejo. *Annu. Meet. Soc. Am. Archaeol., Dallas, Tex.*

3

NUTRITIONALLY SIGNIFICANT INDIGENOUS FOODS INVOLVING AN ALCOHOLIC FERMENTATION

Keith H. Steinkraus

I. Introduction	36
II. Alcoholic Foods in Which Sugars Are the Major Fermentable Substrate	37
A. Mead	37
B. Palm Toddys	38
C. Mexican *Pulque*	39
D. Miscellaneous Alcoholic Beverages from Sugary Substrates	41
III. Alcoholic Foods in Which Saliva Is the Amylolytic Agent	41
A. South American *Chicha*	42
B. Related Fermentations	43
IV. Alcoholic Food Fermentations Involving an Amylolytic Mold and Yeast	43
A. Indonesian *Tapé Ketan*	43
B. Indonesian *Tapé Ketella (Peujeum)*	46
C. Chinese *Lao-chao*	46
D. Indian Rice Beer, *Pachwai*	46
V. Alcoholic Foods in Which Starch Hydrolysis Involves a Malting (Germination) Step	46
A. Kaffir (Sorghum) Beer	47
B. Egyptian *Bouza*	50
C. Ethiopian *Talla*	51
D. Mexican *Tesgüino*	51
E. Russian *Kvass*	52
VI. Alcoholic Fermentations Involving Use of a *Koji* for Starch Hydrolysis	52
A. *Koji*	52
B. Japanese Rice Wine, *Saké*	53
VII. Alcoholic Milk Beverages	54
A. *Busa*	54
B. *Kefir*	54

FERMENTED FOOD BEVERAGES IN NUTRITION
Copyright © 1979 by Academic Press, Inc.
All rights of reproduction in any form reserved.
ISBN 0-12-277050-1

C. *Koumiss* 55
D. *Mazun* 55
VIII. Significance of Indigenous Alcoholic Foods in Human Nutrition 55
IX. Summary 56
References 57

I. INTRODUCTION

Vast numbers of the human population subsist principally on rice, wheat, or corn along with beans, starchy roots and tubers, fruits, and vegetables. Such diets generally provide limited intakes of total protein and very low levels of animal protein. If the rice is highly polished, it is low in thiamine and riboflavin. Such diets are conducive to beriberi. Maize diets are limiting in niacin and can lead to pellagra. Vegetarian diets, in general, are low in vitamin B_{12}.

Some modern methods of food processing such as canning, puffing, clarification, dehydration, and storage result in losses of essential amino acids, protein quality and quantity, and vitamins. Since the majority of people in the affluent Western world are overnourished and eat highly diversified diets, losses of particular nutrients in certain of the foods are probably not too serious, and the losses of specific nutrients can be offset by enrichment. This is within the means of the Western food consumer, but it is beyond the means of the masses in the developing world unless enrichment is heavily subsidized.

Thus it can be illuminating to examine some of the ways indigenous populations have developed to provide proper nutrition for themselves. Their methods frequently demonstrate how nutritive values of basic substrates can be modified and improved at minimum cost through fermentation.

Two fermentation products widely involved in the preservation of foods are acids and alcohol. Generally, food spoilage and disease-producing microorganisms cannot develop in an acid or an alcoholic environment. In a number of foods described in this paper, both acid and alcohol are produced. However, this chapter deals primarily with those foods in which ethanol is an important product. It is well known that consumption of alcoholic products such as beer or wine is generally safe even in environments where the water supply is contaminated. In addition, it is recognized that nearly all humans everywhere, except for those who do not consume alcoholic beverages for religious reasons, incorporate such products into the diet enthusiastically. In most primitive societies, birth, marriage, death, religious and social activities require provision and consumption of appreciable quantities of alcoholic beverages.

Modern food technology uses enrichment or fortification to improve nutritive value of foods (Harris, 1959). It is possible to accomplish much the same enrichment by use of suitable fermentations. Platt (1946, 1964) coined the phrase "biological ennoblement" for the latter. This chapter deals with a number of foods in which there are striking examples of biological ennoblement.

This chapter is not a comprehensive review but rather a selection and classification of indigenous alcoholic fermentations designed to introduce the reader to important aspects of these and related beverages, their processes of manufacture, and their nutritional implications. For details on additional fermentations, readers are referred to the reviews of Hesseltine (1965), Pederson (1971), and Batra and Millner (1976).

II. ALCOHOLIC FOODS IN WHICH SUGARS ARE THE MAJOR FERMENTABLE SUBSTRATE

Man has been producing wines since the beginning of recorded history. Honey was the only concentrated sweet available in prehistoric times and fermented honey may well have provided man's first common alcoholic drink, long before the cultivation of fruit or grain crops (Morse and Steinkraus, 1975). The half-empty honeypot left in the rain may have been the first alcoholic drink (Brothwell and Brothwell, 1969). Fruit wines were probably discovered as soon as man tried to collect and store sweet fruits and berries.

A. Mead

Diluted honey wines have been consumed in Africa for centuries, and primitive, unclarified mead is still produced in Ethiopia as an alcoholic beverage called *tej* (Vogel and Gobezie, 1977). In various parts of Africa, one part of honey is diluted with from three to six parts of water. In the early days, a handful of roasted barley was added to the pot to promote fermentation. Also, wood or bark from the shrub *Rhamnus tsaddo* was added. Fermentation time was 5 or 6 days (Platt, 1955). Today the fermentation pot is "smoked" over an olive wood, hop stem fire to introduce a smokey flavor that is desired. One part of honey is mixed with four parts of water. The pot is covered with cloth and incubated in a warm place for 2 or 3 days. Wax and top scum are removed. Washed, peeled, and boiled hops are added. The pot is again covered and fermented for another 8 days in a warm place or 15 to 20 days in a cold place. It is then coarse filtered through cloth to remove sediment, hops, and spices. *Tej* is yellow, sweet, alcoholic, and cloudy due to the yeasts and other fermenting organisms. It has a very pleasant flavor and also a reputation for causing headaches if one consumes too much.

Nothing appears to have been published regarding its ethanol content, acidity, or nutritive value. Since honey is relatively expensive, consumption of *tej* is restricted generally to special holidays.

B. Palm Toddys

Apparently, sap collected from most palms can be fermented to toddy. However, the most frequently tapped palms include the coconut palm (*Cocos nucifera*), the oil palm (*Elaeis guineensis*), Palmyrah palm (*Borassus flabellifer*), the Nipa palm (*Nipa fruticans*), the wild date palm (*Phoenix sylvestris*), the Raphia palm (*Raphia hookeri* or *R. vinifera*), and the Kithul palm (*Caryota urens*). Sap can be collected from slits along the trunk, but the major source is the unexpanded flower spathes. There is considerable art connected with binding the flower spathes, pounding to cause the sap to flow, properly cutting the spathe tip, and collecting the sap (Faparusi, 1973). During full production, each spathe can yield from 600 to 1000 ml of sap per day. Three spathes can be tapped on a single tree; thus yield can total from 1.8 to 3.0 liters of sap per tree per day.

Generally the sap is collected in earthenware pots which contain yeasts and bacteria and even leftover toddy from previous fermentations. Thus, fermentation commences as soon as fresh sap flows into the pot. Fermentation generally continues until the toddy is collected soon after which it is consumed.

The yeast *Saccharomyces cerevisiae* always seems to be present, whereas *Lactobacillus plantarum, Leuconostoc mesenteroides,* and other bacterial species vary (Okafor, 1972; Faparusi, 1973).

Palm wine is essentially a heavy milky-white, opalescent suspension of live yeasts and bacteria with a sweet taste and vigorous effervescence (Okafor, 1975a, b).

Palm sap is a sweet, clear, and colorless liquid containing about 10–12% sugar and is neutral in reaction (Okafor, 1975b).

Bassir (1967) reported that palm sap contains $4.29 \pm 1.4\%$ sucrose, $3.31 \pm 0.95\%$ glucose, and $0.38 \pm 0.015\%$ NH_3. Yeasts, mainly *S. cerevisiae* and *Schizosaccharomyces pombe,* accumulate in the millions on exudates around the flower stocks and inoculate the sap naturally. *Lactobacillus plantarum* and *L. mesenteroides* lower the initial pH of the juice from 7.4 to 6.8 and by the end of 48 hours the pH is about 4.0. Palm wine is generally drunk when the pH is between 5.5 and 6.5 (after 12 hours fermentation). Bassir (1967) states that the ethanol content never rises above 7.0%. In 24-hour palm wines, the alcohol ranges from 1.5 to 2.1%. In 72-hour wine, the ethanol is from 4.5 to 5.2%. Organic acids present at 24 hours are lactic acid (32.1–56.7 mg/100 ml), acetic acid (18.6–28.6 mg/100 ml), and tartaric acid (11.7–36.0 mg/100 ml).

An enormous amount of palm wine is consumed in southern Nigeria. Four million people drink approximately half a liter per day. Half a liter contains approximately 300 calories from the sugar and alcohol, 0.5–2 gm of protein, and important amounts of water-soluble vitamins. During the first 24 hours, there is as much as 83 mg ascorbic acid per liter. This declines with age. Thiamine increases from 25 to 150 μg per liter. Riboflavin content increases from 0.35 to 0.50 μg per liter in the first 24 hours. Pyridoxine increases from 4 to 18 μg per liter. Since vitamin B deficiencies are particularly widespread among pregnant women and teenagers, palm wine is an important addition to their diets (Bassir, 1967).

van Pee and Swings (1971) reported that palm juice and wines also contain vitamin B_{12} as shown in the following tabulation.

Source	Vitamin B_{12} (ng/ml)
Fresh palm juice (26 samples)	17–180
Palm wine (12-hour fermentation)	140–190
Palm wine (24-hour fermentation)	190–280

The vitamin B_{12} content is very important for people who consume primarily vegetable foods. The fact that the fresh palm juice contains vitamin B_{12} suggests that there is bacterial growth in the fresh juice before it is "fermented." van Pee and Swings also reported discovery in palm wines of a motile rod-shaped high alcohol-producing anaerobic bacterium closely related to *Zymomonas mobilis*.

Platt (1955) reported that in some areas African palm wine is as important in society for working parties and for ceremonial and other social activities as are the "beers." In fact, the palm wines often sell for less than the price of a soft drink. Thus, the palm wines represent an excellent value and play an important role in diets wherever they are consumed in the world.

C. Mexican *Pulque*

Pulque is the national drink of Mexico, where it was inherited from the Aztecs (Goncalves de Lima, 1975). It is a cloudy, white, viscous beverage produced by fermentation of the juices (called *aguamiel*) of certain cactus plants of the genus *Agave*. The *Agave* plants grow on very poor soil and consumption of pulque is of particular importance in the diets of the poor and low-income people.

In the traditional process, the floral stem of mature plants 8 to 10 years old is cut off and the juice accumulates in the cavity. The juices are collected daily by oral suction and the juice is then fermented in wooden, leather, or fiberglass tanks with capacity of about 700 liters. The inoculum is *pulque* from a previous fermentation. Fermentation time is 8–30 days depending upon temperature and the season. Generally 30 to 70 liters of fresh juice are added to a tank and an equal amount is withdrawn. Thus, the process is semicontinuous.

The sugars in *aguamiel* consist of sucrose, glucose, fructose, and pentoses. *Pulque* has an ethanol content of 4–6%, pH between 3.5 and 4.0, and total acidity of 400–700 mg/100 ml (as lactic acid). It also contains aldehydes up to 2.5 mg/100 ml; esters (as ethyl acetate) 20–30 mg/100 ml; and fusel oils from 80 to 100 mg per liter (Sanchez-Marroquin and Hope, 1953).

The essential organisms are homofermentative and heterofermentative *Lac-*

**Table I.   Chemical Changes during Pure Culture
Fermentation of *Agave* Juice (*Aguamiel*)** [a]

Analyses	Sterile aguamiel	*Pulque*
Brix	11.0	6.0
Specific gravity	1.042	0.978
pH	7.0	4.6
Total acidity (as lactic acid) (%)	0.018	0.348
Total reducing sugars as glucose (%)	10.00	0.48
Sucrose (%)	7.6	0.42
Crude protein (%)	0.17	0.17
Dry solids (%)	15.29	2.88
Ash (%)	0.31	0.29
Ethanol (% w/v at 20°C)	0.00	5.43
Higher alcohols (%)	—	0.51
Volatile acidity (as acetic acid) (%)	—	0.02

[a] Adapted from Sanchez-Marroquin and Hope (1953).

tobacillus sp., *Leuconostoc* sp. (closely related to *L. dextranicum* and *L. mesenteroides*), *Saccharomyces carbajali* (closely related to *S. cerevisiae*), and *Pseudomonas lindneri*. The *Leuconostoc* sp. produce bacterial polysaccharides or dextrans which are responsible for the characteristic viscosity of *pulque* (Sanchez-Marroquin, 1953).

Because of problems controlling the natural fermentation, pure culture techniques have been developed in recent years (Sanchez-Marroquin, 1953). Optimum temperature was found to be 28°C (82°F) and optimum pH 5.0. Under pure culture conditions using pasteurized *Agave* juice, fermentation time is

Table II.   Vitamins in *Pulque* [a]

Vitamin	Amount (μg/100 ml)
Thiamine	22–30
Niacin	54–515
Riboflavin	18–33
Pantothenic acid	55–86
Pyridoxine	16–33
Biotin (ng/100 ml)	11–32
p-Aminobenzoic acid	15–29

[a] Adapted from Sanchez-Marroquin and Baldrano (1949) and Sanchez-Marroquin and Hope (1953).

48–72 hours versus 8–30 days for natural fermentations. Chemical changes occurring in aguamiel using pure culture fermentation are summarized in Table I. The changes are similar to those observed in traditional *pulque*.

Traditional *pulque* plays an important role in the nutrition of low-income people in the semiarid areas of Mexico. Per capita consumption has been about 1 liter per day. Nutritional surveys indicate that there is a protein–calorie deficiency in the main *pulque*-producing areas (State of Hidalgo). Children at or under school age receive *pulque* three times a day. This provides 2.2–12.4% of their calories and 0.6–3.2% of their protein requirements (Instituto Nacional de Nutricion, 1976). Vitamin contents of *pulque* are in the approximate ranges shown in Table II. It is clear that *pulque* contributes worthwhile amounts of B vitamins to the diets of the low-income people consuming it.

D. Miscellaneous Alcoholic Beverages from Sugary Substrates

1. Mexican *Mezcal*

Mezcal is a tan-colored, sour beverage with an alcohol content similar to that of beer. The substrate is the core of the *Agave* (century plant), which is stripped of leaves and cooked to yield a sweet pulp. Organisms from a previous fermentation serve as a source of inoculum (Pennington, 1969; Pederson, 1971).

2. Sugar Cane Wines

Wines such as Philippine *Basi* are made by fermenting freshly pressed sugar cane juice. The juice is concentrated by boiling, and dried samac leaves are added. The juice is filtered and transferred to earthenware jars. Yeast powder and steamed rice are added. Fermentation takes 7–10 days followed by aging. Dominant microorganisms are *Saccharomyces, Endomycopsis,* and *Lactobacillus* sp. The wines have a fine flavor with a good balance between sweetness and acidity (Kozaki, 1976).

III. ALCOHOLIC FOODS IN WHICH SALIVA IS THE AMYLOLYTIC AGENT

The discovery of the alcoholic fermentation whereby maize and other plant products are chewed to introduce salivary amylase is lost in antiquity. The use of saliva to hydrolyze starch could very well be the most primitive process of fermentation. It may have developed centuries ago as a result of prechewing grains to be fed to infants, the sick, or elderly toothless individuals. Such prechewed grains mixed with saliva would tend to become sweet and ferment naturally.

A. South American *Chicha*

Chicha is a fermented beverage which is produced traditionally among the Indians of the Andes by prechewing maize, yuca, mesquite, quinoa, or plantain (see also Chapter 2). Well-made *chicha* is an attractive beverage, clear and effervescent, resembling apple cider in flavor. It generally contains 5% or less ethanol with a range of 2–12% (Cutler and Cardenas, 1947). Maize in this region has always had a profound religious and magical significance as well as economic importance (Nicholson, 1960). *Chicha* plays a predominant role in all fertility rites. It has been used to induce the Thunder God to send rain and used in the Sun and Harvest Festivals. *Chicha* is probably derived from the word *chichal,* meaning "with saliva" or "to spit." Generally the home-pounded maize is chewed by older women and children. The ground maize is first moistened with water, rolled into a ball, placed in the mouth and thoroughly mixed with saliva using the tongue. The teeth play little role in the process. The "gob" is then pressed and flattened against the roof of the mouth and removed as a single mass with the fingers. The lumps are sun-dried and stacked.

The salivated flour called *muko* commands a high price on the market; thus, owners of haciendas may put as many young and older persons as are available to the task of producing the *muko.* The brewing process involves filling a wide-mouth, earthenware pot [about 30 inches (75 cm) high and 34 inches (85 cm) in diameter] one-third full with the dried, pulverized, salivated flour. Unsalivated flour and/or sugar may also be added. The pot is then filled with water and heated to 75°C (167°F). It is never boiled as such a process makes the product pasty. The pot contents are well-mixed and heated for 1 hour, then settled and cooled. Three layers are found: (1) The liquid top called *upi*, (2) a jellylike middle layer and, (3) the coarse particles on the bottom. The liquid layer is scooped out with a gourd and placed in another large-mouth pot where it is allowed to stand. The jellylike layer is placed in a shallow pan, simmered, and concentrated to a sugarlike product. More *muko* is added to the coarse particles on the bottom and the process is repeated. As a top layer of liquid is produced, it is added to the *upi* collected earlier. Some of the concentrated sugary jelly layer is also added to the liquid to be fermented to *chicha.* On the third day, the liquid becomes slightly bitter. On the fourth day, fermentation begins and the liquid *upi* bubbles violently. Fermentation is generally complete by 6 days or at latest 10 days depending upon the temperature. Floating froth is removed. Part of this may be used as an inoculum for new *chicha*; however, the earthenware pots are so inoculated with yeasts and microbes, no additional inoculum is usually required (Cutler and Cardenas, 1947).

Although the "saliva" method is still practiced in parts of the Andes region (Robinson, 1977), most *chicha* is now produced using germinated maize (Nicholson, 1960). In this process, red maize kernels are soaked in water in earthenware pots for 12 to 18 hours or overnight. The kernels are then germinated in the dark in layers 2 or 3 inches deep for 3 days. At this stage the plumule is from ¼ to ½ inch (6 to 12 mm) long and the kernel tastes sweet. Optimum germination temperature is 91°F (33°C). The most important consideration is that germination is uniform. The germinated maize is then "heaped up" carefully and covered with burlap. Enough heat is generated to burn a man's hand if placed in the "heap." Within two days, the kernels are white, parched, and covered with a thin layer of ash. The kernels are then sun-dried for 2 to 5 days. The sun-dried kernels, called *jora*, are milled to produce a flour from which *chicha* is then made. The milled *jora,* called *pachucho* is mixed with water and undergoes a series of boilings and processings to separate the hulls from the starch. Eventually it is strained through cloth and the liquid falls into prepared pots which have been used previously for fermentation and therefore contain a natural inoculum. The *chicha* is ready to drink when it has lost its sweetness and is "semi-sharp." Sometimes brown sugar or molasses is added to increase the alcohol content (Nicholson, 1960). Some *jora* is also chewed to make *chicha,* particularly in Bolivia. In this case, it is the young girls who do the chewing (Nicholson, 1960).

B. Related Fermentations

In Brazil, a beer called *kaschiri* is made in which cassava tubers are chewed as a first step. In Mozambique, women chew the yuca plant, spit it out, and allow it to ferment to a beer called *masata* (Pederson, 1971). The fermentations are probably similar to traditional *chicha.*

IV. ALCOHOLIC FOOD FERMENTATIONS INVOLVING AN AMYLOLYTIC MOLD AND YEAST

A. Indonesian *Tapé Ketan*

Indonesian *tapé ketan* is very closely related to primitive rice wine. Saké as produced centuries ago was a cloudy slurry containing the rice residues, the microorganisms, alcohol, and other metabolic products (Casal, 1940). The primary source of amylolytic enzymes in modern saké production is a *koji* which is described later. Under primitive conditions, it would be expected that amylolytic molds and yeasts also would be present in the fermenting mixture.

Indonesian *tapé ketan* is a sweet-sour, alcoholic paste in which an amylolytic mold *Amylomyces rouxii* and at least one yeast of the *Endomycopsis burtonii* type hydrolyze steamed rice starch to maltose and glucose and produce ethanol and organic acids which provide an attractive flavor and aroma (Ko, 1972). If yeasts of the genus *Hansenula* are present, the acids and ethanol are esterified producing highly aromatic esters. Fermentation time is 2 to 3 days at 30°C (86°F). With continued incubation, the product becomes more liquid.

One type of primitive rice wine is made from inocula which are obtainable in the markets of Indonesia as a product called *ragi* (in Thailand, *luk-paeng*) (Fig. 1). *Ragi* is a white, dried rice flour cake about 2.5 cm in diameter containing a variety of molds and yeasts including those described above (Dwidjoseputro and Wolf, 1970). Housewives buy the *ragi,* steam-soaked glutinous rice, inoculate it with the powdered *ragi,* place the inoculated rice into earthenware jugs with added water, and allow the mixture to ferment for 3 to 5 days (Fig. 2). The liquid portion is then consumed and additional water is added. Fermentation continues for 3 to 5 days following which the liquid portion is again drunk. This is repeated until all the rice has been consumed. Thus, there is no loss of nutrients. Any residual dregs are sun-dried and used as a type of *ragi*.

Detailed studies have been made of the biochemical and nutritional changes that occur during *tapé* fermentations (Ko, 1972; Cronk *et al.,* 1977). Most of the

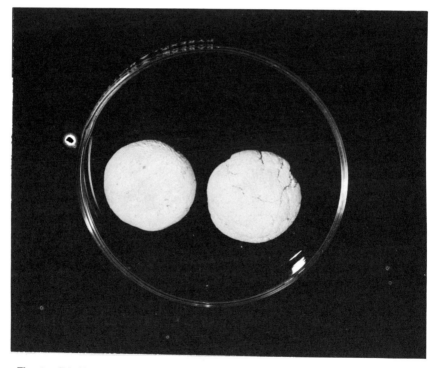

Fig. 1. Primitive rice wine manufacture in Thailand. *Luk-paeng (ragi)* dried yeast/rice flour cakes about 2.5 cm in diameter.

Fig. 2. Earthenware crocks used for the rice wine fermentation. Inoculated rice is placed in the crock and covered with water. Within 3 or 4 days a portion of the rice is fermented to rice wine. The liquid portion is consumed. Additional water is added and the fermentation continues. This is repeated until the rice grains and liquid have been entirely consumed. A portion of the dried dregs may be used in place of the traditional *luk-paeng* for inoculation of new batches of rice wine.

starch is hydrolyzed to sugars which, in turn, are fermented to ethanol and organic acids. Lysine, first limiting amino acid in rice, is selectively synthesized by the microorganisms so that it increases by 15%. Thiamine, which is very low in polished rice (0.04 mg/100 gm), is increased 300% (to 0.12 mg/100 gm) by the action of the microorganisms. Up to 8% ethanol is produced. This serves as calories for the consumers. It also undoubtedly contributes to destruction of food-spoilage and disease-producing organisms that might be present in the water and environment.

Through the loss of total solids due to utilization of the starch combined with the synthesis of new protein by the microorganisms, the protein content of *tapé ketan* is increased to as much as 16% (dry basis) versus 7–9% in the rice.

Thus, we have in the *tapé ketan* process a way of raising the protein quantity and quality in starch substrates and also producing thiamine which may very well be limiting in predominately polished rice diets (Cronk *et al.*, 1977).

B. Indonesian *Tapé Ketella (Peujeum)*

Protein enhancement is all the more important in the case of *tapé ketella,* which is also produced in Indonesia. *Tapé ketella* is a sweet–sour alcoholic food made from cassava tubers. The tubers are peeled, steamed, and cut into pieces about 5 × 5 cm (2 × 2 inches). They are then carefully inoculated on all surfaces with powdered *ragi. Amylomyces rouxii* and yeasts of the *Endomycopsis* or *Hansenula* genera, along with related molds and yeasts, overgrow the cassava utilizing a portion of the starch for energy. Cassava contains as little as 1–2% protein and, by itself, it is clearly unable to support proper protein nutrition even though it can provide sufficient calories. The biochemical changes that occur are similar to those in *tapé ketan*. Protein content can be increased to as high as 8% (Steinkraus and Cullen, 1977). Consumption of even a portion of the cassava as *tapé ketella,* thus can have a beneficial effect on the nutrition of the consumer.

C. Chinese *Lao-chao*

The starter for *lao-chao* is *chiu-yueh* or *peh-yueh* (Wang and Hesseltine, 1970). It is a gray-white ball containing yeasts and fungi grown on rice flour. It obviously is closely related to Indonesian *ragi*. The major species present are *Rhizopus oryzae, R. chinensis,* and *Chlamydomucor oryzae* now classified as *Amylomyces rouxii* (Ellis *et al.,* 1976). The process of manufacture appears to be identical to Indonesian *tapé ketan*. Wang and Hesseltine note that it is considered a good food to help new mothers regain their strength.

D. Indian Rice Beer, *Pachwai*

The starter for Indian *pachwai* is *bakhar*. The *bakhar* contains *Rhizopus* sp., *Mucor* sp., and at least one species of yeast. Ginger and other plant materials are dried, ground, and added to rice flour. Water is then added to make a thick paste and small round cakes about 1.0–1.5 cm in diameter are formed and inoculated with powdered cakes from previous batches. The cakes are then wrapped in leaves, allowed to ferment 3 days, and then sun-dried. *Pachwai* is manufactured by adding powdered *bakhar* to steamed rice and allowing the mixture to ferment 24 hours. The whole mass is then transferred to earthenware jars, water is added, and fermentation continues. The beer develops a characteristic alcoholic flavor and is ready to drink in 1 or 2 days (Hutchinson and Ram Ayyar, 1915). This fermentation also is closely related to Indonesian *tapé*.

V. ALCOHOLIC FOODS IN WHICH STARCH HYDROLYSIS INVOLVES A MALTING (GERMINATION) STEP

Germination as a means of converting starch to sugar and thus sweetening cereal products was most likely discovered soon after man started harvesting and storing grain. Grain stored in open pots and occasionally wet with rain would germinate. The grains became sweet and also underwent other desirable changes such as softening of texture and production of a sprout which was like a fresh vegetable. Sprouted grain left in an excess of water could have led to very early forms of primitive beers.

A. Kaffir (Sorghum) Beer

Kaffir beer is an alcoholic beverage with a pleasantly sour taste and the consistency of a thin gruel. It is the traditional beverage of the Bantu people of South Africa. Alcohol content may vary from 1 to 8% v/v. Kaffir beer is generally made from kaffir corn (*Sorghum caffrorum*) malt and unmalted kaffir corn meal. Maize or finger millet (*Eleusine coracana*) may be substituted for part or all of the kaffir corn depending upon the relative cost (Schwartz, 1956). Even cassava and plantains may be used (Platt, 1964).

The kaffir corn grain is steeped for 6 to 36 hours. It is then drained and placed in layers and germinated with periodic moistening for 4 to 6 days. Germination continues until the plumule is 1 inch (2.5 cm)/or longer in length. It is then sun-dried. Over 80% of the malt is sold for home brewing of kaffir beer. Although it is possible that molds growing on the grains during germination contribute to the amylolytic power of the malted grain, Schwartz (1956) found that the kaffir corn malt itself with diastatic powers of 11.6 to 17.2 (°Lintner) was sufficient to convert the starches for fermentation.

The essential steps in brewing are mashing, souring, boiling, conversion, straining, and alcoholic fermentation. Mashing is carried out in hot (50°C) water. Proportions of malted to unmalted grains vary, but 1:4 is satisfactory. One gallon of water is added for about every 4 pounds of grain. Souring begins immediately because of the presence of lactobacilli. A temperature of 50°C (122°F) favors development of *Lactobacillus delbrueckii*. Souring is complete in from 6 to 15 hours. Water is added and the mixture is boiled. It is then cooled to 40°–60°C (104°–140°F) and additional malt is added. Conversion proceeds for 2 hours and then the mash is cooled to 25°–30°C (77°–86°F).

Yeasts present on the malt are responsible for the natural fermentation. However, *Saccharomyces cerevisiae* isolated from kaffir beer is frequently inoculated in the more modern breweries. Kaffir beer is ready for consumption in 4 to 8 hours. It is drunk while still actively fermenting. This state may continue for up to 40 hours. Ethanol content is generally from 2 to 4% w/v. The beer also contains from 0.3 to 0.6% lactic acid and 4 to 10% solids. Production of acetic acid by *Acetobacter* is the principal cause of spoilage. van der Walt (1956) has described the microbiology of kaffir beer production. Other studies on the process include those by Platt and Webb (1946), Platt and Webb (1948), Aucamp *et al.* (1961), Horn and Schwartz (1961), von Holdt and Brand (1960a,b), O'Donovan and Novellie (1966), and Novellie (1959, 1960a,b, 1962a,b,c, 1966).

Platt (1964) described village processing of kaffir beer. Brewing is a woman's responsibility. Every girl learns the procedure before she marries.

Consumption each day of 5 pints of kaffir beer, which requires approximately 425 gm of grain, is not at all unusual for a working man. How the consumption of the beer diet compares in nutrient intake with a diet in which the grains are consumed directly is shown in Table III (Platt, 1964).

It is seen that caloric content of the two diets is quite similar. Only 37 calories are lost in the diet containing the beer. However, most notable is the doubling of riboflavin and near doubling of nicotinic acid content of the diet containing beer because of synthesis of vitamins during malting and fermentation. Pellagra, which is relatively common in those subsisting on maize diets, is never noted in those consuming usual amounts of kaffir beer (Platt, 1964).

Some people complain about the use of grain for beer making, especially when there is a shortage of grain. However, there is very little loss of calories during the fermentation, and it is believed that workers consuming the beer over a number of hours while working will utilize most of the ethanol for calories. Average daily beer consumption can be as high as 10 pints per worker. Approximately 35% of the caloric intake of the workers comes from beer. Platt (1955) reports that one village with 156 inhabitants consumed over 4000 gallons in one year. Half of the beer was consumed during regular work hours and the additional for "beer parties" during which collective work is done by women and men accomplishing tasks which individually would take much longer.

The infants in the villages are given the dregs which contain very little alcohol,

Table III. Comparison of Diet with and without Maize Beer[a]

	Amount of food eaten	
Food item	Diet without beer	Diet with Kaffir beer
Maize, whole meal	350 gm	137.5 gm
Maize, 60% extraction	350 gm	137.5 gm
Maize beer	—	5 pints (2840 ml)
Vegetables	130 gm	130 gm
Sweet potatoes	470 gm	470 gm
Kidney beans	30 gm	30 gm
Calories	3016	2979
Vitamin B_1	2.00 mg	1.95 mg
Riboflavin	1.13 mg	2.32 mg
Nicotinic acid	11.70 mg	20.30 mg

[a] Adapted from Platt (1964).

Table IV. Effect of Germination on the Vitamin Content of Kaffir Corn and Maize[a]

		Vitamin content (μg/gm) (dry weight basis)		
Sample	Moisture (%)	Thiamine	Riboflavin	Nicotinic acid
Kaffir corn				
Grain	11.6	3.34	1.29	35.1
Malt	10.7	1.73	2.41	34.1
Maize				
Grain	12.1	3.50	1.46	19.6
Malt	8.6	2.32	2.02	29.1

[a] Adapted from Goldberg and Thorp (1946).

are rich in vitamins and protein content, and very likely highly beneficial to infant nutrition.

The effect of germination on vitamin content of kaffir corn and maize is presented in Table IV. Thiamine decreases, riboflavin shows a substantial increase, nicotinic acid remains nearly constant in kaffir corn but increases by about 50% in maize. Thus, the microorganisms involved in the fermentation must be contributing additional thiamine, riboflavin, and nicotinic acid to the product.

Aucamp *et al.* (1961) report the results of chemical analyses of municipal kaffir beers (Table V). According to these figures, the content of ascorbic acid in kaffir beer is negligible. However, consumption of one-half gallon of beer per

Table V. Composition of Municipal Kaffir Beers[a]

	Range	Mean	Number of analyses
pH	3.2–3.7	3.4	10
Alcohol (% w/v)	1.8–3.9	3.0	17
Solids (% w/v)			
Total	3.0–8.0	5.4	17
Insoluble	2.3–6.1	3.7	6
Nitrogen (% w/v)			
Total	0.059–0.137	0.093	16
Soluble	0.010–0.017	0.014	9
Thiamine (μg/100 ml)	20–230	93	21
Riboflavin (μg/100 ml)	27–170	56	21
Nicotinic acid (μg/100 ml)	130–660	315	21
Ascorbic acid (mg/100 ml)	0.01–0.15	0.04	7

[a] From Aucamp *et al.* (1961).

Table VI.   Amounts of B Vitamins Supplied by One-half Gallon of Kaffir Beer[a]

	Thiamine (mg)	Riboflavin (mg)	Nicotinic acid (mg)
Minimum	0.45	0.61	2.9
Maximum	5.2	3.9	15.0
Mean	2.1	1.3	7.2
Minimum daily requirement[b]	1.0	1.2	10.0
Recommended allowance[c]	1.5	1.8	20.0

[a] Calculated from values in Table V.
[b] United States Food and Drug Administration (Harris, 1959).
[c] Adult males; Food and Nutrition Board, National Research Council (1974).

day, which is rather common, would supply more than the minimum daily requirements (Harris, 1959) of thiamine, riboflavin, and nicotinic acid as shown in Table VI. The vitamins supplied, however, would be less than the allowances recommended for American adult males by the Food and Nutrition Board, National Research Council (1974).

Nevertheless, kaffir beer nutritionally is a very important part of the South African village diet.

B. Egyptian *Bouza*

Egyptian *bouza,* a fermented alcoholic wheat beverage, has been known since the Pharaohs (Morcos *et al.,* 1973). In earlier times, *bouza* was consumed by all classes, including children. Most *bouza* today is consumed in villages by the lower income groups.

Bouza is a thick, sour, pale yellow, alcoholic beverage with an agreeable flavor. Wheat (*Triticum vulgaris*) is the preferred substrate; however, maize is sometimes used. Coarsely ground wheat grains are placed in wooden basins and kneaded with water into a dough. The dough is cut into thick loaves and baked lightly. About one-quarter of the total wheat grains used are moistened with water and germinated for 3 to 5 days, then sun-dried and ground. The germinated wheat flour is mixed with crumbled bread loaves and soaked in water in a wooden barrel. *Bouza* from a previous fermentation is added as an inoculum. The mixture is allowed to ferment for 24 hours at room temperature, sieved to remove large particles, and is ready to consume. The pH at 24 hours is 3.9 to 4.0; alcohol content is 3.8 to 4.2%. If allowed to ferment another 24 hours (48-hour *bouza*), the pH falls to 3.6–3.8 and the ethanol content rises from 4.1 to 5.0%. By 72 hours, the pH falls to 3.5–3.7 and ethanol content reaches 4.4–5.4%. Total solids content in *bouza* ranges from 12.4 to 16.2% (24-hour *bouza*). Protein content (dry weight basis) of *bouza* gradually increases as fermentation time increases: 11.8% (24 hours); 12.0–12.5% (48 hours); and 12.7–13.1% (72 hours) (Morcos *et al.*, 1973). The increase in protein content most likely reflects the additional growth of yeasts and bacteria as fermentation progresses. The protein content, the vitamins, and the calories consumed in *bouza* undoubtedly play an important role in the nutrition of the lower income groups of Egypt even today. Further research is needed on the content of specific vitamins in *bouza*.

C. Ethiopian *Talla*

Talla is a tan to dark brown home-processed beer with smokey flavor. Amhara *talla* is hopped and concentrated; Gurage *talla* is delicately aromatized with a variety of spice. Oromo galla *talla* is unhopped, thick, and sweet. *Talla* is used as a solvent for tapeworm medicine and other drugs. The mold formed on the dregs is used as a cure for wounds and amebic dysentery (Vogel and Gobezie, 1977).

The smoked flavor is derived from inversion of the fermentation container over smouldering olive wood. Grains are also toasted until they are partly charred and begin to smoke slightly. Barley or wheat is cleaned and soaked overnight covered with water. The grains are then drained and allowed to sprout for three days, sun-dried, and ground into a flour. Powdered hop leaves and water are placed in the fermentation container and allowed to stand for 3 days. On the third day ground barley or wheat malt is added along with pieces of flat bread baked to a crisp coal black on the outside. On the fifth day, pulverized hop stems are added along with cereal flour made by boiling, toasting, and milling sorghum, millet, teff, barley, wheat, or maize flour. On day 6, water is added, the pot is covered and fermentation continues for 5 to 7 days. It is then strained and served.

Some similarities are seen in the methods of manufacturing *bouza* and *talla*. Both utilize bread in the fermentation. However, the partial charring of the grains and the bread in *talla* manufacture appear to be unique and the reason for such drastic heating is not readily apparent. Certainly heating the grains or bread to the point of charring damages protein and the availability of amino acids such as lysine. Further research is needed on the nutritive value of *talla*.

D. Mexican *Tesgüino*

Tesgüino is an alcoholic slurry prepared by fermentation of germinated maize. This beverage is very important in the everyday life of the native Indian population. It is the preferred beverage for religious and other rituals. Diluted with water it is also fed to infants and is consumed in the home. From one-quarter to several liters may be consumed per day (Pennington, 1969; Ulloa and Ulloa, 1973; Herrera *et al.*, 1972; Ulloa *et al.*, 1974).

Soaked maize kernels are placed in small baskets in the dark until they germinate. The Indians believe that sprouts germinated in the light are green and bitter, while germination in complete darkness yields white, sweet sprouts which will yield a sweet *tesgüino*. The sprouted grains are ground in a hand mill and boiled in water until the mixture is yellow (approximately 8 hours). The liquid is then transferred to a clay pot and certain barks are added which have been previously boiled in water for hours. Fermentation continues for 3 to 4 days. *Saccharomyces cerevisiae* appears to be the most important organism in the fermentation. It apparently is transferred from fermentation to fermentation by the reused utensils and clay pots (Ulloa and Ulloa, 1973).

Little specific information regarding nutritional changes that occur during fermentation has been found in the literature. However, based upon the changes occurring in similar or related fermentations, i.e., kaffir beer and *bouza*, it would be expected that the proteins and B vitamins consumed in *tesgüino* contribute importantly to the nutrition of the Mexican Indians.

E. Russian *Kvass*

Russian *kvass* is prepared by mixing equal parts of barley malt, rye malt, and rye flour, adding boiling water, stirring and permitting the mash to stand and cool for several hours, adding more boiling water, and then cooling and inoculating with yeast. Peppermint is added to the final product for flavoring (Prescott and Dunn, 1959, pp. 170–171).

VI. ALCOHOLIC FERMENTATIONS INVOLVING USE OF A *KOJI* FOR STARCH HYDROLYSIS

A. *Koji*

The *koji* process was developed centuries ago in the Orient (Murakami, 1972). The mold-covered rice containing the enzymes is called *koji*. It serves as a source of enzymes for the saccharification of starch in the saké brewing process much as malt serves a similar purpose in Western beer manufacture. A sporulated mold culture (*tane-koji*) is produced by culturing *Aspergillus oryzae* on soaked, steamed polished rice at 28°–30°C (82° to 86°F) for 5 or 6 days or until there is abundant sporulation. The *tane-koji* is then used to inoculate larger quantities of steamed rice to produce heavy mycelial growth and maximum enzyme production on and through the rice. This requires about 48 hours at 28°–30°C (82 to 86°F). The most important enzymes in *saké-koji* are the amylases, but proteases and lipases are also present. The *koji* should be used as soon as harvested or dried for stability.

B. Japanese Rice Wine, Saké

Saké is a clear, pale yellow wine with an alcoholic content of 15–16% or higher, a characteristic aroma, little acid and slight sweetness (Murakami, 1972; Kodama and Yoshizawa, 1977). Japanese saké is closely related to Chinese rice wine, *shaosing chu*. Both processes utilize a *koji* to saccharify the starches. However *shaosing chu* has a deeper color and a more oxidized "sherry-like" flavor. It usually contains some wheat.

Japanese *amazake* is home-brewed rice wine. Boiled rice is allowed to mold and then mixed with freshly boiled rice and water. In time, it ferments and is consumed (Casal, 1940). In feudal times, saké was a home industry. The product was very likely unclarified and consumed cloudy with the microorganisms and residual solids much as other primitive rice wines are consumed today (Casal, 1940, p. 71; Rabbitt, 1940).

Rice wines are widely consumed in the Orient. About 1.6 million kiloliters of saké were produced in Japan in 1970 (Murakami, 1972). Of the factories producing rice wines 87% are small, producing less than 500 kiloliters of saké per year.

It is likely that wines made from millet predate rice wines in both China and Japan. Rice wine was probably developed in China centuries before it was produced in Japan (Casal, 1940).

In beer brewing, saccharification of the starch is a separate step followed by filtration and then by fermentation of the wort. In saké brewing, saccharification and fermentation proceed simultaneously in the unfiltered mash (called *moromi*). This particular set of circumstances leads to very high populations of yeast cells in the mash and high ethanol contents in the saké. It is not unusual to have the ethanol content reach 20% v/v.

The sequence of microorganisms is as follows. First, growth of *Pseudomonas* produces nitric acid from nitrates in the water. This is followed by growth of *Leuconostoc mesenteroides* and *Lactobacillus sake* which acidify the mash and eliminate the nitric acid. Then the yeast *Saccharomyces sake* (closely related to *Saccharomyces cerevisiae*) propagates to levels of 3 or 4 × 10^8 cells per milliliter of mash at which time it makes up 95 to 99% of the microflora.

High quality rice is used for saké manufacture. It is necessary to remove the bran and 25–30% of the weight of the original kernels during polishing for high quality saké. The brewing water must contain not more than 0.02 ppm iron as it results in a deep color in the saké.

Recent developments include acidification of the mash with lactic acid. This decreases the amount of bacterial activity in the mash and allows a higher fermentation temperature 15°C (59°F), thus shortening the overall fermentation time (Murakami, 1972).

Along with the *koji,* it is necessary to build up the yeast inoculum called *moto* (Kodama and Yoshizawa, 1977). Traditional *ki-moto* undergoes acidification because of lactic acid bacteria in the mash. In *sokujo-moto* 140 ml of lactic acid (75%) is added to 60 kg of *koji,* 140 kg steamed rice, and 200 liters of water to acidify the mash. Yeasts (10^5 to 10^6 cells per milliliter) are then inoculated. While traditional *moto* requires 10 to 15 days to develop, acidified *moto* is ready in a few days.

The main mash is called *moromi.* Proportions of *koji,* water, and steamed rice are the same as for *moto.* Sugar concentration due to starch hydrolysis reaches 20%. Saké brewing is a relatively low-temperature process. The temperature begins at 7° or 8°C and is gradually raised to 15° to 16°C. Fermentation time is about 20–25 days. It is then filtered and clarified like other modern alcoholic beverages. Final sugar concentration can be adjusted by addition of steamed rice to the mash before filtration. There is enough amylolytic activity left to convert the added rice starch to sugar, which then remains in the saké. If necessary the saké can be fortified with added alcohol to bring the ethanol content to 20–22%, thus increasing the stability of the wine.

The saké process is included in this chapter principally because it is an example of an alternative method of saccharifying starch in the production of alcoholic foods. The clarified rice wines are too expensive for wide usage among the lower income groups and they therefore do not contribute much to the basic nutritional requirements of the masses.

VII. ALCOHOLIC MILK BEVERAGES

A. *Busa*

Busa is a fermented alcoholic milk produced in Turkey. It contains up to 0.78% lactic acid and up to 7.1% ethanol by weight. Since the whole product is consumed, it would seem to be very rich in nutrients (Tschekan, 1929).

B. *Kefir*

Kefir is a well-known fermented milk consumed in the Balkans. It is described as an acidic, mildly alcoholic, effervescent beverage (la Riviere *et al.,* 1967). The fermentation is carried out by a yeast *Torulopsis holmii* and *Lactobacillus brevis* growing symbiotically in a polysaccharide matrix called *kefiran.* The complex is called a *kefir* "grain" and doubles in weight every 7 to 10 days if transferred daily into fresh milk (la Riviere, 1969). The concentration of ethanol in *kefir* is very low, generally less than 1% (Steinkraus and Cullen, 1977). Fermentation time is approximately 24 hours at room temperature. Keeping quality and nutritive value are excellent.

C. Koumiss

Koumiss is fermented mare's milk produced and consumed in Russia. The inoculum is *koumiss* from a previous batch. Fermentation time for "new" *koumiss* is 12–24 hours. "Medium" *koumiss* requires another 12–20 hours fermentation. This is the most commonly consumed product. "Strong" *koumiss* requires 56–72 hours fermentation. Medium *koumiss* contains 1–2% ethanol and 0.6–0.8% lactic acid. *Koumiss* is effervescent and contains from 0.6 to 1.3% carbon dioxide. Protein content of *koumiss* ranges from 2.3 to 2.6%. The fermenting microorganisms are a yeast and a bacterium. *Streptococcus lactis* and *Bacillus acidi lactici* are also generally present (Rubinsky, 1910).

D. Mazun

Mazun is an Armenian fermented beverage made from cow, goat, or buffalo milk. It contains both alcohol and acids when fermented at 30°C (Emmerling, 1898).

VIII. SIGNIFICANCE OF INDIGENOUS ALCOHOLIC FOODS IN HUMAN NUTRITION

Alcoholic foods and beverages are highly acceptable the world over except among those groups that forbid the consumption of alcohol for religious reasons. When such a degree of acceptability can be combined with an important contribution to the caloric, protein, and vitamin content of the diet, obviously this is an ideal combination. Such is the case with most, if not all, the indigenous primitive alcoholic foods described in this chapter. Another important point is that all the foods described except perhaps clarified Japanese saké and Chinese *shaosing chu* can be manufactured at the village level at costs within the range of the majority of the world's masses of poor and hungry.

One of the great advances in microbiology over recent years has been the development of technology enabling the world to produce single cell protein (SCP) on inedible substrates such as hydrocarbons with addition of inorganic nitrogen and minerals. Important as it is, this is capital-intensive, sophisticated technology. The SCP must be separated from the substrate and formulated into foods. The products cannot be within the means of the world poor and hungry. In the indigenous alcoholic foods described herein, microorganisms, i.e., single cell proteins are produced on edible substrates and the microorganisms, the residual substrate, and products are all consumed together. This is labor intensive, low-cost technology. And, if the world wishes to improve the nutrition of its poor and hungry, it must look to these processes for the means of achieving its goals.

Malting is a relatively simple, low-cost method of increasing the vitamin content of cereal grains. Malting by itself can approximately double the content of riboflavin and niacin in the food. It can also increase the ascorbic acid content. This does involve some handling of the grains, but it does not involve the expenditure of dollars which would be required to achieve the same degree of enrichment by adding vitamins to the product. Growth of yeasts and bacteria in the indigenous alcoholic foods also results in some selective synthesis of amino acids. The balance of amino acids can be improved as it is in *tapé ketan* and at very little cost. There is a net increase in protein in some of the products which can help the consumer improve his diet.

There are a few words of caution. Most of the primitive alcoholic beverages described are highly nutritious because they have not been refined and Westernized. The Western world wants most of its beverages to be crystal clear. Fortunately, many primitive societies still prefer their beverages heavy and cloudy. Much of the improved nutritional value is in the substrate residues and the microorganisms which develop during the fermentation. It would be a tragedy if primitive societies were persuaded to clarify their fermented alcoholic foods thus losing some of the nutrients they presently enjoy.

IX. SUMMARY

Some important indigenous alcoholic foods and beverages have been reviewed. These have been classified into those in which sugars are the major substrate, those in which saliva is the amylolytic agent, whose in which starch hydrolysis is accomplished by an amylolytic mold and yeast, those in which starch hydrolysis is accomplished by a malting step, those involving use of a *koji* in which an amylolytic mold is grown in a starch substrate as a source of enzymes, and alcoholic foods with a milk base. The world has hundreds of alcoholic beverages, but it is believed that most if not all of them will fit one or more of the classes described. Also, the fermentations included can be used as patterns for producing alcoholic foods from nearly any starch- or sugar-containing substrate using simple technology. Based upon the biochemical and nutritional changes occurring in the indigenous alcoholic foods thus far studied, it is possible to forecast the nutritional improvements in vitamins and protein quality that can be achieved at low cost and in the form of highly acceptable foods.

REFERENCES

Aucamp, M. C., Grieff, J. T., Novellie, L., Papendick, B., Schwartz, H. M., and Steer, A. G. (1961). Kaffircorn malting and brewing studies. VIII. Nutritive value of some kaffircorn products. *J. Sci. Food Agric.* **12**, 449-456.

Bassir, O. (1967). Some Nigerian wines. *West Afr. J. Biol. Appl. Chem.* **10**, 42-45.

Batra, L. R., and Millner, P. D. (1976). Asian fermented foods and beverages. *Dev. Ind. Microbiol.* **17**, 117-128.

Brothwell, D., and Brothwell, P. (1969). "Foods in Antiquity." Thames & Hudson, London.

Casal, U. A. (1940). Some notes on *sakazuki* and on the role of *saké* drinking in Japan. *Trans. Asiat. Soc. Jpn. Ser. 2* **19**, 1-186.

Cronk, T. C., Steinkraus, K. H., Hackler, L. R., and Mattick, L. R. (1977). Indonesian tapé ketan fermentation. *Appl. Environ. Microbiol.* **33**, 1067-1073.

Cutler, H. C., and Cardenas, M. (1947). Chicha, a native South American beer. *Bot. Mus. Leafl., Harv. Univ.* **13**(6), 33-60.

Dwidjoseputro, D., and Wolf, F. T. (1970). Microbiological studies of Indonesian fermented foodstuffs. *Mycopathol. Mycol. Appl.* **41**, 211-222.

Ellis, J. J., Rhodes, L. J., and Hesseltine, C. W. (1976). The genus *Amylomyces. Mycologia* **68**, 131-143.

Emmerling, O. (1898). Ueber Armenisches mazun. *Centralbl. Bakteriol., Abt. 2* **4**, 418-420.

Faparusi, S. I. (1973). Origin of initial microflora of palm wine from oil palm trees (*Elaeis guineensis*). *J. Appl. Bacteriol.* **36**, 559-565.

Food and Nutrition Board, National Research Council. (1974). "Recommended Dietary Allowances," 8th ed. Natl. Acad. Sci., Washington, D.C.

Goldberg, L., and Thorp, J. M. (1946). A survey of vitamins in African foodstuffs. VI. Thiamin, riboflavin and nicotinic acid in sprouted and fermented cereal products. *S. Afr. J. Med. Sci.* **11**, 177-185.

Goncalves de Lima, O. (1975). "Pulque, Balche Pajauaru." Univ. Fed. Pernambuco, Recife, Brasil.

Harris, R. S. (1959). Supplementation of foods with vitamins. *Agric. Food Chem.* **7**(2), 88-102.

Herrera, T., Taboada, J., and Ulloa, M. (1972). Fijacion de nitrogeno en al tesguino y el pulque. *An. Inst. Biol., Univ. Nac. Auton. Mex., Ser. Biol. Exp.* **43**(1), 77-78.

Hesseltine, C. W. (1965). A millenium of fungi, food and fermentation. *Mycologia* **57**, 149-197.

Horn, P. J., and Schwartz, H. M. (1961). Kaffircorn malting and brewing studies. IX. Amino acid composition of kaffircorn grain and malt. *J. Sci. Food Agric.* **12**, 457-459.

Hutchinson, C. M., and Ram Ayyar, C. S. (1915). Microbiology of bakhar. *Mem. Dep. Agric. India* **1**, 137-168.

Instituto Nacional de Nutricion (1976). *Encuestas Nutr. Mex.* **2**, 91-111.

Ko, S. D. (1972). Tapé fermentation. *J. Appl. Microbiol.* **23**, 976-978.

Kodama, K., and Yoshizawa, K. (1977). Saké. *In* "Economic Microbiology" (A. H. Rose, ed.), pp. 423-475. Academic Press, New York.

Kozaki, M. (1976). Fermented foods and related microorganisms in Southeast Asia. *Proc. Jpn. Assoc. Mycotoxicol.* **2**, 1-9.

la Riviere, J. W. M. (1969). Ecology of yeasts in the Kefir grain. *Antoine van Leeuwenhoek; J. Microbiol. Serol.* **35**, Suppl. Yeast Symp. 15-16.

la Riviere, J. W. M., Kooiman, P., and Schmidt, K. (1967). Kefiran, a novel polysaccharide produced in the kefir grain by *Lactobacillus brevis. Arch. Mikrobiol.* **59**, 269-278.

Morcos, S. R., Hegazi, S. M., and El-Damhougi, S. T. (1973). Fermented foods in common use in Egypt. II. The chemical composition of *bouza* and its ingredients. *J. Sci. Food Agric.* **24**, 1157-1161.

Morse, R. A., and Steinkraus, K. H. (1975). Wines from the fermentation of honey. *In* "Honey: A Comprehensive Survey" (E. Crane, ed.), pp. 392-407. Heinemann, London.

Murakami, H. (1972). Some problems in saké brewing. *Ferment. Technol., Proc. Int. Ferment. Symp., 4th, Kyoto,* pp. 639–643.

Nicholson, G. E. (1960). Chicha maize types and chicha manufacture in Peru. *Econ. Bot.* **14,** 290–299.

Novellie, L. (1959). Kaffircorn malting and brewing studies. III. Determination of amylases in Kaffircorn malts. *J. Sci. Food Agric.* **10,** 441–449.

Novellie, L. (1960a). Kaffircorn malting and brewing studies. IV. The extraction and nature of the insoluble amylases of kaffircorn malts. *J. Sci. Food Agric.* **11,** 408–421.

Novellie, L. (1960b). Kaffircorn malting and brewing studies. V. Occurrence of β-amylase in kaffircorn malts. *J. Sci. Food Agric.* **11,** 457–471.

Novellie, L. (1962a). Kaffircorn malting and brewing studies. XI. Effect of malting conditions on the diastatic power of kaffircorn malt. *J. Sci. Food Agric.* **13,** 115–120.

Novellie, L. (1962b). Kaffircorn malting and brewing studies. XII. Effect of malting conditions on malting losses and total amylase activity. *J. Sci. Food Agric.* **13,** 121–124.

Novellie, L. (1962c). Kaffircorn malting and brewing studies. XIII. Variation of diastatic power with variety, season, maturity and age of grain. *J. Sci. Food Agric.* **13,** 124–126.

Novellie, L. (1966). Kaffircorn malting and brewing studies. XIV. Mashing with kaffircorn malt. Factors affecting sugar production. *J. Sci. Food Agric.* **17,** 354–365.

O'Donovan, M. B., and Novellie, L. (1966). Kaffircorn malting and brewing studies. XV. The fusel oils of kaffir beer. *J. Sci. Food Agric.* **17,** 362–365.

Okafor, N. (1972). Palm-wine yeasts from parts of Nigeria. *J. Sci. Food Agric.* **23,** 1399–1407.

Okafor, N. (1975a). Preliminary microbiological studies on the preservation of palm wines. *J. Appl. Bacteriol.* **38,** 1–7.

Okafor, N. (1975b). Microbiology of Nigerian palm wine with particular reference to bacteria. *J. Appl. Bacteriol.* **38,** 81–88.

Pederson, C. S. (1971). "Microbiology of Food Fermentations." Avi, Westport, Connecticut.

Pennington, C. W. (1969). "The Tepehuan of Chihuahua—Their Material Culture." Univ. of Utah Press, Logan.

Platt, B. S. (1946). "Nutrition in the British West Indies," Colon. Rep. No. 195. HM Stationery Off., London.

Platt, B. S. (1955). Some traditional alcoholic beverages and their importance in indigenous African communities. *Proc. Nutr. Soc.* **14,** 115–124.

Platt, B. S. (1964). Biological ennoblement: Improvement of the nutritive value of foods and dietary regimens by biological agencies. *Food Technol.* **18,** 662–670.

Platt, B. S., and Webb, R. A. (1946). Fermentation and human nutrition. *Proc. Nutr. Soc.* **4,** 132–140.

Platt, B. S., and Webb, R. A. (1948). Microbiological protein and human nutrition. *Chem. Ind. (London)* Feb. 7, 88–90.

Prescott, S. C., and Dunn, C. G. (1959). "Industrial Microbiology," 3rd ed. McGraw-Hill, New York.

Rabbitt, A. (1940). Rice in the cultural life of the Japanese people. *Trans. Asiat. Soc. Jpn.* **19,** 187–259.

Robinson, W. B. (1977). Personal communication.

Rubinsky, B. (1910). Studien über den Kumiss. *Centralbl. Bakteriol., Abt. 2* **28,** 161–219.

Sanchez-Marroquin, A. (1953). The biochemical activity of some microorganisms of pulque. *Mem. Congr. Cientif. Mex., 4th, Centenario Univ. Mex.* pp. 471–484.

Sanchez-Marroquin, A., and Baldrano, H. (1949). Studies on the microbiology of pulque. VIII. Presence of some bacterial growth factors. *An. Esc. Nac. Cienc. Biol., Mexico City* **6,** 31–39.

Sanchez-Marroquin, A., and Hope, P. H. (1953). Agave juice fermentation and chemical composition studies of some species. *J. Agric. Food Chem.* **1,** 246–249.

Schwartz, H. M. (1956). Kaffircorn malting and brewing studies. I. The kaffir beer brewing industry in South Africa. *J. Sci. Food Agric.* **7,** 101–105.

Steinkraus, K. H., and Cullen, R. (1977). Unpublished data.

Tschekan, L. (1929). Mikrobiologie der Busa. *Centralbl. Bakteriol., Abt. 2* **78,** 74–93.

Ulloa, M., Salinas, C., and Herrera, T. (1974). Estudio de *Bacillus megaterium* aislado del Tesgüino de Chihauhan Mexico. *Rev. Latinoam. Microbiol.* **16,** 209–211.

Ulloa, S. C., and Ulloa, M. (1973). Alimentos fermentados de maiz consumidos en Mexico y otros paises Latinoamericanos. *Rev. Soc. Mex. Hist. Nat.* **34,** 423–457.

van der Walt, J. P. (1956). Kaffircorn malting and brewing studies. II. Studies on the microbiology of kaffir beer. *J. Sci. Food Agric.* **7,** 105–113.

von Holdt, M. M., and Brand, J. C. (1960a). Kaffircorn malting and brewing studies. VI. Starch content of kaffir beer brewing materials. *J. Sci. Food Agric.* **11,** 463–467.

von Holdt, M. M., and Brand, J. C. (1960b). Kaffircorn malting and brewing studies. VII. Changes in the carbohydrates of kaffircorn during malting. *J. Sci. Food Agric.* **11,** 467–471.

van Pee, W., and Swings, J. G. (1971). Chemical and microbiological studies on Congolese palm wines (*Elaeis guineensis*). *East Afr. Agric. For. J.* **36**(3), 311–314.

Vogel, S., and Gobezie, A. (1977). Personal communication.

Wang, H. L., and Hesseltine, C. W. (1970). Sufu and lao-chao. *J. Agric. Food Chem.* **18,** 572–575.

THE NUTRIENT CONTRIBUTIONS OF
FERMENTED BEVERAGES

W. J. Darby

I. Nature of Nutritional Contribution of Beverages 61
II. Time-Honored Precepts of Use 62
III. Excesses and Nutritional Deficiencies 66
IV. Direct Nutrient Contributions of Fermented Beverages 67
V. Clarification, Filtration, and Distillation 69
VI. Trace Elements.. 71
VII. Organic Compounds in Fermented Beverages 71
VIII. Summary 72
    References 73
    Editorial Comment 75

I. NATURE OF NUTRITIONAL CONTRIBUTIONS OF BEVERAGES

Interest of the physician in the nutritional roles of fermented beverages stems from (a) a widely shared personal enjoyment of moderate amounts of tasteful beverages and (b) concern for the harmful effects on patients of excessive amounts of the beverages. To consider these points one immediately asks: What are "moderate amounts" of alcoholic beverages? What contribution do they make to nutrition or how may they influence nutriture? What are excessive quantities? What are the consequences or nutritional risks of excesses? How does one enjoy the pleasures and benefits of such beverages within the acceptable risk–benefit framework? Such are considerations developed in this volume or for which the authoritative reviews provide a basis for making judgments. Several

FERMENTED FOOD BEVERAGES IN NUTRITION
Copyright © 1979 by Academic Press, Inc.
All rights of reproduction in any form reserved.
ISBN 0-12-277050-1

chapters also identify the voids in scientific information and areas of needed future research.

The roles of fermented food beverages in nutrition (1) are several:

1. They provide a pleasurable aspect to the intake of food; their enjoyment adds a quality that in the value judgments of many throughout centuries of human experience are deemed desirable.

2. They contribute to fulfillment of a subjective desire for something that the individual's past experience has associated with the pleasure of eating.

3. Some satisfy one's thirst.

These properties serve to enhance the pleasure of eating, and are widely held to stimulate appetite as well.

The effects of fermented beverages upon digestion, absorption, metabolism, and utilization of nutrients can have nutritional significance. These effects depend on many considerations, especially quantity and type of beverage, time and circumstance of use, interaction with drugs, and the physiologic or disease state. These matters are considered elsewhere in this volume.

II. TIME-HONORED PRECEPTS OF USE

The nutrient composition of fermented food beverages varies with the type of the beverage, and hence different beverages make differing direct contributions to the nutriture of the individual.

Before considering nutrient composition, however, I would note that for centuries the pleasurable benefits of moderate use of alcoholic beverages have been repeatedly described and also the consequences of excessive use recognized and warned against. Dr. Todhunter (Chapter 5) quotes the "many and singular commodities of wine" as described by Tho. Venner, Doctor of Physick in Bathe, in his "Via Recta ad Longam, or, A Treatise wherein the right way and best manner of living for attaining to a long and healthfull life, is clearly demonstrated and punctually applied to every age and constitution of body" (London, 1650) (2).

It further is of interest to observe the early precepts regarding the appropriate use of fermented beverages at various ages as set forth by Tho. Venner:

Ive: the firſt is, that it be not given unto children, for this will be as if you ſhould adde fire
unto fire : For they being of hot and moiſt
temperature, would thereby become over-hot, and
their heads alſo filled with vapors, whereof enſue many
evils, and ſometimes the falling ſickneſſe. The ſecond
is, that it be not given to Youths, as from 14 yeares
of age unto 25, for wine is unto them moſt repugnant;
becauſe it doth above meaſure heat their haſtie, hot,

and agitating nature, and extimulate them (like mad men)unto enormious and outrageous actions. The third is,that it be very moderately given, and that not too often unto young men,as from 25 years of age un to 35, and that it be alfo of the fmaller forts of wines , as Claret, &c. efpecially if they are of hot conftituti on : for otherwife it will make them prone unto wrath and unlawfull defires, dull the wit, and confound the memory. The fourth is, that it be more liberaliy gi ven unto them that are in their manhood and conftant age, as from 35 yeares unto 50,and let fuch when they are paft forty years of age, begin to make much of the ufe of wine : and yet if they be of hot conftitutions, let them abftain from the ftronger forts of wines,efpe cially from the often ufe of them , becaufe they will be offenfive unto the head and finewes. The fift is, that it be given with a liberall hand unto old men,and that alfo of the ftronger forts of wines , efpecially when they are in the later part of old age , as from 60 yeares upward unto the end of their life. For un to old men there come foure excellent commodities, by the ufe of pure wine. The firft and greateft com modity , feeing that they are cold, and for the moft part alfo without good alimentall bloud, is , becaufe it greatly correcteth the coldnes of their age, and bring eth them unto a better temperature of heat , with in creafe of bloud. The fecond, becaufe it expelleth fad nefle and melancholy, whereunto that age is moft fub ject. The third is, becaufe it maketh them to fleepe well, which by reafon of the ficcity of the braine, and paucity of vapours , many old men oftentimes want. The fourth and laft commodity is , becaufe it remo veth obftructions, whereunto they are very fubject. To conclude, as pure wine is moft unmeet and hurt full for children , and fuch as are young : fo for old men it is moft convenient and wholfome.

Foure princi-pall commodi-ties come unto the aged by the ufe of pure wine.

Venner's precepts are so similar to the advice of Plato concerning improper use of such beverages as well as to their use at different ages as to suggest that he may have based them upon Plato's advice. The translated Plato, as quoted by Darby *et al.* (3) reads:

... to drink to the point of intoxication is not proper to any other occasion except the festivals in honour of the god who gave the wine, and it is not safe; neither is it appropriate at the time when one is seriously engaged in the business of marriage, wherein, more than at any other time, bride and groom ought to be in their sound senses, since they are undergoing no little change in their lives; and at the same time, because their off-spring ought in all cases to be born of sound-minded parents ...

VIA RECTA

AD

John: Cotes his (handwritten) Vitam Longam. (handwritten +)

Booke 1682:) (handwritten)   OR,

A Treatife wherein the right way and beft
manner of living for attaining to a long and
healthfull life , is clearly demonftated and punctually
applied to every age and conftitution of body.

Much more enlarged than the former Impreſsions.

By T HO. V ENNER Doctor of Phyfick in Bathe.

Whereunto is annexed by the fame Authour,
A very neceffary, and compendious Treatife of the famous Baths of B A T H E.

WITH

A Cenfure of the Medicinall faculties of the Water of
St. *Vincents-Rocks* neer the City of *Briftoll.*

As alfo

An accurate Treatife concerning T O B A C C O.

All which are likewife amplified ſince the former Impreſsions.

John: Cotes (handwritten) **LONDON,**

Printed by *James Flesher*, for *Henry Hood*, and are to be fold at his Shop
in Saint *Dunftans* Church-yard in Fleetftreet.

2   **1650.**

Elsewhere Plato wrote:

> ... boys under eighteen shall not taste wine at all; for one should not conduct fire to fire; wine in moderation may be tasted until one is thirty years old, but the young man should abstain entirely from drunkenness and excessive drinking; but when a man is entering upon his fortieth year he, after a feast at the public mess, may summon the other gods and particularly call upon Dionysus to join the old men's holy rite, and their mirth as well, which the god has given to men to lighten their burden—wine, that is, the cure for the crabbedness of old age, whereby we may renew our youth and enjoy forgetfulness of despair ... (Laws, II, 666, A-B)

Again, note the similarity of wording in Avicenna's advice (3)

> ... to give wine to youth is like adding fire to a fire already prepared with matchwood. Young adults should take it in moderation, but elderly persons may take as much as they can tolerate ...

William Wadd (4) in his 1816 book, "Cursory Remarks on Corpulence; Or Obesity Considered as a Disease: With a Critical Examination of Ancient Modern Opinions, Relative to its Causes and Cure," stated:

> The article of *drink* requires the utmost attention. Corpulent persons generally indulge to excess; if this be allowed, every endeavour to reduce them will be vain.

In the subsequent edition of his book (5) "Comments on Corpulency Lineaments of Leanness" (1829), Wald depicted one of his patients standing beside a table

Comments on Corpulency.

with a decanter and glass of port and again counsels the need to restrict such beverages as part of a successful regimen of weight reduction. Caricaturists of the eighteenth and nineteenth centuries commonly pictured alcoholic beverage excesses with obese subjects.

While such older writings contain much wisdom concerning the proper uses of alcoholic beverages, there has arisen—and is perpetuated—a vast lore of unsound mysticism relating to these beverages and health. Bookstores contain many examples of appealing volumes filled with misinformation and fanciful recommendations concerning the use of wines and other beverages in the treatment of numerous diseases and of fanciful claims concerning the "complete nutritive value" of wine. The information in the present volume provides documented scientific background against which such claims may readily be assessed and the absurd ones dismissed.

III. EXCESSES AND NUTRITIONAL DEFICIENCIES

In more recent times, nutritionists and physicians have documented the frequent occurrence of nutritional deficiencies among alcoholics, a relationships neatly summarized in the First Report to the U.S. Congress on Alcohol and Health (6) in 1971 as follows:

> Malnutrition is commonly observed among alcoholic persons. In recent years this has been more true of those found on skid row, but it is by no means rare among those in better circumstances. One of the main reasons for this is the fact that alcohol itself represents an important source of calories. Each gram of alcohol provides 7.1 calories, which means that a pint of 100-U.S.-proof distilled spirits represents more than 1,300 calories, possibly half to two-thirds of the normal daily caloric requirement of many people. Therefore, heavy drinkers need less food to fulfill their caloric needs. Since alcoholic beverages do not contain significant amounts of protein, vitamins, minerals, and amino acids, they provide only "empty calories," and the intake of the vital elements of nutrition by a heavy drinker may readily become borderline or insufficient. Economic factors may also reduce the consumption of nutrient-rich food by the alcoholic person.

> In addition, even in a person consuming a good diet, heavy alcohol intake can result in malnutrition by interfering with the normal processes of food digestion and absorption. As a consequence, there is inadequate digestion of the food actually consumed. Some of the side effects of gastritis also reduce appetite, thereby lessening food intake. Moreover, alcohol appears to affect the capacity of the intestine to absorb various nutrients, including vitamins—especially vitamins B_1, and B_{12} and amino acids. In addition, malnutrition itself further reduces the capacity of the intestine to absorb nutrients.

> Some of the important diseases associated with defective nutrition alcoholic persons have.... been noted in.... dealing with the alcoholic encephalopathies, liver, and heart. Others long known to afflict these individuals are polyneuropathy (associated chiefly with deficiency of vitamin B_1); pellagra, due to deficiency of niacin and other fractions of the vitamin B complex, with symptoms of dermatosis, digestive disorder, and mental dysfunction; anemia, due to deficiency of iron and cobalamin (vitamin B_{12}); and, less often, scurvy,

due to deficiency of vitamin C; and other hypovitaminoses which result in defective bodily functioning of the organism.

In addition, an adequate supply of nutrients, especially protein and vitamins, is needed for the normal maintenance of liver function. The role that protein deficiency may play in the development of liver disease in adult humans has not been clarified. In experimental animals, however, severe protein deficiency has been shown to aggravate the pathological effects on tissues of an alcoholized diet.

One of the significant recent findings concerning the development of injury from alcohol has been the observation that, although nutritional deficiency states can aggravate the effect of alcohol upon the liver, sufficient alcohol—even in the absence of malnutrition—can have a deleterious effect upon that organ.

A particularly dramatic complication of alcohol intoxication is low blood sugar (hypoglycemia) which, if unrecognized, may be responsible for some of the "unexplained" sudden deaths observed in acutely intoxicated alcoholic patients. This complication occurs in individuals whose liver glycogen stores are depleted by fasting or starvation, or in those who have pre-existing abnormalities of carbohydrate metabolism.

Nutritional studies in the 1930s to 1950s highlighted the effects of a deficiency of several nutrients—protein, methionine, choline and other "labile methyl," riboflavin, vitamin B_{12}, selenium—on fatty changes, on necrosis, and/or cirrhosis of the liver in experimental animals. The hypothesis was put forward that similar liver pathology in the alcoholic was of nutritional origin and an era of intensive therapeutic supplementation ensued, including the development of a liver extract concentrate that was used to treat cirrhosis. The pendulum has now swung away from this explanation. However, one may ask whether the relationships between nutrient deficiencies, liver pathology, and alcoholic intake have been clearly resolved.

The nutritional roles of fermented food beverages include indirect contributions through enhancement of appetite, a variety of physiologic effects and, when misused, displacement of nutrients or dilution of the nutrient density of the diet. The relationship of malnutrition per se to certain of the consequences of alcoholism in man remains unresolved. I might add, parenthetically, that the tenuous hypothesis advanced some years ago by Roger Williams (7) that excessive alcohol consumption results from a deficiency of thiamine does not have support, insofar as I can determine, from any critical study in man. Conversely, however, excessive users of alcohol not infrequently suffer varying degrees of thiamine deficiency (Chapter 26).

IV. DIRECT NUTRIENT CONTRIBUTIONS OF FERMENTED BEVERAGES

Against this background let us examine the direct contribution of fermented food beverages to the nutrient intake. The use of fermented drinks containing yeast cells as part of man's diet since earliest times is discussed elsewhere in this volume. Beer, mead, and other beverages were formerly consumed in the freshly prepared state; consumption of comparable freshly fermented beverages is found today in the palm wine, Kaffir beer, maize beer, and other fermented grain products of Africa, in *pulque* (the fermented, unfiltered juice of the century plant) drunk by the Otomi Indians of Mexico, and other unclarified, unstrained products (see Steinkraus, Chapter 3).

The Accessory Food Factors Committee (8) of the MRC noted in 1945 that "In diets of most tropical tribes, various types of crude fermented liquors or gruels play an important part." They further noted that often when these foods have been discarded with the advent of civilization a deterioration in health has been noted and, in some instances, frank deficiency diseases have appeared.

In this relationship, a study of the Nutritional Status and Food Habits of the Otomi Indians (9) in Mexico is of interest. The Otomi Indians live in the Mezquital Valley of Mexico.

> The region is arid and barren and, economically and culturally, one of the most depressed of the country. The inhabitants eat very few of the foods which are commonly considered as essential to a good nutritional pattern. Their consumption of meat, dairy products, fruits and vegetables is exceedingly low. However, through the eating of tortillas, the drinking of *pulque* . . . and the eating of every edible plant available, a fairly good diet is maintained.

Mean blood levels for vitamin C were well above 1.0 mg/100 ml of serum, reflecting in large part the intake of *pulque* that for adults amounted to 1–2 liters or more daily. Anderson *et al.* (9) comment relative to the beverage:

> Pulque is produced by fermentation of the juice of the maguey. Eight to ten years after transplanting, just before the plant is ready to put out the central flower-bearing stalk, this central part is removed, leaving a cup-shaped receptacle. Into this cup juice from the leaves drains and is removed daily with a large pipette made from an elongated gourd. The sides of the cup must be scraped daily in order to keep the flow going. The leaves have stored up a great deal of sugar (chiefly sucrose) for the needs of the rapidly growing flowering stalk, and the juice produced is therefore quite sweet and is known as agua miel (honey water). This agua miel is innoculated with a culture from a previous batch of pulque and allowed to ferment for a variable time, usually about 10 to 12 days. After it reaches the optimum point it should be drunk within 24 to 48 hours, since the fermenting organisms are not removed and the fermentation proceeds unchecked, causing it to spoil. The presence of the fermenting organisms gives pulque a whitish, turbid appearance. It is mildly acid and not particularly unpleasant to the taste. It is usually produced under very unhygienic conditions, but its acidity probably prevents it from being a good culture for pathogenic organisms.

> Because the organisms which cause the fermentation are not removed they contribute some vitamin B and protein to the drink, and the vitamin C content is considerable. The average consumption for adults in our dietary records was 1 to 2 liters a day and this is probably low, since it is difficult to get an accurate report on a substance drunk abundantly at odd times. It was not unusual for a man to drink as much as 10 liters a day. The alcohol content of pulque is low (3 to 5 per cent) but, in spite of this, obvious drunkenness was quite common, particularly on market and fiesta days.

They conclude that in view of the otherwise marginal character of the diet, the nutrient content, especially vitamin C, of *pulque* is important. Of the nutrients in the diet, *pulque* contributed the following: energy, 12%, protein, 6%, thiamine, 10%, riboflavin, 24%, niacin, 23%, vitamin C, 48%, calcium, 8%, and iron, 20%. *Pulque* and other locally produced fermented foods are described by Steinkraus (Chapter 3).

These nutrients are derived from the combined raw material used in fermentation plus the yeast contained in the unfiltered beverage. Yeast per se has been exploited as human food because of its content of protein, B vitamins (niacin, thiamine, riboflavin, folic acid), and, if irradiated, vitamin D activity (8,10). The supplementary value of yeast in preventing or treating deficiency diseases in man is illustrated in its use in management of beriberi and pellagra (Todhunter, Chapter 5). Broquist (Chapter 7) and Krampitz (Chapter 6) discuss means of increasing the nutrient content of yeasts.

As Steinkraus indicates (Chapter 3), the chemical analysis and nutritional studies of other "native beers" and fresh, unfiltered wines similarly reflect the nutrient content of the yeasts and the modified cereals or juices.

V. CLARIFICATION, FILTRATION, AND DISTILLATION

As beverages are filtered, clarified, or distilled, there is a decrease in the content of nutrients other than alcohol, for example, unclarified Kaffir beer (sorghum) (2.8% alcohol) provides (11) per 100 gm: 117 calories; 21.2 gm carbohydrate; 2.6 gm protein; and 0.7 gm ash. Clarified (strained) Kaffir beer provides 31 calories; 3.6 gm carbohydrate; 0.5 gm protein; and 0.2 gm ash.

Comparison of the average content of usually reported nutrients in commercial beers and wines with that of distilled spirits further illustrates this generalization. The recently revised U.S. Department of Agriculture's Table of Food Composition (12) gives average values for nutrient contents of beers, wines, and whiskey, as summarized in Table I.

Rather extensive tabulations of the small amounts of vitamins in wines indicate that the vitamin content of these beverages is of relatively little significance in terms of the total dietary allowances. Amerine (13) summarizes this by stating:

Ascorbic acid is present in small amounts in fresh grapes (1–18 mg/100 ml), where it is an important factor in the reducing system of the ferment; only negligible amounts are found in wines.

Vitamin A is likewise present in negligible amounts. But thiamine is found in amounts of 0 to 50 μg per 100 g. of wine. Riboflavin occurs in amounts of 5 to 120 μg per 100 g. About 70 μg of pyridoxin, 65–120 μg of nicotinic acid, and 70–140 μg of pantothenic acid are also reported. In addition, *p*-aminobenzoic acid, biotin, and inositol occur, the latter in appreciable amounts.

TABLE I. Average Contents of Selected Nutrients in Alcoholic Beverages

	Grams	Calories	Protein (gm)	Fat (gm)	CHO (gm)	Ca (mg)	P (mg)	Fe (gm)	Na (mg)	K (mg)	Vitamin A (IU)	Thiamine (mg)	Riboflavin (mg)	Niacin (mg)	Ascorbic acid (mg)
1. Beer (4.5% alcohol by volume; 3.6% by weight): Can or bottle (12 fl. oz.)	360	151	1.1	0	13.7	18	108	Trace	25	90	—	0.01	0.11	2.2	—
2. Wines: Table (12.2% alcohol by volume: 9.9% by weight): wine glass (serving portion 3.5 fl. oz. or 103 ml)	102	87	0.1	0	4.3	9	10	0.4	5	94	—	Trace	0.01	0.1	—
3. Whiskey: 86-proof (36.0% alcohol by weight): jigger (1.5 fl. oz. or 44 ml)	42	105	—	—	Trace	—	—	—	Trace	1	—	—	—	—	—

VI. TRACE ELEMENTS

The trace element content of wines is likely negligible inasmuch as care is taken to reduce certain inorganic components (copper especially) because of their influence on clarity, taste, or keeping quality. On the other hand, some Spanish red wines have been reported to contain significant levels of iron—a mean content of about 12 mg/liter with ranges from 3 to 25 mg (14). Indeed, this was suggested as a contributing source of iron to siderosis in a region where these clarets were consumed. (Bottled wines were found to contain less—2–5 mg.) It is of interest that EDTA, a chelating substance, in appropriate ratio with iron, enhances the bioavailability of this element (15).

Brewer's yeast, unlike *Torula,* contains glucose tolerance factor (GTF) activity, and serves as a source of the chromium-containing GTF. Doisy *et al.* (16) report that brewers' yeast has the highest GTF activity of any of the foods studied. The quantity of this factor (and of chromium) in commercial beers varies over a wide range. There are no studies of its content in native beers.

The United States Brewers Association has recently developed information on the content of three essential trace elements in beer: Se, Cr, and Cu (17). The selenium content of 15 samples averaged 0.005 ppm, similar to the content of most fresh fruits and vegetables. Chromium content averaged 0.047 ppm. Copper averaged 0.23 ppm, a figure comparable to yellow corn and appreciably lower than tomato juice (average 1.2 ppm).

There is need for additional data on the trace element content of a wide variety of fermented food beverages in view of the increasingly recognized importance of these nutrients in human nutrition, and the effects of levels of alcohol consumption on the metabolism of at least some inorganic elements. For example, alcohol ingestion is reported to increase the urinary loss of zinc and, in some studies, the gastrointestinal absorption of iron.

Detailed data do not appear to exist to permit definitive comparisons of the different nutrient content of various types of beers and wines. The major compositional differences nutritionally appear to be in the caloric content reflecting different levels of alcohol and of carbohydrates. There is relatively little information on the nutrient content of the large number of fruit wines. Obviously, the sweet or dessert wines contain more sugars than the dry wines; strong ale and stout contain more carbohydrate (and calories) than mild ale or light beers. Since these differences can be meaningfully great, there is a need to develop informative tables of composition and nutrient content for use of the nutrition community.

VII. ORGANIC COMPOUNDS IN FERMENTED BEVERAGES

Finally, I wish to direct attention to the vast number of organic compounds that occur in differing amounts in various fermented beverages. Their possible physiologic or toxicologic significances need increased consideration in order to interpret the results of studies that are being conducted relating occurrence of pathologic conditions to different beverage consumption patterns. In 1969, Kahn (18) tabulated some 400 organic compounds that have been identified in whiskey, wine, and beer. These include a long series of fatty acids, dicarboxylic acids, amino acids, aromatic acids, keto and hydroxy acids, aliphatic esters (some 58), aliphatic dibasic esters, aliphatic and aromatic carbonyls (aldehydes), acetals, phenols, hydrocarbons, N- and S-containing compounds and heterocyclics, lactones, sugars, and miscellaneous organic chemicals. (See the appendix to this volume) Maynard Amerine (see Chapter 8) has indicated the occurrence of certain of these in wines. Kahn's tabulation does not represent a systematic comparative study of the distribution of the representatives of each group between the three beverage types. It does, however, suggest that there are some rather striking differences in the occurrence of some of them in different fermented beverages. For example, none of the 11 miscellaneous compounds reported in whiskey were listed as being encountered in wines or beers; of the 11 acetals reported in whiskey, only seven were reported from wine and one in beer; and of the 58 esters listed in whiskey, 53 had been detected in wine, but only 19 in beer.

While many, if not most of these substances, occur in trace quantities, the better definition of the chemical content of beverages with assessment of their potential biologic significance may be fruitful. Modern analytical methodology makes feasible the identification and quantitative determination of these substances.

VIII. SUMMARY

Fermented food beverages enhance the pleasure of eating. Their nutritional roles include indirect contributions through subjective enhancement of appetite, a variety of other physiologic effects, and when misused, dilution of nutrient density of the diet. The nutrient composition varies with the type of beverage. Some crude fermented beverages or gruels make important direct nutrient contributions to the diets of many tropical indigenous populations, and the discard of these foods has resulted in deterioration in health.

The chemical analysis and nutritional studies of native beers and fresh, unfiltered wines reveal that they reflect the nutrient content of the yeasts and the modified cereals or juices from which they are prepared. As the beverages are filtered, clarified, or distilled, there is a decrease in the content of nutrients other than alcohol. The small amounts of vitamins in wines are of relatively little significance in meeting the total recommended dietary allowances. The iron content of some red wine has been held to be significant.

Brewers' yeast contains a variable amount of the chromium-containing glucose tolerance factor activity. Beer varies in its GTF activity; its chromium content averages 0.047 ppm; its selenium content is similar to that of fresh vegetables and fruits, its copper content to yellow corn but lower than tomato juice.

There is need for additional data on the trace element content of a wide variety of fermented food beverages (wines, beers, crude fermented foods). Tables of nutrient composition for different varieties of wines and beers are needed.

A systematic comparative study of the content of the large number of organic compounds occurring in alcoholic beverages should be made. The possible physiologic or toxicologic significance of these compounds deserves greater consideration in the interpretation of epidemiologic findings in relation to beverage consumption patterns.

REFERENCES

1. Oser, B. L., The evaluation of beer as a food. *Wallerstein Lab. Commun. Sci. Pract. Brew.* No. 2, 5–10 (1938).
2. Venner, T. "Via Recta ad Vitam Longam, or, A Treatise Wherein the Right Way and Best Manner of Living for Attaining to a Long and Healthfull Life, is Clearly Demonstrated and Punctually Applied to Every Age and Constitution of Body." James Flesher for Henry Hood, London, 1650.
3. Darby, W. J., Ghalioungui, P., and Grivetti, L. "Food: The Gift of Osiris," Vol. 2, pp. 587–588. Academic Press, New York, 1977.
4. Wadd, W. "Cursory Remarks on Corpulence; or Obesity Considered as a Disease: With a Critical Examination of Ancient and Modern Opinions, Relative to its Causes and Cure," 3rd Ed. J. Callow, Soho, London, 1816.
5. Wadd, W. "Comments on Corpulency: Lineaments of Leanness: Mems on Diet and Dietetics." John Ebers & Co., London, 1829.
6. "First Special Report to the U.S. Congress on Alcohol and Health," Sec. Health, Educ. Welfare, Natl. Inst. Alcohol Abuse Alcoholism. Washington, D.C., 1971.
7. Williams, R. J. "Nutrition and Alcoholism." Univ. of Oklahoma Press, Norman, 1951. "You Are Extra-ordinary." Random House, New York, 1967.
8. Accessory Food Factors Committee. Food yeast: Survey of its nutritive value. *Med. Res. Counc. (G.B.), War Memo.* No. 16 (1945).
9. Anderson, R. K., Calvo, J., Serrano, G., and Payne, G. C. A study of the nutritional status and food habits of Otomi Indians in the Mezquital Valley of Mexico. *Public Health Nation's Health* 36(8), 883–903 (1946).
10. Tannenbaum, S. R., and Wang, D. I. C., eds. "Single-Cell Protein," Vol. 2. MIT Press, Cambridge, Massachusetts, 1975. See esp. Scrimshaw, N. W. Single cell protein for human consumption—an overview. Ch. 2, pp. 24–45.
11. Leung, W.-T. W., Busson, F., and Jardin, C. "Food Composition Table for Use in Africa," pp. 218–219. U.S. Dep. Health, Educ. Welfare, Natl. Cent. Chron. Dis. Control, Bethesda, Maryland; Food Agri. Organ. U.N., Rome, 1968.
12. Table on food composition. *U.S. Dep. Agric., Agric. Handb.* No. 456, p. 31 (1975).
13. Amerine, M. S. The composition of wines. *Sci. Mon.* 77(5), 254 (1953).
14. Herreros, V., and DeCastro, S. El hierro del vino y su posible intervencion en el genesis de la siderosis cirrotica. *Rev. Clin. Esp.* 117, 505–508 (1970).

15. Layrisse, M., and Martinez-Torres, C. Fe(III)-EDTA complex as iron fortification. *Am. J. Clin. Nutr.* **30**(7), 1166–1174 (1977).
16. Doisy, R. J., Streeten, D. H. P., Freiberg, J. M., and Schneider, A. J. Chromium metabolism in man and biochemical effects. *In* "Trace Elements in Human Health and Disease. Vol. 2: Essential and Toxic Elements" (A. S. Prasad, ed.), Nutr. Found. Monogr., pp. 80–81, Academic Press, New York, 1976.
17. "Report of Analyses of American Beers for Selenium, Cadmium and Copper," Unpubl. Rep. U.S. Brew. Assoc., Washington, D.C., 1974.
18. Kahn, J. H. Compounds identified in whisky, wine and beer: A tabulation. *J. Assoc. Off. Anal. Chem.* **52**(6), 1166–1178 (1969).

EDITORIAL COMMENT

The nutrient of greatest consequence in modern wines and beers is alcohol, which is a source of energy. Accordingly, it is important to be familiar with comparative information on the alcohol content of differing beverages. Medicinal and therapeutic qualities have long been attributed to wines and beers with certain varieties of beverages being considered especially efficacious for particular disorders. Analysis of nutrient contents give little if any basis, however, for such therapeutic specificity or use.

Among the more beneficial properties of these beverages is the enhancement of the taste of food. Thus the use of wine or beer may aid the ill or convalescent person to consume a nutritionally more complete meal. The serving of wine or beer can also add a ceremonial touch to a meal and thereby increase its acceptabiliy.

Beer in particular may have an especial virtue in encouraging consumption of generous amounts of water while providing minimal amounts of sodium. To put the sodium content in perspective, a 12-oz. bottle of beer or a half-liter of wine will provide 25 mg of sodium, where a diet moderately restricted in sodium as a part of treatment of hypertension for instance might permit no more than 2000 mg per day, or a more severely restricted diet would limit sodium consumption to 500 mg or less. Thus an "extra" 25 mg of sodium in a bottle of beer or 5 mg in a glass of wine would be a minor contribution to the daily sodium content of the diet, but allow use of a welcome enhancer of an otherwise bland menu.

There may be small amounts of protein, carbohydrate, vitamins, and minerals as nutrients provided by some beverages, but an important therapeutic usefulness is inherent in the alcohol itself. Under convalescent or nursing home circumstances beer and wine in moderation may be effective mild sedatives and tranquilizers (Watkin, Chapter 15). Nurses and physicians must be aware of possible excessive use, however, just as for any sedative. If the person has been dependent on alcohol in younger life, one should be cautious about recommending beer or wine for sedative purposes.

1. Ethanol Equivalents in Selected Amounts of Various Beverages

Table 1 is based primarily on grams of absolute alcohol or ethanol with equivalents in ounces and milliliters of spirits, wine, and beer. Items referred to under footnote *a* are the equivalents of some selected amounts of the various beverages, e.g., 150 gm of ethanol, are contained in approximately 16 oz. or one pint of 80-proof spirits. Conversion factors for other frequently used strengths are also given; for example, 86-proof spirits contains 7.5% more ethanol than the same volume of 80-proof spirits; 100-proof spirits, which now has but a small share of the market in the United States, contains 25% more ethanol than 80-proof.

　　　　　　　　　　　　　　　　　　　　　　　　　　　　　　　　　　　W. J. Darby

Table I.   Ethanol Equivalents of Some Commonly Used Alcoholic Beverages

Grams	Ethanol		80-Proof spirits[b]		Table wine (12.5%)[c]		Beer (3.6%) by weight[d]	
	fl. oz.[e]	ml	oz.	ml	oz.	ml	oz.	No. of 12-oz. bottles
1	0.0428	1.267	0.1071	3.167	0.3427	10.14	0.952	0.08
5	0.2142	6.335	0.5355	15.84	1.714	50.68	4.76	0.40
8.75[a]	0.3750	11.09	0.9372	27.72	3.00	88.72	8.33	0.69
9.34[a]	0.4000	11.83	1.000	29.57	3.20	94.64	8.89	0.74
10	0.4284	12.67	1.071	31.68	3.427	101.4	9.52	0.79
12.6[a]	0.54	15.97	1.350	39.92	4.319	127.8	12.0	1.0
15[a]	0.6427	19.0	1.606[f]	47.61	5.142	152.1	14.3	1.2
20	0.8569	25.34	2.142	63.35	6.855	202.73	19.0	1.6
23.34[a]	1.000	29.57	2.500	73.94	8.000	236.6	22.2	1.9
50	2.142	63.35	5.355	158.4	17.14	506.8	47.6	4.0
80	3.427	101.4	8.568	253.4	27.42	810.9	76.2	6.4
100	4.284	126.7	10.71	316.7	34.27	1014	95.2	7.9
150[a]	6.427	190.0	16.07[g]	475.0	51.42	1520	143	12
200	8.569	253.4	21.42	633.5	68.55	2027	190	16
239[a]	10.24	302.8	25.6[h]	757.1	81.92	2423	228	19
300	12.85	380.1	32.13[i]	950	102.8	3040	286	24
400	17.14	506.8	42.85	1266	137.1	4055	381	32

[a] Indicates grams of ethanol in selected amounts of various beverages; 1 oz. = 29.574 ml; to convert milliliters of ethanol to grams multiply by 0.7893, the specific gravity of ethanol.

[b] In the same volume of 86-proof spirits the amount of ethanol is 7.5% greater, and in 100-proof 25% greater.

[c] Fortified wines have a higher alcohol content. Sherry is commonly 18–20%.

[d] Malt liquors usually have about 4.5% alcohol by weight; low beers about 3% or less.

[e] The British fluid ounce is 4% less than the U.S. ounce.

[f] Approximately the amount in a "miniature."

[g] Approximately one U.S. pint; the British pint is 20 oz., with their quart and gallon being correspondingly higher.

[h] One fifth (of a U.S. gallon) or 4/5 of a quart.

[i] Approximately one U.S. quart.

While these differences are small in one drink, when computed for a week or year the differences are substantial. Care, therefore, should be taken in making such conversions. For example, the amount of ethanol in one bottle of the usual strength beer (3.6% by weight) is 12.6 gm, but the amount of spirits served in the usual 1.6-oz. 80-proof miniature drink on an airplane contains 14.94 gm of ethanol. The size of a drink varies, of course, in hotels, clubs, and the home; in hotels the average drink is probably closer to 1 oz. while in homes often between 1.5 and 2 oz.

Table wines commonly contain about 12.5% or less ethanol, often closer to 12 or 11.5%. Sherry or fortified wines have alcohol added and are usually 18–20% ethanol. Brandy-type liqueurs vary in ethanol strength, some 80-proof, others substantially lower.

The bulk of the beer sold in the United States is about 3.6% alcohol by weight or 4.5% by volume. Malt liquor tends to have higher alcohol content, usually about 4.5% by weight, and low calorie beers less alcohol, in the range of 3% by weight or lower.

Table II. Nutritional Properties of Malt Beverages

Constituent	Range per liter	Average values	
		Per liter	Per 12 oz. (355 ml)
Calories	324–425	395	140
Ethyl alcohol (gm)	35.6–39	37	13
Total protein (gm)	2.5–3.0	2.8	1
Total carbohydrates (gm)	11.7–40	31.5	11
Total lipids (gm)	0.025–0.03	0.028	0.01
Minerals (mg)			
Sodium	20–104	52	18.5
Potassium	171–475	303	108
Magnesium	68–88	76	27
Calcium	45–63	53	18.8
Phosphorus	20–135	76	27
Iron	Trace–0.07	—	—
Zinc	0.04–0.40	0.19	0.07
Chromium	0.01–0.012	0.011	0.004
Copper	0.06–0.10	0.09	0.03
Manganese	0.13–0.020	0.17	0.06
Vitamins (mg)			
Ascorbic acid	Trace–0.20	—	—
Thiamine	0.01–0.08	0.037	0.013
Riboflavin	0.2–0.4	0.32	0.114
Niacin	5.1–7.52	6.58	2.34
Pantothenic acid	0.45–1.16	0.7	0.25
Pyridoxine	0.51–0.68	0.62	0.22
Folacin	0.05–0.06	0.055	0.02
Cyanocobalamin (B12)	0.005[a]	—	0.002[a]
Hop resins (mg)	26–50	38	13.5

[a] One determination only.

2. Nutritive Properties of Malt Beverages

Many of the published references to the nutritional properties of malt beverages are values obtained from foreign beers which tend to be "heavier" than the popular lighter American beers; the former often have a somewhat higher carbohydrate content, and consequently a number of higher calories. A summary of these values was published by Davidson in 1961 (1). Other papers have emphasized various special aspects of malt beverage, such as its relative low sodium content (2–4).

More recently the U.S. Brewers Association in collaboration with the U.S. Department of Agriculture has sponsored analyses of beer by constituent members of U.S.B.A. for inclusion in the Food Data Bank of the Consumer and Food Economics Institute of U.S.D.A. A summary of these values for "regular" beers is shown in Table II. The "dark" American beers tend to have a somewhat higher carbohydrate content, and the so-called low calorie beers lower calorie, alcohol, and carbohydrate values. Beverages designated as malt liquor usually have a higher alcohol content than regular beers, about 4.5% by weight.

Table III.   Free Amino Acid and Fatty Acid Content of Beer[a]

Amino acids	Range (mg)	Average (mg)	Fatty acids	Range (mg)	Average (mg)
Total	247–375	311	Total	5–12	8.5
Tryptophan	8.6	8.6	Caproic	1.3–30	1.9
Leucine	2.0–10.9	6.5	Caprylic	2.5–7.0	4.7
Cysteine	Trace	—	Capric	1.0–3.0	1.8
Valine	2.9–17.8	10.4	Lauric	0.15–2.0	0.18
Alanine	14.5–21.6	18	Myristic	0.13[c]	—
Glycine	8.1–11.5	9.8	Palmitic	0.03	—
Threonine	3.7–4.6	4.2	Stearic	0.06	—
Lysine	0.2–4.4	2.3	Oleic	0.17	—
Phenylalanine	3.1–32.4	17.8			
Arginine	2.0–9.4	5.7			
Aspartic acid	4.5–20.6	12.6			
Proline	151–169	160			
Isoleucine	2.1–6.6	4.4			
Methionine	1.4–2.7	2.1			
Tyrosine	14.7–28.4	21.6			
Histidine	5.9–20.4	13.2			
Glutamic acid	1.2–6.6	3.9			
Serine	—	5.3[b]			

[a] Values per liter.
[b] Average of 11 tests; range not given.
[c] One determination only.

It will be seen that one 12-oz. bottle of regular beer contains about 140 calories, 11 grams of carbohydrate, mostly in the form of dextrins and maltose, small amounts of protein and amino acids, relatively small amounts of sodium, and comparatively high amounts of potassium. The lipid content is negligible; varying amounts of trace minerals and vitamins are present. The so-called light or low calorie beers contain from 70 to 95 calories and 3–5 gm of carbohydrate per 12-oz. bottle. Since brewers' yeast is rich in thiamine, the small amount of this vitamin in finished beer tends to reflect the extent to which yeast is eliminated by filtration and autolysis prevented.

In Table III are shown values for various free amino acids and fatty acids.

REFERENCES

1. Davidson, C. S. Nutrient content of beers and ales. *N. Engl. J. Med.* **264,** 185–186 (1961).
2. Dutchess, C. E. Malt beverages as supplements to medically approved diets. *Ind. Med. Surg.* **28,** 544–546 (1959).
3. Newburg, B. Sodium-restricted diet. *Arch. Intern. Med.* **123,** 692–693 (1969).
4. Olmstead, E. G., Cassidy, J. E., and Murphy, F. D. The use of beer in the low salt diet with special reference to renal disease. *Am. J. Med. Sci.* **230,** 49–53 (1955).

II

Fermentation

5

A HISTORICAL PERSPECTIVE ON FERMENTATION BIOCHEMISTRY AND NUTRITION

E. Neige Todhunter

I.	Introduction	83
II.	Early Fermented Beverages	85
	A. Discovery	85
	B. Types of Beverages	85
III.	Health Value of Fermented Beverages	86
	A. Early Dietary Recommendations	87
	B. Utilization of Alcohol	87
	C. Health Hazards	88
	D. Spruce Beer, an Antiscorbutic	88
	E. Fermented Milks	89
IV.	Early Chemistry of Fermentation	90
V.	Yeast: A Living Organism	91
	A. Identification of Growth of Yeast	91
	B. Pasteur's Contribution	92
VI.	The Discovery of Zymase	93
VII.	Chemistry of Yeast–Juice Action	94
VIII.	Yeast: Its Nutritional Aspects	94
	A. Requirements for Growth	94
	B. Nutrient Synthesis	95
	C. Nutritive Value of Yeast	96
	References	96
	Editorial Comment	98

I. INTRODUCTION

This discussion covers the period from ancient man's discovery of fermentation to the twentieth century biochemist's determination of the intermediary metabolism of carbohydrates. It is the biography of yeast: how yeast the un-

Table I.   Chronology of Major Events in the History of Fermentation

c 3200 B.C.	Egyptians were skilled in the art of preparing fermented beverages and leavened bread
c 1610	Van Helmont tried to explain fermentation in chemical terms and introduced the term "gas"
1680	Leeuwenhoek observed small round or oval particles in beer under his microscope
1754	Joseph Black discovered "fixed air" (carbon dioxide). He distinguished between effervescence of fermentation and gas production by acids acting on carbonates
1766	Cavendish identified "phlogiston" or "inflammable air" (hydrogen). He showed alcoholic fermentation produced fixed air only, and putrefaction produced fixed and inflammable air
1772	Daniel Rutherford discovered "residual air" (nitrogen)
1774	Priestly discovered "dephlogisticated air" (oxygen)
1789	Lavoisier showed fermentation of sugar produced alcohol and carbon dioxide and quantitatively determined the amounts of these products
1810	Appert devised a method of preventing fermentation or putrefaction by heat treatment and hermetic sealing of food in glass containers
1815	Gay-Lussac showed that alcoholic fermentation involved conversion of 1 mole of glucose to 2 moles of ethyl alcohol + 2 moles carbon dioxide
1837	Cagniard-Latour's microscope studies determined yeast was a living organism belonging to the vegetable kingdom
1837	Theodor Schwan's chemical experiments distinguished between fermentation and putrefaction
1837	Kützing, by microscopic examination, showed alcoholic fermentation depended on yeast which grew by budding
1839	Berzelius claimed yeast action was a catalytic one only. Wohler and Liebig denied the living organism concept and ascribed yeast action to instability of the reacting materials
1857–1860	Pasteur's classic research establishing yeasts as living cells responsible for fermentation
1876	Kühne introduced the term "enzyme," from the Greek meaning "in yeast"
1883–1886	Emil Christian Hansen introduced pure cultures of yeast for beer production
1897	Edward Buchner showed cell-free extracts of yeast obtained by dialysis caused fermentation of sugar. He called the active constituent "zymase"
1901	Bios, an unknown factor essential for yeast growth, discovered by Wildiers
1906	Harding and Young discovered a co-ferment (later called a coenzyme) in yeast juice
1908	Phosphate combines with glucose as an ester during fermentation; elucidation of metabolic pathways of carbohydrate progressed from this stage
1920s	Studies began on yeast and the vitamins

known was identified as a living organism, a producer of enzymes and coenzymes, a synthesizer of vitamins and protein, a nutritionally significant food supplement, and a major contributor to the development of microbiology, biochemistry, and the understanding of cell processes. So many centuries and such a broad involvement of a number of developing sciences necessitate that only the major events of this long history can be briefly highlighted here. Table I provides a chronology of these events.

II. EARLY FERMENTED BEVERAGES

A. Discovery

The history of fermentation covers a period of over 5000 years. The earliest records of civilized man including archaeological materials, paintings, drawings, and carvings show that in early Egyptian times men were skilled in the preparation of fermented beverages and leavened bread (Darby *et al.*, 1977). By what accident ancient man discovered this process cannot be clearly identified. It probably happened in the food-gathering era when native fruits were gathered for food. Fruits broken or bruised and allowed to stand would begin to ferment, and thus early man found a pleasing beverage. This could have been the beginning of what early civilized man developed to a fine art.

From recorded historic times it is known that about 3200 B.C. in Egypt, the grapevine was widely found throughout the Middle East and eastern Mediterranean area, though Greek and Roman mythology provide conflicting statements as to its origin. According to legend, Osiris and his son Horis taught the Egyptians (Bacchus taught the Greeks) to cultivate the vine and make wine (Darby *et al.*, 1977). Tradition of the Israelites tells that Noah provided them with the means to make wine.

B. Types of Beverages

Wines of great variety were used by early man. The grape, wherever the climate permitted the vine to flourish, was the most common source of wine. In other areas any available fruit was used. In northern Europe where the colder climate favored the growth of apples, cider was prepared. It has been called England's ancient wine and is one of the oldest fermented drinks. Perry was prepared from crushed pears.

Wild honey left to stand for some time was observed to ferment and with the addition of water an alcoholic beverage was obtained. In Greece it was named *hydromel* and in early Britain it was called *mead* and was the beverage of warriors. The great hall of the Saxon palaces was known as the mead hall.

Metheglin was similar to mead but with aromatic herbs added. These beverages continued in use in England until the eighteenth century and were also among the "home brews" of Americans until the nineteenth century.

Fermented milk drinks were discovered after man had domesticated hoofed animals and sheep, and one of the oldest of these drinks is fermented mare's milk called *koumiss*. Most fermented milks are produced by bacterial action and those that have some yeast fermentation are very low in alcohol content.

Beer became the chief fermented beverage of early man as soon as he settled in one place and developed some agricultural practices. Whatever grain was available, Neolithic man ground it between stones, then cooked and made it into some kind of gruel. If left standing for a few days this gruel would ferment, thus providing an easy source of a beverage. Theophrastus, around 400 B.C. described beer as the wine of barley. In some parts of the world beer was produced from maize and from rice. Some kind of beer has been part of man's diet from earliest times, and was drunk by both rich and poor, though wine was more frequently the beverage of the privileged class.

Beer is the generic term for the product of fermentation of a decoction of cereal malt. Over the centuries it has been used to describe a wide variety of products with different names and widely differing ingredients. The old English term was ale, which came from the name of the Viking's drink. In the fourteenth century hops were added to the wort in the brewing process because of their preservative effect, and then beer became the accepted term in England. At different times beer has been regarded as an essential beverage (especially when an unpolluted water supply was not available), as a food product that was nourishing, wholesome, and stimulating, or as a tonic or medicine. It has also been reviled as a destroyer of health. The first English settlers at Jamestown, Virginia, believed they could not live without beer; it was, to them, an essential for health and well-being. The Puritans also considered it necessary, wholesome, and nourishing.

Until the nineteenth century the flavor and quality of ale and beer depended on the ingredients, the mineral content of the water, the extent to which the malt was roasted, the degree to which the brew was hopped, and the particular methods of each brewer. By about 1870 the nature, function, and behavior of yeast were understood, and in the 1880s pure yeast cultures could be obtained and the technology of beer production rapidly advanced.

III. HEALTH VALUE OF FERMENTED BEVERAGES

Fermented beverages were long accepted simply for the pleasure they gave. Health effects were of little concern until the time of Hippocrates, when much empirical advice was given. Not until the nineteenth century was chemistry sufficiently advanced to allow chemical determination of food and beverage constituents, and thus to determine its nutritive value.

A. Early Dietary Recommendations

Andrew Boorde (c 1490–1549), the English physician and traveler, wrote "A Dyetary of Helth" in 1542. It is believed to be one of the earliest medical works written in English. With regard to beer he wrote: "Ale for an Englishe man is a naturall drynke." He described the health value of wine as follows: "Moderately drunken, it doth acuate and doth quicken a mans wits, it doth comfort the heart, it doth scour the liver; specially, if it be white wine, it doth rejuice all the powers of man, and doth nourish them; it doth ingender good blood, it doth nourish the brain and all the body" (Furnivall, 1870).

One hundred years later Dr. Tobias Venner (1577–1660) of Bath, England published "Via Recta ad Vitam Longam" in 1620. [See also Chapter 4.] He made various health recommendations and notes on food, and he praised wine. His comments are reminiscent of Boorde. Venner says of wine:

> Many and singular are the commodities of Wine: for it is of it selfe, the most pleasant liquor of all other, and was made from the beginning to exhilarate the heart of man. It is a great increaser of the vitall spirits, and a wonderfull restorer of all concoction, distribution, and nutrition, mightily strengtheneth the natural heat, openeth obstruction, discusseth windinesse, taketh away sadnesse, and other hurts of melancholy, induceth boldnesse and pleasant behavior, sharpeneth the wit, abundantly reviveth feeble spirits, excellently amendeth the coldnesse of old age, and correcteth the tetrick [morose] qualities which that age is subject unto; and to speak all in a word, it maketh a man more coragious and lively both in mind and body.

B. Utilization of Alcohol

In the last century there was much debate as to the function and fate of alcohol from fermented beverages. Pereira in 1843 concluded from the available evidence that alcohol did undergo some change within the body. It was absorbed but it was not excreted by the intestines, the kidney, or the skin. It was recognizable by its odor in the breath, but this could account for only a very small portion. It could not form tissues because it lacked nitrogen, sulfur, and phosphorus which were then recognized as the essential tissue-building constituents. The explanation then was that it had to be oxidized and removed by exhalation, and therefore calories must be evolved (Pavy, 1874; Pereira, 1868).

The problem was further resolved with the building of a respiration calorimeter at Wesleyan University in Connecticut and the extensive experiments by Atwater and Benedict in 1902 which showed that "alcohol is similar to the fuel ingredients of ordinary food in that it is oxidized in the body, yields energy for warmth and probably for work, and protects body material from consumption" (Chittenden, 1930). They computed that 4 gm alcohol were isodynamic with 7 gm sugar or starch and 3 gm fat.

C. Health Hazards

All fermented beverages have some dangers if consumed in excess and, therefore, must be mentioned as health hazards. That keen observer and famous diarist Samuel Pepys gave up drinking wine because he went "to bed very near fuddled," and in 1662 he wrote "Since leaving drinking of wine I do find myself much better, and do mind my business better, and so spend less money, less time lost in idle company."

"Devonshire colic" was reported in 1767 by George Baker, a physician in Devonshire, England, as being endemic in that area. He identified it as lead poisoning caused by drinking cider contaminated with lead (Schmidt, 1959). A London physician and one of the early writers on diet gives the following description: "Cyder and Perry are, it is said, generally fermented and kept in leaden vessels, or at least the apples and pears are passed through leaden tubes; and the lead being readily dissolved by the acid, is gradually introduced into the body which produces painful and dangerous colics" (Willich, 1799).

D. Spruce Beer, an Antiscorbutic

From ancient Egyptian times beer has been a yeast fermentation product. Through the centuries there have been many variations in the starting material, the flavorings and herbs that have been added to it, and in the procedures used. Therefore, there have been many very different beverages that were called beer, and spruce beer is an unusual example.

Spruce beer is of particular interest to nutritionists because of its use as an antiscorbutic, particularly in this country in colonial times. The Oxford Universal Dictionary describes spruce beer as a beverage from Prussia and gives the date for it as 1500. Apparently this type of beer was introduced into England around 1640 and called *Mum,* which the Oxford Dictionary describes as a kind of beer originally brewed in Brunswick (Prussia). In 1664 Pepys recorded that he had visited a "mum-house" in London. Lorenz gives a recipe for making *mum* or spruce beer as described in 1682 by an anonymous writer. "The inner rind of the Firr, the tops of Firr and Birch" and a variety of other seeds and berries were added to fermenting mash. This recipe originated in the House of Brunswick, Germany, and was recommended as "very powerful against the breeding of Stones, and against all Scorbutich Distempers" (Lorenz, 1953).

This at once reminds the nutritionist of the experience of Jacques Cartier in his famous second voyage to Newfoundland in 1535–1536. While wintering in the St. Lawrence River, a strange disease (scurvy) "spread among the three ships to such an extent that of the 110 men forming our company there were not ten in good health.'' Cartier learned from the Indians how to prepare an infusion of leaves and bark from an evergreen tree. Reluctantly, one or two men ventured to try it and after drinking it two or three times, they recovered health and strength. When this became known the men all wanted it ''so that in less than eight days a whole tree as large as any I ever saw was used up and all recovered health and strength'' (Biggar, 1924). The tree that produced this cure was not named but was believed to be the hemlock.

The spruce beer of the Germans served the same purpose as Cartier's infusion but was probably a more palatable product because of the fermentation.

When colonial Americans could not get hops, which they were accustomed to use for flavoring their beer, they used sassafras, spruce, or a variety of herbs instead. Spruce beer was mainly a home brew until the Revolution (Baron, 1962). It also was a component of the rations ordered by General George Washington to be issued to all troops of the Continental Army. The order was given on December 24, 1775, that the ration was to include ''one quart of spruce beer per day, or nine gallons of molasses to one hundred men per week'' (Manchester, 1976). Apparently it had been learned in the early wars with the Indians that some preparation or decoction of spruce or fir tips was an effective antiscorbutic, and scurvy was likely to affect soldiers who often were on very restricted rations.

E. Fermented Milks

Although the main emphasis here is on wine (vinous fermentation) and beer which are dependent on yeast fermentation, mention should be made of fermented milk products because of their contribution to health. They were known and recorded in Biblical times and were used by the early Egyptians, Greeks, and Romans. Cheese is an ancient fermented food, but the microorganisms concerned are bacteria rather than yeast. Similarly the fermented milks used by the ancient nomadic tribes of Asia Minor, eastern and southern Europe, and western Asia are mainly the result of bacterial activity, but there is also some yeast action. Yogurt (with its many various spellings) is the best known of these, and originally was prepared locally from the milk of buffaloes, cows, donkeys, sheep, and goats. It contains little or no alcohol. *Kefir* is a fermented milk originating in the Caucasian mountains. The combined fermentation of milk by lactic acid bacteria and yeast produces about 0.8% lactic acid, about 1% alcohol and carbon dioxide and it foams and effervesces like beer (Keogh, 1976). *Koumiss* originated in the Asiatic steppes and was traditionally made from mares' milk. *Kuranga* is another fermented milk prepared by the Mongols and other Asiatic peoples by the action of lactobacilli and a special yeast, *Torula curunga* (Borgstrom, 1968). The Icelandic people have prepared a skim milk product since the tenth century, which uses combination of bacterial and yeast action. It is mildly alcoholic and effervescent (Keogh, 1976).

In general, these fermented milk products retain the major nutrients of the original milk and are believed to be more easily digested. These fermented milks vary in fluidity and usually are viscous or a soft curd.

IV. EARLY CHEMISTRY OF FERMENTATION

Alchemy finally had to give way to more modern chemistry, and physiology had to develop a chemical viewpoint. This change began in the seventeenth century. This was the century that saw the rise of independent thinkers, men such as Harvey, Gilbert, Kepler, Descartes, Boyle, and Newton in science. One man was on the borderline between the old and the new thinking in science; he was Jean Baptiste van Helmont (1577–1644), who practiced medicine in Brussels. He introduced some new concepts and terminology in chemistry but also retained some of the old mystical concepts. He tried to explain the process of fermentation and the chemistry involved. He observed that if whole grapes were allowed to dry they kept indefinitely, but if the skin was broken the grapes fermented. He noted the production of gas by this fermentation, and that the gas produced was different from atmospheric air. Van Helmont introduced the new term ''gas'' to describe this product of fermentation. He also considered gastric digestion as a chemical process rather than a mechanical action. Franciscus Sylvius (1614–1672) of Leyden considered gas bubbles an essential part of fermentation and since the same bubbles arose when acid was poured on chalk he concluded the two processes were the same (Foster, 1901).

By the eighteenth century rapid progress was being made in chemistry, particularly in discovery of the gases of the air. Joseph Black in 1756 discovered what he called ''fixed air'' (carbon dioxide). In 1772 Daniel Rutherford discovered a new type of air ''residual air'' (nitrogen). Cavendish in 1766 identified ''phlogiston'' or ''inflammable air'' (hydrogen), and Priestley discovered ''dephlogisticated air'' (oxygen) in 1774. It is interesting' to note that the English scientists did not accept van Helmont's term gas; throughout all their work in this period in the development of pneumatic chemistry they referred to ''airs.''

Until this time alcoholic fermentation and putrefaction were believed to be the same. The gas produced in each instance was identified by MacBride as fixed air. Two years later, 1766, Cavendish found fixed air was the only gas produced by fermentation, but the gas from putrefaction was a mixture of fixed air and inflammable air (Harden, 1923).

Antoine Lavoisier (1743–1794) resolved these problems of chemical nomenclature and developed quantitative and systematic chemistry. Nutritionists have called Lavoisier ''the Father of nutrition'' because of his basic studies in respiration calorimetry. In 1783, using an ice calorimeter, he measured the body heat of a guinea pig, and in 1785 with Seguin he determined the amount of oxygen absorbed or utilized by a man at rest, after taking food, and when physical work was done. This was followed by chemical studies of vinous fermentation showing that fermentation of fruit juices or any sugar substances was accompanied by production of carbonic acid gas and alcohol. Earlier investigators had referred to "spirit of wine," but Lavoisier restored the old Arabic name of "alcohol" (McKie, 1952).

Lavoisier's experiments on fermentation were quantitative. He determined the amount of carbon, hydrogen, and oxygen in sugar. He used a weighed amount of sugar, water, and yeast, and determined the weight of carbonic acid gas and water produced by fermentation and the weight of the residual liquor and its components. From this he wrote the equation (believed to be the first chemical equation): must of grapes = carbonic acid + alcohol.

Gay-Lussac continued the quantitative investigations begun by Lavoisier and showed that alcoholic fermentation by yeast involved conversion of 1 mole of glucose to 2 moles of ethyl alcohol plus 2 moles of carbon dioxide (McKie, 1952).

V. YEAST: A LIVING ORGANISM

A. Identification of Growth of Yeast

The first recorded observation of a yeast cell was made in 1680 by Leeuwenhoek of the Netherlands. He was a tradesman in Delft but his hobby was grinding lenses, which he did with great patience and skill. With these lenses he made what was referred to as a microscope, but was really a magnifying glass capable of 250–270 magnification. When he placed a droplet of fermenting beer on the lenses he observed globular bodies. These he drew and described in one of his many communications to the Royal Society of London. Leeuwenhoek did not know that these were yeast cells, and his observation attracted little attention until some 50 years later when three investigators working independently in France and Germany used the microscope to study the activity of beer yeast (Lafar, 1910; Harden, 1923).

Cagniard-Latour of France was interested in brewing and studied the fermentation of beer. He examined brewers' yeast under the microscope and saw small globules which had no motion but reproduced by budding; therefore, he concluded they were living organisms. He believed these cells acted on a sugar solution only while they were still living and by their vital activity sugar was changed into alcohol (Lafar, 1910). These observations were made in 1835–1836 and presented to the Academy of Sciences in a paper published in 1838.

Theodor Schwann in Germany had demonstrated that putrefaction of a solution of animal or vegetable matter did not occur if the solution was boiled and entering air was first heated. Schwann attributed the origin of putrefaction to living germs in the air. In 1837 he made similar observations with beer yeast and also studied yeast cells under the microscope and found that the yeast globule reproduced by pushing out from its interior a small nodule which grew to normal size. This reproductive process established the vegetable nature of the cell. Schwann also observed that the reproduction of the globules increased with the rate of fermentation and concluded that the development of fermentation was most probably induced by these organisms. This yeast plant was recognized as a fungus and given the name "Zuckerpilz" or sugar fungus, which has been perpetuated in the generic term *Saccharomyces* (Lafar, 1910).

Independent observations by Kützing of Germany added strong confirmation that yeast was a vegetable organism and was responsible for alcoholic fermentation. By microscopic examination he observed the mode and rate of growth of yeast and wrote in 1837, "It is obvious that chemists must now strike yeast off the role of chemical compounds, since it is not a compound but an organized body, an organism" (Harden, 1923). This the leading chemists of the day were not willing to do. They repudiated the suggestion of a life force in yeast and contended that all activity attributed to yeast could be explained by chemical action. Berzelius believed it was a catalytic type of action. Liebig believed in a theory of motion of the elements within a compound and this caused a disturbance of equilibrium which was communicated to the elements of the substance with which it came in contact thus forming new compounds.

B. Pasteur's Contribution

Pasteur the chemist became a skilled investigator of biological problems, and from 1855 to 1876 his research was concentrated on studies of fermentation. He clearly demonstrated that fermentation was a physiological action associated with the life processes of yeast. The brewer added yeast to treated barley and the yeast grew, reproduced itself, and the resultant beverage was beer. The winemaker did not add yeast to grapes; the yeast was already there on the surface of the grapes as Pasteur could see with his microscope. Pasteur demonstrated that if he took pure grape juice from the center of the grape and exposed it to air purified by heat or by filtering, there was no fermentation. Yeast, the living organism, which he watched budding and growing under the microscope, was the essential agent of fermentation. His classic paper of 1857 dealt with lactic acid fermentation, the production of lactic acid rather than alcohol. He showed this was fermentation also, the action of a living organism, but a different organism from that involved in alcoholic fermentation.

Other papers followed in which Pasteur described his technique of isolating different microorganisms from fermenting material, growing them on a culture medium and determining their nutrient requirements and conditions for growth. Pasteur's life and work have been described many times (Vallery-Radot, 1923; Dubos, 1950; Conant, 1952). His major achievements in fermentation were to establish that (a) fermentation is caused by living organisms, (b) a specific ferment is responsible for each kind of fermentation, and (c) ferments are not born spontaneously.

VI. THE DISCOVERY OF ZYMASE

Liebig and other chemists disputed Pasteur's theories. Other investigators remembered the work of Lavoisier and of Gay-Lussac who showed that the conversion of sugar to alcohol and carbon dioxide was a balanced chemical reaction. The search began for something in yeast that might be the cause of that reaction. Payen and Persoz (1833) had obtained a substance from malt extract that converted starch to sugar, and they named it diastase. Three years later Schwann described a substance in gastric juice that degraded proteins and named it pepsin. These substances which were not recognized at that time as enzymes were called "ferments" and were so classified until 1876. However, it was recognized that diastase and pepsin were different in some way from those ferments which caused fermentation, and so they were called unorganized or structureless ferments.

In 1858 Moritz Traube gathered together all the available evidence and theorized that ferments were present in those living organisms that caused fermentation and that those ferments were definite chemical substances produced within the cells of the organisms (Harden, 1923). These were classified as organized or structural ferments. This terminology was retained until 1876 when Kühne proposed that ferments be called "enzymes" from the Greek words meaning "in yeast."

Many attempts were made to separate an active substance from yeast but all failed until 1897 when Edward Buchner obtained an enzyme from cell-free juice from yeast. This was one of those fortunate discoveries by investigators who were seeking something else at the time. Buchner and his brother Hans, a bacteriologist, were attempting to prepare an extract from yeast for therapeutic purposes. They ground yeast cells with sand but the extract obtained rapidly decomposed. A sugar solution was added as a preservative but the extract fermented producing carbon dioxide and alcohol. They realized the significance of this observation and with repeated preparations verified that an extract from yeast, with all traces of yeast cells removed, contained an enzyme which they named zymase. The living organism was not the cause of fermentation, the action was produced by a ferment secreted by the yeast cell (Harden, 1923). No longer was it necessary to make a distinction between organized and unorganized ferments, all were enzymes (Leicester, 1974).

VII. CHEMISTRY OF YEAST-JUICE ACTION

Following Buchner's discovery that yeast juice contained an active agent, zymase, which fermented sugar solutions, many investigators began to search more deeply into the chemical processes involved.

Harden and co-workers were among the leaders in this group and by 1906 they reported that zymase required the presence of a coferment. They dialyzed yeast juice and found that neither the filtrate nor the residue alone was capable of fermenting glucose. When the filtrate and residue were combined active fermentation resulted. These and other experiments led to the conclusion that fermentation of glucose required not only the presence of the enzyme but also of another substance that was dialyzable and thermostable. The essential substance in the dialyzate they called cozymase (Harden, 1923).

In 1905 Harden also found that addition of inorganic phosphate stimulated fermentation of fresh yeast juice with a measurable increase in evolution of carbon dioxide; subsequently the activity declined, and the phosphate disappeared. By continued intensive investigations of the action of added phosphate, Harden and co-workers were able to show that an organic phosphate ester was formed as an intermediate in the fermentative degradation of the sugar molecule. They were able to isolate a hexose diphosphate and finally to identify it as fructose 1,6-diphosphate.

These studies on alcoholic fermentation along with those of many other investigators led to the determination of the intermediary metabolism of carbohydrates and rapid advances in biochemistry.

It is hardly surprising that modern biochemists make such statements as "biochemistry itself was born of yeast technology" (Rose and Harrison, 1971). Rene Dubos (1950) has said "much of our understanding of the biochemical reactions of living processes has evolved from the study of yeast and of alcoholic fermentation," and Sir Hans Krebs (1968) has stated "looking back at the history of biochemistry one can say that many outstanding discoveries in biochemistry and general biology were made in the course of work on yeast."

VIII. YEAST: ITS NUTRITIONAL ASPECTS

A. Requirements for Growth

Not much attention had been paid to the nutrient requirements of yeast for its maximum growth until the beginning of this century. In 1901, Wildiers working with Ide at the University of Louvain, Belgium, reported that yeast required for its growth an organic substance which he called "bios." He worked with a beer strain of *Saccharomyces cerevisiae* and failed to get any appreciable growth in a solution of sucrose, ammonium chloride, and inorganic salts until he added a filtered extract of yeast cells. The bios theory aroused much controversy and many attempts were made to isolate it. By 1931 two fractions of bios had been separated; neither fraction alone stimulated yeast growth, but a combination of the two fractions was effective. Inositol was found to be associated with bios I but bios II could not be identified as a single substance (Koser, 1968).

Following the early discovery of bios as a growth stimulant there was little progress in understanding the nutrition of yeast until the discovery of the vitamins, fat-soluble A and water-soluble B. Through the 1920s research was focused on vitamin B and its antineuritic properties. The antineuritic activity was lost when yeast was autoclaved, but the heated fraction retained some growth-stimulating capacity. An alcoholic extract of yeast material adsorbed on charcoal cured polyneuritis in pigeons but did not sustain growth in rats unless supplemented by autoclaved yeast. Thus the separation of vitamin B into a thermolabile fraction, called vitamin B-1, and a thermostable fraction was clearly demonstrated by 1926.

The identification of all the other factors now recognized as parts of the vitamin B complex took some 20 more years and followed many different lines of investigation. However, the growth requirements of yeasts and other microorganisms have been a source of much information, leading to the discovery of several of the B-complex vitamins, namely pyridoxine, folic acid, inositol, pantothenic acid, and biotin. The problem was not a simple one, especially because it took time for investigators to realize that there were marked differences in the nutritive requirements of different species of yeasts, and particularly of different strains of the same species. For example, yeasts unable to ferment lactose did not require preformed nicotinic acid but synthesized their own supply of it. On the other hand, more than 100 strains of lactose-fermenting yeasts all required nicotinic acid for their growth (Koser, 1968).

B. Nutrient Synthesis

Some of the B vitamins are essential in the media for growth of yeasts, and others are synthesized by yeasts during their growth. Because of this synthesis of nutrients the fermented beverages, particularly the home brews of our ancestors, often made a significant contribution to the food value of the diet. These beverages contained yeast cells that had not been removed by clearing or refining as used in modern processes. Thiamine, riboflavin, nicotinic acid, and pyridoxine are present in yeast cells in comparatively high amounts. Yeast as a daily therapeutic treatment of beriberi has long been recommended (Williams and Spies, 1938).

Because of their ability to synthesize nutrients and because they grow rapidly on raw materials such as industrial or agricultural washed products, yeasts are being more thoroughly investigated today as commercially feasible sources of food supplements.

C. Nutritive Value of Yeast

The vitamin value of yeast has just been mentioned. Yeast is also an excellent source of protein. During World War I in Germany a *Torula* yeast was cultivated on a sugar medium containing only inorganic sources of nitrogen. This yeast produced little alcohol and when dried was a source of protein for addition to the diet (Accessory Food Factors Committee, 1945). During the 1930s the League of Nations Technical Commission on Nutrition and other groups considered the use of this method for improving tropical diets, and at the beginning of World War II the British Medical Research Council investigated and gathered together all available research data on the value of dried food yeast and found it to be a superior source of thiamine, and riboflavin and high in pyridoxine and niacin. Feeding tests with experimental animals showed yeast proteins were deficient in methionine but of high supplementary value when added to diets consisting mainly of cereals. Tests for tolerance by adults and children when food yeast was incorporated into the food components of the daily diet showed good acceptance and no digestive disturbances. It was concluded that 10 gm daily of food yeast was acceptable (Accessory Food Factors Committee, 1945).

Pellagra was a major disease problem in the Southern states in the early part of this century and dried bakers' or brewers' yeast was introduced by Goldberger in 1925 in the successful treatment of the disease. He used 1 gm dried yeast per kilogram of body weight and some physicians used higher doses without any deleterious effect. Yeast was widely distributed throughout the pellagrous regions; the usual amount was 30 gm/day (Sebrell, 1938).

Food yeasts are widely used today as sources of nutrients and flavor improvers in food. They are classed as "dried yeast" and are obtained from spent brewers' yeast or from yeast cultivated in special mineral-supplemented media using sugar-bearing materials such as molasses, wood hydrolysates, and cheese whey. The product consists of nonfermenting stable whole yeast cells (Peppler, 1970).

REFERENCES

Accessory Food Factors Committee. (1945). Food yeast: A survey of its nutritive value. *Med. Res. Counc. (G.B.) War Memor.* No. 16, 3–16.
Baron, S. (1962). "Brewed in America, a History of Beer and Ale in the United States," pp. 17 and 99. Little, Brown, Boston, Massachusetts.
Biggar, H. P. (1924). "The Voyages of Jacques Cartier," Published from the original with translations, notes and appendices, No. 11, pp. 212–215. Publ. Public Arch. Can., Ottawa.

Borgstrom, G. (1968). "Principles of Food Science," Vol. 2, p. 103. Macmillan, New York.

Chittenden, R. H. (1930). "The Development of Physiological Chemistry in the United States," pp. 59–60. Chem. Catalog Co. (Tudor), New York.

Conant, J. B. (ed.). (1952). "Pasteur's Study of Fermentation," Harvard Case Histories in Experimental Science, No. 6. Harvard Univ. Press, Cambridge, Massachusetts.

Darby, W. J., Ghalioungui, P., and Grivetti, L. (1977). "Food: The Gift of Osiris," Vol. 2, pp. 532–536. Academic Press, New York.

Dubos, R. J. (1950). "Louis Pasteur: Freelance of Science," p. 118. Little, Brown, Boston, Massachusetts.

Foster, M. (1901). "Lectures on History of Physiology during the Sixteenth, Seventeenth and Eighteenth Centuries," pp. 216–217. Cambridge Univ. Press, London.

Furnivall, F. J. (ed.). (1870). "Andrew Boorde's Dyetary," pp. 254–256. Early Engl. Text Soc., London.

Harden, A. (1923). "Alcoholic Fermentation," pp. 1–9. Longmans, Green, London.

Keogh, B. P. (1976). Microorganisms in dairy products—friends and foes. *PAG Bull.* **6**(4), 34–41.

Koser, S. A. (1968). "Vitamin Requirements of Bacteria and Yeasts," pp. 8–13 and 500. Thomas, Springfield, Illinois.

Krebs, H. (1968). Introductory remarks. *In* "Aspects of Yeast Metabolism" (A. K. Mills, ed.), p. xiii. Davis, Philadelphia, Pennsylvania.

Lafar, F. (1910). "Technical Mycology" (transl. by C. T. C. Salter), Vol. 1, pp. 12–14. Griffin, London.

Leicester, H. M. (1974). "Development of Biochemical Concepts from Ancient to Modern Times," p. 182. Harvard Univ. Press, Cambridge, Massachusetts.

Lorenz, A. J. (1953). Some pre-Lind writers on scurvy. *Proc. Nutr. Soc.* **12**, 306–324.

McKie, D. (1952). "Antoine Lavoisier: Scientist, Economist, Social Reformer," pp. 282–284. Schuman, New York.

Manchester, K. E. (1976). General Washington and the patriot soldiers. *J. Am. Diet. Assoc.* **68**, 421–433.

Pavy, F. W. (1874). "A Treatise on Food and Dietetics, Physiologically and Therapeutically Considered," pp. 124–125. Churchill, London.

Payen and Persoz. (1833). Memoire sur le Diastase. *Ann. Chim. Phys.* **53**(2), 73–92. Cited by Balls, A. K. (1942). *In* "Liebig and after Liebig" (F. R. Moulton, ed.), p. 39. Am. Assoc. Adv. Sci., Washington, D.C.

Peppler, H. J. (1970). Food yeasts. *In* "The Yeasts. Vol. 3: Yeast Technology" (A. H. Rose and J. S. Harrison, eds.), pp. 439–441. Academic Press, New York.

Pereira, J. (1868). "A Treatise on Food and Diet" (C. A. Lee and S. R. Wells, eds.), Am. Ed., pp. 25–27. Samuel R. Wells, New York. (Engl. Ed., 1843.)

Rose, A. H., and Harrison, J. S. (eds.). (1971). "The Yeasts. Vol. 2: Physiology and Biochemistry of Yeasts," p. 1. Academic Press, New York.

Schmidt, J. E. (1959). "Medical Discoveries—Who and When," p. 125. Thomas, Springfield, Illinois.

Sebrell, W. H. (1938). Vitamins in relation to the prevention of pellagra. *J. Am. Med. Assoc.* **110**, 1665–1672.

Vallery-Radot, R. (1923). "The Life of Pasteur" (transl. from the Fr. by R. L. Devonshire). Doubleday, Page, New York.

Williams, R. R., and Spies, T. (1938). "Vitamin B$_1$ (Thiamin) and Its Use in Medicine," p. 88. Macmillan, New York.

Willich, A. F. M. (1799). "Lectures on Diet and Regimen," pp. 352–353. Longman, London.

EDITORIAL COMMENT

The traditional view that gout was a consequence of excessive use of port by the well-to-do Englishmen of the eighteenth and nineteenth centuries has been illuminated by recent studies of an epidemic of gout among drinkers of illicit whiskey distilled with the use of car radiators as condensers in the southern United States. It appears that contamination of both the port and the whiskey with lead from the lead solder used in the stills in Portugal at that time and in the car radiators led to a form of nephritis characterized by hyperuricemia. The term saturnine gout has been applied to this form of the disease and is derived from the alchemist's symbol for lead, Saturn.

Although alcohol consumed in a fasting state might be expected to cause ketosis and thereby decrease renal clearance of uric acid, alcohol taken with food would not likely contribute to hyperuricemia.

REFERENCE

Ball, G. V. (1977). Two epidemics of gout. *Bull. Hist. Med.* **45**, 401–408.

6

BIOCHEMISTRY OF FERMENTATION

L. Krampitz

I. Introduction   99
II. The Alcoholic Fermentation: A Look into the Future   101
III. Glycerol Fermentation   103
IV. The Citric Acid Fermentation   104
    Reference   106

I. INTRODUCTION

It is my opinion that there will be a resurgence of interest in fermentation processes. This interest is one necessitated by the rapid dwindling of our petroleum resources and the increased demand of the products of fermentation for nutritional purposes. Prior to our great technological advances in petroleum chemistry and the catalytic processes in which petroleum was converted to products formerly associated with the fermentation industry, much emphasis was placed on fermentation processes. To mention only a few: the alcoholic fermentation, the so-called solvent fermentations wherein butyl alcohol, acetone, and 2,3-butylene glycol were produced by fermentation. To keep pace with the demand for synthetic fibers whose raw materials begin with petroleum conversion, the fermentations industry will have to step in. The biochemistry of many of these fermentations has been fairly well elucidated and is outlined in textbooks of biochemistry or microbiology; therefore, I will not dwell on their metabolic pathways. Figure 1 illustrates a portion of the metabolic pathways concerned with fermentation processes. The advances made in recent years in molecular biology have not as yet been applied to any large extent to fermentation processes. I will outline some personal thoughts on how the application of molecular biology may aid in accomplishing some desired results in fermentation processes.

FERMENTED FOOD BEVERAGES IN NUTRITION
Copyright © 1979 by Academic Press, Inc.
All rights of reproduction in any form reserved.
ISBN 0-12-277050-1

99

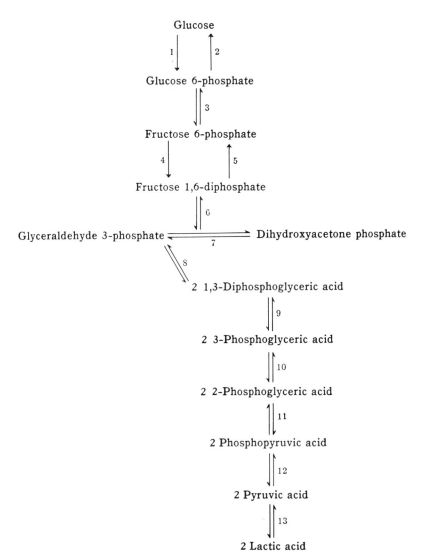

Fig. 1. Metabolic pathways in fermentation processes.

II. THE ALCOHOLIC FERMENTATION: A LOOK INTO THE FUTURE

The alcoholic equation is

$$C_6H_{12}O_6 \rightarrow 2C_2H_5OH + 2CO_2$$

The equation represents the overall conversion of glucose to ethanol and CO_2 by the yeast fermentation. The simplified equation omits the details of how phosphate esters are formed and how energy-rich phosphate bonds can be generated by anaerobic oxidation. The equation is stoichiometric, and from a carbon and electron balance point of view it is impossible to improve the yield of ethanol. The question arises then, if this is the case what is there to improve? The obvious answer is kinetics or rates of reactions.

If methods can be developed whereby one microorganism can accomplish what several do now in the same amount of time, it is obvious that the costs of production will become more attractive and be much more competitive with alternate processes. In recent years we have learned that cells are not merely bags of enzymes that carry out their specific reactions, but rather an organization of these catalytic processes that are well regulated. In some cases we have learned how to manipulate these regulatory mechanisms in order to bring about more desired metabolic pathways and greater abundance of products with reference to time. To illustrate, I will deviate from the alcoholic fermentation and describe a well-documented case with tryptophan synthetase (1). Wild-type *Escherichia coli* has a tryptophan operon located in its chromosome which codes for the biosynthesis of tryptophan synthetase. Mutants lacking genes in this operon have been obtained. They are genes for the structural proteins of the enzyme tryptophan synthetase, and therefore these mutants cannot synthesize the enzyme. These mutant organisms are nutritionally dependent upon tryptophan for growth. Extramicrobial chromosomal elements known as plasmids, which are also capable of transferring genetic information, are known to exist in many bacteria. These are relatively small double helix circular structures of DNA, which, when they enter the appropriate microbial cell, replicate and form more genetic material. The DNA which carries specific genetic markers can code for messenger RNA, and subsequently specific protein biosynthesis may occur. Plasmids have been obtained which replicate only to the extent of two or three copies; however, others can replicate to produce many more, thus enhancing the biosynthesis of messenger RNA and subsequently specific protein or enzyme biosynthesis. These latter plasmids are said to be under relaxed control. By means of recombinant DNA techniques, these circular plasmids can be opened by endonucleases and foreign DNA can be inserted by ligase enzymes *in vitro* to construct a plasmid now containing a new piece of DNA.

Figure 2 diagrammatically outlines the procedure. Plasmid DNA is isolated

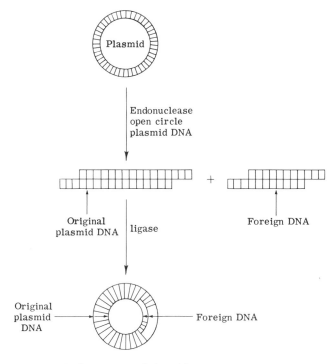

Fig. 2. Recombinant DNA technique.

from disrupted *E. coli* by density gradient centrifugation. Endonucleases have been isolated from *E. coli* which specifically hydrolyze the plasmid circular DNA at specific points, thus opening the circle. The opening occurs in such a manner that there is an overlapping of the two DNA strands. These overlapping areas have been termed "sticky" inasmuch as complementary pieces of foreign DNA can be attached by means of a ligase enzyme, thus closing the circle and creating a newly constructed plasmid. When placed in the microbial cell, the altered plasmid will code for the genetic marker present in the newly inserted piece of DNA, and the metabolic mechanisms will synthesize the specific protein (enzyme) as a final gene product. It is apparent that if the foreign DNA has been inserted into one of the so-called relaxed control plasmids, replication of the DNA will occur and form many copies of the plasmid. Thus, such a cell will have the potential capacity to biosynthesize large quantities of messenger RNA followed by a larger capacity to synthesize the specific enzyme coded for by the DNA.

What has been done in the tryptophan synthetase area is that a plasmid was constructed by enzymatically inserting a piece of DNA carrying the tryptophan operon. The plasmid constructed was one under relaxed control, and therefore when inserted into the mutant which did not possess the tryptophan operon, replicated to the extent that several copies of plasmid were formed. Consequently, increased messenger RNA for the synthesis of tryptophan synthetase, and hence increased amounts of tryptophan were formed as compared to the wild type.

While plasmids of the type found in bacteria have not been found in yeast, forms of DNA have been isolated which bear close resemblance to these plasmids.

The demand for industrial ethanol will remain, but its source from petroleum will be severely curtailed; therefore, I predict that active basic research into the fermentation process along the lines of recombinant DNA with plasmids described above with tryptophan will be actively pursued to increase the rate of ethanol formation by fermentation.

Until recently much of the research performed to increase the yield of products of fermentation was done by empirical screening techniques. Now that we are learning more about the molecular biology of all cells, these problems can be approached in a more rational manner. We may find that in the whole array of enzymes responsible for the alcoholic fermentation in yeast, there are only a few that are under strict regulatory control, and by removing these regulatory mechanisms, we can obtain strains of yeast capable of forming ethanol at a much greater rate.

III. GLYCEROL FERMENTATION

Mutant forms of microorganisms are readily obtained by the techniques of molecular biology. With few exceptions this technique has not been fully exploited by the fermentation industry. By illustration I refer to the glycerol fermentation. During World War I Germany was extremely short of glycerol for munition purposes inasmuch as the supply of triglycerides was extremely limited. The famous German biochemist, Carl Neuberg, was asked to investigate the problem. The mechanism of fermentation of glucose by yeast to ethanol was beginning to be elucidated. Figure 3 illustrates Neuberg's thinking in an attempt to use the yeast alcoholic fermentation for the production of glycerol. He reasoned that if he could prevent pyruvate from being decarboxylated to acetaldehyde and CO_2, the reduced form of diphosphopyridine nucleotide (DPNH) formed from the oxidation of 3-phosphoglyceraldehyde to 3-phosphoglyceric acid would then reduce dibydroxyacetone phosphate to glycerol instead of reducing acetaldehyde to ethanol. He added sodium bisulfite to the normal yeast fermentation of glucose which made a bisulfite complex with pyruvate, preventing it from decarboxylation to acetaldehyde and CO_2, thus favoring the reduction

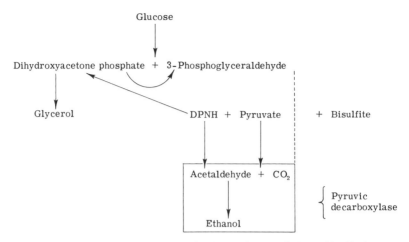

Fig. 3. Yeast alcoholic fermentation for the production of glycerol by Neuberg.

of dihydroxyacetone phosphate to glycerol. The yields were not stoichiometric, since the binding of pyruvate by bisulfite was not complete, permitting ethanol to be formed. In view of this half-century-old information and a greater need for glycerol for industrial purposes, it would appear that the selection of a mutant yeast devoid of the pyruvate decarboxylase gene would be a mechanism to obtain yields of glycerol as high as 50% of the glucose fermented. A "spinoff" of the fermentation would be pyruvate, the other portion of the fermentation. A mutant lacking pyruvic decarboxylase would be much more efficient than the bisulfite binding technique employed by Neuberg, since the latter technique was not a total block of the conversion of pyruvate to acetaldehyde and CO_2.

IV. THE CITRIC ACID FERMENTATION

Finally, I want to mention the citric acid fermentation—a compound so important to the beverage industry. I am not acquainted with the economics of this fermentation, but it would appear to me that the initial substrate in the fermentation should be cheap, perhaps cheaper than glucose. Another objection to glucose as the initial substrate in the citric acid fermentation is that by-products other than citric acid are formed. I also am not privy to the processes now being used by the major citric acid producers, and perhaps they have already employed some of the procedures I will discuss. In thinking about this fermentation, it became apparent that acetate would be the convenient substrate whose initial cost would be relatively lower. Figure 4 outlines a mechanism of the total biosynthesis of citrate from acetate after being primed with a small quantity of glucose. Certain yeasts

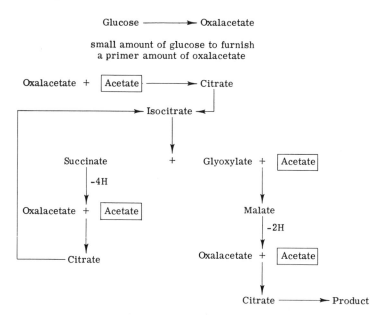

Fig. 4. Biosynthesis of citrate from acetate.

have the full complement of enzymes illustrated in this figure. Citrate synthetase is responsible for the biosynthesis of citrate from oxalacetate plus acetyl-coenzyme A (CoA). For the sake of simplicity the CoA derivatives are not depicted in the figure. To initiate citrate biosynthesis from acetate, a primer amount of oxalacetate must be present. A small amount of glucose serves as the source of the primer oxalacetate through oxidative mechanisms. The term fermentation in this case is a misnomer inasmuch as the process depicted is aerobic. Acetyl-CoA condenses with the oxalacetate to form citrate. The latter is converted to isocitrate.

A key enzyme in the series of reactions is isocitrate lyase, which cleaves isocitrate into succinate and glyoxylate. Note that a C_4 compound, succinate, has been formed which by aerobic oxidation is the precursor for another mole of oxalacetate. The latter is constantly being regenerated, providing a mechanism for the regeneration of citrate and isocitrate and permitting another cleavage to succinate and glyoxylate. The fate of glyoxylate is important since it provides for a net synthesis of citrate. The glyoxylate condenses with another mole of acetate through catalysis by malic synthetase to form malate. The latter, by aerobic oxidation, forms oxalacetate which condenses with a second mole of acetyl-CoA to form citrate. As a result, three moles of acetate have condensed to form one mole of citrate and a fourth mole of acetate is used to regenerate isocitrate to regenerate the cycle. The original C_4 compound, i.e., oxalacetate which serves as a primer, arose from glucose and is maintained in the cyclic events. In the mechanism outlined for citrate formation, known enzymatic reactions found in some microorganisms have been presented. These reactions are no doubt under strict regulatory control, and by molecular biological research increased yields of citrate ought to be obtained. For example, since the process is aerobic, oxidative mechanisms may be present which will remove the accumulated citrate by oxidation to carbon dioxide and water. Mutants should be readily obtainable that are devoid of isocitric dehydrogenase, thus preventing the first step in this oxidative process.

These modern techniques are available and more will be discovered through basic research and all will be applicable to vast numbers of biological conversion processes which form valuable commercial products.

REFERENCE

1. Hershfield, V., Boyer, H. W., Yanofsky, C., Lovett, M. A., and Helinski, D. R. 1974. Plasmid ColE1 as a molecular vehicle for cloning and amplification of DNA. *Proc. Natl. Acad. Sci. U.S.A.* **71,** 3455–3459.

7

NUTRIENT SYNTHESIS BY YEAST

Harry P. Broquist

I. Introduction 107
    A. Food Yeasts 108
    B. Yeast Single Cell Production (SCP) 109
II. Production of Nutrients by Yeast 110
    A. Gross Composition of Yeasts 110
    B. Vitamin-Coenzyme Forms 110
    C. Riboflavin Biosynthesis 111
    D. Yeast Protein 114
    E. Amino Acid Biosynthesis—General 115
    F. Lysine Biosynthesis in Yeast 116
III. Conclusions 118
    References 119

I. INTRODUCTION

Although the art of brewing is a centuries' old practice, it has only been since the turn of this century that the residual brewer's yeast has been recognized as having utility in itself as a nutritional supplement of human food and animal feed. This practice was stimulated by early studies in this century of such nutritionists as Wildiers and Funk, who demonstrated that yeast was a rich source of the newly discovered "vitamines" and of Osborne and Mendel that yeast protein could replace protein of vegetable and animal origin for animal growth.

The development of a whole industry to specifically produce food yeast in this and other countries throughout the world has been well reviewed elsewhere (Peppler, 1970; Baird, 1963). From these reviews it is apparent that such development was due to a number of factors as well as nutritional considerations including, for example, the status of war and peace, famine or plenty in the world, economics, availability of fermentable substrates. For example, in World War I the German war effort demanded new sources of glycerol for the manufacture of gun powder. The result was the "Neuberger fermentation," wherein bisulfite was added to a *Torula* yeast fermentation which shunted the glycolytic pathway at the triose stage to yield glycerol. The yeast was then utilized by the starving populus for food or as fodder in the feeding of livestock. Now, some 60 years later, again the specter of starvation looms ominously on the horizon as we seriously consider the Malthusian proposition that population growth may ultimately exceed the available food supply necessary to sustain such growth. But again man's ingenuity is in evidence as schemes for developing huge fermentation facilities for the production of single cell protein from hydrocarbons are emerging as one approach to meet world protein needs (Tannenbaum and Wang, 1975). What is the nutritional rationale for such activity? And what are the research frontiers essential for back-up of such efforts? This chapter will endeavor to answer, at least in part, these questions.

A. Food Yeasts

Food yeasts may be defined as "nonfermenting stable whole yeast cells, carefully prepared and dehydrated to yield flakes and powders, which are intended for the nutritious and flavorous improvement of foods in human dietary" (Peppler, 1970). They are either species of the genus *Saccharomyces* or the genus *Candida*. The most common food yeasts manufactured in this country are *Saccharomyces carlsbergensis* from beer, *Saccharomyces cerevisiae* strains of bakers' yeast grown on molasses, *Saccharomyces fragilis* grown on cheese whey, and *Candida utilis* propagated on waste sulfite liquor. About 86,000 metric tons (dry weight basis) of such yeast was estimated to have been produced in the United States in 1968.

The nutritional requirements for growth of most yeasts are quite simple. With the exception of the B complex vitamins usually provided by the crude medium, most yeasts synthesize all other cell constituents from simple sugars, ammonium salts, and inorganic salts. Hence a cheap source of carbohydrate is the prime consideration for yeast propagation, and materials such as molasses, grain, fruit products, whey, and cellulosic waste including spent sulfite liquor are the most frequent substrates for yeast growth. There is also a high oxygen requirement necessary to obtain aerobic conditions necessary for a high yield of yeast cells rather than ethanol which is formed under anaerobic conditions. An example of a medium suitable for the propagation of food yeast is as follows: blackstrap molasses 4.6%, beet molasses 6.4%, ammonium sulfate 0.35%, magnesium sulfate heptahydrate 0.25%, ammonium hydrogen phosphate 0.05%, and phosphoric acid 0.04%, pH 5.0.

B. Yeast Single Cell Production (SCP)

In the case of single cell production in which *Candida lipolytica* is a favored microorganism, hydrocarbons serve as the primary carbon and energy source for cell growth which occurs mainly at the interface between the oil and water. The water phase provides the other nutrients, including ammonium salts, phosphate, and trace minerals. Table I taken from some data reviewed by Johnson (1969) illustrates a number of common substrates considered in yeast single cell protein production from which it can be seen that the conversion of such hydrocarbons to yeast cell mass is very efficient indeed. The dissimilation of hydrocarbons such as octadecane proceeds via oxidative processes: alcane → alcohol → acid and then via β-oxidation to yield acetyl-coenzyme A (CoA) units and adenosine triphosphate (ATP). Acetyl-CoA then either enters anabolic pathways for the production of yeast cells substance or is catabolized to CO_2 and H_2O. The oxygen requirement for the overall fermentation is very high, and some of the ATP that is formed is evolved as heat, all of which poses engineering and economic problems.

The case for yeast SCP production as one approach to close the protein gap on the worldwide scene can be summarized as follows (Kihlberg, 1972):

1. Yeasts have a very short generation time (4–12 hours) and can thus provide a rapid mass increase.

2. Yeast can be easily modified genetically such that mutants might be produced with more desirable physical and nutritional characteristics, e.g., faster growth rate, adaptability to higher temperatures, altered amino acid composition.

3. The protein content of yeast is of the order of 50% on a dry weight basis, which is much higher than that for most common food stuffs.

4. SCP production is based upon such substrates as cellulosic wastes, coal petroleum, natural gas, methanol, ethanol, which hopefully will be readily available for a long time to come.

Table I. Yeast Cell Yields from Diverse Substrates[a]

Yeast strain	Substrate	Cell yield (gm/gm/substrate)
Torula sp.	Hexadecane	0.82
Candida lipolytica	Octadecane	0.85
Candida sp.	"*n*-Paraffins"	0.85–1.10
Candida intermedia	Docosane	0.89
Candida utilis	Glucose	0.51
Candida utilis	Ethanol	0.68

[a] From Johnson (1969), reprinted by permission.

5. SCP production can be carried out in a continuous fermentation process independent of climatic changes with only small land area and water requirements.

II. PRODUCTION OF NUTRIENTS BY YEAST

A. Gross Composition of Yeasts

The gross composition of two common food yeasts and two yeast strains grown on alkanes is shown in Table II, which illustrates the nutritional merit of these products as rich sources of protein and B complex vitamins. In addition, alkane-grown yeast strain G is high in lipid content. As pointed out earlier, it has been recognized for over half a century that yeast is a rich source of the B complex vitamins, and this has had both basic and applied application. In addition to the B vitamins listed in Table II, other vitamins found in yeast include folacin, calcium d-pantothenate, biotin, p-aminobenzoic acid, choline chloride, and inositol. Of these vitamins, thiamine, riboflavin, niacin, pyridoxine, and folacin figure importantly in meeting the recommended daily allowances for these vitamins when consumed either in enriched foods or taken directly in capsule or pill form.

B. Vitamin–Coenzyme Forms

Because of the ready availability of yeast and because it can be easily grown under defined cultural conditions, it has been a popular biological source material

Table II. Gross Composition of Food Yeasts and Alkane-Grown Yeasts[a]

Analysis	Saccharomyces cerevisiae (molasses)	Candida utilis (sulfite liquor)	Alkane-grown yeast	
			Strain G (pure alkane)	Strain L (middle distillate)
Moisture (%)	9.2	5.8	<7.0	<8.0
Protein (N × 6.25) (%)	40.6	52.0	60	68–70
Fat (%)[b]	4.8	4.8	8–10	1.5–2.5
Ash (%)	9.7	9.0	6.0	7.9
Thiamine (μg/gm)	28	5		
Riboflavin (μg/gm)	62	45		
Niacin (μg/gm)	283	415		
Pyridoxine (μg/gm)	34	29		

[a] Reprinted from "Single-Cell Protein," Vol. 2, by S. R. Tannenbaum and D. I. C. Wang (eds.), by permission of The M.I.T. Press, Cambridge, Massachusetts. Copyright © 1975 by The M.I.T. Press.
[b] Ether extract of acid-hydrolyzed sample.

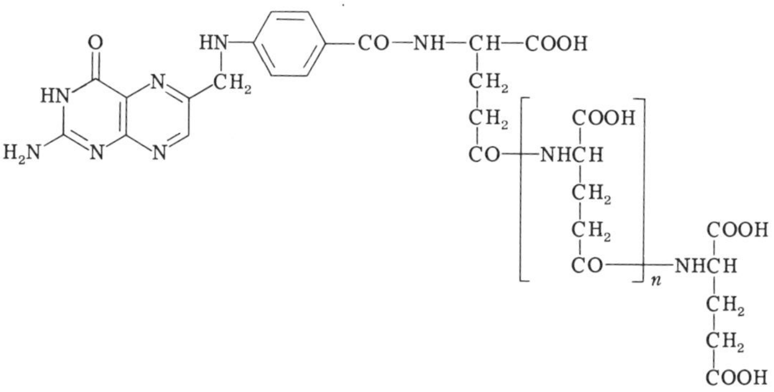

Fig. 1. Polyglutamate forms of folic acid.

for the isolation of the B complex vitamins and for study of their coenzyme forms. Such studies contributed, for example, to the recognition that thiamine existed in the cell as its pyrophosphate ester and that such vitamins as riboflavin and niacin are linked to nucleotides flavin adenine dinucleotide and nicotinic adenine dinucleotide, respectively. Another example of such research that is still attracting attention today was the discovery of Pfiffner *et al.* (1946) that folacin exists in yeast almost entirely in a polyglutamate form, pteroylheptaglutamate ($n=5$, Fig. 1). This subsequently led to findings that folacin may exist in natural materials in various conjugated forms having from 1 to 7 glutamyl residues in γ-glutamyl-peptide linkage (Fig. 1). Such conjugated folic acid forms have varying degrees of folic acid activity per se depending on species differences, but for man the polyglutamates must be hydrolyzed in the gut by folic acid conjugase to release pteroylmonoglutamate ($n=1$, Fig. 1), which can then be absorbed by the tissues. In the tissues the vitamin is then used, e.g., reduced and/or formylated and may then be reconjugated to generate specific coenzyme forms. Thus, it is clear that early studies of the nature of folic acid in yeast led to broad concepts bearing on both storage and functional forms of this important vitamin.

C. Riboflavin Biosynthesis

The data of Table II indicate that the two food yeasts *Saccharomyces cerevisiae* and *Candida utilis* contain about 50 μg of riboflavin, which is probably a reflection both of uptake of the vitamin from the rich medium and synthesis *in vivo*. Certain yeasts and related yeastlike molds which synthesize riboflavin in amounts far beyond their metabolic needs have been the subject of much investigation to elucidate the details of riboflavin biosynthesis. The requirement of most microorganisms for riboflavin is of the order of 0.1 mg/liter, and hence the ability of yeastlike molds *Eremothecium ashbyii* and *Ashbyii gossypii* to produce on the order of 5000 mg/liter of riboflavin is a "fermentation wonder" and poses interesting questions about the relase of control mechanisms in this metabolic pathway. The delineation of the pathway of riboflavin biosynthesis in these fungi has occupied the attention of research groups for many years (for reviews, cf. Demain, 1972; Plaut and Smith, 1974) and may be summarized with reference to Fig. 2.

Techniques used to elucidate the metabolic pathway included incorporation of radioactive test precursors into riboflavin formed in growing cultures of *E. ashbyii* or *A. gossypii,* accumulation of metabolites by *S. cerevisiae* riboflavin auxotrophs, demonstration of postulated transformations in cell-free fungal enzyme systems, etc. There are still some uncertainties about some of the transformations shown in Fig. 2, but an overview of the biosynthetic pathway is as follows: A purine nucleotide guanosine triphosphate (GTP) is first formed and initiates events of riboflavin synthesis. Carbon-8 of the purine ring is removed together with a loss of ribose phosphate, giving a pyrimidine, 6-hydroxy-2,4,5-triaminopyrimidine (HTP). Ribitol, the pentose moiety of riboflavin, is then added to the N_4 atom of HTP in a reaction involving cytidine diphosphate to give the pyrimidine nucleoside 2,5-diamino-6-hydroxy-4-ribitylaminopyrimidine (DHRAP). The latter is then deaminated at C-2 yielding 5-amino-2,6-dihydroxy-4-ribitylaminopyrimidine (ADRAP). The question of the nature of the C_4 unit which is then added to ADRAP to form the lumazine 6,7-dimethyl-8-ribityllumazine (DMRL) has long been sought. Although there have been a number of candidates proposed (e.g., acetoin, acetate units), the nature of the ADRAP → DMRL conversion is still shrouded in mystery. The conversion of DMRL to riboflavin was also a puzzling matter until the brilliant discovery from Plaut's (1960) laboratory that DMRL could react with itself in a complex reaction catalyzed by riboflavin synthetase forming a molecule of riboflavin and regenerating a molecule of ADRAP for reentry into the pathway of riboflavin synthesis. The enzyme riboflavin synthetase has been purified some 2000-fold from bakers' yeast. Binding studies support the view that one site on the enzyme binds DMRL in such a way that it functions as the donor of the C_4 moiety transferred in the riboflavin synthetase reaction, while a second site binds the lumazine that serves as the acceptor of the C_4 fragment.

As mentioned above, *S. cerevisiae* riboflavin auxotrophs have been useful in establishing the events shown in Fig. 2. The riboflavin-deficient mutants were divided according to accumulation products into five groups corresponding to the five enzymatic steps shown in the figure. It is thought that the two genes, *rib* 3 and *rib* 4, involved in the conversion of ADRAP to DMRL may be concerned with the synthesis of the C_4 moiety and the ring closure step, respectively.

The synthesis of riboflavin in the yeastlike molds discussed above represents perhaps the most outstanding example of the microbial overproduction of a vitamin yet discovered. Yet an overview of the biosynthetic pathway reveals a marked economy of devices employed to fashion the isoalloxazine ring system of the

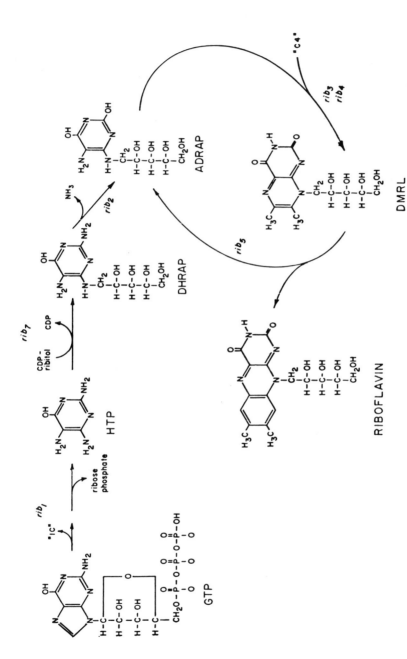

Fig. 2. Probable pathway of riboflavin biosynthesis. GTP, Guanosine triphosphate; HTP, 6-hydroxy-2,4,5-triaminopyrimidine; DHRAP, 2,5-diamino-6-hydroxy-4-ribitylaminopyrimidine; ADRAP, 5-amino-2,6-dihydroxy-4-ribitylaminopyrimidine; DMRL, 6,7-dimethyl-8-ribityllumazine. (From Demain, 1972, reproduced by permission of Annual Reviews, Inc.)

riboflavin molecule. Thus simple substrates readily deriving from carbohydrate and nitrogen metabolism such as CO_2, formate, glycine, aspartate, and the amide nitrogen of glutamine are utilized in the formation of GTP. The transformations GTP → HTP → DHRAP → ADRAP all have their counterpart in other metabolic reactions in living cells, and the final step in the synthesis generates both product riboflavin and precursor ADRAP, which reenters the biosynthetic pathway. Such detailed biosynthetic studies are worthwhile not only in revealing pathway details but now permit a study of the relevant enzymes and the regulation of their activity, which may bear importantly on factors controlling riboflavin overproduction and which might have useful application in the production of other nutrients by fermentation processes.

D. Yeast Protein

The data of Table II indicate that yeast is a rich source of protein, although the values shown must be considered approximate in view of the empirical methods used in the determination. The more important matter is the nutritional quality of yeast protein. This is illustrated by the bar graphs of Fig. 3, which compares the amino acid compositions of yeast, wheat, and rice with an FAO (1957) reference pattern. It can be seen that yeast compares favorably with the FAO standard in

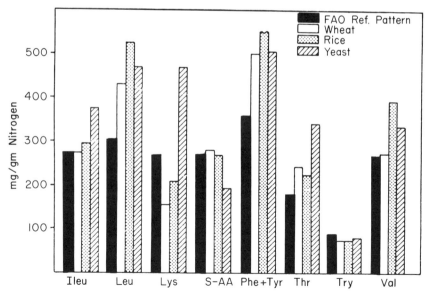

Fig. 3. Essential amino acid composition of wheat, rice, and yeast compared to the Food and Agriculture Organization (1957) reference pattern.

essential amino acid content with the exception of its sulfur amino acids. In contrast, the cereal grains are limited in lysine. Supplementation of the latter grains, which are adequate in methionine, by yeast which is ample in lysine, would yield nutritional products of high quality. This serves to illustrate the merit of food yeast as a protein supplement. Further enrichment of such yeasts by amino acids would only enhance their attractiveness in food supplementation, as will be discussed.

E. Amino Acid Biosynthesis—General

The elucidation of the pathways of amino acid biosynthesis which occurred following World War II was due in important part to (a) the availability of isotopes, which permitted the tracing of metabolic patterns in living systems, (b) a knowledge of microbial nutrition, which provided a simple biological system for experimentation, and (c) an ever-increasing sophistication of biochemical methodology. Much of such pathway knowledge which is widely summarized (cf. Greenberg, 1969; Rodwell, 1969) came from studies with *Escherichia coli, Neurospora crassa,* and yeast amino acid auxotrophs. Thus in early studies of Abelson and Vogel (1955) and Ehrensvärd *et al.* (1951), for example, the labeling patterns of [^{14}C]glucose and [^{14}C]acetate into the amino acids of *Torula utilis* were established.

From these and subsequent studies, the general concept of the grouping of the amino acids into families to denote common biosynthetic origins evolved. In yeast these families may be divided into five groups on the basis of the parent amino acids or carbohydrate precursors (Moat and Ahmad, 1966).

1. The glutamate or α-ketoglutarate family: glutamate, arginine, lysine, proline, and hydroxyproline
2. The aspartate family: aspartate, threonine, methionine, and isoleucine
3. The pyruvate family: alanine, valine, isoleucine, and leucine
4. The serine–glycine family: serine, glycine, cysteine, and cystine
5. The aromatic family: histidine, phenylalanine, tyrosine, and tryptophan

The relationships outlined above were shown in general to exist also in bacteria with the notable exception that lysine was derived in bacteria and in higher plants from aspartic acid rather than α-ketoglutarate as shown in Fig. 4. Thus aspartic semialdehyde serves as a branch point leading either to lysine or to homoserine, which then serves as a precursor for the synthesis of methionine, threonine, and isoleucine depending on conditions in the cell. The synthesis of the amino acids derived from aspartate is governed by an elaborate system of control mechanisms principally elucidated in George Cohen's laboratory (Truffa-Bachi and Cohen, 1968). Manipulation of such fine control can have enormous effects on the elaboration of lysine via the aspartate pathway (Fig. 4).

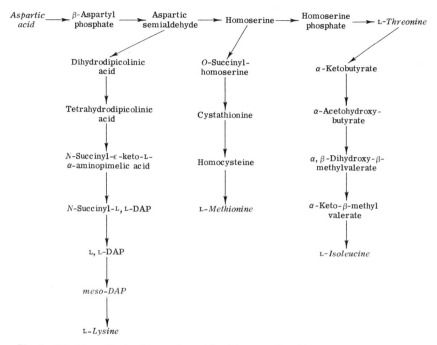

Fig. 4. The biosynthesis of the amino acids of the aspartic acid family. (From Truffa-Bachi and Cohen, 1968, reproduced by permission of Annual Reviews, Inc.)

For example, a mutant of *Corynebacterium glutamicum* which lacks homoserine dehydrogenase can produce yields of lysine ranging from 30 to 40% of the proffered sugar when the concentrations of methionine and isoleucine in the medium are carefully controlled (Nakayama, 1972). Indeed this fermentation serves as the primary commercial source of lysine today.

F. Lysine Biosynthesis in Yeast

In yeast, however, lysine is formed in a process in which α-ketoadipic acid and α-aminoadipic acid are prominent intermediates (Fig. 5). Work from a number of laboratories (cf. references in Broquist, 1971) has established that yeast has a series of enzymes that carry out certain reactions quite analagous to the TCA cycle to form α-ketoadipic acid. Thus as is shown in Fig. 5, acetate (as its CoA ester) condenses with α-ketoglutarate yielding homocitrate, which following dehydration, rehydration, dehydrogenation, and decarboxylation yields α-ketoadipic acid. The latter is then transaminated to α-aminoadipic acid, which is then converted to α-aminoadipic-δ-semialdehyde in a process requiring activation by ATP and reduction by reduced pyridine nucleotide. Aminoadipic semialdehyde is then aminated to give lysine in a unique transamination in which saccharopine is a stable intermediate.

Early evidence for the homocitrate–aminoadipate pathway of lysine by biosynthesis (Fig. 5) came from the authors' findings that bakers' yeast and *Torula* yeast could convert α-ketoadipic acid and α-aminoadipic acid to lysine in high yield (Broquist *et al.*, 1961). Subsequently, many of the other intermediates of the pathway were established by appropriate techniques including the use of yeast lysine auxotrophs, which permitted the accumulation and subsequent identification of products. Figure 5 indicates the location of the genetic block in a series of these yeast mutants that were extensively employed in these pathway studies.

Recently Florentino and Broquist (1974) prepared a high lysine bakers' yeast from aminoadipate fermentation, which contained about 10% of its dry weight as lysine and evaluated its nutritional quality in the rat. At least 90% of such lysine was available for growth of weanling rats fed a 20% wheat gluten diet limited in lysine. Supplementation of wheat flour with 2% of this high lysine bakers' yeast doubled the protein efficiency ratio of the flour, thus illustrating that this lysine-

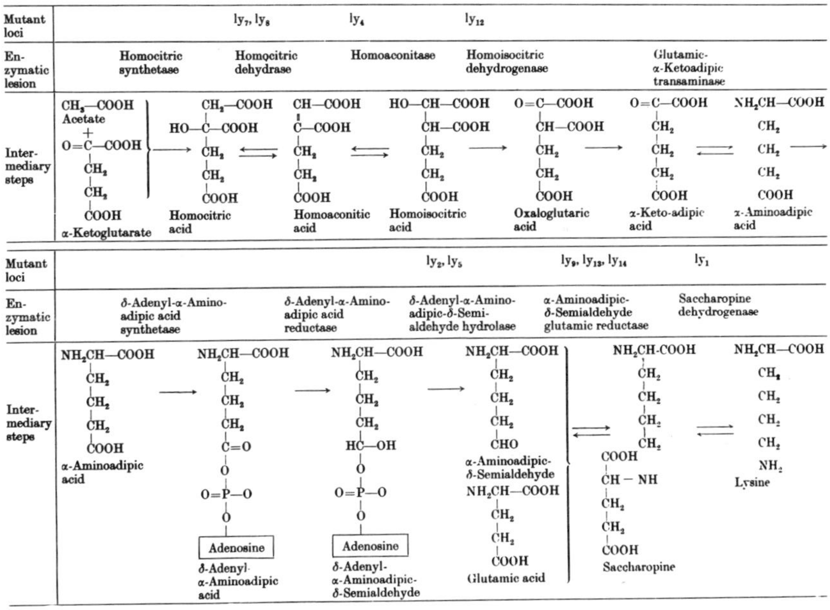

Mutant loci	ly_7, ly_8		ly_4	ly_{12}	
Enzymatic lesion	Homocitric synthetase	Homocitric dehydrase	Homoaconitase	Homoisocitric dehydrogenase	Glutamic-α-Ketoadipic transaminase
Intermediary steps	CH₃—COOH Acetate + O=C—COOH / CH₂ / CH₂ / COOH α-Ketoglutarate	CH₃—COOH / HO—C—COOH / CH₂ / CH₂ / COOH Homocitric acid	CH—COOH / C—COOH / CH₂ / CH₂ / COOH Homoaconitic acid	HO—CH—COOH / CH—COOH / CH₂ / CH₂ / COOH Homoisocitric acid → O=C—COOH / CH—COOH / CH₂ / CH₂ / COOH Oxaloglutaric acid → O=C—COOH / CH₂ / CH₂ / CH₂ / COOH α-Keto-adipic acid	NH₂CH—COOH / CH₂ / CH₂ / CH₂ / COOH α-Aminoadipic acid

Mutant loci	ly_2, ly_5			ly_9, ly_{13}, ly_{14}	ly_1
Enzymatic lesion	δ-Adenyl-α-Aminoadipic acid synthetase	δ-Adenyl-α-Aminoadipic acid reductase	δ-Adenyl-α-Aminoadipic-δ-Semialdehyde hydrolase	α-Aminoadipic-δ-Semialdehyde glutamic reductase	Saccharopine dehydrogenase
Intermediary steps	NH₂CH—COOH / CH₂ / CH₂ / CH₂ / COOH α-Aminoadipic acid	NH₂CH—COOH / CH₂ / CH₂ / CH₂ / C=O / O / O=P—O / O / Adenosine δ-Adenyl-α-Aminoadipic acid	NH₂CH—COOH / CH₂ / CH₂ / CH₂ / HC—OH / O / O=P—O / O / Adenosine δ-Adenyl-α-Aminoadipic-δ-Semialdehyde	NH₂CH—COOH / CH₂ / CH₂ / CH₂ / CHO α-Aminoadipic-δ-Semialdehyde NH₂CH—COOH / CH₂ / CH₂ / COOH Glutamic acid	NH₂CH-COOH / CH₂ / CH₂ / CH₂ / COOH / CH—NH / CH₂ / CH₂ / COOH Saccharopine	NH₂CH—COOH / CH₂ / CH₂ / CH₂ / NH₂ Lysine

Fig. 5. Relationship among the genes, enzymes, and intermediates of the biosynthetic pathway of lysine in *Saccharomyces*. (From J. K. Battacharjee and A. K. Sinha, 1972, *Mol. Gen. Genet.* **115**, 26. Reproduced by permission.)

enriched yeast has merit as a protein supplement in improving the nutritional quality of cereal grain protein.

In certain of our early experiments (Broquist *et al.,* 1961) as much as 4 gm of lysine was formed from 5 gm of α-ketoadipate per liter. This again is a flagrant example of overproduction of a nutrient when it is considered, for example, that the requirement of the lactobacillus, *Pediococcus cerevisiae,* for lysine is only about 10 mg per liter. We have shown in *N. crassa* lysine auxotroph 33933 that the first enzyme of the pathway, homocitric synthetase, is subject to both repression and feedback inhibition by lysine (Hogg and Broquist, 1968); and in yeast Maragoudakis *et al.* (1967) reported that $5 \times 10^3 M$ lysine inhibited the activity of this enzyme. It is apparent, however, that the lysine pathway (Fig. 5) is not subject to end-product control when lysine is being synthesized from preformed α-aminoadipate or α-ketoadipate. A cheap synthesis, either chemically or biologically for either of these adipic acid derivatives, would provide an attractive route for the production of a yeast enriched in lysine, which would intrinsically contain other desirable nutrients as well (Table II). Furthermore, the addition of high lysine yeast as an amino acid supplement to cereal grains, for example, at the 2% level as in the study of Florentino and Broquist (1974), corrects the lysine deficiency and minimizes the amount of yeast nucleic acid ingested and ultimately excreted as uric acid. Yeast consumption by man has been criticized on this basis, but clearly, if yeast products being considered as food or feed supplements can be enriched to meet specific nutritional needs, less yeast is then required as a supplement, and the uric acid problem can be obviated.

The biosynthesis of lysine in yeast was presented in some detail as but one example of amino acid pathway knowledge in yeast that can have application to the improvement of the nutritional qualities of yeast. It is apparent from Fig. 3 that an increase in the methionine content of yeast would be a major contribution toward improving the nutritional quality of yeast protein. Relationships for the biosynthesis of amino acids including methionine from aspartate in yeast is as that shown in Fig. 4 with the exception that there is no branch point from aspartic semialdehyde leading to lysine. Study of the regulation of methionine synthesis in yeast might well yield useful information that could be applied to improving the methionine content of yeast.

III. CONCLUSIONS

It is apparent that an impressive yeast technology exists to produce a high quality yeast product for food or animal feed, either via conventional substrates or more novel carbon sources such as petroleum-based substrates. The yeast cell is outstanding in its ability to produce vitamins and proteins and much is known about the pathways of biosynthesis of vitamins and amino acids and their control in yeast. Present knowledge of the biosynthesis of riboflavin in certain yeastlike molds and of lysine via the aminoadipate–homocitrate pathway in yeast was discussed both from the standpoint of specific metabolic transformations and as examples of the overproduction of nutrients. A combination of present knowledge of fermentation, engineering, yeast genetics, and intermediary metabolism as applied to the problem of yeast production should lead to yeast products of even superior nutritional quality, which will find their place in meeting nutritional needs of the world's burgeoning population.

REFERENCES

Abelson, P. H., and Vogel, H. J. (1955). Amino acid biosynthesis in *Torulopsis utilis* and *Neurospora crassa. J. Biol. Chem.* **213,** 355–364.

Baird, F. D. (1963). "The Food Value and Use of Dried Yeast," pp. 1–36. Cerevisiae Yeast Inst., Chicago, Illinois.

Broquist, H. P. (1971). Lysine biosynthesis (yeast). Metabolism of amino acids and amines. *In* "Metabolism of Amino Acids and Amines" (H. Tabor and C. W. Tabor, eds.), Methods in Enzymology, Vol. 17B, pp. 112–129. Academic Press, New York.

Broquist, H. P., Stiffey, A. V., and Albrecht, A. M. (1961). Biosynthesis of lysine from α-ketoadipic acid and α-aminoadipic acid in yeast. *Appl. Microbiol.* **9,** 1–5.

Demain, A. L. (1972). Riboflavin oversynthesis. *Annu. Rev. Microbiol.* **26,** 369–388.

Ehrensvärd, R. L., Saluste, E., and Stjernholm, R. (1951). Acetic acid metabolism in *Torulopsis utilis.* III. Metabolic connection between acetic acid and various amino acids. *J. Biol. Chem.* **189,** 93–108.

Florentino, R. F., and Broquist, H. P. (1974). Production and nutritional evaluation of a high lysine baker's yeast (*Saccharomyces cerevisiae*) in rats. *J. Nutr.* **104,** 884–893.

Food and Agriculture Organization of the United Nations (FAO). (1957). Protein requirements. *FAO (F.A.O. U.N.) Nutr. Stud.* No. 16.

Greenberg, D. M. (1969). Biosynthesis of amino acids and related compounds (Part I). *In* "Metabolic Pathways" (D. M. Greenberg, ed.), pp. 238–315. Academic Press, New York.

Hogg, R. W., and Broquist, H. P. (1968). Homocitrate formation in *Neurospora crassa*—Relation to lysine biosynthesis. *J. Biol. Chem.* **8,** 1839–1845.

Johnson, M. J. (1969). Microbial cell yields from various hydrocarbons. *In* "Fermentation Advances" (D. Perlman, ed.), pp. 833–842. Academic Press, New York.

Kihlberg, R. (1972). The microbe as a source of food. *Annu. Rev. Microbiol.* **26,** 427–466.

Maragoudakis, M. E., Holmes, H., and Strassman, M. (1967). Control of lysine biosynthesis in yeast by a feedback mechanism. *J. Bacteriol.* **93,** 1677–1680.

Moat, A. G., and Ahmad, F. (1966). Reprinted from *Wallerstein Lab. Commun.* **28,** No. 96, 111–136. Biosynthesis and Interrelationships of Amino Acids in Yeast.

Nakayama, K. (1972). Lysine and diaminopimelic acid. *In* "The Microbial Production of Amino Acids" (K. Yamada, S. Kinoshita, T. Tsunoda, and J. Aida, eds.), pp. 369–397. Kodansha Ltd., Tokyo.

Peppler, H. J. (1970). Food yeasts. *In* "The Yeasts. Vol. 3: Yeast Technology" (A. H. Rose and J. S. Harrison, eds.), pp. 421–462. Academic Press, New York.

Pfiffner, J. J., Calkins, D. G., O'Dell, B. L., Bloom, E. S., Brown, R. A., Campbell, C. J., and Bird, O. D. (1946). On the peptide nature of vitamin B conjugate from yeast. *J. Am. Chem. Soc.* **68,** 1392–1393.

Plaut, G. W. E. (1960). Studies on the stoichiometry of the enzymic conversion of 6,7-dimethyl-8-ribityllumazine to riboflavin. *J. Biol. Chem.* **235,** PC41–PC42.

Plaut, G. W. E., and Smith, C. M. (1974). Biosynthesis of water-soluble vitamins. *Annu. Rev. Biochem.* **43**, 899–922.

Rodwell, V. W. (1969). Biosynthesis of amino acids and related compounds (Part II). *In* "Metabolic Pathways" (D. M. Greenberg, ed.), pp. 317–373. Academic Press, New York.

Tannenbaum, S. R., and Wang, D. I. C. (eds.). (1975). "Single-Cell Protein," Vol. 2. MIT Press, Cambridge, Massachusetts.

Truffa-Bachi, P., and Cohen, G. N. (1968). Some aspects of amino acid biosynthesis in microorganisms. *Annu. Rev. Biochem.* **37**, 79–108.

8

THE BIOCHEMICAL PROCESSES INVOLVED IN WINE FERMENTATION AND AGING

Maynard A. Amerine

I.	Introduction	121
II.	Products	122
III.	By-Products of Fermentation	123
IV.	Indirect By-Products	126
V.	Higher Alcohols	126
VI.	Heat	129
VII.	By-Products of Processing	129
VIII.	The Malo-Lactic Fermentation	130
IX.	Research Needs	131
	References	132

I. INTRODUCTION

The mid-nineteenth century marks the emergence of microbiology and biochemistry as formal scientific disciplines. If not their father, Louis Pasteur was surely their midwife. From simple studies on the variation in alcohol yield of different molasses fermentations spring Pasteur's studies on yeasts and bacterial contamination and later to the biochemical pathway of alcoholic fermentation. Pasteur (1866) also made significant studies on the aging of wine and on the role of oxygen in aging.

Since 1866 (the date of the first edition of his ''Études sur le Vin''), great progress has been made in the study of the biochemical pathways involved in alcoholic fermentation. Today in the classic Embden–Meyerhof–Parnas pathway for the conversion of sugar to alcohol and carbon dioxide, we have a clear picture of the dozen or more enzymes involved, the metals required, the multiple by-

FERMENTED FOOD BEVERAGES IN NUTRITION
Copyright © 1979 by Academic Press, Inc.
All rights of reproduction in any form reserved.
ISBN 0-12-277050-1

products, and the energy changes accompanying this pathway. The regulation of the end products during vinification is still not so clear—at least from the practical point of view. The basic problem of enology is to determine where the 300 plus organic compounds present in wines come from and what the relation of each to quality is. Control of the amounts present is the objective.

My purpose here is to outline the main biochemical pathways involved in alcoholic fermentation of wines and in their processing and aging. I will relate the end products to the types and quality of the products. However, the no less important biochemical pathways involved in the ripening of grapes must be omitted. The total and relative amounts of the various sugars and acids produced by different varieties of grapes and under varying climatic conditions is, of course, very important to the quality of the final wine.

II. PRODUCTS

Early in the nineteenth century Gay-Lussac established his equation for alcoholic fermentation

$$C_6H_{12}O_6 \rightarrow 2C_2H_5OH + 2CO_2 \tag{1}$$
$$180 \qquad 92 \qquad 88$$

Theoretically, then, 51.1% of the sugar should appear as ethanol and 48.9 as carbon dioxide. In fact, the yield of alcohol in molasses or grape fermentation does not exceed 47% and Pasteur showed that this was due to various by-products and to the use of sugar in the metabolism of the yeasts.

The yield of alcohol is obviously of great importance to the winemaker in producing enough alcohol for a stable wine. This is critical when the sugar content of the grapes is limited (as it is in many areas in the world). Empirical studies have shown that the alcohol yield depends on sugar concentration, the temperature of fermentation, strains and activity of the yeast, aeration, acidity, and other factors. In general, alcohol yields are higher at lower sugar content (within the limitation of about 47%) and at lower temperatures. Oxygen deficiency leads to accumulation of high-energy compounds such as ethanol and the higher alcohols. The higher the sugar content the less efficient the production of ethanol. Winemakers utilize this information in various ways. If the sugar content is high, the percentage of sugar converted to alcohol is less and retention of residual sugar easier.

Yeast strain, ammonia content, and temperature are also important. Ough (1964) in the University of California enology laboratory showed that the rate of fermentation from 20° to 0° Brix at 21.1°C could be approximately predicted from an equation involving sugar, pH, and NH_3:

$$0.117 + 0.00085 \ NH_3 + 0.1066 \ pH - 0.0169° \ Brix \tag{2}$$

From this, about 80% of the variation in fermentation rate of white musts was predictable. The inhibitory effect of ethanol was greater at higher temperatures and at lower pH values.

Alcoholic fermentation ceases at about 16% ethanol, though by special techniques and with ethanol-tolerant strains of yeast higher levels can be achieved. This justifies the distinction between low alcohol and high alcohol wines—set at 14% in this country.

In practice ethanol yield also varies with variety of grape, but this may be an artifact based on differences in sugar, ammonia, and amino acid contents, as well as pH, and possibly on their varying natural yeast flora. The pentose cycle constitutes an alternative form of hexose metabolism, and this explains the formation of essential pentose cell constituents.

III. BY-PRODUCTS OF FERMENTATION

The Embden–Meyerhof–Parnas (EMP) scheme revealed all of the direct by-products. It does not show all of the possible by-products, since many result from reactions of the by-products with each other. The next-to-the-last step of the EMP pathway is the decarboxylation of pyruvic acid to acetaldehyde. If the further participation of the acetaldehyde in the scheme is blocked (by SO_2, for example) the aldehyde cannot act as a hydrogen acceptor. Under these conditions, dihydroxyacetone acts as the acceptor, and glycerol and acetic acid accumulate. Some acetaldehyde always remains—more when the fermentation is conducted completely anaerobically.

During fermentation, pyruvic acid and acetaldehyde accumulate in the early stages and decrease thereafter (Fig. 1). α-Ketoglutaric acid accumulates in the later stages. This is important because these three compounds fix SO_2 (whether added or as a by-product of alcoholic fermentation). This means that if they are allowed to accumulate, the winemaker is faced with a new wine with a high fixed SO_2 content or with a wine which will fix a considerable amount of SO_2 without increase in the free SO_2 content. This is not critical in red wine production, but in some white table wines where a little free SO_2 is desirable to maintain the wine in a reduced state, this can be important. Some acetaldehyde is also produced by autolysis of yeasts. Different strains of yeast produce more or less acetaldehyde (Amerine and Kunkee, 1968).

Other carbonyls that have been reported are formaldehyde, propionaldehyde, butyraldehyde, isobutyraldehyde, valeraldehyde, 2-methylbutyraldehyde, hexanal, furfural, acetone, 2-butanone, and 2,3-butanedione. The latter is probably the result of condensation of two molecules of acetaldehyde (by carboligase) to acetoin and its reduction (by other enzymes) to the diol. Acetaldehyde also reacts with ethanol to produce acetal (in brandies) and with acetic acid to produce

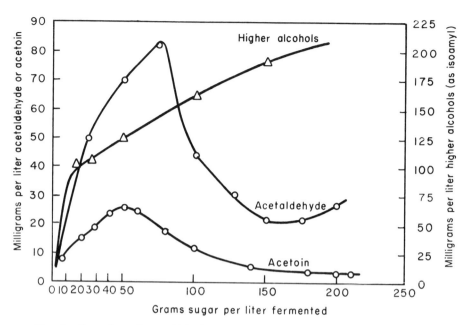

Fig. 1. Formation of acetaldehyde, acetoin, and higher alcohols during alcoholic fermentation. (From Amerine and Joslyn, 1972, reproduced by permission.)

diacetyl (noted in some red wines). 2,3-Pentanedione and 2,3-hexanedione are probably formed in similar reactions.

Glyercol and formic, acetic, lactic, and succinic acids are also constant by-products (Fig. 2). The amounts produced are small and, except in rare cases, of little importance with regard to quality. The same is true of acetoin, 2,3-butanedione, and other direct by-products.

Pasteur first reported the constant production of lactic and succinic acids and glycerol during alcoholic fermentation. The mechanism has been noted above. The amount produced is greater at 30°C than at higher fermentation temperatures.

Lactic acid is probably produced by reduction of pyruvic acid. Succinic acid no doubt results from condensation of pyruvic acid and carbon dioxide to form oxalacetic acid which is reduced to succinic acids. A little may come from glutamic acid.

Acetic acid is produced during alcoholic fermentation by dismutation of acetaldehyde (oxidation of one molecule of acetaldehyde to acetic acid and the simultaneous reduction of another to ethanol). Different strains of yeast vary in the amounts produced. Unless care is taken, some acetic acid may be formed as a result of bacterial synthesis. Moreover, there is an extraordinary lability among the by-products of alcoholic fermentation. Using C^{14}-labeled compounds, Dur-

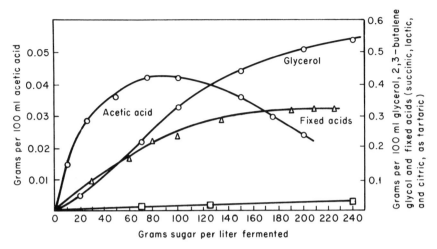

Fig. 2. Formation of acetic acid, glycerin, and fixed acids during alcoholic fermentation. (From Amerine and Joslyn, 1970, reproduced by permission.)

mishidze (1966) found labeled acetic acid converted to acetaldehyde, ethanol, 2,3-butanedione, lactic acid, and glycolic acid; labeled lactic acid was found as ethanol, acetic acid, glycolic acid, 2,3-butanedione, and glycerol; and glycerol appeared in acetaldehyde, ethanol, glycolic and lactic acids, and 2,3-butanedione. Obviously, the tricarboxylic (Krebs) and glyoxylic cycles are intervening with the glycolytic Embden–Meyerhof–Parnas cycle (if not in the complete anaerobic mitochondrial cycle, in those parts of the cycle involving soluble enzymes of the yeast). The pyruvic acid produced enters directly into the tricarboxylic acid cycle, etc.

One further problem is the inhibiting effect of some of the products of alcoholic fermentation on the process. Ethanol, acetic acid, propionic acid, butyric acid, and acetaldehyde all have a marked inhibiting effect and, of course, more at higher concentrations. Under aerobic conditions, carbon dioxide has a distinct inhibitory effect on yeast growth. It is worth noting that carbon dioxide is utilized by yeasts. The apparent slightly greater efficiency of conversion of glucose to ethanol in pressure tank fermentations may result from this. It has also been attributed to less yeast growth.

The primary mechanism of formation of ethyl acetate is not by simple esterification. The acetate is formed aerobically by alcoholysis of ethanol and coenzyme A (CoA) in the following reaction:

$$CH_3Co\text{-}S\text{-}CoA + C_2H_5OH \rightarrow CH_3COOC_2H_5 + CoASH \tag{3}$$

Howard and Anderson (1976) have recently demonstrated this in cell-free yeast extracts. Small amounts of many other esters are formed, probably by similar mechanisms. Polyhydroxy acids form both neutral and acid esters—more neutral at lower compared to higher pH. More total esters are formed at lower pH's.

IV. INDIRECT BY-PRODUCTS

Most strains of *Saccharomyces* release some hydrogen sulfide and/or, somewhat reciprocally, sulfur dioxide. The source of both seems to be from reduction of the sulfate naturally present in the must, from elemental sulfur present on the grapes (used for mildew control), or to a lesser extent from sulfur-containing amino acids. Some sulfide may come from reduction of sulfur dioxide. The hydrogen sulfide (and mercaptans) and ethanethiol (formed from the reaction of acetaldehyde and hydrogen sulfide) have very undesirable odors. Prevention is the best policy. Should they be formed, the amount present is reduced by reactions with any cupric ion present or by aeration followed by addition of SO_2 and close filtration.

Small amounts of methanol are found in wines, particularly in fruit wines. This alcohol is not formed by alcoholic fermentation, but results from the de-methoxylation of pectins and (to a very limited extent) of polyphenols and flavanols. Wines made by a second- fermentation of the skins and stems contain more methanol, and to prevent this practice legal limits on the methanol content are enforced in Italy and elsewhere. Wines made from moldy grapes are also higher in methanol.

V. HIGHER ALCOHOLS

A number of alcohols besides ethanol are found in all alcoholic beverages, including wine (Table I). For many years these were believed to be produced by the Ehrlich mechanism: deamination and decarboxylation of amino acids followed by reduction to yield an alcohol containing one carbon acid less than the amino acid. Higher alcohols may also be produced by the Strickland reaction: an amino acid is oxidized (serves as a hydrogen donor) and another is reduced (serves as a hydrogen acceptor). The greater amount, however, seems to come from the carbohydrate source, utilizing the enzymes involved in the normal sequence of the corresponding branched-chain amino acids. By one or the other mechanisms, *1-propanol,* 2-propanol, *1-butanol, 2-methyl-1-propanol* (isobutanol), *2-butanol,* 2-methyl-2-propanol (*t*-butanol), *1-pentanol* (*n*-amyl), *3-methyl-1-butanol* (isoamyl), *2-methyl-1-butanol* (act-amyl), *1-hexanol,* 1-heptanol, 2-phenylethanol, 2-(4-hydroxy-phenyl)ethanol (tyrosol), 2-(3-indole)ethanol (tryptophol), and 1,2-ethanedial (glycol) may be formed. (Compounds itali-

Table I. Source of Alcohols in Alcoholic Fermentation [a]

Alcohol[a]	Common name	Aldehyde found	Keto acid found	Corresponding amino acid
Ethanol	Ethanol	Acetaldehyde	Pyruvic	Alanine
1-Propanol	n-Propanol	n-Propionaldehyde	α-Ketobutyric	α-Aminobutyric
2-Propanol	Isopropanol	—	—	—
1-Butanol	n-Butanol	n-Butyraldehyde	α-Ketovaleric	Norvaline
2-Methyl-1-propanol	Isobutanol	Isobutyraldehyde	α-Ketoisovaleric	Valine
2-Butanol	sec-Butanol	—	—	—
—	—	—	α-Keto-γ-methiobutyric	Methionine
2-Methyl-2-propanol	tert-Butanol	—	—	—
1-Pentanol	n-Amyl	Valeraldehyde	α-Ketocaproic	Norleucine
3-Methyl-1-butanol	Isoamyl	Isovaleraldehyde	α-Ketoisocaproic	Leucine
2-Methyl-1-butanol	act-Amyl	act-Valeraldehyde	α-Keto-β-methylvaleric	Isoleucine
1-Hexanol	n-Hexanol	n-Hexanal	—	—
1-Heptanol	n-Heptanol	n-Heptanal	Oxalacetic	Aspartic
—	—	—	α-Ketoglutaric	Glutamic
2-Phenylethanol	β-Phenethyl	β-Phenylacetaldehyde	β-Phenylpyruvic	Phenylalanine
2-4-Hydroxyphenylethanol	Tyrosol	p-Hydroxyphenylacetaldehyde	p-Hydroxyphenylpyruvic	Tyrosine
2-(3-Indole)-ethanol	Tryptophol	Indolacetaldehyde	Indolpyruvic	Tryptophan
1,2-Ethanediol	Glycol	Glyoxal	—	Serine

[a] Source of data: Amerine and Joslyn (1970).

[b] In this text we have followed the IUPAC nomenclature for the higher alcohols. The more common names are listed in column 2. For discussion of the rules, see Hodgman (1976).

cized are always present. The largest amounts are of 3-methyl-1-butanol and 2-methyl-1-butanol.)

Very variable amounts of the higher alcohols are formed by different strains of yeasts, three times as much in some cases. This depends on the environmental conditions and on the intensity of metabolism of the yeasts. Temperature and growth factors (thiamine, etc.) have complex effects. The nitrogen source obviously has a profound effect on the amount of higher alcohols found. The fraction of amino acids (valine, leucine, isoleucine, phenylalanine) transformed to higher alcohols appears to be inversely related to the total nitrogen content of the substrate. Generally, anaerobic conditions result in more of the higher alcohols, at least of 3-methyl-1-butanol.

Too much of the higher alcohols, particularly of the amyl alcohols, is objectionable with regard to flavor. However, in practice this rarely occurs. In fact, at low concentrations (i.e., near threshold levels) they may add to the complexity of the odor and be desirable. In the presence of some bacteria more is reported to form, so one method of reducing the amounts is relatively bacteria-free fermentations, i.e., use more SO_2 and cleaner (mold-free) grapes.

In the production of dessert wines, wine spirits are added to raise the alcohol to 17% or more. In the distillation of wine in continuous stills to produce wine spirits, it is possible to remove most of the higher alcohols if the still is properly operated.

While the cause of the hangover is still a mystery, the higher alcohols are often considered the culprit. The available data do not support this claim. Excess consumption of ethanol has enough bad effects without incriminating the higher alcohols.

At least nine lactones and one lactane are found in wines, according to Webb (1977): γ-butyrolactone, 4-carboethoxy-γ-butryolactone, pantolactone, two isomers of 4,5-dihydroxyhexanoic acid-γ-lactone, 4-acetyl-γ-butyrolactone, 4-ethoxy-γ-butyrolactone, 4-keto-5-methyl-δ-valerolactone, 3-methyl-4-butyl-γ-butyrolactone, ethyl pyroglutamate, and 2-pyrrolidone-5-carboxylic acid. He also notes the probable presence of higher homologues of the acyl and alkoxy series of the butyrolactones. Of these, γ-butyrolactone has been found in many fermented products. It could come from glutamic acid or from related compounds (succinic, 2-oxoglutaric, or aminobutyric acids). Other lactones are produced during fermentation of flor sherries (see Section VII) by the action of yeast reductases on 2-ketopantoyl lactone. Others arise during fermentation by reactions involving 4-oxobutyric acid (succinic semialdehyde) or β-formylpropionic acid and to a lesser extent 2-oxoglutaric acid: It is also known that wines and brandies aged in oak contain a lactone from the oak—the trans form of the γ-lactone of 4-hydroxy-3-methyloctanoic acid. These compounds have potent tension-reducing properties as measured by the open field testing procedure with animals.

VI. HEAT

One by-product of alcoholic fermentation is frequently forgotten—heat. The amount of heat evolved is about 23.5 calories per 180 grams of sugar fermented. Unless the heat is dissipated the temperature of the fermenting mass will rise until the yeast cells are killed and fermentation ceases (i.e., "sticks"). Some heat, of course, is lost with the carbon dioxide (about one-fifth). With small size fermenting vessels in a cool room much heat is lost by radiation from the open surface or by conduction from the walls of the fermenter (about half). However, if the grapes are warm when crushed, the fermenters large, and the ambient temperature high, then an undue rise in temperature can be expected, resulting in "stuck" fermentations. To prevent this, various methods of cooling the fermenters are used, or the fermenters are made small and placed under cool conditions (caves),* or the rate of production of heat is reduced by fermenting under pressure (a greater percent is lost from the fermenter), and SO_2 is used.

VII. BY-PRODUCTS OF PROCESSING

Two methods are used to produce what is called sherry in this country. One is made by baking white wine of about 17% ethanol for 2 to 4 months at 50°-60°C. Depending on the amount of sugar present and the length and temperature of heating, the products have more or less of a caramel odor. A number of compounds are responsible, some very complicated. Hydroxymethyl furfural is one by-product. It is produced by the dehydration of fructose.

The other product is produced by a secondary yeast fermentation on the new wine. The same yeast that was responsible for the alcoholic fermentation can develop a film stage in which the yeast grows as a thick film on the surface of the wine. Many reactions take place, but acetaldehyde is a major product. From time to time the yeast film sinks and settles in the bottom of the container. There further complicated changes in odor occur. These are associated with autolysis of the yeast. The process is somewhat tricky. Not all yeasts will form the necessary film. The alcohol concentration is critical—if much below 15%, acetification may occur and, if above 16%, the film will not form.

The process has been made simpler by saturating the wine with oxygen under pressure and stirring the active yeast culture (to prevent settling). Under these conditions the yeast multiples and acetaldehyde is formed.

Wines with excess carbon dioxide are produced by two procedures: direct carbonation (now seldom used) or a secondary fermentation in a closed container

---

*Not recommended without adequate forced ventilation.

in which the carbon dioxide produced from added sugar is retained. Direct carbonation has little effect on the odor content of the wine. The secondary fermentation in a closed container adds only ethanol, carbon dioxide, and traces of the normal by-products of alcoholic fermentation. However, if the wine and yeast remain in contact for 9–36 months, subtle changes in odor occur. The nature of this change has been much studied in the Soviet Union. Apparently yeast autolysis and the release of the necessary enzymes involved in flavor formation take several months and the reactions that develop the special odor compounds are slow. Traditionally, this process is done in bottles where only a thin layer of yeast cells is present. When it is done in large tanks, a thick layer of yeast develops. Yeast autolysis in these thick yeast deposits produces highly reducing conditions and hydrogen sulfide is produced. The tank process is much less expensive. This is why wines produced by both processes are found on the U.S. market.

VIII. THE MALO-LACTIC FERMENTATION

In many areas of the world, grapes do not receive sufficient heat during the growing season to ripen properly. This results in a deficiency of sugar and an excess acidity in the fruit when it is harvested. The sugar deficiency can be made up by adding sugar or concentrated grape juice. However, the excess acidity has proven more troublesome. Direct neutralization with calcium carbonate has undesirable side effects (delayed precipitation of calcium tartrate, etc.). Ion exchange is feasible but again there are side effects (increased sodium, etc.). In many cases the double salt precipitation of calcium and potassium tartrate malate has been used successfully. The most interesting from the biochemical point of view has been the malo-lactic fermentation.

Certain bacteria produce enzymes by which malic acid is decarboxylated to lactic acid. This reduces the acid directly (one hydroxium ion is lost) and the pH increases (the pK_a of lactic is much higher than that of malic acid and one carbonyl is lost as CO_2). The process is successfully used for red wines in the Burgundy and Bordeaux regions of France and in the cooler regions of Australia and California. Pure bacterial cultures are available to induce the fermentation in cases where it does not occur spontaneously. The biochemical pathway has been studied in several countries, especially by Kunkee (1974) and his students. They note that the reaction [at least in one organism (*Leuconostoc oenos*, ML 34) which is widely used by California winemakers] is

$$\text{NAD} + \text{malate} \rightarrow [\text{pyruvate} + \text{NADH} + CO_2] \rightarrow \text{lactate} + \text{NAD} \qquad (4)$$

with a "spill-off" of the intermediate material (shown in brackets). They believe the formation of this small amount of pyruvate has an important effect on the growth of the malo-lactic bacteria by providing hydrogen acceptors which "spark" the early stages of the fermentation. This in turn brings about the observed stimulation of initial growth of the organism in the presence of malic acid.

As to the desirability of the process, many wine producers believe it essential, not only for the reduction in total acidity and the higher pH, but also because they believe it produces odor by-products, which they welcome. Other producers, particularly of white table wines, find some of the freshness of flavor is lost by the malo-lactic fermentation and wish to avoid it (relatively easy to do in acid wines by early clarification and discrete use of SO_2; not so simple in less acid wines, as in parts of California). Some producers want to avoid it because it is too uncertain or because they do not like its odor. In some cases the malo-lactic stench may be more from poor cooperage than from an excessive malo-lactic fermentation. Bacteriological stability is the only really important use of the malo-lactic fermentation according to Kunkee (1977).

Another microbiological process of importance is infection of grapes with *Botrytis cinerea*. This results in loss of water, a higher skin to volume ratio, and development of a special flavor. There is considerable interest in the wines of these "late harvest" grapes in Germany, Sauternes, the Loire and in certain areas of California.

IX. RESEARCH NEEDS

To complete the conversion of fermentation from an "art" to a science, much more information will be needed on the biochemical pathways involved. Unless the biochemical pathways are known it is difficult to devise experiments destined to produce new products, to produce a better level of certain products, or to secure a better balance between products.

This research is needed to determine the odor and taste factors influencing quality, particularly those which establish varietal character. Next we need to know the biochemical pathways involved. Then the factors influencing the production of such compounds must be determined.

While most of the fermentation by-products and the biochemical pathways of their production are known, the control of the extent of the different reactions is far from accomplished. This is particularly important since temperature–pressure recorder controllers and simple computers capable of multiple operations are available. Some manual and some automatic analyses would be necessary.

Finally, if we knew the biochemical pathways involved in aging, we could possibly change the environmental conditions so that more or less of various products were produced.

REFERENCES

Amerine, M. A., and Joslyn, M. A. (1970). "Table Wines: The Technology of Their Production," 2nd Ed. Univ. of California Press, Berkeley.

Amerine, M. A., and Kunkee, R. E. (1968). Microbiology of winemaking. *Annu. Rev. Microbiol.* **22,** 323–358.

Durmishidze, S. V. (1966). Contribution à l'étude de la formation et de l'évolution chimique des produits secondaires de la fermentation alcoolique. *Bull. OIV* **39,** 465–481.

Hodgman, C. D. (1976). "Handbook of Chemistry and Physics," 57th Ed. Chem. Rubber Publ. Co., Cleveland, Ohio.

Howard, D., and Anderson, R. G. (1976). Cell-free synthesis of ethyl acetate by extracts from *Saccharomyces cerevisiae. J. Inst. Brew., London* **82,** 70–71.

Kunkee, R. E. (1974). Malo-lactic fermentation and winemaking. *Adv. Chem. Ser.* No. 137, 151–170.

Kunkee, R. E. (1977). Personal communication.

Ough, C. S. (1964). Fermentation rates of grape juices. I. Effect of temperature and composition on white juice fermentations rates. *Am. J. Enol. Vitic.* **15,** 167–177.

Pasteur, L. (1866). "Études sur le Vin." Imprimerie Impériale, Paris.

Webb, A. D. (1977). Some physiologically active lactones in wines. *In* "Alcohol, Industry and Research" (O. Forsander, K. Erikson, E. Oura, and P. Jounela-Eriksson, eds.), pp. 206–219, 249–250. Alko, Helsinki.

THE BREWING OF BEER

W. A. Hardwick

I. Brewing Constituents	133
A. Malt	133
B. Cereal Adjunct	134
C. Hops	134
II. Steps in Brewing	134
A. Mashing	134
B. Lautering	136
C. Kettle Boil	136
D. Wort Cooling and Trub Removal	136
E. Yeast Pitching and Primary Fermentation	137
F. Yeast System	137
G. Lagering	137
H. Finishing and Packaging	137
I. Beer Flavor	138
III. Terminology	138
Editorial Comment	140

I. BREWING CONSTITUENTS

Brewing can be defined as the production of beer from malted barley, hops, and water, with or without the addition of other carbohydrate materials.

A. Malt

Malted barley, or malt, as it is generally referred to, is the principal brewing ingredient. It provides carbohydrate that is the source material for alcohol, and it provides nitrogenous compounds, some of which nourish the brewers' yeast while others contribute to flavor of the final product.

FERMENTED FOOD BEVERAGES IN NUTRITION

133

Copyright © 1979 by Academic Press, Inc.
All rights of reproduction in any form reserved.
ISBN 0-12-277050-1

Malt is prepared by soaking barley in water long enough to stimulate germination and to allow the grains to germinate until both rootlets and acrospire are in evidence. When germination has progressed to this point, the grain has produced high levels of diastatic and proteolytic enzyme activity. Considerable proteolysis and some degradation of barley starch granules have already begun to take place. At this juncture the germinating grain is gently dried to about 4% moisture so that the enzymatic activity is arrested but is not destroyed. Such malt can be stored for months or even years and still be suitable for brewing.

B. Cereal Adjunct

When malt alone is used to brew a beer, that beer is rather full-flavored and heavy on the palate owing to the relatively high concentration of flavorful compounds in the malt. American brewers prefer to brew a beer that is somewhat lighter in palate. They do this by including with the malt varying quantities of unmalted cereal grains or parts of cereal grains that contribute essentially nothing to the brewing system but carbohydrates. These are generally referred to as cereal adjunct materials, and they come from either rice or corn.

C. Hops

The brewers material referred to as hops, are, in reality, the clones or blossoms taken from the hop plant *Humulus lupulus,* which is a member of the hemp family. These clones are harvested from the hop vines, dried in special drying sheds, and compressed into bales for convenient shipment and storage. The hop clones contain various essential oils, which contribute to beer flavor to some extent. Most significantly, they contain certain bittering substances called ''humulons'' which give beer its characteristic bitter quality. Most of the hops used to flavor American beers are grown domestically in the Pacific Northwest, although a significant quantity of hops is still imported from the hop districts in Europe, especially Bavaria.

The following description of the various brewing steps is summarized in Fig. 1. Both the outline and the description attempt to give representative brewing steps as carried out by brewers in the United States.

II. STEPS IN BREWING

A. Mashing

The brewer grinds the malt, or more specifically, he crushes it with rollers that do minimal damage to the grain husk but which crush and release the enclosed

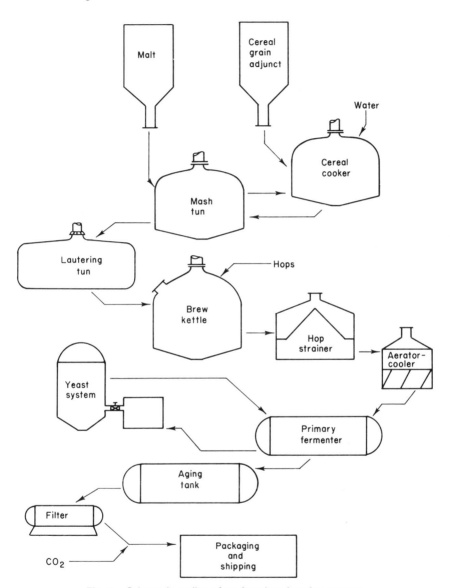

Fig. 1. Schematic outline of an American brewing process.

kernel. Too much breaking up of the husk material allows the release of excessive amounts of polyphenolic compounds from the husk during mashing and lautering. These can exert adverse effects on both beer flavor and beer stability. The crushed grain is mixed with water in the mash tun (tank). The amylolytic enzymes, which have lain dormant in the malt in its dried state, now hydrate and begin to break down the starch material present in the malted barley grain. These enzymes will also break down any carbohydrate adjunct that a brewer might add. Both α- and β-amylases are present, and normal brewing practice is to program the mashing temperatures so that the proper activity of each is brought into play. The result is the conversion of 80% of the starch to maltose with some glucose and trisaccharide material present. The remainder of the starch is broken down to intermediate fragments or chains of glucose units that range from four units to over twenty units in length; these are not fermentable by yeast.

Cereal adjunct is ground and sent to a cereal cooker, where it is gelatinized and mixed with the malt mash in a variety of ways. All of these are designed to allow the saccharifying enzymes present in the malt to break down the grain starch-producing fermentable sugars and carbohydrate fragments.

B. Lautering

The grain residue–liquid mixture produced by the mashing operation is passed through a straining device called a lauter tun (tank), where the insoluble spent brewers grains are removed. These residual grains are high in protein and have a value as a feed supplement.

C. Kettle Boil

The clear liquid coming from the lauter tun is called wort. It is pumped into the brew kettle where either hops or extracts prepared from hops are added to give the characteristic hop aroma and bitter taste to beer. Atmospheric boiling is carried out in the kettle for slightly more than an hour, generally. During this time the characteristic beer color develops, and a significant portion of the soluble barley protein is heat coagulated. Some brewers add corn syrup at this juncture rather than use a cereal grain adjunct in the mashing operation.

D. Wort Cooling and Trub Removal

The freshly boiled wort is sent through a hop strainer to remove the spent hops and is then allowed to stand until the coagulated protein, commonly called hot trub, settles out. The wort is then pumped through a cooling system which aerates the wort and rapidly cools it down to about 10°C. This cooling causes more protein and protein–polyphenolic conjugated substances to fall out of solution. This material, called cold trub, must be removed by settling, filtration, or centrifugation.

E. Yeast Pitching and Primary Fermentation

Yeast is added to this cooled, aerated wort where it begins to grow and to ferment the sugars present. Active growth (budding) of the yeast takes place for about 48 hours, after which budding practically ceases but fermentatioin of the sugars to alcohol continues actively.

F. Yeast System

Production brewers' yeast is recovered from the primary brewing fermenter and is pumped to a yeast-handling area where it is washed, reconditioned, and prepared for reuse.

In most modern breweries a pure culture propagation system is employed to provide a constant supply of fresh brewers' yeast. This is used to replace older yeast in the brewing system. The older yeast is dried and sold as a feed supplement, or it is washed free of hop-bittering substance, dried, and sold into the human food market.

G. Lagering

At the end of primary fermentation, which generally lasts from 4 to 7 days, the yeast will have settled to the bottom of the fermenter. The young beer, called ruh beer, is pumped into an aging tank; the yeast is recovered and is sent back to the yeast handling area. Brewers vary in the lengths of time that they age their beers; some age for only a few days, while other will age for a month or more. This aging step, called lagering, is generally done in a cold (0–10°C) environment and is employed to improve flavor and stability of the beer.

H. Finishing and Packaging

When lagering is completed, the beer is given a brief chill-proofing treatment, filtered, carbonated (if the natural carbonation has not been retained), and packaged. Most packaged beer (over 80%) is put into cans or bottles, which are then most generally pasteurized. Draft beer is not pasteurized. It is pressed into the kegs when it is quite cold and the kegs are maintained in this condition until the beer is sold.

I. Beer Flavor

Beer gets its characteristic flavor from three sources: malt, hops, and yeast. Grain character and malty-to-caramel flavors come from the malt. Some fruity character, pleasing aromatics, and bitterness comes from the hops. The pleasing sourness, fruity aromatics, and general aroma character of beer come from yeast fermentation by-products, as do alcohol and carbon dioxide. Of the more than 200 chemical compounds identified in beer to date, more than half are the result of yeast metabolism.

III. TERMINOLOGY

Beer is one of several malt beverages sold in this country. Some confusion exists about the names of some of them as well as some of the terms used to describe how they are made. Definitions listed below of a number of common terms should provide clarification.

Beer. Beer is a beverage obtained by the alcoholic fermentation of malted barley to which hops have been added; it may also contain other starchy materials.

Lager beer. Lager, or lager beer, is a beer which has been subjected to a typical "bottom" fermentation and then has been subjected to storage or aging for some period of time. Essentially all the beers sold in the United States are lager beers.

Ale. Ale, a popular drink in England and in Canada, is fermented by a special ale yeast which carries out a "top" fermentation in the presence of air. It is more hoppy and more tart in flavor than lager beer and is slightly higher in alcohol content.

Bock beer. Bock beer is a "fest" beer or special occasion beer, generally brewed from roasted malt which is much darker in color and richer in taste than lager beer.

Porter. Porter is a heavy malt beverage, fermented much in the fashion of beer except that darker malts are used so that a malty flavored and sweeter beverage is produced, which also contains less hop bitter.

Stout. Stout is similar to porter except that it has an even stronger malt flavor and a much stronger hop character.

Malt liquor. Malt liquor is brewed as beer. It is brewed from less malt and more fermentable carbohydrates than beer and it contains less hop. It is generally more fruity in flavor than is regular beer.

Brewers' yeast. Brewers' yeast is a unicellular organism, *Saccharomyces cerevesiae*. Some taxominists give it the classification of *Saccharomyces carlsbergensis*. This group of yeast is divided into two major classes, top fermenting and bottom fermenting. The top fermenting yeasts are employed to produce ale, porter, and stout; the bottom fermenting yeasts are used to produce lager beers and malt liquors. The top fermenting yeasts remain suspended in the ales and utilize oxygen during fermentation. Bottom yeasts ferment in the absence of oxygen and settle in the bottom of the fermentation vessel.

Pasteurization. Although classical pasteurization is defined as the heat treatment of beer in order to kill or inhibit yeast and bacteria, a more modern definition of beer pasteurization would be treatment by any process during manufacturing and packaging of beer which effectively destroys, inactivates, or prevents further microbial development in the product.

Chillproofing. Chillproofing is the term generally ascribed to measures taken by a brewer to render his product stable during its shelf-life storage and during chilling to drinking temperatures. A malt beverage characteristically tends to develop a precipitate unless certain measures are taken to prevent this. In some instances, the enzyme papain is employed to digest large molecular weight protein fragments during finishing stages of the beer. This renders these fragments more soluble and prevents them from precipitating later on. Other procedures involve the addition of substances which will specifically adsorb certain of the grain husk-related polyphenols present in the beer. The adsorbed complex then settles to the bottom of the finishing tank and the beer is moved away from it. Chillproofing is brought about by the removal of either large polypeptides or of polyphenolic materials, or both.

EDITORIAL COMMENT

In the nature of brewing, the complex interacting biological system results in a product which varies somewhat in its composition from brewery to brewery and from time to time in the same brewery. Some of these variations, especially in taste and certain constituents, are the result of the brewing art as practiced by individual companies. Different types of beer as described by Hardwick differ from one another in certain constant characteristics. The darker beers, for example, tend to have a somewhat higher carbohydrate content then the lighter beers; malt liquor has a higher alcoholic content (see pp. 79–80) than regular brews, while low calorie (light) beers, not referred to by Hardwick, are lower in both carbohydrate and alcoholic content.

III

Consumption of Beer and Wine

10

NATIONAL PATTERNS OF CONSUMPTION AND PRODUCTION OF BEER

Philip C. Katz

I. Introduction 143
II. History of Beer 144
    A. Ancient Times 144
    B. United States 145
III. Production Trends 146
    A. Geographic Changes 147
    B. Packaging Shifts 147
    C. Purchasing Habits 148
IV. Consumption Trends 148
    A. Historical and National 148
    B. Geographical Region 149
    C. Consumption in Specific States 151
    D. Imports 151
V. World Production and Consumption 152
    A. Production 152
    B. Consumption 152
    C. Total Alcohol Consumption 153
VI. Conclusions 153
    Economic Contributions 153
    References 154
    Supplementary Reading 154

I. INTRODUCTION

The story of malt beverages throughout the world and the United States is of great interest not only to the historian, the scientific community, and those engaged in the manufacture and distribution of its various products, but also to the consumer as well.

The development of the modern brewery and the current distribution techniques have evolved in two distinct periods in the United States, pre-Prohibition and post-Prohibition. This chapter emphasizes the changes that have occurred since Prohibition in the production and distribution of malt beverages. These changes are reflected in packaging innovations, purchasing habits, and geographic shifts in per capita consumption. They are a reflection not only of the changes that have occurred in malt beverage consumption but in society as well.

The relative position of malt beverages in the United States to the rest of the world is discussed as well as some of the economic contributions of the industry to this country.

II. HISTORY OF BEER

In 1976 the sale of domestic and imported malt beverages in the United States reached an all-time high of 152,811,000 barrels, which represents 4,737,000,000 U.S. gallons. The calendar year 1976 was the nineteenth consecutive year in which sales of malt beverages exceeded the previous year and established all-time highs (1).

The malt beverage industry has undergone vast changes in the United States in the last 30 or 40 years since beer has become a more popular beverage. In 1976, the total per capita consumption of malt beverages was 21.8 gallons per person, and while this figure ranks approximately thirteenth throughout the world, it compares with a figure of 16.4 ten years ago, 15.7 twenty years ago, and 17.8 in 1946 (2). Beer presently ranks fourth in the United States behind soft drinks, coffee, and milk (3). This will be discussed later in this chapter.

Information concerning the brewing industry and its sales performance in this country is readily available and dates back over 110 years or to the 1860s, when the first excise tax was applied to beer in the United States.

A. Ancient Times

The history of beer, of course, dates back thousands of years and throughout the world.

The first written history of beer production is recorded on a clay tablet, from ancient Mesopotamia (4), where 4000 years ago brewing was a highly respected profession and women were the master brewers of the day. This was also true in Babylon. There, women brewers actually were temple priestesses (5). For temple ceremonies, a special beer was reserved which was not available to the general public. In those ancient times, persons drank beer through reeds or tubes. One king, in fact, used a straw of gold, long enough to reach from his throne to a large container of beer. (One golden straw, belonging to Queen Shu-bad of Mesopotamia, was unearthed and is presently in the University of Pennsylvania Museum.)

In Egypt, where people revered moderation, beer was present in moderate amounts. Beer was so precious that it was offered as libation for the gods (6), so worthy that Rameses III, for example, sacrificed 30,000 gallons a year (7), and so highly regarded that it was used as an ingredient in some 100 medical prescriptions.

Other cultures and other people displayed a similar dependence on beer. To the Syrians, it was a prescription for patients to induce relaxation. To the Saxons, it was a cure for hiccups, simply by steeping medicinal roots in hot beer. To Sophocles, the beverage was part of his diet of moderation, "bread, meat, vegetables and beer." To the Arabs, it was a recipe ingredient as an agent to leaven bread. To Roman soldiers, beer was a morale booster. Julius Caesar, for instance, toasted his officers in beer after crossing the Rubicon in northern Italy during their march in Rome shortly before the first century.

In later centuries, beer followed the path of Western civilization. It is not surprising, therefore, to find that the hearty Vikings ate six meals a day and drank a soup of bread and beer with every one of them.

King Arthur served his Knights of the Round Table "bragget"—a beer highly seasoned with spices. King Henry VIII offered a breakfast for three that consisted of a roast beef, a loaf of bread, and a gallon of ale. England's Queen Elizabeth I insisted on good beer when she traveled. [She sent couriers ahead to test the beverage in the next town. If it was not satisfactory, she had her own favorite brew brought from London in time for her arrival (7, p. 15).] Her rival, Mary, Queen of Scots, drank beer even while a prisoner in Tutbury Castle. The Silver Jubilee year of the coronation of Queen Elizabeth II, marking the occasion of her ascension to the throne, was distinguished by special brews.

The Germans were the first to flavor their beer with hops. In their country, hop gardens were commonplace in the eighth century (8). In southern Germany, the Irish monk St. Gallen headed a vast religious establishment that included a brewery, housed in the same building as the bakery. From earliest recorded times, beer was an important part of life in Germany.

B. United States

The importance of beer was also recognized in America. On his fourth and final voyage to Central America in 1502, Christopher Columbus was served a beer, brewed from maize (9).

Beer was brewed in the tragic lost colony of Virginia in 1587 and on Manhattan Island as early as 1612. Had it not been for beer, Plymouth rock might not be a tourist shrine. A diary kept by a Mayflower passenger tells us the landing was made because "we could not now take time for further search, our victuals being much spent, especially our beere . . ." (4).

Many illustrious Americans (among them, Samuel Adams and George Washington) favored beer as a beverage (10). Thomas Jefferson went on record as saying, "I wish to see this beverage become common."

By 1800, beer was indeed common. It was being brewed in every one of the original states and was as important to their economy as it is today.

III. PRODUCTION TRENDS

Some of the production trends in the United States and the changes that have occurred are considered below.

First of all, the number of brewers in this country has declined drastically, although in concentration the brewing industry is not unique. It would, in fact, be difficult to name any industries that have not become consolidated in our country. In 1935, there were 750 brewing plants throughout the United States, representing 750 different companies. Today, there are 48 companies with some 90 plants which commercially produce beer (11).

Table I presents the total beer output in the United States in 1976 and the amount produced relatively in the various sections of the country. Owing to the decline in the number of breweries in the country, it has become necessary to combine regions; thus, individual data are not disclosed.

Table I. Malt Beverage Output Share by Region, 1946–1976 [a]

	Output (%)	
Region	1946	1976
New England	4	2
Middle Atlantic	34	14
Northeast central	33	27
Southwest central	3	10
South Atlantic and southeast central	6	19
Mountain, Pacific and northwest central	20	28
	100	100

[a] From U.S. Treasury Department, Bureau of Alcohol, Tobacco, and Firearms (11).

A. Geographic Changes

While we will consider trends in malt beverage consumption later in this chapter, our present analysis will be devoted to brewery shipments or output.

As indicated in Table I, about 2% of all beer produced in the United States is derived from the New England area, while approximately 29% comes from the vast area of the Mountain, Pacific, and northwest central states.

What is of greater significance, however, is the changes that have occurred over time. In 1946, the Middle Atlantic states of New Jersey, New York, and Pennsylvania produced almost 34% of all beer in this country, while in 1976 the figure had declined to 14%. Similarly, in the south Atlantic and southeast central states, which represented only 6% of output in 1946, today accounts for almost 19%. The southwest central region similarly increased from 3.3% to almost 10%.

Thus, while the trend in this country has been toward concentration, in the sense of fewer brewers selling more product, there has been a shift away from centralization in the geographic sense as the output becomes more spread out throughout the nation.

B. Packaging Shifts

The reasons for this change are related basically to the packaging process and the shifts in consumer preference as it relates to beer consumption habits. Over the past 40 years beer has increasingly become a product more apt to be served and consumed in the home than in the tavern or ''on-premise'' as it is called.

Immediately after Prohibition, about 75% of the beer sold in this country was draught while the balance was sold in reusable bottles (12). By comparison, today about 12% of the beer consumption is draught and 12% is in reusable bottles— the remaining 76% is sold in convenience containers, cans and one-way bottles.

The development of the convenience container has completely revolutionized the marketing of malt beverages in this country (13). The development of the beer can in the mid-1930s was followed by the one-way bottle in the 1950s and 1960s. Thus the share of canned beer sales in the past 30 years went from 7% to its present 53% while the one-way bottle increased from 2% to its current 23% (13, p. 33).

Other significant packaging changes and innovations were the easy open or pull tab can in the 1960s and the twist-off bottle cap. All these changes tended to increase the competition among metal and glass suppliers to the brewing industry and to keep prices down to the benefit of the consumer (12, pp. 20–21). As an example, between 1956 and April of 1977, the retail price of beer as measured by the U.S. Department of Labor, Bureau of Labor Statistics increased by 61% while the price index for all items rose by 121% (14).

The development of the convenience container also allowed great increases in the productivity of the brewers. As an example, employee production increased by 4½ to 5 times or 2400 barrels per year up to 11,000 per worker (12, p. 26).

C. Purchasing Habits

Along with the shift in packaging quite naturally came a change in purchasing habits. Initially beer was sold for off-premise use in wooden cases of 24 bottles, and as acceptance of convenience containers generated, the six-pack grew more prominent, so that by 1955—only 20 years after initial development—over two-thirds of all canned beer was sold in six-packs (15).

As we indicated previously, after Prohibition, beer was essentially an on-premise drink—it was not purchased in food stores. With the development of packaging, beer became more prominent in the supermarket and, more recently, in the convenience store.

Today only about 30–35% of all beer is consumed in the tavern or restaurant, while as recently as 30 years ago the comparable figure was about 50–55%. The food store alone probably accounts for close to 40% of all beer sold in this country, while practically none was sold in the 1930s (15, p. 12).

With this change in product convenience and purchasing habits in mind let us now consider what has happened to the overall acceptance and use of the product.

IV. CONSUMPTION TRENDS

A. Historical and National

A very long-term prospective of malt beverage per capita consumption in this country indicates a growth in popularity of the product, and even recent trends show this to be continuing.

We indicated previously that official records go back to 1863 and in that year per capita consumption of beer in the United States was 1.7 gallons. This figure represents total population, not just adults. When one views this over the years, we see an increase up to the time just prior to Prohibition when the per capita rose to 21.0 gallons in the years 1911, 1913, and 1914 (2).

Immediately after Prohibition, per capita consumption was at the 10 gallon level and as indicated in Table II, it rose to 12.1 in 1940 and 18.7 in 1945. The post-Prohibition period high point was reached in that year and then the decline started. By 1950, the per capita had dropped to 17.0 gallons, and in 1955 it stood at 15.9. In 1960, the figure was 15.1, but in 1962, industry per capita trend began to reverse itself.

Looking at the most recent trends (Table III), we see that the per capita

Table II.   **Per Capita Consumption of**
Malt Beverages[a]

Year	Consumption (U.S. gallons)
1935	10.1
1940	12.1
1945	18.7
1950	17.0
1955	15.9
1960	15.1

[a] From U.S. Brewers Association (2), based on state tax reports.

consumption has increased every year since 1962 and finally in 1971 has surpassed the previous post-Prohibition high. Growth has continued since that time and 1974 saw a new all-time high of 21.1 gallons per capita consumption in this country and, as we noted previously, a high of 21.8 in 1976.

B. Geographical Region

The regional differences in beer consumption in the United States have drastically changed over the years with the change in purchasing habits and shipping patterns. As a result, the vast differences which used to exist among the various areas of the United States are no longer present. The increases in income in the

Table III.   **Per Capita Consumption of**
Malt Beverages[a]

Year	Consumption (U.S. gallons)
1962	15.1
1965	15.9
1970	18.6
1971	19.0
1972	19.4
1973	20.2
1974	21.1
1975	21.6
1976	21.8

[a] From U.S. Brewers Association (2), based on state tax reports.

Table IV.   Regional Per Capita Consumption of Malt
Beverages, 1950 and 1968[a,b]

Region	1950	1968
New England	19	18
Middle Atlantic	23	19
Northeast central	22	20
Northwest central	16	16
South Atlantic	10	14
Southeast central	8	11
Southwest central	12	17
Mountain	15	17
Pacific	16	17
Total	17	17

[a] From State Tax Reports (2) and U.S. Department of
Commerce, Bureau of the Census (2).
[b] In U.S. gallons.

Southern and Western states, the mobility of the population, and the additional
retail outlets for beer have all contributed toward the changing regional consumption of beer.

We can see the evidence of this in two ways. First, Table IV shows regional
per capita consumption in 1950 and 1968; in both years the national average was
17.0 gallons. The spread from the national average in 1968 is considerably less
than it was in 1950. In 1950, the highest per capita consumption was 23 gallons
and the lowest was 8, while in 1968 the high was 20 and the low was 11.

Thus while there has been a smoothing out, we really have seen the previously
lower regions gain sharply, and those which had been high declined modestly in
terms of the national average.

To confirm this point, we compared the share of total U.S. consumption in
1956 and 1976 and can see a pattern similar to that viewed previously for
shipping trend, as shown in Table V.

In 1956, the two largest regions in beer consumption were the Middle Atlantic
and northeast central states, which represented over one-half the beer consumed
in the country. The two smallest areas were the Mountain states and the southeast central section, which together accounted for only about 7%.

In 1976, the two largest regions (Middle Atlantic and northeast central states)
accounted for less than 37% of all beer consumption, while the two smallest
(Mountain states and southeast central states) now represent over 10% of the
total.

Table V.   Malt Beverage Consumption Share by
Region[a]

Region	Consumption (%)	
	1956	1976
New England	7	6
Middle Atlantic	25	16
Northeast central	26	20
Northwest central	8	8
South Atlantic	9	15
Southeast central	4	5
Southwest central	8	11
Mountain	3	5
Pacific	10	14
	100	100

[a] From U.S. Brewers Association (2), based on state tax reports.

C. Consumption in Specific States

There has, however, been very little change in the share of consumption represented by the ten largest consuming states. Twenty years ago, the ten leading states represented 61% of all consumption, while in 1976 this figure was 56%.

In the case of other alcohol beverages, the statistics are higher. For example, the three leading states for beer—California, New York, and Texas—account for 25% of total consumption, while the leading states for wine represent almost 40% and distilled spirits 28% (2, pp. 58,60,61).

Finally, on a total population basis, Nevada shows the highest per capita consumption, approximately 36 gallons, followed by New Hampshire with 33.1. Wyoming stands at 32.0 gallons per person, Wisconsin at 31.1, and Montana rounds out the list at 30.9.

D. Imports

One final element of beer consumption in this country remains—that of imports. These have increased significantly, although they still represent only 1.6% of all beer consumed. In 1976, imports were 2,385,000 barrels, which was a 42% increase over 1975. Other recent increases are shown in Table VI, and we might add that the leading countries shipping beer into the United States are The Netherlands, Canada, and Germany.

**Table VI. Shipments of Malt Beverages to the United
States, Percentage Change, 1970–1975[a]**

1970	12%
1971	3%
1972	1%
1973	22%
1974	22%
1975	21%

[a] From U.S. Department of Commerce (2).

V. WORLD PRODUCTION AND CONSUMPTION

A. Production

The United States ranks first in the production of malt beverages, and in 1975 its figure of 160 million barrels compares with West Germany at 79,630,000 barrels, followed by the United Kingdom, Russia, Japan, Czechoslovakia, France, Canada, East Germany, and Mexico, which are the ten top producers (Table VII).

It is further estimated that the world production of beer was approximately 683,809,000 barrels or 21.2 billion U.S. gallons.

B. Consumption

While the United States is the leader in production, it stands only thirteenth in per capita consumption. The most recent figures available for 1975 (Table VIII)

Table VII. 1975 World Production of Malt Beverages[a,b]

United States	160,572,000
Germany	79,630,000
United Kingdom	55,057,000
Russia	51,130,000
Japan	33,483,000
Czechoslovakia	19,066,000
France	19,018,000
Canada	18,099,000
East Germany	17,214,000
Mexico	16,510,000

[a] From *Brewers Digest* (15a). Reprinted by permission.
[b] In U.S. barrels.

Table VIII. 1975 World Per Capita Consumption of Malt Beverages[a]

Country	Consumption (U.S. gallons)	Country	Consumption (U.S. gallons)
1. West Germany	39.0	8. United Kingdom	31.1
2. Czechoslovakia	37.6	9. Denmark	31.0
3. Australia	37.5	10. Austria	27.4
4. Belgium	37.0	11. Canada	22.7
5. New Zealand	35.2	12. Ireland	22.3
6. Luxembourg	34.1	13. United States	21.6
7. East Germany	31.1	14. The Netherlands	20.8

[a] From The Brewers' Society, London (15b).

reveal that West Germany was the leader with 39.0 gallons per capita, followed closely by Czechoslovakia and Australia. Others ahead of the United States are Belgium, New Zealand, Luxembourg, East Germany, United Kingdom, Denmark, Austria, Canada, and Ireland.

Thus, while per capita consumption in the United States has been steadily advancing, it has not yet approached that of the world leaders.

C. Total Alcohol Consumption

In terms of total alcohol consumption, that is, beer, wine, and distilled spirits, the United States ranks seventeenth in the world, the leaders being France, Portugal, Spain, Italy, and West Germany (16).

VI. CONCLUSIONS

This chapter has been a brief description of the brewing industry in the United States—its history, development, changing patterns, and relationship to the rest of the world. Malt beverages of varying types in the United States are products which have a number of factors in common; they are moderate beverages, unmatchable in product quality and integrity, and they combine the talents of advanced science with the art of brewing.

Economic Contributions

The American brewing industry is one which contributes over 2 billion dollars yearly in excise taxes to federal and state governments (17) (2, p. 108), purchases over 900 million dollars worth of the finest of agricultural products, and uses over 2.5 billion dollars in packaging materials plus providing employments for thousands of workers.

The brewers of this country and the world look forward to continued product improvement and distribution so that the ultimate beneficiary and judge—the consumer—will be able to enjoy its product properties to the fullest extent.

REFERENCES

1. U.S. Department of Commerce, "Malt Beverage Imports." Washington, D.C. 1976. See also U.S. Department of Labor, Bureau of Labor Statistics, "Monthly Labor Review," May. U.S. Gov. Print. Off., Washington, D.C., 1977.
2. U.S. Brewers Association, "The Brewing Industry in the United States," p. 13. Washington, D.C., 1976.
3. Maxwell Associates, "Maxwell Consumer Services Reports," March. Richmond, Virginia, 1977.
4. U.S. Brewers Association, "The Story of Beer," Washington, D.C., n.d.
5. W. C. Firebaugh, "The Inns of Greece and Rome," pp. 18-19. Pascal Coviei, Chicago, Illinois, 1928.
6. J. P. Arnold, "Origins and History of Beer and Brewing," p. 65. Alumni Assoc. Wahe-Henius Inst. Fermentol., Chicago, Illinois, 1911.
7. G. Donaldson and G. Lampert, eds., "The Great Canadian Beer Book," p. 11. McClelland & Stewart, Toronto, 1975.
8. H. L. Rich & Co., "One Hundred Years of Brewing," p. 32. Chicago, Illinois, 1903.
9. M. S. Weiner, "The Tasters Guide to Beer; Brews and Breweries of the World," p. 52. Collier, New York, 1977.
10. S. Baron, "Brewed in America," p. 39. Little, Brown, Boston, Massachusetts, 1962.
11. U.S. Department of the Treasury, Bureau of Alcohol, Tobacco, and Firearms, "Alcohol, Tobacco, and Firearms Summary Statistics," p. 41. U.S. Gov. Print. Off., Washington, D.C., 1976.
12. F. J. Sellinger, "Statement Before the Panel on the Material's Policy of the Subcommittee of Environmental Pollution of the Committee of Public Works," p. 13. U. S. Senate, July 11. Washington, D.C. 1974.
13. R. S. Weinberg, "The Effects of Convenience Packaging on the Malt Beverage Industry." U. S. Brewers Assoc., Washington, D.C. 1971.
14. U.S. Department of Labor, Bureau of Labor Statistics, "Monthly Labor Review." U.S. Gov. Print. Off., Washington, D.C., 1977.
15. American Can Company, "A History of Packaged Beer and Its Market in the United States," p. 10. New York, 1969.
15a. *Brewers Digest,* Aug., p. 14 (1976).
15b. The Brewers' Society, London.
16. Produktschap Voor Gedistilleerde Dranken, "Hoeveel Alcoholhoudende Dranken Worden er in de Wereld Gedronken?" Schiedam, Nederland, 1975.
17. U.S. Department of the Treasury, Bureau of Alcohol, Tobacco, and Firearms, "Commissioner of Internal Revenue's Annual Report." U.S. Gov. Print. Off., Washington, D.C., 1975.

SUPPLEMENTARY READING

American Can Company (1969). "A History of Packaged Beer and Its Market in the United States." New York.

Arnold, J. P. (1911). "Origins and History of Beer and Brewing." Alumni Assoc. Wahe-Henius Inst. Fermentol., Chicago, Illinois.

Baron, S. (1962). "Brewed in America." Little, Brown, Boston, Massachusetts.

Bickerdyke, J. (1889). "The Curiosities of Ale and Beer." Spring Books, London.

Donaldson, G., and Lampert, G., eds. (1975). "The Great Canadian Beer Book." McClelland & Stewart, Toronto.

Emerson, E. R. (1908). "Beverages Past and Present." Putnam's Sons, New York.

Firebaugh, W. C. (1928). "The Inns of Greece and Rome." Pascal Covici, Chicago, Illinois.

Maxwell Associates (1977). "Maxwell Consumer Services Reports." Richmond, Virginia.

Porter, J. (1975). "All About Beer." Doubleday, Garden City, New York.

Produktschap Voor Gedistilleerde Dranken (1975). "Hoeveel Alcoholhoudende Dranken Worden er in de Wereld Gedronken?" Schiedam, Nederland.

H. L. Rich & Co. (1903). "One Hundred Years of Brewing." Chicago, Illinois.

Sellinger, F. J. (1974). "Statement Before the Panel on the Material's Policy of the Subcommittee of Environmental Pollution of the Committee of Public Works," U.S. Senate, July 11. Washington, D.C.

U.S. Brewers Association (1976). "The Brewing Industry in the United States." Washington, D.C.

U.S. Brewers Association (n.d.). "The Story of Beer." Washington, D.C.

U.S. Department of Commerce (1976). "Malt Beverage Imports." Washington, D.C.

U.S. Department of Labor, Bureau of Labor Statistics (1977). "Monthly Labor Review." U.S. Gov. Print. Off., Washington, D.C.

U.S. Department of the Treasury, Bureau of Alcohol, Tobacco, and Firearms (1975). "Commissioner of Internal Revenue's Annual Report." U.S. Gov. Print. Off., Washington, D.C.

U.S. Department of the Treasury, Bureau of Alcohol, Tobacco, and Firearms (1976). "Alcohol, Tobacco, and Firearms Summary Statistics." U.S. Gov. Print. Off., Washington, D.C.

U.S. Department of the Treasury, Bureau of Alcohol, Tobacco, and Firearms (1976). "Statistical Release—Beer." U.S. Gov. Print. Off., Washington, D.C.

Weinberg, R. S. (1971). "The Effects of Convenience Packaging on the Malt Beverage Industry." U.S. Brewers Assoc., Washington, D.C.

Weiner, M. A. (1977). "The Tasters Guide to Beer; Brews and Breweries of the World." Collier, New York.

11

PRODUCTION AND CONSUMPTION OF WINE: FACTS, OPINIONS, TENDENCIES

Werner Becker

I. Introduction 157
   A. General Comments 157
   B. Definition of Wine 158
   C. The Place of Wine Compared to Other Beverages 161
II. Production and Distribution 163
   A. Countries of Wine Cultivation and Wine Harvests 163
   B. Types of Wine and Qualities 165
   C. Trade (or Commerce) 166
III. Consumption 170
   A. Quantities 170
   B. Per Capita Consumption 170
   C. Occasions for Consumption: Habits and Behavior
      of Consumption 178
   D. Influence of Sex, Age, Profession, and Income 181
   E. Consumer and Health Protection 182
IV. Conclusions 183
   A. Production and Consumption 183
   B. The Product 183
   C. Tendencies 184
   References 186

I. INTRODUCTION

A. General Comments

In June 1976, an international symposium on world wine consumption took place in the Papal Palace at Avignon, in the heartland of one of the largest French

FERMENTED FOOD BEVERAGES IN NUTRITION

wine-growing areas (Office International de la Vigne et du Vin, 1977). One hundred participants from 25 countries of the world were present. I had the honor to report on the relationship between wine import and consumption.

This chapter deals with the production and consumption of wine and the national differences which exist in this area. However, this necessitates qualification. A worldwide description in all aspects would go beyond the fixed framework. For this reason, this topic is discussed primarily from the viewpoint of my country (Germany) and the European economic community, with only occasional reference to other continents. No usable data exist from some countries with regard to many of the points considered.

B. Definition of Wine

It is useful to consider the question, What is wine?

If one inquires about the world's understanding of what constitutes wine, the conclusion that may be reached is that until now there has not existed a general valid universally applicable definition. The most meaningful definition is that offered by the Office International de la Vigne et de Vin (OIV, 1975) in Paris and which has been adopted by the majority of member states:

> *VIN* (Definition de base): Le vin est exclusivement la boisson résultant de la fermentation alcoolique complète ou partielle du raisin frais foulé ou non, ou du moût de raisin. Son titre alcoométrique acquis ne peut etre inférieur à 8° 5. Toutefois, compte tenu des conditions de climat, de terroir ou de cépage, de facteurs qualitatifs spéciaux ou de traditions propres à certains vignobles, le titre alcoométrique total minimal pourra être ramené à 7° par une législation particulière à la région considérée.*

On the other hand, the European Economic Community (EEC, 1970) in Brussels defines wine as follows:

> Wine, the product which is exclusively produced through complete or partial alcoholic fermentation of the fresh crushed wine grapes or grape juice.

The EEC regulations are legally valid in the nine countries of the European Economic Community. In contrast to the general definition of the OIV, no specification of alcohol content is contained in the EEC definition of wine. However, minimum alcohol contents are specified in individual paragraphs of the text of the basic regulations (EEC) No. 816/70 and 817/70.†

Table I presents the very complicated determination of the criteria by wine-

---

*Translation: Wine is exclusively the drink that results from complete or partial alcoholic fermentation of grape, freshly pressed or not, or of must of grape. The alcohol content cannot be less than 8.5%. However, with certain variations in climatic conditions, the soil, the species of vine, and the special conditions peculiar to certain vineyards, the minimal alcohol content may be reduced to 7% by regulations particular to the region that is surveyed.

†Official paper of the EEC No. L 99 of April 28, 1970 finally changed through regulation (EEC) No. 1167/76 of May 17, 1976 in the official gazette No. L 135 of May 24, 1976.

Table I. Wine-Growing Areas in the EEC: Minimum and Maximum Alcohol Contents According to EEC Regulations[a][b]

Zone	Natural alcohol content (minimum vol %)	Total alcohol content after enrichment (maximum vol %)	True alcohol content (minimum vol %)
Table wines			
A White wine	5.0	11.5 ⎫	
Red wine		12 ⎬	8.5[c]
B White wine	6.0	12 ⎬	
Red wine		12.5 ⎭	
C Ia	7.5	12.5 ⎫	
Ib	8.0	13 ⎬	9.0
II	8.5	13.5 ⎭	
III	9.0		

Zone	Natural alcohol content		Total alcohol content after enrichment (maximum vol %)	Total alcohol content (minimum vol %)
	According to EEC (minimum vol %)	According to national determination (final)		
Quality wines				
A	6.5	—	⎫	
B	7.5	—	⎬ Not regulated by	
C Ia	8.5	—	⎬ the EEC.	9.0
Ib	9.0	—	⎬ National deter-	
II	9.5	—	⎬ minations are	
III	10	—	⎭ void	

[a] See Fig. 1.
[b] From Becker (1968, 1973), reprinted by permission.
[c] By analysis, not enrichment, of white wine of 8.5% content.

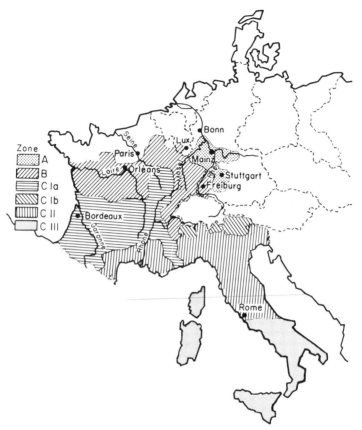

Fig. 1. The wine-growing zones of EEC definition (From Becker, 1968, 1973, reprinted by permission.)

growing zones (Fig. 1), types of wine (table wine or quality wine), etc. In addition to the general definitions of wine, the EEC standards also contain more than 20 additional definitions concerning the pre-product of wine (grapes, crushed grapes, grape juice, etc.) or with sub- or subsequent products (e.g., table wine, quality wine, sparkling wine, slightly sparkling wine, dessert wine, or brandy).

To recapitulate: The "Code international de traitements oenologique" of the OIV as well as the EEC regulations and—as far as is known, most legal standards for wine of the different countries of the world—define wine as well as its pre-, by-, and secondary products. The general definition of wine appears to be inexact. This "inexactness" reflects different climatic conditions under which wine growing takes place and is a result of different concepts as to what the alcohol content of wine should be in order to still be wine!

The general definition does not include sugaring the wine. (The EEC regulations use the word "enrichment" instead of the word "sugaring".) By using the word "exclusively" in the general definition, it is intended to prevent the manufacture of artificial wine.

The initial base for the manufacture of wine always has to be the *grape* and alcoholic fermentation—to a defined percentage—has to have occurred! Otherwise world legislation either prohibits or allows certain treatment procedures and additional or supplementary materials, a fact which tends to make the general definition of wine somewhat similar.

The "Codex Oenologique International" published by the OIV contains the description of more than 30 "produits chimiques, organiques ou gas utilisés dans l'élaboration ou la conservation du vin."

Next to the legal wording, the popular definition plays a special role. However, literature and poetry provide a more simple description. Wine is described in song and verse throughout the world in incomparable terms. A few popular descriptions of wine make clear how much wine has become part of our culture:

> "The dispenser of joy"
> "The milk of old age"
> "Captured sunshine"
> "A gift of God"
> "Gift from heaven"
> "Wine is a second life"
> "Le vin es un puissant rectificateur de l'hérédité"
> "Le vin est l'ensemble de la civilisation"

C. The Place of Wine Compared to Other Beverages

The complexity of this subject is too great to treat in detail here. The following simple concepts are relevant.

1. Physiologically man can only absorb a certain amount of liquids during a certain unit of time. Wine is an alcoholic beverage (the alcohol content per liter is, in general, between 7 and 15%, except for the alcohol-fortified dessert wines), the limits of intake of which are imposed by alcoholic content. Wine as an alcoholic beverage competes with other alcoholic beverages (as, e.g., beer, brandy, etc.) as well as with nonalcoholic beverages (e.g., mineral water, milk, coffee, tea, etc.) and in, most recent times, increasingly with pop or mixed drinks. Beverage intake, then, is a blend of alcoholic and nonalcoholic drinks.

2. As complicated as relationships might be with regard to the consumption of different beverages, one can be reasonably sure of a few trends (at normal usage): (a) the more alcoholic beverages other than wine a person consumes, the less remains for wine drinking; (b) wine drinkers often consume other alcoholic drinks also; (c) one can consume more of the light, dry wines of lower alcohol

Table II. World Consumption of Alcoholic Beverages[a]

Country	Consumption per capita (liters)	Population	Consumption (hl)
Russia	3.3	254,380,000	8,304,540
U.S.A.	3.13	213,121,000	6,670,687
Japan	3.05	110,950,000	4,382,525
Federal Republic of Germany	3.04	61,830,000	1,879,632
Poland	4.6	34,020,000	1,564,920
France	2.6	52,910,000	1,375,660
Italy	2	55,810,000	1,116,200
Spain	2.6	35,471,900	922,269
Canada	3.61	22,830,000	824,163
Great Britain	1.46	55,962,000	817,045
Yugoslavia	3	21,350,000	640,500
German Democratic Republic	3.4	16,850,000	572,900
Rumania	2.4	21,250,000	510,000
Holland	3.39	13,660,000	463,074
Czechloslovakia	2.98	14,686,300	437,652
Mexico	0.7	60,150,000	421,050
South Africa	1.38	25,470,000	351,486
Hungary	3	10,540,000	316,200
Sweden	2.97	8,176,691	242,848
Peru	1.4	15,870,000	222,180
Turkey	0.5	39,180,000	195,900
Belgium	1.99	9,813,200	195,283
Australia	1.3	13,500,000	175,500
Bulgaria	2	8,720,000	174,400
Finland	2.8	4,727,000	132,356
Switzerland	1.94	6,400,000	124,080
Austria	1.65	7,520,000	124,160
Cuba	1	9,090,000	90,900
Denmark	1.79	5,060,000	90,574
Portugal	0.9	8,257,000	74,313
Norway	1.84	4,010,000	73,784
Ireland	2.03	3,086,000	62,646
New Zealand	1.7	3,090,000	52,530
Israel	0.9	3,493,000	31,437
Luxembourg	3.5	358,000	12,530
Cyprus	1.8	640,000	11,520
Iceland	2.4	216,600	5,198
	2.74	1,232,448,691	33,752,642

[a] From *Journ. Vinic.* (1977), reprinted by permission.

content than he can of the heavy, sweet, or overly sweet wines of high alcohol content.

Next to the physiological situation the comparative cost of various beverages plays a significant role in determining the consumption level. Furthermore, there appears to be considerable national differences with regard to the amount of alcohol consumption (Table II).

II. PRODUCTION AND DISTRIBUTION

A. Countries of Wine Cultivation and Wine Harvests

From a grape growing area of more than 10 million hectares* (mio ha) more than 300 million hectoliters† (mio hl) wine (7920×10^6 gallons) are produced annually in the world. In 4 of the last 6 years world harvest was considerably larger than 300 mio hl; in the year 1973 production was even larger than 350 mio hl.

If one compares the distribution and production by continent, the following results, shown in the tabulation, are obtained (Mauron, 1976):

	Percentage of harvest 1964–1973	Producing area (%)	Percentage of wine production[a] 1975	Production in mio hl
Europe	80	72	79	253.653
Asia	0.5	13	1	2.046
America	14	9	14	46.698
Africa	5	5	5	15.335
Oceania	0.5	1	1	3.786
	100	100	100	321.518

[a] Because of other types of use (e.g., table grapes, raisins), the grape production shows the following numbers (1975): Europe, 67%; America, 16%; Asia, 11%; Africa, 4.5%; Oceania, 1.5%.

At 253,653 mio hl, Europe is therefore considered the largest producer, the major share belonging to the EEC (see tabulation below):

Country in EEC	Wine production 1975 (mio hl)
France	65.975
Italy	69.814
Federal Republic Germany	9.04
Luxemburg	0.157
Total	144.986 = 45% of world production

*1 Hectare $= 10,000 \text{ m}^2 = 2.47$ acres.
†1 Hectoliter $= 100$ liter $= 26.53$ gallons (U.S.).

Table III.   World Wine Production, 1927–1973[a]

Year	Production	Year	Production
1927–1936	181,307	1959	240,110
1937	177,852	1960	235,369
1938	204,962	1961	212,999
1947	164,762	1962	274,601
1948	167,849	1963	252,616
1949	166,260	1964	281,753
1950	192,961	1965	283,659
1951	193,395	1966	268,933
1952	179,862	1967	279,733
1953	215,162	1968	280,874
1954	210,865	1969	271,523
1955	226,896	1970	305,907
1956	218,601	1971	288,672
1957	173,846	1972	283,834
1958	224,264	1973	354,213

[a] Values in thousands of hectoliters. From OIV (1975).

According to calculations of the OIV the wine production of the Eastern Block Countries which are united in the Comecom amounts to a total of about 60 mio hl or about 17% of the world production (Mauron, 1976). Table III shows the development of wine production in the world since 1927.

Because of the progress that has been made in the growing technique, because of tendencies to expand the grape area in some countries, and because of the higher yield obtained per hectare, one can assume that the world production in the course of the next decade will reach 400 mio hl.

B. Types of Wine and Qualities

There is no internationally agreed upon formula for designating "type of wine." We shall refer here to the color of the wine and its type of use. According to color we can differentiate between white and red wines (and, in addition, to in-between colored wines, rosés); and according to use to table drinking wines and base wines.* To the table wines one can add the sparkling wines and slightly sparkling wines as well as dessert wines. The latter are also referred to as "special wines."

As far as the qualitative nature of the wines is concerned, there are two types,

---

*Base wine is a wine base for the manufacture of sparkling wine, slightly sparkling wine, vermouth wine, or vinegar. "Brennwein" is a wine base for the manufacture of brandy. Processing wines are not discussed here.

quality, and simple or ordinary wines. The limiting factor is largely a matter of legislation within individual countries or of custom, which varies somewhat internationally.

A very clear delineation, however, has taken place in the legislation of the EEC, which determined that in the sector of wine (in addition to special wines) destined for immediate human consumption there are two categories: ordinary wine and quality wine of certain growing areas. This classification has been influenced not only by qualitative points of view, but also by questions of market regulations (market and price intervention). Among quality wines of certain growing areas are the following:

France: wines with an "appellation d'origine" or a "Vin dèlimité de qualité supérieure."

Italy: wines with an "denomiazione di origine controllata" or a "Denominazione di origine controllata e garantita."

Federal Republic of Germany: the "Qualitätsweine" as well as "Qualitätsweine mit Prädikat." The latter always coupled with the concept "Kabinett," "Spätlese," "Auslese," "Beerenauslese," "Trockenbeerenauslese," as well as "Eiswein."

Luxemburg: wines with the "Marque national du vin Luxemburgeois."

With regard to the legislation of other European countries (e.g., Austria), it can be said that these are in part in the process of approximating those of the EEC regulations.

An attempt to divide the world wine harvest according to the color or quality of the wine was not possible because of the lack of sufficient data. However, an exact division of harvest by steps of quality, based on the strict organization of the EEC, is presented below for 1976 (in mio hl).

Total harvest (146,800)		Ordinary wine (100,250 = 68.3%)		Quality wine (29,900 = 20.4%)		Other wines (16,650 = 11.3%)	
White	Red	White	Red	White	Red	White	Red
54.500	92.300	26.600	73.650	15.500	14.400	12.400	4.250
(37.1%)	(62.9%)	(26.5%)	(73.5%)	(51.8%)	(48.2%)	(74.5%)	(25.5%)

Through the above summary we can come to the following conclusion: the production of ordinary wine dominates in the EEC. Also, more red of all qualities is produced. A similar tendency, with few exceptions, appears worldwide. The production of red wine takes place mainly in the Mediterranean region, as well as in hot countries of the other continents. Obviously, the hotter climate is less suitable for the production of white wine.

A detailed examination of the EEC area shows the following, from which the differences from country to country become quite evident. The tabulation below presents the wine production in the ECC in 1976 (in mio hl).

Country	Harvest		Red wine		White wine		Table wine		Quality wine	
	hl[a]	%	hl	%	hl	%	hl	%	hl	%
France	73.035	100	50.750	69.2	13.247	18.1	48.073	65.8	15.914	21.8
Italy	64.978	100	41.652	64.1	23.326	35.9	55.178	84.9	6.500	10.0
Germany	8.659	100	1.127	13.0	7.532	87.0	0.027	0.3	8.631	99.7
Luxembourg	0.128	100	—	—	0.128	100	0.055	43.0	0.073	57.0

[a] Tentative numbers. Totals differ from 100% of wines produced because of the omission of base wines.

C. Trade (or Commerce)

In the political economy of the grape-growing countries, wine, depending on the country and region, plays a more or less large role. In many regions the landscape widely reflects the viniculture. Impressive examples are in France (region of Bordeaux and Langedoc-Roussilion), in Italy (Apulia, Trentione/South Tyrol), in Spain (Rioja or La Mancha), in the United States (California), and in the Federal Republic of Germany (Rhine-Moselle valley or the Baden).

Since wine as a typical argicultural product is largely characterized through the enological conditions of its home or area of production (type of grape, ground, climate, and weather conditions), the potential for comparisons is limited. A white Moselle, for instance, is a different product from a white Bordeaux of even a port wine! This fact creates a lively international trade, especially in the area of quality wines. Insofar as there are not hinderances to this trade, some wine-growing countries import wines from other countries.

Countries that do not grow wine can obtain it from wine-growing countries throughout the world. Furthermore, in the area of table wines the importation of certain wines for the purpose of blending (which is generally forbidden in the area of quality wines) plays an important role in some countries. A typical example of this has been represented for many decades by the purchase of wine by France from Algeria. Because of political reorientation, the large wine-growing country of Italy has, in recent years, become one of the largest suppliers of blending wines for France. In the course of the last 2 years France has purchased 6 to 7 mio hl from Italy each year.

Even though in some countries or in some wine-growing areas exportation plays an important role, this should not overestimated. It is true that the world wine export has risen greatly since 1949, but it amounts to only 12% of the world production. In 1949, the export of wine totalled 14,055 (thousand hectoliters); in 1959, it had risen to 25,325. Table IVA gives the figures from 1968 to 1973 and shows the distribution of these exports in the individual countries.

Some grape-growing countries, and especially those where for religious reasons little wine is consumed (a fact that possibly might change somewhat in the course of the next several decades), a major portion of wine is exported. Algeria is such an example.

Germany, a typical grape-growing country with considerable importation (in 1976 a total of 7.9 mio hl was imported; of this 5.7 mio hl was ordinary table wine and 2.2 mio hl was base wine), has increased its exports in recent years. In 1976 a total of 1.04 mio hl was exported to approximately 100 countries, which corresponds to 12% of its own harvest.

Within the EEC the border-crossing trade is, in general, not as large as one would assume, since customs duties and import quotas are completely lacking (That means no trade hinderances within the area of the nine member countries!). Within the EEC the border-crossing inland trade amounts to 13.7 mio hl or about 9% of the EEC production.

From these export–import movements one observes that with few exceptions the major amount of wine produced remains in its country of origin and is consumed there.

Exportation from one country logically has to be balanced by importation in one or several other countries. The following comparison gives some interesting insights into these balances as of 1973 (Mauron, 1976):

	Export (mio hl)	Import (mio hl)
Europe	31.635	38.363
America	0.128	3.160
Africa	11.524	1.907
Asia	0.550	0.268
Oceania	0.062	0.147
World total	43.899	43.845

Obviously by far the largest international traffic takes place within Europe, whereby import (including movements within Europe) is considerably larger than is export. Noteworthy is the large export from Africa and the relatively small (3.160 mio hl) import of America (North and South America combined). Of this, the United States with 2.088 mio hl represents roughly two-thirds of the volume. The largest roles in international trade in Europe are played by France, Italy, USSR, and Germany. The import and export of some selected European wine-growing countries in 1973 are given in Table IVB.

Table IVA. Wine Exports [a,b]

Country	1968	1969	1970	1971	1972	1973
World total	**25,241**	**32,461**	**37,399**	**34,086**	**41,794**	**43,899**
Europe	**16,085**	**18,047**	**22,193**	**26,660**	**33,174**	**31,635**
Albania	38	55	40[c]	—	—	—
Federal Republic of Germany	244	288	346	451	524	677[c]
Austria	13	28	49	157	226	181
Belgium [b]	54	88	97	59	70	122
Bulgaria	1,708	1,864	1,941	2,095	2,300[c]	1,945
Denmark					34	35
Spain	2,430	2,629	3,439	3,744	3,862	4,069
France	3,505	3,873	4,019	4,798	5,754	7,223
Greece	818	914	1,639	935	780	972
Hungary	795	867	984	1,108	1,280	1,456
Italy	2,661	3,028	5,550	9,307	14,247	10,468
Luxemburg [b]	57	57	62	67	71	79
Malta	140	100	154	37[c]	97[c]	111
Low countries [b]	159	377	395	350	108	102
Portugal	2,413	2,410	2,050	2,019	1,965	2,096
Rumania	579	675	688	801	789	999
United Kingdom	19	25	80	98	97	116
Switzerland	6	8	7	8	8	8
Czechoslovakia	21	—	—	—	2	7
Russia	98	241	230	210	280	280
Yugoslavia	327	520	468	616	680	689

America	**95**	**132**	**107**	**118**	**117**	**128**
Argentina	29	44	23	63	30	64
Canada	1	1ᶜ	1ᶜ	1ᶜ	1ᶜ	1ᶜ
Chile	53	72	67	40	37	35
USA	12	15	16	14	19	28
Africa	**8,560**	**13,865**	**14,678**	**6,816**	**7,916**	**11,524**
South Africa	150	130	110	123	115	120
Algeria	6,920	11,978	12,548	5,980ᶜ	6,382	9,754
Egypt	20	26	27	14	27	33
Morocco	790	1,055	959	440ᶜ	684	795
Tunisia	680	676	834	238	698	812
Other countries	—	—	200ᶜ	21ᶜ	10ᶜ	10ᶜ
Asia	**417**	**335**	**362**	**426**	**507**	**550**
Cyprus	304	239	270	342	406	437
Israel	22	24	26	28	33	30ᶜ
Turkey	83	59	54	43	45ᶜ	45ᶜ
Other countries	8ᶜ	13ᶜ	12ᶜ	12ᶜ	23ᶜ	38ᶜ
Oceania	**84**	**82**	**59**	**66**	**80**	**62**
Australia	84	82	59	66	80	62

ᵃ Values in thousands of hectoliters.

ᵇ The figures for Belgium, Luxemburg, and the Low Countries are not related to the calendar year but to the wine production season 1962–1963; 1963–1964, etc. From OIV (1975).

ᶜ Estimation based on earlier years.

169

Table IVB.   Import and Export of Some European
Wine-Growing Countries, 1973

	Export (mio hl)	Import (mio hl)
Italy	10.468	1.124
France	7.223	8.784
Spain	4.069	0.832
Federal Republic of Germany	0.677	7.547[a]
German Democratic Republic	—	1.180
USSR	0.280	6.720
Portugal	2.069	0.002
Bulgaria	1.945	0.152
Hungary	1.456	0.227

[a] Includes base wine.

III. CONSUMPTION

A. Quantities

There is as much variety in the amounts consumed of the several types of wine as there is variety in the quantity and types produced. The preceding section indicated that roughly 90% of the wine produced remains in the country of origin, so that wine consumption is highest in the wine-producing countries.

For several years the OIV compiled the available data on the consumption of wine and has published this in its most recent memento (OIV, 1975). Table V shows the consumption by amount distributed over continents and countries. The world wine consumption is given at 278,214 mio hl for 1973. This is but 78.5% of production (Fig. 2). In the EEC area the degree of self-sufficiency in the last 3 years was above 100%. The trend toward a greater production than consumption appears to be a worldwide phenomenon. There are also considerable increases in grape areas in California, USSR, and several other countries.

B. Per Capita Consumption

More interesting than the quantitative estimates of production in each country is the comparison of per capita consumption. The statistics of the OIV provides this information (Table VI). The enormous differences in the annual per capita consumption (lowest consumption in Europe is in Norway, 2.95 liters, the highest consumption is in Italy with 108 liters; in the Americas the highest consumption is in Argentina with 73 liters and lowest is in Mexico with 0.17 liters) can be readily explained. Consumption is highest in large wine-producing countries; countries with a low per capita income have a lower consumption.

Table V. Wine Consumption[a,b]

Country	Population in 1973 (thousands)	1963	1969	1970	1971	1972	1973
Total World		**227,258**	**257,641**	**259,091**	**263,373**	**265,073**	**278,214**
Europe		**189,122**	**212,997**	**214,285**	**216,701**	**217,625**	**231,097**
Federal Republic of Germany	61,970	7,369	9,620[c]	9,793	10,650	11,965	13,388
German Democratic Republic	16,980	705	884	850	867	905	985
Austria	7,520	1,600	2,500	2,800	3,000	3,150	3,250
Belgium	9,760	710	1,010	1,155	1,262	1,275	1,397
Bulgaria	8,620	1,141	1,820	1,584	1,641	1,664	1,724
Denmark	5,030	156	250	289	330	373	540
Spain	34,860	19,100	20,742	20,800	20,100	22,859	25,500
Finland	4,640	93	174	190	204	226	211
France	51,500	59,432	57,462	55,634	54,389	56,632	56,469
Greece	8,950	2,501	3,500	3,500	3,500	3,500	3,145
Hungary	10,410	1,930	3,964	3,883	3,927	4,160	4,008
Ireland	3,030	34	48	98	96	97	90
Italy	53,500	55,101	59,048	59,547	59,596	55,305	59,980
Luxembourg	350	85	111	125	139	142	140[c]
Malta	30[c]	30[c]	30[c]	30[c]	30[c]	30[c]	30[c]
Norway	3,960	50	84	90	103	107	116

(Continued)

Table V. *(Continued)*

Country	Population in 1973 (thousands)	1963	1969	1970	1971	1972	1973
Europe (*Continued*)							
Low Countries	13,448	304	664	690	752	853	850[c]
Poland	33,360	1,443	1,763	—	2,018	2,135	2,168
Portugal	8,830	10,179	8,723	6,750	8,551	6,142	8,000
Rumania	20,830	5,246	—	4,678	—	4,500	6,800
United Kingdom	55,887	1,103	1,533	1,609	1,918	2,209	2,889
Sweden	8,140	304	462	513	564	642	594
Switzerland	6,440	2,096	2,424	2,476	2,551	2,646	2,729
Czechoslovakia	14,580	894	1,300	1,548	1,500	1,543	1,579
USSR	247,460	11,828	24,000	27,410	28,790	29,000	29,000[c]
Yugoslavia	20,960	5,508	5,764	5,480	5,545	5,515	5,515[c]
America		**32,573**	**37,641**	**38,574**	**39,239**	**39,841**	**39,133**
Argentina	24,290	18,123	21,211	21,447	20,215	19,076	17,754
Brazil	101,710	1,560	1,566	1,566	1,566	2,000	2,034
Canada	22,130	508	709	430	892	976	1,186
Chile	10,230	4,318	4,194	3,957	—	—	—
USA	205,400	7,000[c]	8,919	10,119	11,553	12,755	13,144
Mexico	54,300	73	125	—	—	—	98
Paraguay	2,670	—	—	53	53	53	—
Peru	14,910	104	139	149	150	145	—
Uruguay	2,990	837	728	728	728	754	762

Africa		**4,203**	**4,846**	**5,123**	**5,143**	**5,191**	**5,345**
South Africa	23,500	1,317	2,158	2,467	2,467	2,525	2,690
Algeria	15,770	380	138	96	96	—	75
Morocco	15,830	361	200ᶜ	200ᶜ	200ᶜ	300ᶜ	300ᶜ
Tunisia	5,510	145	150ᶜ	160ᶜ	180ᶜ	170ᶜ	180ᶜ
Other countries		2,000	2,200	2,200	2,200	2,100	2,100
Asia		**724**	**1,034**	**848**	**975**	**1,055**	**1,064**
Cyprus	660	69	65	65	65	64	64
Israel	3,254	110	111	105	112	115	120
Japan	108,350	140	200	180	200	386	390
Libya	3,060	—	104ᶜ	—	—	—	—
Turkey	37,930	262	469	309	409	300ᶜ	—
Other countries		150	150	150	150	150	150
Oceania		**636**	**1,128**	**1,261**	**1,315**	**1,361**	**1,575**
Australia	13,130	571	1,001	1,108	1,111	1,157	1,300
New Zealand	2,980	65	127	153	204	—	275

[a] Values in thousands of hectoliters.
[b] From OIV (1975).
[c] Estimation based on earlier years.

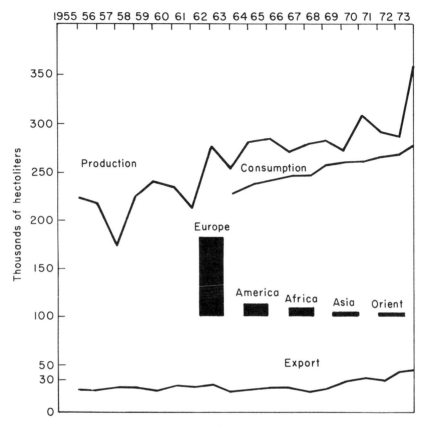

Fig. 2. World wine production, export, and consumption. From OIV (1975).

In agreement with the findings of the OIV (Mauron, 1976) one can divide the countries into three groups: First are those with large consumption, namely, Italy, France, Spain, Portugal, and Argentina. The per capita consumption in these countries is between 70 and 108 liters. It is astonishing that these 5 countries, with only 5% of the world population, consume 60% of the world's wine production! The average per capita consumption of these high-consuming countries is 97 liters per year! Second are countries with medium per capita consumption, which is between 10 and 50 liters. These include Federal Republic of Germany, Austria, Belgium, Bulgaria, Greece, Hungary, Luxemburg, Rumania, Switzerland, Czechoslovakia, Russia, Yugoslavia, Chile, Uruguay, South Africa, Australia, New Zealand, and Cyprus. These countries with 14% of the world population consume some 30% of the world wine production. The arithmetical mean of consumption is about 17 liters annually. The third group is

Table VI. Consumption of Wine Per Capita[a,b]

Country	Population, 1973 (thousands)	1963	1969	1970	1971	1972	1973
Europe							
Albania							
Federal Republic of Germany	61,970	12.90	15.90	16.20	17.50	19.40	21.70
German Democratic Republic	16,980	4.00	5.20	5.00	5.10	5.40	5.80
Austria	7,520	22.50	34.00	37.80	39.80	42.70	44.00
Belgium	9,760	7.15	10.10	12.00	12.80	12.90	12.90
Bulgaria	8,620	14.20	21.20	18.60	19.30	19.34	20.00
Denmark	5,030	3.32	5.13	5.91	6.66	7.48	10.73
Spain	34,860	63.60	60.90	61.50	60.00	67.00	73.00
Finland	4,640	2.07	3.71	4.07	4.35	4.80	4.40
France	51,500	127.27	112.44	109.13	108.04	106.88	106.88
Greece	8,950	29.50	40.00	40.00	40.00	40.00	35.73
Hungary	10,410	29.30	38.50	37.70	38.50	40.00	38.50
Ireland	3,030	1.20	1.60	3.30	3.20	3.20	3.00
Italy	53,500	107.00	112.00	111.00	102.00	100.00	108.46
Luxemburg	350	25.75	33.40	37.87	39.74	41.50	40.90
Norway	3,960	1.35	2.20	2.34	2.61	2.74	2.95

(Continued)

Table VI. (*Continued*)

Country	Population, 1973 (thousands)	1963	1969	1970	1971	1972	1973
Europe (*Continued*)							
Low Countries	13,448	2.60	4.92	5.15	6.23	7.81	8.87
Poland	33,360	4.70	5.60	5.60	6.20	6.40	6.50
Portugal	8,830	114.00	98.50	76.70	97.15	69.77	90.90
Rumania	20,830	27.50	—	23.10	—	21.80	32.60
United Kingdom	55,887	2.07	2.76	2.89	3.46	4.06	5.18
Sweden	8,140	3.90	5.80	6.40	7.00	7.90	7.30
Switzerland	6,440	38.10	39.00	39.00	40.80	42.00	42.65
Czechoslovakia	14,580	6.40	9.00	10.60	10.40	10.60	10.80
USSR	247,460	5.63	10.00	11.40	11.80	12.00	9.10
Yugoslavia	20,960	27.50	38.30	26.90	27.60	27.50	27.10
America							
Argentina	24,290	86.00	88.50	91.80	85.30	79.70	73.10
Brazil	101,710	2.20	1.80	1.80	1.80	2.00	2.00
Canada	22,130	2.36	3.36	2.20	4.23	4.63	5.00
Chile	10,230	51.60	46.60	43.90	—	—	—
USA	205,400	3.65	4.43	4.97	5.60	6.12	6.24
Mexico	54,300	0.19	0.16	0.25	0.22	0.25	0.17
Paraguay	2,670	—	—	2.00	2.00	2.00	2.00
Peru	14,910	0.79	1.05	1.13	1.14	—	—
Uruguay	2,990	29.80	26.00	26.00	26.00	25.00	25.40

Africa							
S. Africa	23,500	6.03	10.02	11.20	11.20	11.10	11.42
Algeria	15,770	3.80	1.30	0.70	0.70	—	0.55
Morocco	15,830	3.00	1.50	1.50	1.30	2.00	2.00
Tunisia	5,510	3.20	—	3.10	3.40	3.20	3.40
Asia							
Cypress	660	12.00	8.20	8.20	8.20	8.20	8.20
Israel	3,254	4.60	3.88	3.35	3.50	3.59	3.75
Japan	108,350	0.40	0.40	0.32	0.35	0.36	0.48
Libya	3,060	3.00	3.70	—	—	—	—
Turkey	37,930	0.70	0.82	1.24	1.08	0.78	—
Oceania							
Australia	13,130	4.37	7.70	8.52	8.54	8.90	9.90
New Zealand	2,980	2.24	4.54	5.45	7.14	—	9.40

[a] Values in liters.
[b] From OIV (1975).

the "small consumers," including all countries where per capita consumption is less than 10 liters annually. In this group we find the north European countries, Sweden, Denmark, Finland, Ireland, Norway, The Netherlands, German Democratic Republic, Belgium, Poland, Great Britain; in the Americas, Canada. United States, Brazil, Mexico, and Paraguay; in Africa, Algeria, Morocco and Tunesia; in the Middle East, Israel, Lebanon, and Turkey, and Japan in the Orient. These and other low-consuming countries have a tremendously high proportion of the world's population (more than 80%) but consume at the most about 10% of the world wine produced!

Global changes in the per capita consumption reveal that per capita consumption in some countries has stagnated or has even been slightly reduced. This is the case in Italy, France, Spain, and Switzerland. Slight increases are found in the United States, England, Belgium, Denmark, Sweden, the USSR, and in the Federal Republic of Germany.

C. Occasions for Consumption; Habits and Behavior of Consumption

Two very basic differences in consumption behavior can be identified. One group drinks wine as an accompaniment during meals. For this group wine assumes the characteristic of a food. Dining without wine is unthinkable. The national production and habits of consumption complement each other well. In this pattern of consumption red wine is usually dominant. For daily consumption simple wines are used; for these the concept "vin de consommation courant" is applied in France. High-quality wines accompany holiday feasts. It can be assumed that a consumer belonging to this group has an annual consumption of 200 to, at times, 300 liters. Outside the wine-producing regions, however, such consumption is the exception.

A second group of people consumes wine only occasionally during a meal. This group enjoys wine at other than mealtime, e.g., after the meal, during TV viewing, in good company at home or in the tavern. They do not consider wine as food; rather it assumes the characteristics of a luxury. This pattern of consumption is largely prevalent in Germany. While the consumers of the first group are always counted among the "intensive users," the second group is dominated by the "regular" or "occasional" users.

A recent study of the attitude concerning beverages of citizens of the Federal Republic Germany included 5000 adults (representative of roughly 33 million citizens between the ages of 20 and 65). It contained 200,000 items about the buying and consumption behavior (Spiegel Documentation "Percentages," 1977). The following tabulation provides some of the information accumulated in this study.

Wine consumption	20-65-year-olds (%)	Estimate of the year per capita use (liters)[a]
Several times a week	13	120
About once a week	14	40
Two to three times a month	20	20
Less often	26	5
Total	73[b]	

[a] Personal estimate.
[b] No wine is consumed by 27%.

The following is an estimate of a subdivision of wine drinkers and nonwine drinkers in the Federal Republic of Germany*: Intensive wine drinkers, 15%; regular wine drinkers, 15%; occasional wine drinkers, 20%; sporadic wine drinkers, 25%; and non-wine drinkers, 25%. From this estimate it appears that less than 30% of the wine drinkers consume about 60% of the total.

What kinds of wine does the consumer drink? This question cannot be answered on a worldwide basis, but only from country to country. The differences are enormous. A German study (Michel, 1976) shows that the share of German wines in the total consumption (domestic and foreign) in Germany has varied between 67 and 82% in the course of the last 10 years. The foreign amount share of consumption in Germany is therefore between 18 and 33% (see following tabulation).

The fluctuating proportions of domestic wine versus foreign wine are mainly the result of the annual fluctions in German harvests. Comparison of the amount

	Share of amounts						
Origin	1965	1968	1969	1971	1973	1974	1975
Germany	82	74	72	71	67	72	75
Foreign	18	26	28	29	33	28	25
Total	100	100	100	100	100	100	100
	Share of value						
Origin	1965	1968	1969	1971	1973	1974	1975
Germany	83	78	76	77	76	79	80
Foreign	17	22	24	23	24	21	20
Total	100	100	100	100	100	100	100

*Personal estimate.

with the value share reveals that as far as price is concerned more is invested in German wines than foreign wines. The same study (Household Panel, 1975) also shows which types of wines the German consumer prefers:

	1969			1973		
	German	Foreign	All	German	Foreign	All
Amounts (%)						
White wine	84	37	71	85	37	69
Red wine	13	60	26	12	59	28
Rosé wine	3	3	3	3	4	3
Total	100	100	100	100	100	100
Value (%)						
White wine	87	39	75	87	39	76
Red wine	10	58	22	10	57	21
Rosé wine	3	3	3	3	4	3
Total	100	100	100	100	100	100
	1974			1975		
	German	Foreign	All	German	Foreign	All
Amounts (%)						
White wine	88	33	73	86	32	74
Red wine	10	65	25	12	65	24
Rosé wine	2	2	2	2	3	2
Total	100	100	100	100	100	100
Value (%)						
White wine	89	35	79	88	38	79
Red wine	9	62	19	10	58	18
Rosé wine	2	3	2	2	4	3
Total	100	100	100	100	100	100

In accord with the predominant type of production in Germany, the German wine drinker prefers white wine (minimum 84%, maximum 88%). As far as foreign wines are concerned, red wine is preferred (minimum 59%, maximum 65%). With rosé there is no significant difference between domestic and foreign wine preference (between 2 and 4% each).

Another very interesting question is, How many other wines are consumed in addition to the genuine drinking wines (table wines)? This question has been somewhat neglected in many places, since attention is almost always directed only to the still wines when speaking of wine consumption.

An analysis of the Wine Institute in California ("La Vigne et le Vin en Californie," 1974) may be used as an example. This shows that in the United

States in 1973 the following wines were consumed: table wine, 55%; dessert wine, 20%; aromatic wine, 16%; sparkling wine, etc., 5%; and vermouth, 3%. This special question deserves to be examined on a worldwide basis.

The question of which quality grade the consumer prefers can only be answered country by country and the answer differs considerably according to the share of production, import practices, level of income, and development of income, as well as the degree to which a wine is known through advertising and public relations.

Since, as has been shown several times, the major amount of wine is usually consumed in the land of its production, the relation of quality wine to table wine is a matter of custom and consumer preference. In the genuine consumer countries and in countries with a production below consumption level the relation of quality wine to standard table wine is influenced greatly by import taxes, and trade politics, as well as the purchasing power of the citizens. In the monopoly countries (e.g., Sweden, Norway, and Finland) the decision as to which wines are imported is made by state offices. However, these state offices cannot completely disregard the wishes of the consumers.

In the Federal Republic of Germany from 1973 to 1975 the market shares of simple "Quality wines" (QbA) and "Quality wines with Prädikat" are about equal (each at about 50%). The market share of standard wines was only 1 to 2%. In the large production and consumption countries of France and Italy the share of the quality wines in consumption is about 30% in France and about 15% in Italy.

An important determining influence in wine consumption is site of residence—urban or rural. Wine consumption is greatest in growing areas. In the Federal Republic of Germany the average per capita consumption in 1975 was 23.6 liters; consumption in the grape-growing area of Baden-Württemberg, however, was 37 liters. On the other hand, in areas of the Federal Republic distant from grape-growing regions it was only 16.5 liter per capita. Data from France and Spain, however, indicate that migration to the large cities does not appear to increase wine consumption.

D. Influence of Sex, Age, Profession, and Income

Wine consumption is influenced by various cultural, historical, and provencial factors and habits, and by the politics of importation, as well as by many other considerations. The roles of sex, age, education, family status, occupation, and income all are significant.

For the Federal Republic of Germany the Spiegel Documentation "Percentages" (1977) presents an extensive study of such factors. This leads to the following conclusions, some of which are surprising.

For individuals between the ages of 20 and 65 years, wine is drunk by 70% of men and 76% of women. The highest consumption rate, 78%, is among 30–39-year-olds; the lowest, 69%, is among 50–65-year-olds. Educational level is a factor—of those with a high school or college education, 82% drink wine, while persons with only an elementary school education and persons who have finished an apprenticeship (equivalent of vocational technical school), 71% are wine drinkers.

Marital status is of little importance here: 74% of married people and 71% of single people drink wine. Whether or not the consumer is employed or not or is in training for a job, plays a less important role than many assume: 74% of employed persons, 73% of unemployed, and 71% of those in training for a job drink wine.

Income is a considerable factor and is related to wine drinking as follows in the study (Spiegel Documentation "Percentages," 1977) in the Federal Republic of Germany.

1000 DM	2000 DM	3000 DM	4000 DM	4000 + DM
69%	70%	77%	78%	81%

E. Consumer and Health Protection

All grape-growing countries and most consuming countries have extensive special regulations that serve to protect the consumer and his health. These pertain to the proper labeling (or designation) of the wine, storage conditions, permitted additives, bottling, and distribution practices. Since the inception of the Common wine market organization in the EEC, extensive regulations as indicated above have been issued. Through these regulations that determine labeling and other practices the consumer is offered a high degree of quality protection. Anything that is not explicitly permitted is forbidden.

Still not regulated are the enological procedures (except for the use of SO_2) for which an EEC regulation exists. Supposedly this situation will shortly be changed. For other existing enological procedures national regulations remain valid.

In the Federal Republic of Germany and in Luxemburg a special quality control, in addition to the wine state control, is maintained on all quality wines. Only after official affixing of a proof for examining number does the producer gain the right upon special application to label his product in question as quality wine. In some countries a similar quality control is conducted in certain regions or wine groups (for instance VDQS in France).

IV. CONCLUSIONS

A. Production and Consumption

The grape production of the world is on the increase, although it may vary from country to country. About 90% of the grapes grown are made into wine; the rest is consumed as table grapes or raisins. In addition to expansion of the growing area the yield per hectare is slowly increasing.

Estimates of wine consumption of the world based on production statistics are too low. Specialists estimate that, at present, world production of wine is 20 to 30 mio hl greater than consumption. Other authors calculate the excess to be, at most, 8%.

About 90% of the world wine harvest is consumed in the country of production. The international trade, however, imported and exported about 44 mio hl in the year 1975, which amounts to 13% based on the 1975 world wine harvest.

In the large wine-producing countries, where at the same time wine consumption is greatest, wine consumption has reached a plateau or has even decreased.

In countries with medium or low per capita consumption, increasing rates of consumption can be expected for a variety of reasons. This is true for some wine-producing countries as well as for those wine-importing countries that do not make wine. It is estimated, according to a study done by Bank of America, that the sale of wine in the United States will about double by 1980 (based on the year 1973) (24 against 13.2 mil hl). It is also possible that a shift in the present relationships between the types of wine may occur as well as a shift in the relationship between domestic and foreign wines.

B. The Product

Wine is sometimes considered as food for daily consumption (i.e., with meals), but in some countries it is considered a luxury (at other times than meals). The latter is generally true in countries with low per capita consumption.

Wine has a good image with the consumer. Many people are becoming increasingly interested in wine. Interest in wine has greatly risen in several countries (i.e., United States, Great Britain, Federal Republic of Germany). However, some countries (France) report the opposite trend. Wine is considered by the consumer to be not only a pleasant but also a healthful product. The enjoyment of wine is regarded as enhancing the quality of life. Wine drinking is "in" in many places.

The tremendous variations of the product (wine is the child of its respective homeland and is the product of nature subject to yearly differences in quality) always makes enjoyment exciting. However, while this is helpful to promotion, it also curtails consumption.

C. Tendencies

Increasingly stricter traffic legislation, such as blood alcohol limits for drivers, and the growing use of automobiles may deter an increase in wine consumption. Even the fact that wine, in the language of the consumer, is regarded as a beverage of moderation does not alter this fact.

It appears that the world population is decreasing or is stable in those countries where buying power (as prerequisite for appreciable wine consumption) is high. In the developing countries, where there are as yet no dedicated wine drinkers, population is increasing. This will have an influence on future sales.

The tastes of the public are not fixed or unchanging. A large number of consumers love dry wines (especially with meals), while others (e.g., in Germany) prefer rather mild and lovely wines with an appealing residual sweetness; women especially prefer lovely wines. Changes in the taste of the public cannot be excluded during a generation, but they do not occur abruptly.

Wine competes with a number of other beverages. One of the main rivals is beer in some countries; this is apparent from market data (Table VII). The fact

Table VII. Popularity of Alcoholic Beverages, Federal Republic of Germany[a]

	Total	Men	Women	Age (years)			
				20–29	30–39	40–49	50–65
	32.94[b]	15.28[c]	17.66[c]	5.79[c]	8.68[c]	8.72[c]	9.76[c]
Whisky	11	17	6	22	14	7	6
Brandy	15	19	11	16	17	13	13
Cognac	10	13	8	11	13	9	8
Vodka	4	5	2	6	4	3	2
Gin	3	3	3	7	4	2	1
Clear schnaps	21	31	12	13	22	22	23
Fruit schnaps	8	11	5	6	8	7	9
Rum	9	9	9	13	8	8	9
Bitters	7	9	5	6	4	9	8
Herb liqueur	8	7	9	8	7	9	8
Liqueur	15	6	22	15	15	16	14
Vermouth/aperitif	8	5	10	12	9	8	4
Sherry	3	3	4	4	4	2	4
Port	4	4	4	5	4	4	4
Champagne	24	18	28	26	25	23	21
Wine	36	34	38	35	40	35	33
Beer	51	70	34	54	58	50	44

[a] From Spiegel Documentation "Percentages" (1977).
[b] All values in percent.
[c] In millions.

Table VIII. Other Alcoholic Beverages Wine Drinkers Consume (Federal Republic of Germany)[a]

Preference	Population (in millions)	Wine (%)
Whisky	3.65	49
Brandy	4.82	41
Cognac	3.34	56
Vodka	1.17	56
Gin	1.00	57
Clear schnaps	6.83	34
Fruit schnaps	2.49	55
Rum	3.01	50
Bitters	2.27	42
Herb liqueur	2.56	49
Liqueur	4.93	50
Champagne	7.79	62
Wine	11.80	100
Beer	16.80	43
Total	32.94	36

[a] Population surveyed, 20- to 65-year-olds. From Spiegel Documentation "Percentages" (1977).

that in the two largest wine-growing countries of the world the annual wine consumption seems to settle at about 10 liters per capita indicates that the saturation point (or limit) is much lower for wine than for beer (in the Federal Republic of Germany it is about 146 liter).

There are evidently only a few who exclusively drink wine; many wine drinkers consume other alcoholic beverages also (Table VIII).

The elasticity of demand for wine (income and price elasticity) is evidently lower than commonly assumed. According to studies of the Federal Republic of Germany the demand for amount of wine changes almost simultaneously with changes in income (Schmitt, 1969).

A recent French study came to the same conclusions (*Bull. OIV*, 1977): "The income elasticity of the demand for wine is relatively low in all countries." There are, however, certain differences according to the category of wine: the elasticity of demand is low for simple and cheap wines; for quality wines, which in general are more expensive than ordinary table wines, it is higher.

For the continued development of wine sales a positive development of the income is necessary. From the latter items it would follow that consumers in large parts of the world are becoming more and more interested in higher quality wines.

REFERENCES

Becker, W. (1968). "Winegrowing and Wine in the Federal Republic of Germany," Handbook for the Beverage Industry, pp. 47–79. Gabler, Wiesbaden.

Becker, W. (1973). "Possibilities and Obstacles toward Bringing into Harmony the Wine Legislation in the Common Market." Publications concerning the History of Wine, Soc. Hist. Wine, Wiesbaden. Germany

European Economic Community (1970). Official papers, diverse issues.

German Winegrowing, diverse issues. Wiesbaden, Federal Republic of Germany.

Household Panel of the Society for Market Research. (1976). Langelohstr. 134, D 2000 Hamburg.

Journ. Vinic. (1977). No. 14, 957 (Mar. 26).

Mauron, P. (1976). *Assem. Gen. OIV, Ljubljana; Congr. Doc.* 1976.

Michel, F. W. (1975). "The German Wine Market in the Year 1975, Market Data from Statistics and Research." Study Soc. Market Res., Hamburg.

Bull. OIV (1977). Numéro Spécial, p. 58.

Office International de la Vigne et du Vin (OIV) (1975). "Mémento de l'OIV." Paris.

Office International de la Vigne et du Vin (1977). *Symp. Int. Consommation Vin Monde (Int. Symp. Wine Consumption World), Avignon, 1976.*

Schmitt, H.-R. (1969). Prerequisites and "ansatzstellen" for production formation and advertisements for promotion of the consumption of wine in the Federal Republic of Germany. Inaug. diss., Inst. Agrarian Politics Market Res., Bonn.

Spiegel Documentation "Percentages" (1977). "A Study of the Citizens of the Federal Republic about Beverages as well as about Buying and Consumption Behavior." Spiegel Publ. House, Hamburg.

"La Vigne et le Vin en Californie" (1974). Brochure. I.T.V./Sicarex, Paris.

The Wine Economy. diverse issues. Neustadt an der Weinstrasse, Federal Republic of Germany.

12

THE WINE INDUSTRY AND THE CHANGING ATTITUDES OF AMERICANS: AN OVERVIEW

J. A. De Luca

I. In the Beginning 188
II. Prohibition Attitudes 189
III. Recovery following Repeal 190
IV. The Present 191
V. The American Wine Consumer 191
    References 193

Louis Pasteur characterized wine as "the most healthful, the most hygenic of beverages," but he also declared that wine "is the one [beverage] a person prefers to all others, if only he is given the chance to accustom himself to it" (1).

Wine is the most storied (or poemed or epigrammed, if you will) of beverages. Well-known enologist Maynard Amerine (2) once declared, "Wine is a chemical symphony," Robert Louis Stevenson (3) called it "bottled poetry." Given the nature of wines, no one can say it all, but those descriptions nicely illustrate the esthetic posture of wine.

Unlike most Americans, I came to wine early in life. When I was a boy in Los Angeles, my father used to buy Zinfandel grapes to make wine at home, and it was very much a family project, particularly the crushing of the grapes. Of course, wine is the natural beverage on dinner tables of those of us with our roots in the Mediterranean. I am pleased to see that it is finding its proper place in American life.

FERMENTED FOOD BEVERAGES IN NUTRITION

187

Copyright © 1979 by Academic Press, Inc.
All rights of reproduction in any form reserved.
ISBN 0-12-277050-1

I. IN THE BEGINNING

Colonists to America's Eastern seaboard attempted early to make wine from native American grapes, *Vitis labrusca* and others; but the wild varieties did not make a palatable European-type wine. The obvious solution, it seemed to them, was to import European vines.

Some of the most distinguished names in early American history—among them, Captain John Smith, George Washington, Thomas Jefferson, William Penn, and James Monroe—failed in this worthy project. It is reported that Thomas Jefferson even went so far as to import soil from Europe (4), thinking that that was the difficulty.

Actually a complex series of factors (winter killing, fungus and bacterial diseases, the root louse *Phylloxera*, to name a few) beyond the resources of that day (5) were responsible.

Although wine was served in the White House from the inception of the Republic, Jefferson was the first, and perhaps greatest, wine connoisseur among our Presidents. His concern for wine was broad (6). He once wrote: "I rejoice, as a moralist, at the prospect of a reduction of the duties on wine by our national legislature. No nation is drunken where wine is cheap; and none sober where the dearness of wine substitutes ardent spirits as the common beverage . . . [Wine's] extended use will carry health and comfort to a much enlarged circle" (6).

At about the same time, Father Junipero Serra was leading an inspired band of men, women, and children into California from Mexico—our other colonists. The Franciscans eventually established a chain of missions extending from San Diego to Sonoma, grew grapes, and made wine at most of the missions (7).

Other Europeans followed: Germans, Swiss, French, Hungarians, and Italians filled the melting pot and became Californians. They may have come for God, gold, or glory, but many stayed for grapes, recognizing that the climate and soil of California could reward their diligence. They dared to dream, worked hard, and succeeded. By the turn of the twentieth century, California wine was used around the country and the world, and winning prizes in European expositions. A wandering Scot, Robert Louis Stevenson, spent the happiest year of his life in California as "The Silverado Squatter," and declared prophetically of the wines, "The smack of California earth shall linger on the palate of your grandson" (3).

Despite these happenings one must not assume that America had become a wine-drinking nation, for then, as now, there were formidable obstacles.

For one thing, although California in the decade 1890–1900 produced an average of 25 million gallons of wine annually, nearly 18 million gallons of that was shipped East in bulk, to be bottled locally (8). And, one may add, to be labeled according to the seller's fancy. Thus, many such California wines had little or no identity. As Robert Louis Stevenson pointed out in 1880, many

Americans could honestly say they had never had a glass of California wine—if they went by the labels (3, p. 28).

From 1870 to 1900, U.S. per capita consumption of wine hovered around four-tenths of a gallon. The total consumption of wine only grew as did the population. In the next decade per capita consumption nearly doubled again (8). It would not be until 1941 that the figure would again be that high.

II. PROHIBITION ATTITUDES

The Prohibition movement spread, albeit piecemeal; first towns and counties, then states, enacted laws against alcoholic beverages. Initially, the movement's objective was ardent spirits. Wine was not included. But, as the movement widened, so did its targets. Oliver Wendell Holmes' injunction that "Wine is a food" (9) was ignored, and wine was swept along with the Prohibition tide.

During the years of this national myopia (1920–1933), a few wineries survived by making sacramental wines; most, however, fell into ruin. Many wine-grape vineyards were abandoned in a rush to plant raisin varieties or thick-skinned types of grapes which could be shipped to home winemakers (4, p. 25).

The disastrous social consequences of Prohibition are apparent; we have only recently recovered from some of them. But at Repeal the wine industry of America was worse off than at almost any time in its existence. Rehabilitation of the industry in America was slow. Wine Institute was formed to help the struggling infant California industry. It constructed stringent industry-wide standards for wine. A state-wide system of farm advisors and other state agencies, particularly the Agricultural Experiment Station of the University of California, provided valuable input into grape growing. Gradually, winemakers with vision insisted on quality control. Federal agencies, other states, and their universities acted together to make an almost national effort.

But something had happened to reshape the wine market. A generation of potentially new wine drinkers did not immediately appear. The Jazz Age ran headlong into the Depression and neither was hospitable to the unhurried pleasures of table wines with meals. For the first time in the history of the industry, fortified sweet wines were produced in larger volume than traditional table wines, and for yet another generation that ratio persisted. Dessert wines kept better, traveled better, and soothed nerves rubbed raw by Depression and war. As Leon Adams wrote in his book "The Wines of America" (4, p. 25), the wine industry was reborn in ruins: "It was making the wrong kinds of wines from the wrong kinds of grapes for the wrong kind of consumers in a whiskey-drinking nation with guilt feelings about imbibing in general and a confused attitude toward wine in particular."

III. RECOVERY FOLLOWING REPEAL

During this time, nevertheless, some people were daring again to dream. Equipped with recent university degrees in enology and viticulture, they became teachers of the next group, literally and figuratively breaking ground for the future. Investigation and experimentation with differences in climatic conditions, fermentation, viticultural practices, and a whole range of enological technology began and have continued. Despite widespread ignorance concerning wine (Some of my colleagues recall going into a restaurant in 1934, ordering a bottle of wine, and finding that there was not a corkscrew in the place!), the industry made progress. In California alone production increased from 26 million gallons in 1934 to 58 million gallons in 1937. Distribution channels were opened, and people were learning which wines they liked and which were reliable. By 1941, sales of California wine stood at 89 million gallons (10).

With the advent of World War II many things, including the growth of the American wine market, were interrupted. After the war, the industry was on the move again. By 1949, over 117 million gallons of California wine were shipped to market in America. By 1960, more than 129 million gallons were shipped, dessert wine sales had leveled off, and sales of table wines were greatly on the rise (10).

But these are mere statistics. Winemaking is still partially an art as well as a science. And perhaps more: As Clifton Fadiman (11) put it, "To take wine into our mouths is to save a droplet of the river of human history." The river kept flowing. As more Americans learned about wine, many of them decided they wanted to be part of it. The older established families of California winegrowers were joined by a vigorous new group of people returning to the soil, building something lasting for their children. Just as the solution to the worldwide ravages of the root louse *Phylloxera* was to graft *Vitis vinifera* vines to resistant native American rootstock, so has the wine industry thrived with successive new groups of scions, from the family of Jean-Louis Vignes in 1834 (7, p. 8) to the Class of 1977.

If many of the new producers were starting small (some had been home winemakers), so was the American public. Rosés sold well, saving people from the problem of choosing between white or red with their meals. Most winery advertising was educational, reassuring, and supportive, to increase appreciation of the product. The tide slowly turned. Foreign travel helped accustom people to taking wine a little more for granted, especially with meals. President Eisenhower established an American wine cellar at the White House. Some European winemakers even sent their sons to study at the University of California at Davis, which had become a premier seat of wine education; the professional American Society of Enologists was founded there. All in all, there was a new beginning (12).

IV. THE PRESENT

The 1960s were as much a time of turmoil in the wine world as in other spheres of activity. The trend was up, but not steadily. Lyndon Johnson instituted a "drink American wine" policy at official functions at home and abroad and was castigated for chauvinism—until people tried the wines. It helped the growing wine industry.

In 1969 table wine outsold dessert wine for the first time since the repeal of Prohibition. Shipments of California wine to market also increased 10% to 172 million gallons (13), the first of several years of double-digit increases that would later come to be known as the American "wine boom." The wine boom did not last long enough to deserve the characterization, but it resulted in establishing the appreciation for wine once and for all. With this quickening pace, winery facilities were upgraded. New and substantial acreages of fine wine grapes were planted (of some varieties more were planted in one year than had ever existed in the whole state). Enology classes at the University of California, Davis and Fresno, were crowded. New books on wine and wine production proliferated, some of them highly technical.

It seemed to be the best of times. However, in 1972 the rate of growth slipped to 5.5%. Over the years since, it has averaged around 4% (13). It continues building on a constantly widening base, and grew even during the 1973–1974 recession. Clifton Fadiman (11) opinioned: "Wine is a civilizing agent, one of the few dependable ones, one that again and again has proved its life-enhancing properties." So there is room to hope that we will become a wine-civilized nation.

I use the word "hope" not only because of my professional concern with wine, but also because I believe it is, simply, such a good thing in so many ways. Its use in moderation contributes to the enjoyment of food and life to millions of people. The role of wine in human nutrition is still being studied, but we know a good deal about the importance of wines in nutrition. We know that alcohol is utilized by the body as an energy source, and that wine does contain some important inorganic elements and various vitamins (14). Above all, however, it is simply a most enjoyable complement to most foods and contributes to a relaxed style of living.

V. THE AMERICAN WINE CONSUMER

Who is the American wine consumer? Until the 1960s, the American wine consumer was mostly male, educated, affluent, living in a metropolitan area, and what the demographers call "upscale"—upwardly mobile on the social ladder.

He was living "the good life," and wine has always been thought of, correctly or not, as part of that life.

But as we have seen, the use of wine is not isolated from the changing events and the social reverberations of the world at large. Wine buyers (and wine connoisseurs) are no longer exclusively male. In thirty states, wine can be sold in supermarkets and food stores, and the housewife has begun to exercise her wine options.

This is no small matter, philosophically or in the marketplace. During the last year there has been at least a "boomlet," perhaps more, in white table wine. It has become a popular pre-lunch and dinner aperitif, even a general beverage of wide use. There are a number of reasons for this—it is light, tasty, competitively priced, and served cold, the way most Americans like such beverages. And of great importance, women like it.

Another change in the American wine market has reflected the emergence of young adults as the major part of the population. The post-World War II "baby boom" bequeathed America in the 1960s a new group who had not been subjected to the myths and hysteria of Prohibition. When they came of age, wines in variety, quality, and value were readily available. Accordingly, many of that generation simply accepted wine for what it is, following the dictates of their own taste and good judgment.

This is not to say that wine is sweeping the entire country, however. Wine in America continues to be largely an urban and suburban beverage. Our ten largest metropolitan areas account for 36% of wine sales; the top 50 urban areas account for 65%. In fact, New York and California, with 18.5% of the population, make up one-third of the American wine market. The increase in consumption in all areas of this country gives rise to the expectation that sometime in the future there may be a truly national wine market (15).

Prohibition left the country with 48 sets of state laws upon Repeal, and that state-by-state situation prevails. Besides a bewildering welter of laws and regulations confronting the industry, many states impose discriminatory taxes on out-of-state wine to protect their local fledgling wine industries. Unlike any other commodity, the wine industry is faced with interstate barriers to trade as well as international ones.

Despite these and other barriers, seven out of ten bottles of wine sold in America come from California. This success has occurred in the freest wine market in the world; almost every winemaking nation is allowed to sell its wares here, unhampered. Over 40 currently do so. The United States industry is not accorded the same privilege in most other countries, however.

In 1976, over 272 million gallons of California wine were shipped to market in the United States.* Per capita consumption is now a little more than 1.7 gallons

*Note: shipments totaled 300 million gallons in 1978.

(16). This is brought into perspective by noting that if you were to consume one glass of wine with dinner every night, your annual consumption would be 17 gallons. This compares with an average of 29 gallons for Italy and 27 for France (17). An area of industrial growth has been the number of wineries. In the state of California over 125 were founded in the last decade, 34 in 1976 alone (18). The total number is about 260, and this will undoubtedly increase in 1977.*

A few years ago people began referring to this as ''The Golden Age of California Wine.'' As flattering as it is, it merely reflects progressive changes in quality, taste appreciation, enjoyment of life and food, and affluence.

The world of wine encompasses the arts, science, commerce, and agriculture. I believe that the wine industry well may become the main force for ideas in California agriculture. I hope it will become a touchstone in America's growing awareness of the principles of good nutrition. The coming of wine to our lives has been considered a benchmark of our progressive civilization and, in the best sense, of advancing culture. Americans have adopted many of the great wine traditions, and at the same time, reinforced those which best fit this country: hospitality, a good life, and a sense of history matched with a view of the future.

REFERENCES

1. Pasteur, L., ''Etudes sur le Vin,'' p. 53. F. Savy, Paris, 1886.
2. Amerine, M. A., Wine. *Sci. Am.* (Reprint) (1964).
3. Stevenson, R. L., ''The Silverado Squatters,'' p. 27. Chatto & Windus, London, 1920.
4. Adams, L. ''The Wines of America,'' p. 19. Houghton, Boston, Massachusetts, 1973.
5. Amerine, M. A., Berg, H. W., and Cruess, V. W., ''Technology of Winemaking,'' p. 60. Avi, Westport, Connecticut, 1972.
6. Jefferson, T., ''The Writing of Thomas Jefferson'' (A. E. Bergh, ed.), p. 177. Jefferson Memorial Found., Washington, D.C., 1907. See also Lawrence, R. T., ''Jefferson and Wine,'' Vol. 3. Vinifera Wine Growers Assoc., The Plains, Virginia, 1976.
7. Carosso, V., ''The California Wine Industry,'' p. 41. University of California Press, Berkeley and Los Angeles, 1951.
8. Blout, J. S. ''Brief Economic History of the California Winegrowing Industry.'' Wine Inst., San Francisco, California, 1943.
9. Holmes, O. W., ''Address to Massachusetts Medical Society.'' Boston, Massachusetts, 1860, *In* ''Dictionary of Quotations'' (H. L. Mencken, ed.).
10. Wine Institute, ''Statistical Panorama: The California Winegrowing Industry.'' San Francisco, California, 1965.
11. Fadiman, C., ''The Joys of Wine,'' p. 21. Abrams, New York, 1975.
12. Wine Institute Bulletins, San Francisco, California, 1950–1960.
13. Wine Institute, ''Wine Industry Statistical Report.'' San Francisco, California, 1975.
14. Morgan, A. F., Wine in the normal diet. *In* ''Table Wines; the Technology of Their Production'' (M. A. Amerine and M. A. Joslyn, eds.), p. 481. Univ. of California Press, Berkeley. See also, ''Uses of Wine in Medical Practice.'' Wine Advis. Board, San Francisco, California, 1975.

*Note: the total by the end of 1978 was 285.

15. "Wine Marketing Handbook," pp. 24, 49. Gavin-Jobson Associates, New York, 1976.
16. Wine Institute, "Wine Industry Statistical Report." San Francisco, California, 1976.
17. *J. Off. Int. Vins Vignes* (1976).
18. Wines & vines. *In* "Directory of the Wine Industry in North America." San Francisco, California, 1977.

IV

Metabolism and Therapeutic Use of Alcoholic Beverages

13

ABSORPTION, METABOLISM, AND EXCRETION OF ETHANOL INCLUDING EFFECTS ON WATER BALANCE AND NUTRITIONAL STATUS

Robert E. Olson

I.	Introduction	197
II.	Absorption of Ethanol	198
III.	Distribution of Ethanol in the Body	199
IV.	Metabolism of Ethanol	199
V.	Excretion of Ethanol	202
VI.	Endogenous Biosynthesis of Ethanol in Animals	202
VII.	Effect of Ethanol Intake on Fluid Balance	205
VIII.	Effect of Ethanol Intake on Nutritional Status	206
IX.	Research Needs	209
X.	Summary	209
	References	210

I. INTRODUCTION

Alcoholic beverages have been used by man since the dawn of history, principally to promote social intercourse and assuage anxiety, but also as a significant component of the diet in some cultures and as a vehicle for political domination in others. The principal active compound in these beverages, which include wine, beer, and distilled spirits, is ethyl alcohol. Although ethanol is the quantitative end product of glycolysis in yeast, the principal organism responsible for the fermentation of alcoholic beverages, it can also be detected as a trace metabolite of glycolysis in mammals and, hence, is not a compound foreign to the animal body. Ethyl alcohol is thus a metabolite, a food, and a drug for man.

One cannot discuss alcohol without touching on alcoholism, the social and medical disorder which sometimes results from the intake of large amounts of alcohol. Although ethanol is the agent of alcoholism, it is an essential but not sufficient cause. Many other factors contribute to the development of the full-blown disease. It is estimated that 90 million Americans drink alcohol, but only 9 million can be considered alcoholics. Although alcoholism is considered a medical disease, the criteria for its diagnosis are principally social, not medical ones. Lack of proper behavioral adjustment to family, friends, job, and law enforcement agencies in association with ethanol intake are primary guideposts to the diagnosis. Medical complications of drinking, although frequent in alcoholics, are also seen in heavy drinkers with an orthodox behavioral pattern (Olson, 1973).

In this chapter I should like to review the absorption, metabolism, biosynthesis, and excretion of ethanol, and discuss some of its effects on selected physiological functions and on nutritional status. Since all alcoholic beverages, particularly distilled spirits, are low in protective nutrients, alcohol taken in large amounts may displace other sources of calories in the diet. It will dilute these "protective calories" with "empty calories" and thus may induce serious nutritional disease. In addition to its function as a diluent of protective nutrients, alcohol has an intrinsic toxicity for certain tissues, which will be discussed.

II. ABSORPTION OF ETHANOL

Alcohol is rapidly absorbed from the stomach, small intestine, and colon. It is absorbed across biological membranes by simple diffusion. It has been shown that the rate of alcohol absorption *in vivo* is affected by alcohol concentration, regional blood flow, absorbing surface, and the presence of other foods. The most rapid absorption is from the duodenum and jejunum, with slower absorption from the stomach, ileum, and colon. There is minimum absorption from the mouth and esophagus. The rapidity of overall absorption in the human subject depends, therefore, principally upon gastric emptying time. Several factors influence this rate, including the concentration of ethanol ingested, the type of beverage used, the presence or absence of food in the stomach, and the effect of drugs or other factors affecting motility and blood flow. Depending on these factors, complete absorption may require from 2 to 6 hours. Subtotal gastric resection greatly enhances alcohol absorption, which again highlights the role of the stomach in controlling the rate of entry of alcohol into the bloodstream. The type of alcoholic beverage ingested also affects the rate of absorption. When diluted to the same ethanol concentration, beer is absorbed more slowly than whiskey, whiskey more slowly than gin, gin more slowly than red wine, and pure ethyl alcohol is absorbed most rapidly (Wallgren and Barry, 1970).

III. DISTRIBUTION OF ETHANOL IN THE BODY

Once absorption has occurred, distribution occurs rapidly. Alcohol diffuses rapidly across capillary walls and through plasma membranes of cells, allowing prompt equilibration of intra- and extracellular concentrations. It has been demonstrated that alcohol is ultimately distributed in total body water. The lipid/water partition coefficient of ethanol is quite low, so that uptake of alcohol by fatty tissue is minimal. Alcohol enters membranes, however, and does change their properties (Chin and Goldstein, 1977)

Alcohol also diffuses readily from blood to alveolar air so that the ethanol content of samples of expired air bears a constant relationship to pulmonary arterial blood levels. The partition coefficient is 2100:1, i.e., 2100 ml of alveolar air saturated with water at 34°C contains the same quantity of alcohol as 1 ml of circulating pulmonary arterial blood. If the blood contains 100 mg per 100 ml of ethanol, the corresponding concentration in the alveolar air will be 232 ppm. These principles provide the basis for most medical legal estimations of alcohol intoxication. In 1939 the American Medical Association recommended that for medical–legal purposes, a concentration of 150 mg per 100 ml of blood signifies alcohol intoxication in man. On November 3, 1960, the AMA revised this recommendation and adopted the view that "blood alcohol levels of 100 mg per 100 ml be accepted as *prima facie* evidence of alcoholic intoxication." Most states have accepted this recommendation, and prosecute automobile drivers for drunken driving if their blood alcohol levels exceed 100 mg%.

IV. METABOLISM OF ETHANOL

Ninety to 98% of the alcohol that enters the body is completely oxidized to CO_2 and water. The remaining small amount is excreted via the lungs and kidneys. The metabolism of alcohol differs from that of most substances in that the rate of oxidation above levels of 10–20 mg per 100 ml is fairly constant per unit time, indicating that it has saturated the relevant enzyme system and represents a zero-order reaction. In the adult, the average rate at which alcohol can be metabolized is about 10 ml/hour (7 gm/hour). Alcohol in amounts contained in 4 oz. of whiskey or 1.2 liters of beer requires 5 to 6 hours to metabolize. The initial oxidation of alcohol occurs chiefly in the liver as shown in Fig. 1. A number of enzyme systems participate in the oxidation of alcohol, the primary one being the cytoplasmic alcohol dehydrogenase (ADH), which is a zinc-containing enzyme (MW about 80,000) that utilizes NAD^+ as the hydrogen acceptor. The acetaldehyde generated is then further oxidized by a cytoplasmic aldehyde dehydrogenase in man. Aldehyde dehydrogenase generates acetic acid, some of which diffuses into the blood and may produce significant acidosis. Some of the cyto-

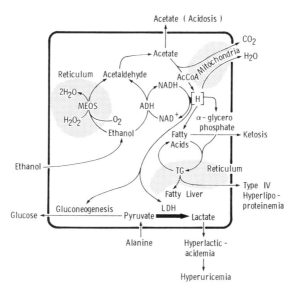

Fig. 1. Metabolic consequences of ethanol ingestion. Modified from Lieber (1973).

plasmic acetate, however, is activated to acetyl-CoA by the enzyme, thiokinase, and then is transferred via acetylcarnitine to mitochondria and oxidized to CO_2 and water. If alcohol is metabolized via cytoplasmic alcohol dehydrogenase and aldehyde dehydrogenase and the resulting acetyl-CoA is metabolized via mitochondrial enzymes, the ATP yield will be 16 moles per mole of alcohol. This allows for the reactivation of acetic acid to acetyl-CoA (which requires 2 moles of ATP) and represents an efficiency of ATP synthesis slightly less than that for glucose. In addition to the classical cytoplasmic, alcohol dehydrogenase which has a K_m of 2 mM, there are other systems known to metabolize alcohol in the hepatic reticulum. There are two enzyme systems in the microsomes that apparently oxidize alcohol and do so at a K_m of about 8 mM. These are the so-called microsomal ethanol oxidizing systems (MEOS). Lieber and DiCarli (1974) proposed the MEOS is a cytochrome P-450 type enzyme employing oxygen and NADPH to generate H_2O plus "active oxygen," which combines with ethanol to produce acetaldehyde and water. The other system in microsomes for which there is good evidence is catalase. The catalase–H_2O_2 complex can oxidize ethanol to acetaldehyde with the production of 1 mole of water. The hydrogen peroxide, which drives this reaction, is formed by a number of NADPH-dependent systems that funnel electrons through flavin enzymes. All pathways of ethanol oxidation generate acetaldehyde, which in turn is converted to acetic acid, and ultimately to acetyl-CoA. Failure to generate 1 mole of NADH in the MEOS oxidation results in the loss of 3 moles of ATP or reduction from 16 to 13 moles of ATP per mole of ethanol. The MEOS pathway thus causes a small drop in the overall efficiency of energy conservation from the administered ethanol. Cytoplasmic ADH provides the main pathway for alcohol oxidation, although it is likely that the higher capacity of alcoholics to metabolize alcohol as compared to nonalcoholics is related to the induction of MEOS in microsomes.

Alcohol dehydrogenase and aldehyde dehydrogenase are both linked to NAD^+-requiring enzymes which generate NADH, which in turn influences a large number of systems in the liver that are sensitive to the $NADH:NAD^+$ ratio. Normally this ratio is kept constant because NADH synthesis is linked to phosphorylation in glycolysis and mitochondrial shuttle systems. After alcohol ingestion, however, NADH is formed almost immediately through alcohol dehydrogenase and causes sudden shifts in the ratio of $NADH:NAD^+$. The reduction of pyruvate to lactate is a prime result. Pyruvate is generated in glycolysis but also appears in liver as a result of transamination of alanine. Increases in lactate are found after alcohol intake and lactic acidosis is sometimes a serious complication of alcoholism. Lactate acidosis also depresses uric acid excretion by competing for the same transport system in the renal tubule and is responsible for the hyperuricemia seen after alcohol intake. It is the physiological event that explains the association of gouty attacks with wine drinking.

Another effect of high NADH levels is inhibition of gluconeogenesis. Although gluconeogenesis requires NADH, it has been observed that the activity of pyruvate carboxylase, which is a rate-determining step in gluconeogenesis, is inhibited by high $NADH:NAD^+$ ratios. Under these conditions pyruvate is converted to lactate, which results in a reduction in the level of substrate for the enzyme (Krebs, 1968).

The high NADH supply also promotes serum liver triglyceride synthesis by virtue of two effects: (1) the biosynthesis of fatty acids from acetate, which requires presence of NADPH formed by transhydrogenase from NADH, (2) the generation of α-glycerophosphate from dihydroxyacetone phosphate, which provides one of the required substrates for triglyceride biosynthesis and hence enhances neutral lipid synthesis. These effects, together with enhanced phospholipid and cholesterol synthesis due to the accumulation of acetyl-CoA and acetoacetyl-CoA will result in the synthesis and secretion of very low-density lipoproteins (VLDL), which may result in a type IV hyperlipidemia. Alcohol administration also leads to triglyceride accumulation in the liver. Because of damage to membranes, the lipoproteins are not secreted through the Golgi apparatus but accumulate as intracellular vacuoles that are visible in the microscope. It is also postulated that fatty acid oxidation is reduced because of injury to the mitochondrial membranes. Both factors, enhancement of synthesis and reduction of β-oxidation, lead to accumulation of intracellular triglyceride. The high level of NADH also depresses citric acid cycle activity because many Krebs cycle dehydrogenases are inhibited by NADH.

Chronic alcohol ingestion leads to proliferation of smooth endoplasmic reticulum, since the enzymes involved in detoxication of many drugs are localized in this organelle (Iseri *et al.,* 1966). It is not surprising that enhanced metabolism of such drugs as warfarin, phenobarbital, and tolbutamide can be demonstrated in patients with chronic alcoholism. Alcohol also influences protein synthesis and results in the decrease of the synthesis of serum albumin and transferrin by the liver. It appears that alcohol, like starvation, results in disaggregation of the polysomes and reduction of the rate of messenger RNA translation (Oratz and Rothschild, 1975).

In some patients increased alcohol intake provokes a hypermetabolic state due to the uncoupling of membrane-bound ATPases with a consequent loss rather than gain in weight (Bernstein *et al.,* 1973). Chronic alcohol intake increases the oxygen consumption in rats, and this is suspected to result from an increased and partially uncoupled potassium sodium ATPase activity. This is, no doubt, one of the toxic effects of ethanol on a variety of cell membranes. Furthermore, it has already been pointed out that the oxidation of alcohol to acetic acid is wasteful because it must be reactivated to acetyl-CoA which utilizes 2 of the 18 moles of ATP synthesized per mole of ethanol oxidized. The MEOS system for ethanol dehydrogenation wastes an additional 3 moles of ATP since no NADH is formed in this initial oxidation. This is not to say that alcohol calories are entirely wasted but simply are not as useful as those from fat and carbohydrate. Further, anabolic reactions like gluconeogenesis and protein synthesis are inhibited by ethanol, which has an overall catabolic effect.

V. EXCRETION OF ETHANOL

Ninety to 98% of the alcohol that enters the body is completely oxidized to CO_2 and water. Since alcohol is not concentrated in the urine, the amount removed by voiding usually amounts to less than 1% of the quantity ingested. Elimination of ethanol through the lung is determined by the diffusion of alcohol from lung capillaries into the alveolar air, but this also accounts for a very small proportion of the alcohol ingested. The resting lung ventilation in man is 8 liters per minute. A man with 100 mg per 100 ml of alcohol in his blood would eliminate approximately 240 mg per hour by this route, a very small fraction of the 50 gm present in his body. If he were hyperventilating, or had an even higher blood alcohol content, the amount lost via the lungs would rise, but still not to levels above 5% of the ingested load.

VI. ENDOGENOUS BIOSYNTHESIS OF ETHANOL IN ANIMALS

Dr. Ivy McManus and I carried out some novel experiments some years ago at the University of Pittsburgh to demonstrate unequivocally that alcohol is not only

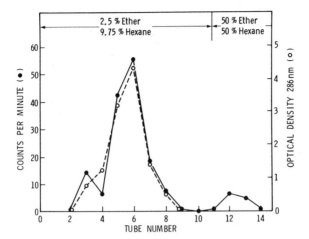

Fig. 2. Chromatographic identification of radioactive ethyl-3,5-dinitrobenzoate prepared from rat liver slices incubated with [2-^{14}C]pyruvate. Radioactivity (CPM), (●). Absorbance at 286 nm, (○). (From McManus *et al.,* 1966, reproduced by permission.)

a nutrient in man, but is in fact a normal metabolite (McManus *et al.,* 1966). Ethanol has been detected in the breath of teetotalers by gas–liquid chromatography. We identified ethanol in mammalian tissue by criteria which included (1) the reduction of NAD^+ in the presence of alcohol dehydrogenase (ADH) by a tissue sublimate, (2) chromatographic identification of ethyl dinitrobenzoate after reaction of the sublimate with 3,5-dinitrobenzoyl chloride (Fig. 2), and (3) formation of [1-^{14}C]ethanol from sodium [2-^{14}C]pyruvate during incubation of liver slices under anaerobic conditions (Fig. 3). Finally, as shown in Table I, we showed that pig heart pyruvate dehydrogenase could be coupled to ADH to form

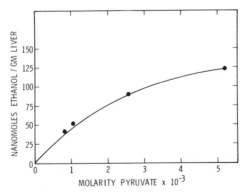

Fig. 3. Effect of pyruvate concentration on ethanol accumulation in rat liver slices. Incubations were carried out for 1 hour under N_2 at 37°C. (From McManus *et al.,* 1966, reproduced by permission.)

Table I.   Dependence of Ethanol Formation on Pyruvate Dehydrogenase System[a]

Conditions	Ethanol formed (nmoles)
Complete incubation mixture	342.0
Minus thiamine pyrophosphate	37.6
Minus pyruvate dehydrogenase	0
Minus NADH	1.8
Minus Mg^{2+}	29.0

[a] The incubation mixture contained 10 μmoles of sodium [2-^{14}C]pyruvate (0.2 μCi per μmole); 8.5 μmoles of NADH; 0.87 μmole of thiamine pyrophosphate; 40 μmoles of $MgCL_2$; 20 units of pyruvate dehydrogenase; 0.80 mg of alcohol dehydrogenase; 100 μmoles of phosphate buffer, pH 6.5, in a final volume of 3.0 ml. Incubated for 1 hour at 37°C under nitrogen.

ethanol from pyruvate. In this system the formation of ethanol required NADH, thiamine pyrophosphate, and magnesium, in addition to the two enzymes. The reaction was inhibited by semicarbazide and bisulfite. A double reciprocal plot of substate concentration vs. velocity from the coupled system showed the K_m to be 5 millimoles. This value is much higher for ethanol synthesis than for pyruvate oxidation and reflects the multiple pathways involved. As shown in Fig. 4 we

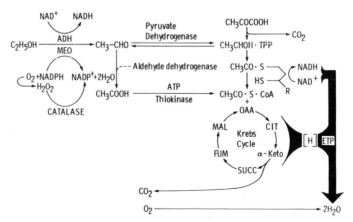

Fig. 4.   Multiple pathways of ethanol oxidation. ADH, Alcohol dehydrogenase; MEO, microsomal ethanol oxidase; TPP, thiamine pyrophosphate; R, Lipoic acid; NADH, nicotinamide adenine dinucleotide (reduced); NADPH, nicotinamide adenine dinucleotide phosphate (reduced); OAA, oxalacetic acid; CIT, citrate; α-keto, α-ketoglutamic acid; SUCC, succinate; FUM, fumarate; MAL, malate.

postulated that hydroxyethylthiamine pyrophosphate formed from pyruvate decomposes in small amounts to generate acetaldehyde which, in turn, is reduced to alcohol by ADH. In other words, there is a small leak in the normal pyruvic dehydrogenase system that permits some ethanol to be formed, whereas in yeast, acetaldehyde is freely disassociated from pyruvate decarboxylase and is quantatively reduced to ethanol.

In some rare cases, it has been observed that gastrointestinal flora may produce enough alcohol to generate a pharmacological effect on the subject. It has been observed in Japan that extensive fungus infection in the stomachs of achlorhydric patients can generate enough alcohol to provide intoxication after ingestion of carbohydrate (Kaji *et al.,* 1976). It has been recently reported that increased ethanol production from bacterial flora is observed in patients with jejunoileal bypass from morbid obesity, although no intoxication was observed in their cases (Mezey *et al.,* 1975).

VII. EFFECT OF ETHANOL INTAKE ON FLUID BALANCE

All of us have had the experience of awakening after a night of revelry, aided by alcoholic beverages, feeling very dry and thirsty. This disability is due to the fact that ethanol promotes diuresis. The mechanism is similar to that which occurs with ingestion of water, although magnified by ethanol. The inhibition of pituitary antidiuretic hormone (vasopressin) causes increased urine flow while chloride excretion falls, as shown in Fig. 5. Haggard *et al.* (1941) observed that

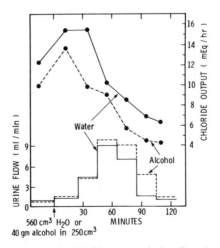

Fig. 5. The average changes in chloride output during diuresis and alcohol diuresis in three subjects. (From Eggleton and Smith, 1946, reproduced by permission.)

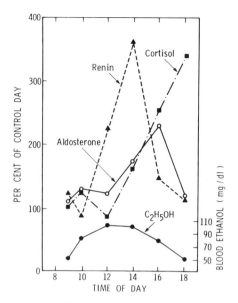

Fig. 6. Effect of acute ethanol consumption on plasma aldosterone, cortisol, and renin activity in normal young men consuming vodka. (From Farmer and Fabre, 1975, reproduced by permission.)

diuresis occurs only when the concentration of alcohol in the blood is rising. Farmer and Fabre (1975) observed that the endocrine response to alcohol ingestion involves an increase in renin, aldosterone, and cortisol, as shown in Fig. 6. The earliest event is a rise in renin levels, which, through angiotensin synthesis, stimulates aldosterone synthesis and release and leads to sodium retention. With the fall in renin and aldosterone levels, sodium and water diuresis occurs. The effects of alcohol intake on cortisol levels are more delayed, and tend to be the cumulative result of many effects of ethanol on tissues which leads to activation of the pituitary–adrenocortical axis. In a group of alcoholics in Pittsburgh, we found that plasma corticoids were universally elevated after a drinking spree (Olson, 1973).

In summary, the effects of acute ethanol intake on water balance are to produce initial modest retention of water and sodium followed by a marked water, sodium, and chloride diuresis. Ethanol-induced phasic changes in vasopressin and renin levels appear to be responsible for this effect.

VIII. EFFECT OF ETHANOL INTAKE ON NUTRITIONAL STATUS

Chronic ingestion of large amounts of ethanol generally leads to malnutrition. There are numerous reasons. One is the antianabolic effects of ethanol itself, relating to the increase in uncoupling of ATPases. Another is the displacement of essential nutrients by ethanol. If a person consumed 800 ml a day of 80 proof whiskey, nearly 1600 calories would be supplied from ethanol, and other sources of calories containing protective nutrients would be reduced. McDonald and Margen (1976) have observed that isocaloric substitution of wine for carbohydrate and fat resulted in weight loss and negative nitrogen balance. A wide range of nutrient deficiencies has been documented in chronic alcoholics, including protein deficiency, thiamine, riboflavin, niacin, folic acid, and pyridoxine depression, and deficiencies of the minerals magnesium, potassium, and zinc (Olson, 1973). The relative contributions of nutrient lack and alcohol toxicity to the causation of alcoholic hepatitis, fatty liver, and hepatic cirrhosis have been debated for some years. The view of Hartroft and Porta (1968) that fatty liver and subsequent liver injury were due solely to dilution of ''protective calories'' has been replaced by the view of Lieber and his co-workers (Lieber, 1973; Rubin and Lieber, 1974) that ethanol has a specific hepatoxicity in addition to its displacement of protective nutrients. It has been suggested (Isselbacher, 1977) that in addition to its generation of reducing equivalents which hasten accumulation of hepatic triglycerides, ethanol interferes with microtubule formation from tubulin. This latter effect appears to give rise to the telltale Mallory bodies in hepatocytes. Collapse of lipocytes, formed under the influence of ethanol, leads to increased collagen synthesis—another prodrome of cirrhosis.

Table II. Average Nutrient Intake as Determined by Diet History of Alcoholic Persons from Food, Alcoholic Beverages, and Vitamin Supplements for 1 Month Prior to the First Vitamin Excretion Test[a]

Intake	Males (26 subjects) (Mean ± SD)	Females (8 subjects) (Mean ± SD)
Calories	2710 ± 933	2578 ± 1459
Protein (gm)	68 ± 32	72 ± 47
Fat (gm)	68 ± 45	79 ± 58
Thiamine (mg)	4.0 ± 5.8	1.4 ± 1.4
Riboflavin (mg)	3.4 ± 3.3	2.1 ± 1.7
Niacin (mg)	24 ± 17	18 ± 15
Alcohol (gm)	141 ± 97	81 ± 72
Distribution of calories		
Protein (%)	10.0	11.2
Fat (%)	22.6	27.6
Carbohydrate (%)	31.0	39.2
Alcohol (%)	36.4	22.0

[a] SD, standard deviation.

Table III. Excretion of Vitamin Metabolites, Fasting and 4 Hours after Ingestion of a Test Dose by Controls and Alcoholics at Admission to the Research Ward[a]

| | No. | Thiamine | | Riboflavin | | N'-Methylnicotinamide | |
		Fasting, μg/gm creatinine	Load, μg/4 hr	Fasting, μg/gm creatinine	Load, μg/4 hr	Fasting, mg/gm creatinine	Load mg/4 hr
Males							
Controls	5	349 ± 76	243 ± 83	746 ± 229	647 ± 172	5.07 ± 1.0	2.74 ± 0.4
Alcoholics	26	265 ± 68	221 ± 41	988 ± 134	1427 ± 323	7.99 ± 1.1	5.39 ± 1.2
Difference[b]		NS	NS	NS	NS	NS	NS
Females							
Controls	7	304 ± 45	420 ± 15	262 ± 35	1090 ± 32	5.58 ± 0.3	5.33 ± 0.2
Alcoholics	8	143 ± 23	180 ± 39	585 ± 118	985 ± 53	3.20 ± 0.4	3.78 ± 0.4
Difference[b]		NS	S	NS	NS	NS	NS

[a] Values are means ± standard error.
[b] Difference between means by t test. Significant (S) when P ≤ 0.05. NS, not significant.

It is somewhat remarkable that many chronic alcoholics escape the ravages of malnutrition and cirrhosis despite heavy drinking. Only about 8% of chronic alcoholics develop cirrhosis. A study by Dr. Janice Neville and me on a group of chronic alcoholics at the University of Pittsburgh showed less evidence of malnutrition than we expected (Neville *et al.*, 1968). As shown in Table II, the mean intake of calories, protein, and three B-complex vitamins was not different from that expected for a normal population. Fat calories, however, were decreased and alcohol calories were markedly increased. Using ICNND guidelines (Interdepartmental Committee, 1960) only 15% of them were ingesting severely restricted amounts of thiamine, riboflavin, and niacin.

The results of fasting and load tests for thiamine, riboflavin, and niacin for alcoholics and controls are shown in Table III. Only thiamine excretion by female alcoholics after a load test was significantly depressed. These data, though limited in number, suggest that periodic heavy drinking in the chronic alcoholic patient does not invariably lead to severe malnutrition.

IX. RESEARCH NEEDS

The mode of absorption, distribution, excretion of ethanol, and the major pathways of ethanol metabolism are well established. The precise contributions of the cytoplasmic ADH pathway and the microsomal ethanol oxidizing systems to the oxidation of ethanol under all conditions in normal subjects and in alcoholic patients, however, are not known. The relative contributions of the cytochrome *P*-450 system and the catalase system to the MEOS are also controversial, but continuing studies should settle this matter. The pathogenesis of alcoholic hepatitis and cirrhosis and the mechanisms of alcoholic toxicity in brain, heart, and liver need much additional incisive investigation. The molecular pharmacology of ethanol in affecting membrane behavior is in its infancy. This remarkable substance, ethanol, which is a metabolite, food, and drug in man, and which has been part of man's acculturation to this planet for millennia, deserves further study on a broad front.

X. SUMMARY

Ethanol is absorbed and distributed like a drug, i.e., by more or less nonspecific mechanisms of diffusion, but is metabolized like a food. Ethanol is not only a food and a drug, it is also a terminal metabolite of glycolysis in animals. This latter pathway represents a vestigial system carried up through the eons of evolutionary time from lowly microorganisms. Although yeast can tolerate the accumulation of ethanol to levels of 10% by volume, the more highly differentiated central nervous system of animals is vulnerable to alcohol concentrations above 50 mg per 100 ml of blood, at which aberrant behavior begins. Ethanol levels above 500 mg per 100 ml are generally lethal. The endogenous alcohol formation in animals generates less than 1 mg per 100 ml, which is below the threshold of effect on biological membranes. The oxidation of alcohol, made possible by the persistence of the enzyme alcohol dehydrogenase in animals, produces wide gyrations in the $NADH:NAD^+$ ratio in liver. The enhanced reducing power in the adenine nucleotide system is responsible for most of the metabolic effects of alcohol in the body including lactic acidosis, ketoacidosis, hyperlipemia, and hyperuricemia. Alcohol, like water, induces diuresis due to changes in pituitary and adrenal function. The degenerative changes that accompany prolonged alcohol intake are a result of the intrinsic toxicity of the drug on biological macromolecules, particularly membranes and intracellular organelles and displacement of protective nutrients in the diet.

REFERENCES

American Medical Association (1970). ''Alcohol and the Impaired Driver: A Manual on the Medical Legal Aspects of Chemical Tests for Intoxication,'' pp. 145–149. Am. Med. Assoc., Chicago, Illinois.

Bernstein, J., Videla, L., and Israel, Y. (1973). Metabolic alterations produced in the liver by chronic alcohol administration. *Biochem. J.* **134**, 515–521.

Chin, J. H., and Goldstein, D. B. (1977). Drug tolerance in biomembranes: A spin label study of the effects of ethanol. *Science* **196**, 684–685.

Eggleton, M. G., and Smith, I. G. (1946). The effects of ethyl alcohol and some other diuretics on chloride excretion in man. *J. Physiol. (London)* **104**, 435–442.

Farmer, R. W., and Fabre, L. F. (1975). Some endocrine aspects of alcoholism. *Adv. Exp. Med. Biol.* **56**, 277–289.

Haggard, H. W., Greenberg, L. A., and Carroll, R. P. (1941). Studies in the absorption, distribution and elimination of alcohol. *J. Pharmacol. Exp. Ther.* **71**, 349–357.

Hartroft, W. S., and Porta, E. A. (1968). Alcohol, diet, and experimental hepatic injury. *Can. J. Physiol. Pharmacol.* **46**, 463–473.

Interdepartmental Committee on Nutrition for National Defense (1960). Suggested guide for interpreting dietary and biochemical data. *Public Health Rep.* **75**, 687.

Iseri, O. A., Lieber, C. S., and Gottlieb, L. S. (1966). The ultra structure of the fatty liver induced by prolonged ethanol ingention. *Am. J. Pathol.* **48**, 535–545.

Isselbacher, K. J. (1977). Metabolic and hepatic effects of alcohol. *N. Engl. J. Med.* **296**, 612–616.

Kaji, H., Asanuma, Y., Ide, H., Saito, N., Hisamura, M., Murao, M., Yoshida, T., and Takahashi, K. (1976). The autobrewery syndrome—the repeated attacks of alcoholic intoxication due to the overgrowth of *Candida (albicans)* in the gastrointestinal tract. *Mat. Med. Polona* **4**, pp. 1–7.

Krebs, H. A. (1968). The effects of ethanol on the metabolic activities of the liver. *Adv. Enzyme Regul.* **6**, 467–480.

Lieber, C. S. (1973). Metabolic adaptation to alcohol in the liver and transition to tissue injury, including cirrhosis. *In* ''Alcohol and Abnormal Protein Biosynthesis'' (M. A. Rothschild, M. Oratz, and S. S. Schreiber, eds.), pp. 321–342. Pergamon, Oxford.

Lieber, C. S., and DeCarli, L. M. (1974). Alcoholic liver injury: Experimental models in rats and baboons. *Adv. Exp. Med. Biol.* **59**, 379–393.

McDonald, J. T., and Margen, S. (1976). Wine versus ethanol in human nutrition. *Am. J. Clin. Nutr.* **29,** 1093–1103.

McManus, I. R., Contag, A. O., and Olson, R. E. (1966). Studies on the identification and origin of ethanol in mammalian tissues. *J. Biol. Chem.* **241,** 349–356.

Mezey, E., Imbrembo, A. L., Potter, J. J., Rent, K. C., Lombardo, R., and Holt, P. R. (1975). Endogenous ethanol production and hepatic disease following jejunoileal by pass for morbid obesity. *Am. J. Clin. Nutr.* **28,** 1277–1283.

Neville, J. N., Eagles, J. A., Samson, G., and Olson, R. E. (1968). Nutritional status of alcoholics. *Am. J. Clin. Nutr.* **21**(11), 1329–1340.

Olson, R. E. (1973). Nutrition and alcoholism. *In* "Modern Nutrition, Health and Disease" (R. S. Goodhart and M. E. Shils, eds.), 5th ed. pp. 1037–1050. Lea & Febiger, Philadelphia, Pennsylvania.

Oratz, M., and Rothschild, M. A. (1975). The influence of alcohol and altered nutrition on albumin synthesis. *In* "Alcohol and Abnormal Protein Biosynthesis" (M. A. Rothschild, M. Oratz, and S. S. Schreiber, eds.), pp. 343–372. Pergamon, Oxford.

Rubin, E., and Lieber, C. S. (1974). Fatty liver, alcohol hepatitis and cirrhosis produced in alcohol by primates. *N. Engl. J. Med.* **290,** 128–135.

Wallgren, H., and Barry, H. (1970). "Actions of Alcohol," Vol. 1, pp. 39–41. Elsevier, Amsterdam.

14

THE ENERGY VALUE OF ALCOHOL

R. Passmore

I. Introduction 213
II. Experiments of Atwater and Benedict 214
III. Utilization of Alcohol by Normal Man 220
IV. Utilization of Alcohol by Alcoholics 222
    Enzymatic Mechanisms 221
    References 223

I. INTRODUCTION

At 7 A.M. on May 5, 1897, at the Wesley Institute, Middletown, Connecticut, Atwater and Benedict (1899, 1902) started the first of their 30 metabolic experiments using the human calorimeter, previously designed and constructed by Atwater and Rosa (1898). In each experiment a subject lived in the calorimeter chamber for a period, usually 3 days, and both his input and output of energy were measured; the results were expressed in terms of heat (calories). These experiments were designed to demonstrate that "the metabolism of both matter and energy in the body could be quantitatively measured and the action of the law of the conversation of energy demonstrated, if practical."

In 13 of the experiments, the subjects were given about 75 gm of alcohol daily as part of their diet. These experiments were "undertaken at the instigation of the Committee of Fifty for the Investigation of the Drink Problem for the purpose of securing more accurate and scientific knowledge of the physiological action of alcohol" (Atwater and Benedict, 1899).

The state of knowledge of physiology as a whole at the end of the nineteenth century was admirably presented by Schäfer (1898) in the two large volumes of his "Textbook of Physiology." In the first volume he writes:

the nutritive value of alcohol has been the subject of considerable discussion and not a few experiments. Some of these tend to show that in moderate non-poisonous doses it acts as a non-protein food in diminishing the oxidation of protein, doubtless by becoming itself oxidised.

Such was the background of ignorance against which Atwater and Benedict started their work, which subsequently showed unequivocally that alcohol could be substituted isoenergetically for carbohydrate and fat as a source of fuel for the human body.

The first part of this chapter gives in detail some of the results of Atwater and Benedict's experiments. They are arguably the best designed and best executed experiments in human physiology. The second part of this chapter describes briefly, but in quantitative terms, how alcohol is utilized by the organs and tissues of the body and gives the limits to which it can serve as a fuel for a normal man. This important information has been available for many years. My excuse for presenting it again is that it is a message that we as teachers of nutrition have failed to put across. In my experience most educated laymen and regrettably large numbers of the medical profession are ignorant of the role of alcohol in the diet. This is a matter of great practical importance in contemporary life. The third part of this chapter deals briefly with the problem of how some chronic alcoholics daily dispose of quantities of alcohol much greater than those a normal person can deal with. A useful working hypothesis provides an explanation for this. In my opinion the mechanism it suggests is almost certainly correct, but many details of how the mechanism develops and how it operates await further research.

II. EXPERIMENTS OF ATWATER AND BENEDICT

Three men acted as subjects for these experiments. They were aged 25, 29, and 32 years. One was born in Canada, one in Sweden, and one in New England, and all three normally worked in Atwater's laboratory. One had ''since boyhood been accustomed to the moderate use of alcohol beverages.'' The other two had always been total abstainers. No differences were found in their utilization of alcohol.

Their diets when in the chamber consisted mainly of beef, skimmed milk, bread, and butter, and were simplified to facilitate analysis. The actual diet in three experiments is shown in Table I. About 72 gm (2½ oz.) of alcohol were given daily, ''as much as would be contained in a bottle of claret or 3 or 4 glasses of whiskey.'' In most cases pure ethyl alcohol was given, but in some whisky or brandy was used. The alcohol was mixed with water or coffee and divided into six doses; three were taken with the meals and three between meals. The amount of food provided was approximately sufficient to meet the subject's requirements of energy; their weights changed little during the course of each experiment.

Table I. Weight, Composition, and Heat of Combustion of Foods[a]

Lab. No.	Food material	Weight per day (gm)	Water (gm)	Protein (gm)	Fat (gm)	Carbohydrate (gm)	Nitrogen (gm)	Carbon (gm)	Hydrogen (gm)	Heat of combustion (calories)
3176	Beef	85.0	53.1	28.7	2.4	—	4.60	16.62	2.30	187
3177	Butter	30.0	3.0	0.5	25.8	—	0.08	19.51	3.01	240
3179	Milk, skimmed	1000.0	900.0	42.0	3.0	47.0	6.70	46.30	6.30	462
3180	Bread	200.0	78.6	17.8	3.2	97.8	2.84	55.52	7.98	561
3181	Ginger snaps	60.0	2.5	3.7	5.0	47.9	0.60	26.59	3.97	266
3168	Parched cereal	50.0	2.8	5.9	0.9	39.5	0.94	21.10	2.97	207
—	Sugar	15.0	—	—	—	15.0	—	6.31	0.97	59
	Total for 1 day	1440.0	1040.0	98.6	40.3	247.2	15.76	191.95	27.50	1982
Supplemental ration										
	Exp. No. 26: Butter	63.5	6.3	1.0	54.5	—	0.16	41.29	6.36	508
	Total for 1 day	1503.5	1046.3	99.6	94.8	247.2	15.92	233.24	33.86	2490
	Exp. No. 27: Alcohol	72.0	—	—	—	—	—	37.56	9.39	509
	Total for 1 day	1512.0	1040.0	98.6	40.3	247.2	15.76	229.51	36.89	2491
	Exp. No. 28: Sugar	128.0	—	—	—	128.0	—	53.88	8.29	507
	Total for 1 day	1568.0	1040.0	98.6	40.3	375.2	15.76	245.83	35.79	2489

[a] From Atwater and Benedict (1902). An experimental enquiry regarding the nutritive value of alcohol.

Measurements of the heats of combustion of the food, feces, and urine were made, and the total heat output of the subjects when in the calorimeter was determined. Chemical analyses of the food, feces, urine, and expired air enabled the nitrogen, carbon, hydrogen, and water balances to be determined. These balances were used to calculate the small amounts of protein and fat lost or gained by the body in each experiment (Table II). It was assumed that there was no change in the carbohydrate content of the body. The complete energy balance could then be derived (Table III).

Tables I, II, and III summarize the results of three experiments on one subject and have been selected as examples of 125 tables in the appendix to the 1902 paper which gives the full protocols of all the experiments. Table III demonstrates the application of the law of the conservation of energy in the human body. It also shows that 72 gm of alcohol, 63.5 gm of butter, and 128 gm of sugar are interchangeable as a source of dietary energy.

The 13 experiments in which alcohol formed part of the diet besides demonstrating that isoenergetic amounts of ethanol, fat, and carbohydrate are equivalent in human diets illustrated several other points in nutrition.

1. Less than 2% of the ingested alcohol appeared in the urine and expired air and the average figure for the availability of dietary alcohol was 98.1%.

2. The ingestion of alcohol did not reduce the availability of dietary protein, fat, and carbohydrate. Indeed, the average figure for availability of protein in the alcohol experiments was 93.7%, as against 92.7% without alcohol.

3. The nitrogen balances indicated that alcohol was as effective as fat and carbohydrate in protecting the body from loss of protein; thus Table II shows that in three comparable experiments in which there was a small negative nitrogen balance, the value for this in the alcohol experiment (No. 27) lay between the values in the fat and carbohydrate experiments. However, the results were not always consistent and are qualified by the statement "that the daily nitrogen balance is a much less reliable indication of the effects of diet, or of drugs, or of muscular work, or of medical treatment than is commonly supposed."

4. The doses of alcohol given in these experiments did not by their vasodilator action increase losses of heat by radiation and so lower body temperature. However, the authors add that

> if the alcohol in these experiments had all been taken at one dose . . . especially if it had sufficed to induce the comatose condition for which the expression "dead drunk" is used, and if the men had at the same time been exposed to severe cold, the production of heat in the body might have been retarded, and the radiation increased so as to lower the body temperature by several degrees.

5. There was no evidence that alcohol raised either heat production or output of carbon dioxide. Thus, it had no specific dynamic action or, in modern terms, thermogenic effects.

Table II. Gain or Loss of Protein (N × 6.25), Fat, and Water[a]

Date and period (in 1900)	(a) Nitrogen gained (+) or lost (−) (gm)	(b) Protein gained (+) or lost (−) (a) × 6.25 (gm)	(c) Total carbon gained (+) or lost (−) (gm)	(d) Carbon in protein gained (+) or lost (−) (b) × 0.53 (gm)	(e) Carbon in fat, etc., gained (+) or lost (−) (c) − (d) (gm)	(f) Fat gained (+) or lost (−) (e) ÷ 0.761 (gm)	(g) Total hydrogen gained (+) or lost (−) (gm)	(h) Hydrogen in protein gained (+) or lost (−) (b) × 0.07 (gm)	(i) Hydrogen in fat gained (+) or lost (−) (f) × 0.118 (gm)	(k) Hydrogen in water, etc., gained (+) or lost (−) (g) − [(h)+(i)] (gm)	(l) Water gained (+) or lost (−) (k) × 9 (gm)
Experiment No. 26											
Feb. 14–15, 7 AM to 7 AM	−1.8	−11.2	+17.0	−5.9	+22.9	+30.1	+8.0	−0.8	+3.6	+5.2	+46.8
Feb. 15–16, 7 AM to 7 AM	−0.3	−1.9	+14.5	−1.0	+15.5	+20.4	−24.4	−0.1	+2.4	−26.7	−240.3
Feb. 16–17, 7 AM to 7 AM	+0.4	+2.5	+18.5	+1.3	+17.2	+22.6	−2.9	+0.2	+2.7	−5.8	−52.2
Total	−1.7	−10.6	+50.0	−5.6	+55.6	+73.1	−19.3	−0.7	+8.7	−27.3	−245.7
Average per day	−0.6	−3.5	+16.7	−1.8	+18.5	+24.4	−6.4	−0.2	+2.9	−9.1	−81.9
Experiment No. 27											
Feb. 17–18, 7 AM to 7 AM	+0.1	+0.6	+12.2	+0.3	+11.9	+15.6	−19.8	—	+1.9	−21.7	−195.3
Feb. 18–19, 7 AM to 7 AM	−0.9	−5.6	+11.8	−3.0	+14.8	+19.4	−12.2	−0.4	+2.3	−14.1	−126.9
Feb. 19–20, 7 AM to 7 AM	−2.1	−13.1	+7.9	−6.9	−14.8	+19.5	+4.2	−0.9	+2.3	+2.8	+25.2
Total	−2.9	−18.1	+31.9	−9.6	+41.5	+54.5	−27.8	−1.3	+6.5	−33.0	−297.0
Average per day	−1.0	−6.0	+10.6	−3.2	+13.8	+18.2	−9.3	−0.4	+2.1	−11.0	−99.0
Experiment No. 28											
Feb. 20–21, 7 AM to 7 AM	−1.3	−8.1	+13.4	−4.3	+17.7	+23.3	+13.8	−0.6	+2.7	+11.7	+105.3
Feb. 21–22, 7 AM to 7 AM	−0.8	−5.0	+12.1	−2.7	+14.8	+19.4	+2.5	−0.3	+2.3	+0.5	+4.5
Feb. 22–23, 7 AM to 7 AM	−0.1	−0.6	+17.0	−0.3	+17.3	+22.7	+15.0	—	+2.7	+12.3	+110.7
Total	−2.2	−13.7	+42.5	−7.3	+49.8	+65.4	+31.3	−0.9	+7.7	+24.5	+220.5
Average per day	−0.7	−4.5	+14.2	−2.4	+16.6	+21.8	+10.4	−0.3	+2.5	+8.2	+73.5

[a] From Atwater and Benedict (1902). An experimental enquiry regarding the nutritive value of alcohol.

Table III. Income and Outgo of Energy[a]

Date and period (in 1900)	(a) Heat of combustion of food eaten (kcal)	(b) Heat of combustion of feces (kcal)	(c) Heat of combustion or urine (kcal)	(m) combustion combustion of alcohol eliminated (kcal)	(d) Estimated heat of combustion of protein gained (+) or lost (−) (kcal)	(e) Estimated heat of combustion of fat gained (+) or lost (−) (kcal)	(f) Estimated energy of material oxydized in the body (a) −[(b)+(c)+ (m)+(d)+(e)] (kcal)	(g) Heat determined (kcal)	Heat determined greater (+) or less (−) than estimated	
									(h) (f) − (g) (kcal)	(i) (h) ÷ (f) (%)
Experiment No. 26										
Feb. 14–15, 7 AM to 7 AM	2490	106	125	—	−64	+287	2036	2077	+41	+2.0
Feb. 15–16, 7 AM to 7 AM	2490	106	125	—	−11	+195	2075	2100	+25	+1.2
Feb. 16–17, 7 AM to 7 AM	2490	106	135	—	+14	+216	2019	2078	+59	+2.9
Total	7470	318	385	—	−61	+698	6130	6255	+125	—
Average per day	2490	106	128	—	−20	+233	2043	2085	+42	+2.0

Experiment No. 27										
Feb. 17–18, 7 AM to 7 AM	2491	97	111	7	+3	+149	2124	2116	−8	−0.4
Feb. 18–19, 7 AM to 7 AM	2491	97	121	6	−32	+185	2114	2126	+12	+0.6
Feb. 19–20, 7 AM to 7 AM	2491	97	140	5	−75	+186	2138	2128	−10	−0.5
Total	7473	291	372	18	−104	+520	6376	6370	−6	—
Average per day	2491	97	124	6	−35	+174	2125	2123	−2	−0.1
Experiment No. 28										
Feb. 20–21, 7 AM to 7 AM	2489	112	119	—	−47	+222	2083	2097	+14	+0.7
Feb. 21–22, 7 AM to 7 AM	2489	112	133	—	−29	+185	2088	2075	−13	−0.6
Feb. 22–23, 7 AM to 7 AM	2489	112	132	—	−3	+217	2031	2065	+34	+1.7
Total	7467	336	384	—	−79	+624	6202	6237	+35	—
Average per day	2489	112	128	—	−26	+208	2067	2079	+12	+0.6

[a] From Atwater and Benedict (1902). An experimental enquiry regarding the nutritive value of alcohol.

6. Whereas in most of the experiments the subjects led a sedentary life in the chamber, in eight (three with and five without alcohol) they worked for up to 8 hours a day on a bicycle ergometer. In these experiments the daily energy exchanges were of the order of 3500 kcal as opposed to a little over 2000 kcal when they were sedentary. Dietary intakes were adjusted accordingly, but the alcohol intake was unchanged. From these experiments the authors concluded that

> the utilization of the energy of the whole ration was slightly less economical with the alcohol than with the ordinary diet, especially when the subjects were at hard muscular work, but the difference in favour of the ordinary diet was very small indeed, hardly enough to be of practical consequences. ... That the alcohol contributed its share of energy for muscular work is a natural hypothesis and very probable, but not absolutely proven.

I got some impression of the work of Atwater and Benedict in the summer of 1929, when, as instructed by K. J. Franklin, my Oxford tutor, I read Lusk's "Science of Nutrition." This book is now again available as a reprint, thanks to the initiative of W. J. Darby and the generosity of the United States Nutrition Foundation. When between 1950 and 1965 I was actively engaged in measuring energy balances in man, albeit by indirect and not direct calorimetry, there were several occasions when their work had to be consulted for technical details. However, only when preparing this chapter did I study their work as a whole and was able to appreciate their true genius. Their analytical and technical skills were of a very high order. Their ability to plan and organize experiments in human physiology has perhaps never been surpassed. The description of their experiments, the tabular presentation of the data, and the discussion of the results are each set out in an orderly and logical manner; they write in simple, clear English, which is a pleasure to read.

III. UTILIZATION OF ALCOHOL BY NORMAL MAN

Ethanol, being soluble in both lipids and water, rapidly crosses cell membranes and enters the cells. Hence when taken by mouth it is quickly absorbed and distributed in both the cellular and extracellular fluids. The concentration in the blood is thus a measure of the total amount in the body. As the total body water in an adult man is about 40 liters, a dose of 32 gm of ethanol consumed rapidly on an empty stomach may raise the blood concentration to 80 mg/100 ml. This is the concentration which interests the police in the United Kingdom. Anyone who has drunk rapidly more than two pints of beer or three ounces of Scotch whiskey is at risk of having such a concentration in his blood.

Ethanol is removed from the body by metabolism at a fixed rate of about 75 mg/kg hourly by a person who is not an alcoholic (Barnes *et al.*, 1965). The rate varies from individual to individual, and values between 50 and 100 mg/kg are probably normal. The rate is not increased when metabolism is raised by muscular exercise, and it is independent of the concentration of ethanol in the blood.

A man weighing 65 kg can metabolize about 117 gm of ethanol daily. As the available energy is 7 kcal/gm (29 kJ/gm), this provides 820 kcal (3.4 MJ), which is just over half the energy required to meet the needs of basal metabolism. The energy utilized in the body is finally dissipated as heat; 3.4 MJ/24 hour is the equivalent of 40 W. Ordinarily ten people at a party produce heat equivalent to that given out by a 1-kW bar of an electric fire. About 40% of this heat may come from the metabolism of alcohol.

If some of the guests turn the party into a binge, they may consume more than 117 gm of alcohol, then, unless they are alcoholics, they may not clear their blood of alcohol for more than 24 hours. Indeed, 24 hours after ceasing drinking, the blood alcohol concentration may be so high as to preclude their driving a car. One hundred and seventeen grams of ethanol are provided by about 7 pints of British beer (ethanol 3–4%) or just over 1 liter of wine (ethanol 10%) or 12 oz. of Scotch whiskey (ethanol 31.5%). These considerations were no doubt responsible for the posters that appeared in the Paris Metro. "Les prescriptions de l'Academie de Medicine. Jamais plus d'un litre du vin par jour." The calculations indicate the limits which are imposed on social drinkers by their capacity to metabolize ethanol.

In medical practice alcohol consumption should always be considered as a source of energy in obtaining a dietary history. Excessive consumption can be a cause of obesity. Alcohol in moderate doses by sparing the utilization of fat and carbohydrate can contribute as a source of energy to the rehabilitation of a patient who has become undernourished for any reason. In dietary surveys consumption of alcoholic beverages should be recorded and utilized in the calculation of the total energy. This has seldom been done in the past. National and international figures of energy consumption and of available energy supplies have usually been underestimated for this reason. In 1975, in the United Kingdom alcohol provided 5.6% of the available energy as against 3.0% in 1955. In these 20 years annual per capita consumption of beer has risen from 140 to 206 pints, from 2.7 to 11.3 pints of wine, and of proof spirits from 1.8 to 4.8 pints. These are hard figures supplied by Customs and Excise. They do not include home-brewed beer, the consumption of which is rising rapidly. To what extent this extra energy has increased the incidence of obesity in the country is uncertain. One effect of the increased consumption is sure. Convictions in the courts for drunkenness have increased from 12.4 to 20.2 per 10,000 of the population over these 20 years.

Enzymatic Mechanisms

Frank Crewe, formerly Professor of Genetics and later of Social Medicine at Edinburgh University, was the finest conversationalist with whom I have had the good fortune to talk. He once said: "It is always possible to distinguish homo sapiens from other species of primates because, where two or three of them are gathered together, there you will find either fermented liquor or theological argument.'' Since man is the only species which has to metabolize large quantities of ethanol, it is proper to ask how the mechanism to do this was acquired. The answer is that the first stage, oxidation to acetaldehyde, is affected by a nonspecific enzyme, alcohol dehydrogenase, present in the liver of all mammals. Its natural substrates are alcohols produced in intermediary metabolism (for example, in steroid and bile acid metabolism). It also metabolizes methanol.

The acetaldehyde formed is converted into acetyl-CoA by an aldehyde dehydrogenase. The acetyl-CoA may then enter the citric acid cycle and be used as a source of energy or enter a synthetic pathway, e.g., for fatty acids or cholesterol. As the acetyl-CoA may pass into the blood and be metabolized elsewhere, ethanol may serve as a source of energy for other tissues, including muscle. However, since the rate-limiting reaction is that carried out in the liver by alcohol dehydrogenase, the rate of clearance of ethanol from the blood cannot be increased by muscular exercise.

Alcohol dehydrogenase is present in the cell cytoplasm. Its cofactor is NAD^+, and the oxidation is linked to the formation of high energy bonds in ATP. Thus all the energy in the ethanol consumed is first either transduced into ATP and utilized in the tissues or incorporated into larger molecules by organic synthesis. None is directly dissipated as heat.

IV. UTILIZATION OF ALCOHOL BY ALCOHOLICS

Many heavy drinkers consume daily amounts of alcohol far in excess of the amounts that a normal man can metabolize. Clearly they must develop an additional mechanism. It is now known that ethanol and many other drugs, normally foreign to the body, may be metabolized and detoxicated in the smooth endoplasmic reticulum of the liver. The enzymes responsible are known as microsomal enzymes and are probably derived under the action of the drugs from physiological enzymes concerned with steroid metabolism. The enzyme system responsible for oxidizing ethanol is the hepatic microsomal ethanol-oxidizing system (MEOS). It is included in the mixed function oxidase system (MFOS) or the monooxygenases. These enzymes are dependent on $NADPH_2$, cytochrome P-450, and cytochrome P-450 reductase, which catalyzes the direct incorporation of molecular oxygen. As the oxidation of ethanol to acetaldehyde by this system is not linked to the formation of ATP, the energy liberated is not utilized by the tissues but dissipated directly as heat.

A convincing hypothesis that microsomal enzymes are responsible for energy wastage in alcoholics has been put forward by Pirola and Lieber (1976). The biochemical details have been presented in a previous symposium (Cederbraum and Rubin, 1975; Lieber *et al.*, 1975). Lieber's experimental model was based on rats fed a diet in which 36% of the energy was provided by alcohol. Within 3 to 4 weeks, this regime produced increased activities of hepatic microsomal enzymes, raised the animal's oxygen consumption, and retarded their growth, presumably because of a lack of available energy. Lieber has been unduly modest in describing his views as a hypothesis. His own work on rats is a convincing demonstration of the relation between energy wastage and induced MEOS activity. However, the findings appear not to have been validated in man. It is not known what dose of alcohol has to be consumed and for what length of time before this mechanism is induced. The use of other drugs almost certainly affects induction of MEOS activity in man.

It is possible, perhaps even probable, that some people who are in no sense alcoholics may have some MEOS activity in their livers. This would account for the occasional finding of a specific dynamic effect of alcohol (Perman, 1962). Perman's subjects were seven healthy young Swedish males, "all with modest ethanol habits."

In conclusion, there is now a need for a comprehensive study of the utilization of ethanol by chronic alcoholics, comparable to the classic study in normal men. The main difficulty of any successors of Atwater and Benedict will be to find subjects among alcoholics as willing, reliable, and co-operative as the three men who contributed so much to the success of the early experiments.

REFERENCES

Atwater, W. O., and Benedict, F. G. (1899). Experiments on the metabolism of matter and energy in the human body. *U.S. Dep. Agric., Bull.* No. 69, p. 112.

Atwater, W. O., and Benedict, F. G. (1902). An experimental enquiry regarding the nutritive value of alcohol. *Mem. Natl. Acad. Sci.* **8**, 231–397.

Atwater, W. O., and Rosa, E. B. (1898). Description of a new respiration calorimeter and experiments on the conservation of energy in the human body. *U.S. Dep. Agric., Bull.* No. 63, p. 94.

Barnes, E. W., Cooke, N. J., King, A. J., and Passmore, R. (1965). Observations on the metabolism of alcohol in man. *Br. J. Nutr.* **19**, 485.

Cederbraum, A. I., and Rubin, E. (1975). Molecular injury to mitochondria produced by ethanol and acetaldehyde. *Fed. Proc., Fed. Am. Soc. Exp. Biol.* **34**, 2045.

Lieber, C. S., Teschke, R., Hasumura, A., and De Carli, L. M. (1975). Differences in hepatic and metabolic changes after acute and chronic alcohol consumption. *Fed. Proc., Fed. Am. Soc. Exp. Biol.* **34**, 260.

Perman, E. (1962). Increase in oxygen uptake after small ethanol doses in man. *Acta. Physiol. Scand.* **55**, 207.

Pirola, R. C., and Lieber, C. S. (1976). Hypothesis: Energy wastage in alcoholism and drug abuse: Possible role of hepatic microsomal enzymes. *Am. J. Clin. Nutr.* **29**, 90.

Schäfer, E. A. (1898). "Textbook of Physiology," Vol. 1, p. 882. Pentland, London.

15

ROLE OF ALCOHOLIC BEVERAGES IN GERONTOLOGY

Donald M. Watkin

I. Introduction	226
II. Alcohol Abuse and Gerontology	226
A. Early Life	226
B. Middle Age	226
C. Old Age	227
III. Prevention of Alcohol Abuse	228
A. Societal Control	228
B. A Public Health Approach: Nutrition Program for Older Americans (NPOA)	229
IV. Results of Carefully Controlled Alcoholic Beverage Consumption Studies among Institutionalized Elderly	230
A. Red Port Study	231
B. Studies on Beer	231
C. Deficiencies of Studies	233
D. Five-Wine Study	233
E. Study Allowing Choice of Beverage among Elderly with High and Low Functional Capacities	234
V. Discussion	235
A. Deficiencies in Studies to Date	235
B. Factors Precluding Governmentally Subsidized Use of Alcohol in Gerontology	235
C. Factors Precluding Unsubsidized Use of Alcohol in Gerontology	235
D. Alternative Strategies	236
VI. Research Needs	239
A. Understanding the Situation	239
B. Understanding the Causes	239
VII. Conclusion	240
References	240
Editorial Comment	243

I. INTRODUCTION

Gerontology, the study of aging as a process from conception to death, must be distinguished from geriatrics, the care of sick persons who happen to be old. The role of alcoholic beverages in gerontology, therefore, is an intrinsic part of many other sections in this volume that deal with the basic, clinical, social, economic, and political sciences and with the religious aspects of human life. All of these are elements of gerontology, nutrition, and health, forming an integrated, inseparable triad to be considered as an entity, not as component parts.

Alcoholic beverages in gerontology must be examined, therefore, in light of the role they play in nutrition and in health from conception to death, and not merely their role during old age. The scope of each agglomerate comprising this triad and the complexities of their interrelationships are so potentially vast that consideration must be focused on but a few aspects deserving of highest priority.

II. ALCOHOL ABUSE AND GERONTOLOGY

A. Early Life

Since aging begins at conception, the influence of alcoholic beverages on the fetus (Jones *et al.,* 1976) and the young child (Rosett, 1976; Streissguth, 1976) deserve brief attention. Aside from the role alcohol abuse may play in the conception of unwanted progeny, the abuse of alcoholic beverages by parents and others who enter a child's environment during early life can be devastating to a child's physical health, as evidenced by the growing national concern over child abuse. A rare influence of the impact of toxic excess on the child's development may be alcohol-induced hypoglycemia (West, 1979) when a child consumes accidentally or is deliberately given an overdose of alcohol. The seeds of successful aging are sown very early in life so that to the extent that alcohol abuse diminishes the child's probabilities of successfully aging, alcohol abuse in the *ambiente* of children is a concern of gerontologists.

B. Middle Age

Abuse of alcohol in one generation may beget similar abuse in subsequent ones. The impact of such geneology on aging patterns in American males is illustrated in findings reported by Linn *et al.* (1969). Using a multivariant analysis on 40 variables (Cumulative Illness Rating Scale, Linn *et al.,* 1968), diagnostic classifications, and various demographic and background factors applied to 100 consecutive deaths in each of three age groups (55-64, 65-74, and 75 years and over), Linn *et al.* found no differences in total pathology scores, number of previous hospitalizations, and total number of past illnesses. However, two of the individual item scores on the Cumulative Illness Rating Scale—hepatic and psychiatric—were significantly higher among the youngest group. Analysis of causes of death showed cirrhosis to be the only cause of death highly related to age, striking heavily at those in the youngest age group. Linn (1968) and Linn *et al.* (1967, 1969) conclude that survival to over 75 years of age is related, first, to membership in a biologic elite with respect to longevity in general but, second, to an ability to avoid or resist diseases which are leading causes of death. In this context, death from cirrhosis and the relation of that disease to abuse of alcohol in early and middle maturity are clearly indicted as risks.

In another study conducted at the Veterans Administration Outpatient Clinic in Boston, during which 77 variables acquired by interviewing next-of-kin listed on 500 death certificates of 500 former participants in the Longevity Study (Enslein *et al.*, 1967; Rose *et al.*, 1967; Rose and Bell, 1971), alcohol abuse appeared to summarize other and perhaps unmeasured variables, rather than operating as a predictor in its own right. Among the variables not measured was nutrition, a significant oversight in view of the work of Lieber *et al.* (1975), suggesting that alcohol is a toxin whose impact is direct as well as other views that its health effects are mediated through nutritional deficiencies.

C. Old Age

Other experiences both clinical and administrative within the framework of county and municipal hospitals, the Clinical Center at the National Institutes of Health, and the Veterans Administration Department of Medicine and Surgery have left little doubt that alcohol abuse contributes significantly to the shortening of life. It also depreciates the quality of life among those excessive imbibers who have avoided the impact of cirrhosis and psychiatric disorders, including suicide, earlier in life. This last-mentioned problem is receiving more prominence as its prevalence among the traditionally silent and ignored older American is brought into the national spotlight.

The 1976 Hearings on Alcohol and Drug Abuse among the Elderly before two Subcommittees of the Committee on Labor and Public Welfare of the United States Senate (Eagleton and Hathaway, 1976) gave prominence to this national problem. More recently, an Interagency Committee on Federal Activities on Alcohol Abuse and Alcoholism has given priority to alcoholism among the nation's elderly (DeLeon, 1977). Preliminary investigations suggest that more than loneliness and feelings of worthlessness may be involved in the pathogenesis of this condition. Retirement-enforced reductions in living standards create "Hobson's choice" situations for the otherwise well-compensated alcoholics. They must either endure the torment of alcohol withdrawal or reduce their expenditures for medical and dental care, nutrition, housing, recreation, and socialization. For many, as was the case with the clients of liveryman Hobson, there is no choice. All other elements of life succumb to the demands of the body for solace from alcohol.

Considering the fact that persons 60 years of age and older in the United States account for over half of the $160 billion health bill (Humphrey, 1977) and are still underutilizers of health services, the needs among older Americans for health care and nutrition are enormous. Failure to fulfill these needs in order to meet more impelling demands of alcohol leads directly to a plethora of health-related problems, many of which could be effectively managed through appropriate health and nutritional care.

III. PREVENTION OF ALCOHOL ABUSE

A. Societal Control

The clear implication that alcohol abuse shortens life and leads to life of poor quality among those who survive to old age is certainly not news to anyone associated with the health industry. Nor is it news that the problem is unresolved in the United States and is increasing in magnitude as the population grows. Obviously, treatment of established alcohol abuse is possible through such well-organized operations as Alcoholics Anonymous. But what about its prevention? What about using health promotion throughout life as a means of converting alcohol from a menace into a virtue?

Comparisons of various societies in their own settings reveal that attitudes toward alcohol vary widely, many of them far different from those of middle-class Americans (Wine Advisory Board, 1975). That rates of alcoholism differ has been shown repeatedly (Bales, 1946; Chafetz, 1964). In addition, the circumstances and purposes of use may vary, even among contiguous countries (Bruun, 1967) or subpopulations of the same country (Snyder, 1962). Alcohol consumption may be regulated or prohibited by religious protocols (Chafetz, 1964) or limited to use as or with food by social custom (Blum and Blum, 1964; Lolli *et al.*, 1958). In the latter situation moderation is regarded as a sign of maturity and overindulgence is not tolerated; alcoholism seldom appears to offer difficulties. Such encouraging observations suggest an alternate approach to the problem of alcoholism. Instilling in younger people more propitious attitudes toward alcohol might have favorable impact on the role alcoholic beverages frequently play in curtailing effective longevity by eliminating among the young and those in early and middle maturity causes of death, including accidents, which may be attributable directly to alcohol abuse.

Unfortunately, temperance-oriented societies are far from ubiquitous (Bales, 1962; Sadoun *et al.*, 1965; Sariola, 1954). Furthermore, the United States descendents, siblings, and cousins of such populations seem to have strayed from the temperate paths followed in their original homelands (Barnett, 1955; Glad, 1947; Lolli *et al.*, 1952).

B. A Public Health Approach: the Nutrition Program for Older Americans (NPOA)

Unfortunately, the needs of older persons in terms of income after retirement, health care, nutrition, housing, or opportunities for recreation and socialization too often are not met. Hence, even among those elderly who did not acquire a problem drinking habit early in life, the seduction of alcohol as a release from various real or imagined miseries of old age is ever present. Obviously, one means of preventing alcohol abuse from developing in this particular group is to devise more effective ways to meet the needs of the elderly and remove the motivation to seek solace from alcohol abuse.

One such choice is the Nutrition Program for Older Americans (NPOA) (Administration on Aging, 1976a) authorized by Title VII of the Older Americans Act of 1965, as Amended (Administration on Aging, 1976b), operational in the United States since September, 1973. Conceptually, dining together creates a center of gravity attracting older Americans through the mechanisms shown in the funnel at the top of Fig. 1 to meal service sites where they have access not only to appropriate nutrition but also to a variety of health services, an array of education and counseling activities, and recreation and socialization (Watkin, 1977). The popularity and political success of this program, as evidenced by the ever-increasing appropriations voted it by Congress, and the innumerable anecdotal reports of virtual rejuvenation of persons who were regarded as lost causes by their children and peers, suggest that comprehensive packages of this type improve the life satisfaction of older Americans. Doing so may diminish the need for succor from alcohol.

Among the successes of NPOA is the rehabilitation of elderly alcoholics, presumably by enabling them again to compensate as they had in earlier stages of life for what may still be a considerable intake of alcoholic beverages.

NPOA serves as an example of what can be done by a caring society to meet the needs of the elderly and, in so doing, to reduce the pressures of circumstances leading to alcohol abuse. NPOA now serves only 25% of those elderly whom the Select Committee on Nutrition and Human Needs of the United States Senate has classified as in desparate need (and these on the average of only twice every three weeks). Since NPOA is a tax-supported activity, subsidized by federal, state, and intra-state governments, attitudes which led to the Eighteenth Amendment

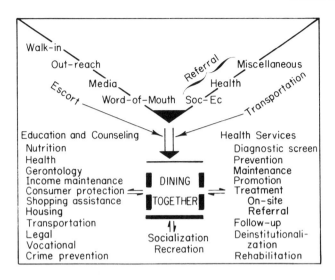

Fig. 1. $\alpha \rightarrow \Omega$ conceptualization of the Nutrition Program for Older Americans (NPOA) authorized by Title VII of the Older Americans Act of 1965, as Amended.

and to the Volstead Act prevail, even though the Twenty-first Amendment will soon have its forty-fourth birthday. NPOA, therefore, has not been a vehicle through which to test the virtues of alcoholic beverages in making a very successful operation even more effective. It is doubtful, therefore, that in the foreseeable future NPOA resources could be used to demonstrate that use of alcoholic beverages in moderation, in conjunction with dining together, could improve the overall success of the program. The matter is further complicated by the fact that many participants in NPOA and some members of various advisory groups that support its program belong to subsets of the total population for whom alcohol in any form is anathema.

Fortunately, data are available from studies elsewhere that have in the past two decades examined the possible role of alcoholic beverages in the management of institutionalized aged. A critical look at some of these studies is instructive.

IV. RESULTS OF CAREFULLY CONTROLLED ALCOHOLIC BEVERAGE CONSUMPTION STUDIES AMONG INSTITUTIONALIZED ELDERLY

The use of alcoholic beverages for social purposes in hospitals and other institutions catering to the well-to-do is not new. Private pavilions in many of the leading hospitals have long included alcoholic beverages via room service or even bartender service for a patient's room. Public institutions and those catering to the less well-to-do not only have avoided such service but have enforced, or at least have tried to enforce, total abstinence on the part of patients and staff. Nonetheless, in some such institutions demonstration programs have been established that provided relatively controlled opportunities to evaluate the influence of consuming alcoholic beverages in moderation on elderly patients or subjects and staff of the institutions.

A. Red Port Study

Perhaps the dean of authorities in this field is Robert Kastenbaum whose studies have been conducted in or near Boston. Kastenbaum and Slater (1964) compared one group of 20 elderly men (median age 76; mean age 76.6) with another group of similar size (median age 73; mean age 75.4) in an investigation with the following design:

1. A period of gathering baseline information
2. Division of the 40 men into two equal groups, Group A began with wine (red port, 1.5–3.0 oz. daily) and Group B began with grape juice
3. A cross-over period of equal length in which the beverages were reversed for Groups A and B
4. A critical free-choice period
5. An extended free-choice period

Participants in the study agreed only to come to a specific place in the hospital where the beverages were available. No patient was required to accept either beverage or even to stay in the area if he chose to do otherwise.

The results and staff impressions included the following: (1) patients for the first time began to relate meaningfully to one another and to the staff. (2) The "club" atmosphere continued long after the study was terminated. (3) Wine was decisively preferred to grape juice in both the critical free-choice session and the extended free-choice phase that followed. (4) Group involvement was significantly greater during the 30 sessions at which wine was used in contrast to those in which grape juice was available. (5) There was, however, little carry-over of the benefits of the group sessions when the men were returned to their home wards.

B. Studies on Beer

Perhaps the most impressive results in Kastenbaum's series were obtained in another study of 34 seriously impaired patients who were transferred from a conventional to a larger ward attractively decorated and equipped with a phonograph, bulletin board, games, cards, checkers, and puzzles as well as with tables for meals and activities (Kastenbaum, 1972). Beer from the pharmacy was served six afternoons a week, along with crackers and cheese. Within a month, the atmosphere of the ward in general and the behavior of individual patients had undergone a remarkable transformation.

At the initiation of the study, 76% of the 34 male patients were incontinent and required safety restraints. After one month, the rate of incontinence was down to 50% while restraints were needed by only 27%. At the end of two months, incontinence was present in only 27% and restraints were necessary in only 12%. The percentage of ambulatory patients increased from 21 to 74% and persisted. Group activity more than doubled, going from 21 to 50% and then to 71% in the second month. Music responsivity showed a slightly different pattern, not changing from 48% in the first month but jumping to 88% in the latter part of the study. Off-ward social activities showed a dramatic rise, from 6 to 15% at the end of the first month and up to 48% at the end of the second.

Improvement was also indicated by the substantial reduction in the amount of psychotropic drugs prescribed. At the start of the study, three-quarters of the group were receiving Thorazine (50 mg qid) and Mellaril (75 mg qid). At the end of one month, only one-quarter of the patients were receiving Thorazine and these at only half the original dosage; one-tenth were receiving Mellaril at about one-fourteenth of the original dosage. By the second month, no Thorazine was used, and the Mellaril dosages and the number of patients receiving them remained at the much lower levels.

Among the remarkable changes observed was the spontaneous decision of ward personnel to give up their accustomed lunch break away from the unit and stay on the ward to eat lunch with the patients.

Similar findings associated with beer consumption by patients in a nursing home in Seattle have been reported by Black (1969). Reduction in the patient demand for tranquilizers, sedatives, and diuretics, a marked decrease in patient irritability, and substantial performance improvements in social competence, social interest, cooperation, neatness, grooming, and physical condition were observed. Patients making the most rapid progress were those most impaired when the test began. In this study the contributions of the effects of social activity and the specific effects of beer could not be differentiated. There was no question in Black's mind, however, that beer produced results under identical conditions which were superior to those produced by fruit juice. Another investigation with similar results was conducted by Becker and Cesar (1973). This one focused on the use of beer in elderly psychiatric patients in a state mental hospital. Beer and fruit juice were compared in two groups, each containing eight male and eight female patients. When social interaction ratings from week 1 to week 11 were compared, the group receiving beer showed increased social activity during the sessions, while the group given fruit juice remained unchanged. In the Kastenbaum and Slater (1964) study, no improvement in ward behavior outside the group sessions was noted in either group. Psychotropic medication remained unaltered. Becker and Cesar suggest that providing beer in a social setting may induce behavioral improvement independent of the influence of beer alone.

Beer therapy was compared with other forms of therapy by Chien (1971) in a study of 40 patients in a state hospital geriatric ward. Three groups received a beverage (beer, fruit punch, or fruit punch plus thioridazine) in a "pub" set up in the hospital, while a fourth group received thioridazine on a ward. The beer sociotherapy group showed the greatest improvement and had the best attendance and the greatest social interaction in the pub. Both groups receiving thioridazine showed improvement but less so than the beer group. Those who received only punch in the pub showed the least change. The author attributed the success of the beer sociotherapy to the total psychosociophysiological effect of the program rather than to any specific nutritional factor in beer itself.

C. Deficiencies of Studies

All of these studies lack sufficient sophistication in research design and methodology to give conclusive support to the efficacy of alcoholic beverages per se in the management of those already old. Most were performed in patients living in deprived institutional environments where there were no longer adequate controls to distinguish the innovation of introducing alcoholic beverages from a generalized Hawthorne effect, i.e., the beneficial effect of interest shown in the subjects by the investigator. A study by Mishara and Kastenbaum (1974) and one by Mishara *et al.* (1975) are exceptions to this generalization.

D. Five-Wine Study

Mishara and Kastenbaum (1974) investigated the therapeutic effects of wine (five different varieties of California wine were used) in quantities of 1.5 to 3.0 oz. daily in 80 elderly patients, half men and half women. One ward containing 20 men and 20 women was run on a Token Economy Program (TEP) and the other on a Free Enrichment Program (FEP). Under TEP, wine was purchasable with tokens earned for good behavior; under FEP, wine was available regardless of behavior. The study lasted for 10 weeks with a 6-month follow-up phase.

The dispensing of wine produced a dramatic reduction in the amount of chloral hydrate used to induce sleep. Interpersonal communication increased. The earning of tokens for wine purchases was a factor in improving the behavior of some patients on the TEP ward, although it was a less effective behavioral reinforcement than cigarettes. Both tranquilizing and stimulating properties of wine were observed, better sleep, on the one hand, and far more interpersonal relations, on the other. The patients seemed to be reminded of previous experiences outside the hospital when wine or other alcoholic beverages were present. More wine was consumed on the FEP ward, suggesting that free access to the beverage is the most productive way of reaching elderly patients. The TEP experiment, however, was more productive in revealing individual differences among patients. Staff were positively affected by the appreciation manifest by wine recipients.

E. Study Allowing Choice of Beverage among Elderly with High and Low Functional Capacities

By far the most impressive study of the beneficial effects of these beverages to date is that of Mishara *et al.* (1975). Specifically designed to compensate for deficiencies of previous studies, this investigation evaluated in a careful, comprehensive fashion the psychological and physical effects of alcoholic beverages in two different populations: (1) elders of a low functional status in a nursing home; and (2) more capable and independent elders in a residence, i.e., a protected environment providing a minimum of supervision and physical care.

In this study, unlike those previously reported, two servings of a wide variety of alcoholic beverages, each containing 0.4 fluid oz. of alcohol (or the equivalent of 1 oz. of 80 proof whiskey or vodka) were available to each experimental subject 5 days a week for 18 weeks. Control subjects had nonalcoholic beverages only during the first 9 weeks and then alcoholic beverages for the remaining 9 weeks.

Medical and psychologic data were recorded by physicians and interviewers who did not have knowledge of the subjects' experimental groups or consumption rates. Data were acquired prior to the study, at the end of 9 weeks, and at the end of the study at 18 weeks. Medical examinations included electrocardiograms, blood tests, and special physical evaluations which concentrated on parameters which might be related to alcohol consumption. The psychological evaluations included three specific tests standardized on elderly populations as well as questions involving daily living.

No negative effects of alcohol consumption were observed. Positive effects of alcohol consumption included improved cognitive performance, increased morale, less worrying, and less difficulty falling asleep. Fears that the experiment might produce alcoholism proved unfounded.

Only after the full 18 weeks did data acquired during interview evaluations indicate that the group consistently consuming alcoholic beverages had more positive findings than did those who participated in social hours not characterized by alcoholic beverage consumption. The different scatter of findings for residence and nursing home suggested to the authors that functional status (high or low, determined by the institutional setting) may interact with effects of alcohol availability and consumption. Alcohol may facilitate certain changes in low functional status groups but quite different changes in groups with higher functional capacity.

As far as beverage choice is concerned, blended whiskey proved most popular (63.4% of all beverages served) with brandy/cordials second (14.6%), rum drinks next (11.5%), and wine and beer combined for the balance.

V. DISCUSSION

A. Deficiencies in Studies to Date

In spite of the care exercised in the planning and conduct of the study by Mishara *et al.* (1975), the health effectiveness, not to mention the cost effectiveness, of alcoholic beverage consumption as opposed to other inducements to socialization, such as NPOA, have not been distinguished. The influence of the Hawthorne effect is not explicitly eliminated. Alcoholic beverage consumption has been restricted in all studies to relatively small quantities daily and has been closely supervised; all subjects were institutionalized elderly persons with varying levels of functional impairment. Hence, extrapolation of findings to make generalized recommendations of public health significance is not justified.

B. Factors Precluding Governmentally Subsidized Use of Alcohol in Gerontology

As pointed out earlier, although the Twenty-first Amendment has been in effect for almost 44 years, the use of alcoholic beverages is still a matter of controversy. The frailty of marriage, the dissolution of the family as an institution, the ever-diminishing number of meals served in the home in a family setting, and the general availability of alcoholic beverages to youth beyond family influences suggest that in the United States and many other societies the family-oriented approaches to the elimination of abuse of alcohol are of limited value. The virtual obsession with alcoholic beverages of many members of the labor and business communities takes its inevitable toll as measured in absenteeism, accidents, and alcohol-induced diseases (National Safety Council, 1976). The alcohol abuse-related mortality rates among Americans in late maturity are appalling (National Center for Health Statistics, 1977).

Among the elderly, both the decompensated former social drinker and the consumer seeking a substitute for companionship and socialization may put their craving for alcohol ahead of other more constructive needs. Hence, when aging is viewed as a lifelong process, it is little wonder that many groups within society are adamantly opposed to any subsidy that would make available alcoholic beverages to those who might even have a clearly indicated need for such intervention.

C. Factors Precluding Unsubsidized Use of Alcohol in Gerontology

In view of the lack of unequivocal evidence that alcoholic beverages per se fulfill a need that is not met by nondrug-related measures, the opposition to subsidized alcoholic beverage consumption cannot be successfully countered.

The high cost of true necessities and the dire economic straits of fixed-income, inflation plagued older Americans make it unlikely that general recommendations for nonsubsidized use of alcoholic beverages will be made by authorities who must view the big picture of aging from conception to death.

D. Alternative Strategies

Professionals from medicine and public health have recently proposed before a congressional committee (McGovern, 1977) constructive approaches to lifelong aging. Approaches controlled by individuals themselves are capable of implementation whenever they become sufficiently motivated, such motivation being derived from genuine desire for effective longevity. This desire may be kindled by examples of vigorous, productive, happy elderly persons; it is not fostered by the poor health and social and economic status of those already old. Improvement in the status of many oldsters can and must be made. Figure 2

Fig. 2. A view of present treatment of the elderly by participants in milieu therapy training at the Institute of Gerontology at the University of Michigan. From Metzelaar (1975).

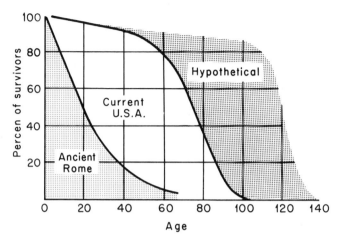

Fig. 3. Hypothetical increase in the technical life span of man. From Bylinsky, see Walford (1976).

illustrates the image that participants in milieu therapy training hold of present-day treatment of the elderly (Metzelaar, 1975). Small wonder that deprivation, loneliness, and despair are rampant among older citizens. Creation of a new image is needed.

Popular articles emphasize the desire of many for this to occur. As illustrated in Fig. 3, Walford (1976), quoted in *Fortune,* believes one can increase the technical life span of man from its present 113 years to close to 140. According to Walford, we should be able to change the lifelong aging process from that depicted at the bottom of Fig. 4 to that shown along the top. Much basic research will be required before even the infrastructure of such a development is in place, if indeed it is ever possible.

Nonetheless, much can be accomplished with what is known. Seven basic characteristics lifestyles have been found by Belloc and Breslow (1972) to lead to effective longevity in population groups. These involve (1) avoidance of gluttony; (2) eating regularly; (3) abstinence from tobacco; (4) use of alcoholic beverages in moderation; (5) regularly scheduled exercise; (6) regularly scheduled hours of sleep; and (7) regularly scheduled recreation and relaxation. When all seven are pursued throughout life, as shown in Fig. 5, persons of 80 years of age may have a "ridit" (Relative to an Identified Distribution) equivalent to persons of 35 who pursue only from zero to two. (The lower the ridit, the better the health status). Estimates of average age at death were increased by 30 years among those who practiced such lifestyles throughout life, and by 11.5 years among those who defer implementing them until age 45 (McGovern, 1977).

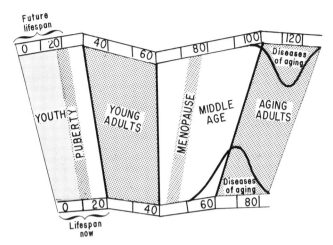

Fig. 4. Hypothetical postponement of life's milestones by age-retarding agents, as proposed by Walford (1976). From Bylinsky, see Walford (1976).

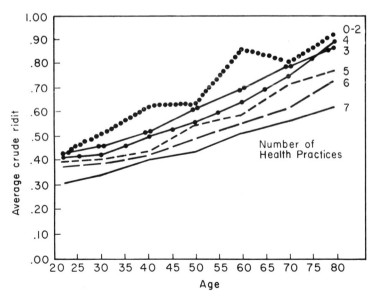

Fig. 5. Average physical health ridit by age group by number of health practices. From Belloc and Breslow (1972).

VI. RESEARCH NEEDS

A. Understanding the Situation

The research needs in this area are extensive and cover a variety of subdisciplines. As implied earlier, gerontology itself draws of many areas of expertise. Thus, any problem within it does likewise.

Probably the largest gap in knowledge is current information on the kinds and amounts of alcoholic beverages consumed by noninstitutionalized older persons. This information is critical because the vast majority of older people are not in institutions. Hence, studies conducted in nursing homes and other protected environments, while valuable, provide data based on very skewed samples. In addition, alcoholic beverage consumption tends to be under-reported and patterns of drinking undefined. The missing information is needed in order to develop valid models of the present situation. The magnitude of alcohol abuse needs to be quantified to indicate the priority its management should recieve in consideration of other pressing needs of older Americans and the limited resources available to fulfill those needs.

For planning purposes it is crucial that a profile of the drinking habits of future elders be developed. Cohort studies of attitudes and practices, a long and complex process, could reveal whether behaviors are likely to change as new cohorts reach age 60 and over. Behaviors presumably may change, but the direction is not clear. Aging persons born during the post-World War II baby boom, the children and adolescents of the 1960s, may not adopt or might even abandon alcohol as a supportive substance in their elder years.

B. Understanding the Causes

Prospective, longitudinal studies of the determinants of alcoholism also are indicated. Present elderly alcoholics need to be identified and characterized to provide a better understanding of which Americans are most vulnerable; cultural factors affecting the current elderly alcoholics should be reassessed. Much of the research of alcoholic beverage consumption among different ethnic groups was done well over a decade ago and may no longer be applicable.

It may be impossible at present to distinguish among the social, economic, and biological factors in the pathogenesis of alcohol abuses. However, qualitative description offers a useful starting point. Are the abusers former chronic drinkers who simply have grown old drinking? Are they responding to the effects of loneliness and fear of old age or to some life crisis, some triggering mechanism? Or are they ''executive drinkers,'' compensated alcoholics who do not heed and have not heeded symptoms and signs of adverse effects which suggest the cessation of drinking? The last-mentioned type may have been perfectly functional when their resources were not limited and may become identifiable only when the constraints of retirement strike. Throughout the world, many types of alcoholism have been described. The types characteristic of older Americans need identification and description.

It is necessary as well to place alcohol in perspective among other problems. Is there a correlation between alcohol abuse and the abuse of other drugs; what is the nature and scope of alcohol–drug interactions on the aged?

With regard to controlled, therapeutic use of alcohol, these questions take on greater significance. Quantifiable parameters should be developed, the use of which will yield data on the value or lack thereof of the use of alcoholic beverages in moderation among the aging and the aged.

In general, research should be directed toward investigating who is using alcohol, how much they are using, and why they are using it. Given this information, the next step would involve exploring constructive ways of using alcohol that would be positive, supportive, and, if the evidence indicates, conducive to better physical and mental health for all Americans.

VII. CONCLUSION

It is appropriate to quote the wisdom of former Supreme Court Justice Oliver Wendell Holmes who lived to be 93: "Some people, as they approach 70, begin to read the Bible and prepare to die; others prepare to live until they are 90. Now if the ones who prepare to live to 90 die at 70, they won't know the difference. But if the ones who prepare to die at 70 die at 90, the last 20 years would be hell" (Corcoran, 1976).

As implied by Holmes, much of man's difficulty in old age is self-imposed. If effective longevity is to be achieved, each person must prepare for a long, vigorous life and lead it accordingly. If alcoholic beverages are desired, and can be afforded without depriving the individual of any necessities, their use in moderation may add greatly to the joy of living and help assure that the last 20 years will not be hell.

ACKNOWLEDGMENT

The author is grateful for the assistance of Cynthia Craig Olds, University of Pennsylvania, class of 1978, in authenticating the references and in performing many other tasks associated with the preparation of this chapter for publication.

REFERENCES

Administration on Aging (1976a). "Older Americans Act of 1965, as Amended, and Related Acts." DHEW Publ. No. (OHD) 76-20170. U.S. Gov. Print. Off., Washington, D.C.

Administration on Aging (1976b). "National Nutritional Program for Older Americans." DHEW Publ. No. (OHD) 76-20230. U.S. Gov. Print. Off., Washington, D.C.

Bales, R. F. (1946). Cultural differences in the rates of alcoholism. *Q. J. Stud. Alcohol* **6,** 480–499.

Bales, R. F. (1962). Attitudes toward drinking in the Irish culture. *In* "Society, Culture and Drinking Patterns" (D. J. Pittman and C. R. Snyder, eds.), pp. 157–187. Wiley, New York.

Barnett, M. L. (1955). Alcoholism in the Cantonese of New York City: An anthropological study. *In* "Etiology of Chronic Alcoholism" (O. Diethelm, ed.), pp. 179–227. Thomas, Springfield, Illinois.

Becker, P. W., and Cesar, J. A. (1973). Use of beer in geriatric patient groups. *Psychol. Rep.* **33,** 182.

Belloc, N. B., and Breslow, L. (1972). Relationship of physical health status and health practices. *Prev. Med.* **1,** 409–421.

Black, A. L. (1969). Altering behavior of geriatric patients with beer. *Northwest Med.* **68,** Part II, No. 5, 453–561.

Blum, R. H., and Blum, E. M. (1964). Drinking practices and controls in rural Greece. *Br. J. Addict.* **60,** 93–108.

Bruun, K. (1967). Drinking patterns in the Scandinavian countries. *Br. J. Addict.* **62,** 257–266.

Chafetz, M. E. (1964). Consumption of alcohol in the Middle and Far East. *N. Engl. J. Med.* **271,** 297–301.

Chien, C. P. (1971). Psychiatric treatment for geriatric patients: Pub or drug. *Am. J. Psychiatry* **127,** 1070–1075.

Corcoran, T. G. (1976). The village elders. (As told to Ann Pincus.) *Washington Post* (Potomac Magazine), Aug. 15, p. 14.

DeLeon, J., chm. (1977). "Treatment and Rehabilitation Working Group. Interagency Committee on Federal Activities for Alcohol Abuse and Alcoholism." Public Health Serv., Rockville, Maryland. (Unpublished activities.)

Eagleton, T. F., and Hathaway, W. D., chm. (1976). "Subcommittees on Aging and on Alcoholism and Narcotics of the Committee on Labor and Public Welfare, United States Senate. Joint Hearings: Examination of the Problems of Alcohol and Drug abuse among the Elderly," Comm. Print No. 75-687 0. U.S. Gov. Print. Off., Washington, D.C.

Enslein, K., Rose, C. L., and Lisanti, V. (1967). Nonlinear multivariate techniques in aging research. *Gerontologist* **7**(3), Part II, p. 29. (Abstr.)

Glad, D. D. (1947). Attitudes and experiences of American-Jewish and American-Irish male youth as related to differences in adult rates of sobriety. *Q. J. Stud. Alcohol* **8,** 406–472.

Humphrey, H. H. (1977). Keynote address. *Proc. Natl. Conf. NASA Shelf-Stable Meal Syst., Washington, D.C.* Lyndon B. Johnson Sch. Public Aff., Austin, Texas.

Jones, K. L., Smith, D. W., and Hanson, J. W. (1976). The fetal alcohol syndrome: Clinical delineation. *Ann. N.Y. Acad. Sci.* **273,** 130–137.

Kastenbaum, R. (1972). Beer, wine and mutual gratification in the gerontopolis. *In* "Research, Planning and Action for the Elderly: The Power and Potential of Social Science" (D. P. Kent, R. Kastenbaum, and S. Sherwood, eds.), pp. 365–394. Behavioral Publ., New York.

Kastenbaum, R., and Slater, P. E. (1964). Effects of wine on the interpersonal behavor of geriatric patients: An exploratory study. *In* "New Thoughts on Old Age" (R. Kastenbaum, ed.), pp. 191–204. Springer, New York.

Lieber, C. S., Teschke, R., Hasumura, Y., and DeCarli, L. M. (1975). Differences in hepatic and metabolic changes after acute and chronic alcohol consumption. *Fed. Proc. Fed. Am. Soc. Exp. Biol.* **34,** 2060–2074.

Linn, B. S. (1968). Viewpoint: Elite physical resistance seen as key to longevity. *Geriatrics* **23,** 48, 55, 58.

Linn, M. W., Linn, B. S., and Gurel, L. (1967). Physical resistance in the aged. *Geriatrics* **22,** 134–138.

Linn, B. S., Linn, M. W., and Gurel, L. (1968). Cumulative and current illness rating scale. *J. Am. Geriat. Soc.* **16**, 622–626.

Linn, B. S., Linn, M. W., and Gurel, L. (1969). Physical resistance and longevity. *Gerontol. Clin.* **11**, 362–370.

Lolli, G. E., Serianni, E., Banissoni, F., Golder, G., Mariani, A., McCarthy, A., and Joner, M. (1952). The use of wine and other alcoholic beverages by a group of Italians and Americans of Italian extraction. *Q. J. Stud. Alcohol* **13**, 27–48.

Lolli, G. E., Serianni, E., Golder, G., and Luzzatto-Fegiz, P. (1958). "Alcohol in Italian Culture: Food and Wine in Relation to Sobriety among Italians and Italian-Americans," pp. 1–140. Free Press, Glencoe, Illinois.

McGovern, G., chm. (1977). "Diet Related to Killer Diseases with Press Reaction and Additional Information. Select Committee on Nutrition and Human Needs of the United States Senate," Comm. Print No. 79-283 O. U.S. Gov. Print. Off., Washington, D.C.

Metzelaar, L., comp. (1975). "A Collection of Cartoons: A Way of Examining Practices in a Treatment Setting." Inst. Gerontol., Univ. of Michigan, Ann Arbor.

Mishara, B. L., and Kastenbaum, R. (1974). Wine in the treatment of long-term geriatric patients in mental institutions. *J. Am. Geriatr. Soc.* **22**, 88–94.

Mishara, B. L., Kastenbaum, R., Baker, F., and Patterson, R. D. (1975). Alcohol effects in old age: An experimental investigation. *Soc. Sci. Med.* **9**, 535–547.

National Center for Health Statistics (1977). "Division of Vital Statistics, Data for 1975." Public Health Serv., Rockville, Maryland.

National Safety Council (1976). "Accident Facts," Publ. No. 021-56, p. 40. Natl. Saf. Counc., Chicago, Illinois.

Rose, C. L., and Bell, B. (1971). "Predicting Longevity: Methodology and Critique," pp. 101–121, 173–193. Heath Lexington Books, Lexington, Massachusetts.

Rose, C. L., Enslein, K., Nuttall, R. L., and Lisanti, U. (1967). Univariate and multivariate findings from a longevity study. *Gerontologist* **7**(3), Part II, p. 42. (Abstr.)

Rosett, H. L. (1976). Effects of maternal drinking on child development: An introductory view. *Ann. N.Y. Acad. Sci.* **273**, 115–117.

Sadoun, R., Lolli, G., and Silverman, M. (1965). "Drinking in French Culture." Rutgers Cent. Alcohol Stud., New Brunswick, New Jersey.

Sariola, S. (1954). "Drinking Patterns in Finnish Lapland." Finn. Found. Alcohol Stud., Helsinki.

Snyder, C. R. (1962). Culture and Jewish society: The ingroup–outgroup factor. *In* "Society, Culture, and Drinking Patterns" (D. J. Pittman and C. R. Snyder, eds.) pp. 188–229. Wiley, New York.

Streissguth, A. P. (1976). Psychological handicaps in children with the fetal alcohol syndrome. *Ann. N.Y. Acad. Sci.* **273**, 140–145.

Walford, R. (1976). Cited in Bylinsky, G. (1976). Science is on the trail of the fountain of youth. Chart by George Nicholson for Fortune Magazine. *Fortune* **94**(1), 134–139.

Watkin, D. M. (1977). The Nutrition Program for Older Americans: A successful application of current knowledge in nutrition and gerontology. *World Rev. Nutr. Diet.* **26**, 26–40.

West, K. (1979). Incorporating beverages containing alcohol into therapeutic diets: Some potentialities and problems. This volume, Chapter 17, pp. 257–262.

Wine Advisory Board (1975). "Uses of Wine in Medical Practice: A Summary for Distribution Only to the Medical Profession." Wine Advis. Board, San Francisco, California.

EDITORIAL COMMENT

Doctor Watkin clearly recognizes the risks of alcoholism and contrasts these with the benefits of moderate alcohol use to the elderly in nursing home and hospital circumstances where use of beer and wine appears to facilitate social interchange in groups. Lessening of use of tranquilizers and sedatives seems to be an additional clear benefit.

He also recognizes the practical difficulties of exploiting the potential of beer and wine for helping our elderly toward a more comfortable life in their later years. For planning purposes, Doctor Watkin calls for collection of data on alcohol use by the elderly in different geographic areas and of different ethnic origin. Such data will change in time since the patterns of alcohol use will likely be different in those who were children and adolescents before World War II and in those who were growing up after that time.

The studies showing beneficial effects of the use of beer and wine in institutionalized elderly persons suggest that older persons not in hospitals or nursing homes may be helped by the judicious use of wine or beer, particularly in aiding socialization and thus alleviating loneliness—and as a substitute for sedatives and tranquilizers.

16

MEDICATIONS, DRUGS, AND ALCOHOL

Frank A. Seixas

    I. Absorption Effects 246
   II. Congeners 246
  III. Metabolism 247
   IV. Microsomal Ethanol Oxidizing System 247
    V. Disulfiram (Antabuse) 248
   VI. Interaction of Alcohol with Other Psychoactive Drugs 249
  VII. Depressants 250
 VIII. Conclusions 251
       References 251
       Editorial Comment 255

The recent description of the fetal alcohol syndrome (1) and the recognition that obstetrical residents and physicans were not in the practice of inquiring about alcoholic beverage intake of pregnant women illustrate the scant attention paid to the common use of alcohol. The ten million alcoholics in the country receive greater attention than do the over 200 million other people in the United States, with a result that information is accumulating on various harmful effects of alcohol consumed in excess.

The effects of alcohol in amounts consumed by the nonalcoholic are important owing to the larger number of persons concerned. The author earlier reviewed one aspect of this subject, namely alcohol and its drug interactions (2). It is timely to examine changes that may influence the selection of drugs and the modifications of alcohol use which may be indicated during their administration. The better understanding of the effects of alcohol can lead us to its more intelligent use. For the small portion of the population that is alcohol-dependent or alcoholic, the different considerations that come into play should also be considered.

I. ABSORPTION EFFECTS

Delayed absorption of ethanol following the administration of epinephrine, amphetamine, atropine, and related drugs has been noted (3). Ethanol enhances the degradation of penicillin with the resultant effect that ingestion of considerable amounts may decrease the quantity of orally taken penicillin available for absorption (4). The influence of alcohol on penicillin and on the absorption of kanamycin are based on an opinion given by a consultant in the question and answer section of the *Journal of the American Medical Association*. This is a matter that requires substantiation by investigation.

Carbonated beverages are said to relax the pyloric sphincter, thus allowing carbonated alcoholic beverages to pass more rapidly into the duodenum. Since absorption of alcohol is more efficient from the duodenum than the stomach, this effect may result in an earlier rise in blood alcohol levels. Food has the opposite effect, i.e., it retards the passage of alcohol from the stomach.

The influences of alcoholism on absorption and the commonly observed associated malnutrition are detailed by Mezey and Halsted in Chapter 19.

Ingestion of aspirin with ethanol combines the irritant effects of the two agents on the mucous membrane, along with clotting difficulties involving factor 7 and proconvertin; these may precipitate gastrointestinal bleeding of serious import (5). Due to its solvent action, ethanol may serve as a vehicle for poorly absorbed substances, including carcinogens, (6). It enhances the absorption of nitroglycerine (7) and of some industrial toxins, such as mercury vapor and lead (8).

II. CONGENERS

The congeners in alcoholic beverages include several longer chain or higher alcohols, differing from beverage to beverage [Appendix I]. Richardson's law states that the intensity of the pharmacologic action of straight chain primary alcohols is directly proportional to the number of carbon atoms in the chain. Thus, one might expect a somewhat higher degree of intoxication with a congener-rich beverage. The most usual effect ascribed to these beverages, however, is an increase in intensity of hangover (not, however, of withdrawal symptoms (9). The presence of free tyramine in some chianti wine can lead to acute hypertensive crises when these wines are drunk by an individual taking monoamine oxidase inhibitors. Tyramine, the metabolism of which is blocked, causes a release of norepinephrin in this circumstance. According to Tacker *et al.* (10), ethanol itself may cause tyramine release.

III. METABOLISM

Alcohol is oxidized in the liver by alcohol dehydrogenase (ADH), a zinc-containing enzyme. Nicotinamide adenine dinucleotide (11) serves as the acceptor of hydrogen produced by this oxidation. ADH is widely thought to have a kinetics of zero order, although some consider that the limiting factor in its reaction rate is the availability of NAD (12), which also is the cofactor for the subsequent conversion of acetaldehyde to acetic acid. These metabolic reactions are discussed by Olson in Chapter 13. ADH is involved in the metabolism of other alcohols, including methanol and ethylene glycol. Methanol (13–14a) is oxidized by ADH to formic acid and formaldehyde, which produce an intense acidosis and optic nerve injury. Large doses of ethanol will saturate the ADH and can prevent the oxidation of methanol to its toxic by-product and be lifesaving. Likewise, large quantities of ethanol can prevent ethylene glycol (15) from being changed to oxalic acid, the crystals of which precipitate in the kidney and produce renal failure.

Ethacrynic acid (16) and phenylbutazone (17) both inhibit ADH. Thus, they "make a little alcohol go a long way" by maintaining persistent blood alcohol levels.

Other metabolic effects of alcohol can change drug reactions and alter the action of drugs. Digitalis sensitivity and arrythmias (18) can be induced by the decrease in potassium and magnesium ion produced by alcohol (19); phosphate concentration may also be decreased by alcohol in conjunction with the use of aluminum-containing antacids with even fatal results (20). Uric acid levels can increase, suggesting gout, but requiring alcohol deprivation rather than allopurinol for relief (21). Puromycin C can be reinforced by alcohol in producing fatty liver (22). Alcoholic hypoglycemia [pp. 260, 446] is well known. Alcohol can interfere with hepatic gluconeogenesis and in conjunction with fasting cause the syndrome of alcoholic hypoglycemia, or it may favor diabetic ketoacidosis or intensify insulin-induced hypoglycemia (23).

IV. MICROSOMAL ETHANOL OXIDIZING SYSTEM

Dr. Charles Lieber's discovery of microsomal ethanol oxidizing system (24) (MEOS) has been vigorously challenged from his first paper. In this regard it is interesting to observe that one of its chief challengers, Dr. Ronald Thurman, said "I retract my previous contention that catalase can explain the entire increase in ethanol metabolism at high levels of consumption" (25). That there is a higher rate of metabolism at higher level of consumption is undisputed, although at least traditionally ADH with its zero order of kinetics would not be responsible for it.

The concept of MEOS is attractive when one is considering drug interactions of alcohol since so much is explained by it. For instance, four medically important drugs, warfarin, dilantin, tolbutamide (26), and isoniazide (27), which are metabolized by the MEOS, are much more speedily metabolized in dry alcoholic subjects. This is readily explainable by postulating that MEOS has been induced by alcohol. The MEOS metabolic pathway is significant in determining doses of the four drugs for dry alcoholic patients.

Similarly, synergism exists when the effect of two drugs in combination is greater than either alone. Alcohol and barbiturates show such synergism. In humans, a blood level of alcohol of 100 mg/100 ml, combined with a barbiturate level of only 0.5 mg/100 ml, has proved fatal (27). This compares to lethal levels of alcohol alone of 500–800 mg% (28) and of phenobarbital alone of 10–29 mg/100 ml (29). Since alcohol and phenobarbital compete for the same MEOS system, the rates of degradation of both are slowed, and thus toxicity is enhanced. They are also synergistic or additive in that both contribute to respiratory depression. In contrast, a person who has developed tolerance to alcohol requires larger than usual doses of barbiturate or other anesthetics when alcohol-free to produce surgical anesthesia.

The microsomal ethanol oxidizing system (30–33) (mixed function oxidizing system) appears to explain this phenomenon. After it is induced and if none of the competitive ethanol is present, barbiturates and other anesthetics are metabolized more readily. If ethanol is present and interfering with the available MEOS, small quantities of the sedative would have a greater effect because it is not broken down by the already engaged MEOS. The observation that chronically administered ethanol enhances the toxicity of carbon tetrachloride (CCl_4) is compatable with the thesis that MEOS is involved in the production of toxic metabolites of CCl_4 and that its capacity to produce them has been enhanced by the long-term alcohol consumption (34).

V. DISULFIRAM (ANTABUSE)

The chance discovery of Hald and Jacobsen in 1948 that alcohol taken after premedication with disulfiram (Antabuse) produced a brisk reaction with flushing, tachycardia, dyspnea, and (after high doses) shock, has been used to advantage to discourage drinking by alcoholics (35). The disulfiram–alcohol reaction is associated with accumulation of acetaldehyde due to interference with the action of acetaldehyde dehydrogenase.

Disulfiram has been used for treating alcoholics successfully by many practitioners (35a–h). However, an occasional toxic side effect has been noted. Particularly distressing is the induction of psychotic or depressive syndromes, thought to be related to the property of disulfiram of inhibiting dopamine β-hydroxylase in the brain (36). Ewing suggests that it may be possible to test for persons likely to react in this manner by measuring dopamine β-hydroxylase. His communication, based on a small sample, reports that persons with low levels of dopamine β-hydroxylase experienced more depression and other central nervous system signs when taking Antabuse than did those with normal levels (37).

A reaction similar to the disulfiram–alcohol reaction (38) can be induced by calcium carbamide (Temposil). This effect lasts only 12–24 hours, whereas that of Antabuse persists for 4–5 days. Calcium carbamide is not approved by the Food and Drug Administration for use in the United States.

Many other drugs have been reported to have an Antabuse-like effect when alcohol is taken. These include metronidazole (39), tolbutamide (40–43) and other antidiabetic drugs, phentolamine (44), chloramphenicol, furazolidine (45), griseofulvin (46), and quinacrine (47). Kalant and associates (48) have expressed doubt that the reaction occurs with metronidazole. Patients, whether alcoholic or not, taking any of these medications should be warned of the possibility of what may be termed the "acetaldehyde reaction" after an alcoholic beverage or alcohol-containing medication.

VI. INTERACTION OF ALCOHOL WITH OTHER PSYCHOACTIVE DRUGS

Alcohol is primarily a depressant of the central nervous system. No amethystic* agent (49) has been found to reverse this depression rapidly in the sense that nalorphine is a specific receptor blocker and antidote for morphine poisoning. However, proponents of the condensation product theory postulate that acetaldehyde condenses with biogenic amines to form morphine-like substances. They have observed that morphine and/or naloxone can block alcohol withdrawal (50) symptoms in mice. Other researchers consider that there is no specific receptor site for alcohol in the brain and that the general cell membrane, neurotransmitter, or calcium effects will account for alcohol's depressant action (51). Myers and Melchoir have challenged this concept by successfully inducing alcohol-seeking behavior in rats by the repetitious delivery of tetrahydropapaveroline in minute amounts to the ventricles in the brain (52). As corroborative new evidence of the action of alcohol on the brain grows, new chemical interrelationships may be found which will be of great value in dealing with alcoholism.

Of the other psychoactive drugs, general analeptics such as picrotoxin and pentelenetrazol (metrazol) are contraindicated for alcohol overdose because of their tendency to produce convulsions. Their stimulant action will only add to the

*Amethystic, an agent which counters drunkenness.

psychomotor agitation of the withdrawal syndrome that follows cessation of alcohol intake (53).

Milder stimulants, such as amphetamine, methylphenidate (Ritalin), caffein, and nicotine, exhibit a more limited degree of pharmacological antagonism to the depressant effects of alcohol, but synergistically increase deterioration of performance with smaller doses of ethanol (54).

A serious overdose of alcohol requires general metabolic support and, if sufficient, even the use of dialysis. To speed the decrease of blood alcohol level, the intravenous introduction of large amounts of fructose (55) has been advocated. The acceleration of alcohol disappearance has been attributed to the resultant reoxidation of NADH. Fructose administration is not without danger, however, owing to the depletion of adenosine triphosphate in the liver (56–59).

VII. DEPRESSANTS

The synergistic actions of alcohol and barbiturates have been discussed. All other sedative hypnotic drugs have either synergistic or additive effects, including bromides, paraldehyde, chloral hydrate, the so-called minor tranquilizers, reserpine, meprobamate, glutethimide (Doriden), methylquaalude, and propoxyphene (Darvon). Alcohol and the sedative drugs exhibit cross-dependence, They not only have similar withdrawal syndromes (differentiated by a longer time of onset) but the administration of one will alleviate the symptoms of withdrawal of the other. This interaction is used to advantage in the treatment of the alcohol withdrawal syndrome (60). Adequate doses are given to counteract the tremor, convulsions, hallucinations, or delerium, the dose level being slowly decreased until the drug can be dispensed with. The unmonitored prescription of large quantities of sedative drugs to patients with alcohol dependency is contraindicated because of the tendency of such patients to relapse into drinking, the development of tolerance to the other sedatives, and the ineffectiveness of this measure in dealing with the underlying problem of alcoholism. There is, as well, a danger of a synergistic lethal reaction (61).

Of the sedatives, the benzodiazipene class is the most used in detoxification because of the large margin of safety between the effective dose and the lethal dose. However, one should not assume that chlordiazepoxide is suicide-free, as the report of Rada *et al.* (62) shows.

Phenothiazines (63) have a different mode of action than the sedatives and act on different sites in the brain. They are associated with greater morbidity and mortality in withdrawal syndrome treatment than are sedatives. However, they are indicated for alcoholics with schizophrenia (64).

In patients taking alcohol, tricyclic antidepressants potentiate sedation, producing central nervous system depression and hypothermic coma. At lower doses tricyclic antidepressants and alcohol taken together impair driving skills (65,66). Antihistamines with alcohol have also been shown to impair driving performance (72,73). The implication of these interactions for highway accidents is evident.

Lithium has been effective in treating the depression of alcoholics according to Kline (67) and Reynolds *et al.* (68). Sellers (69,70) found lithium and propanalol effective ancillary medications in the alcohol withdrawal syndrome, primarily in the control of tremor. Critics of the use of propanalol in this situation predict morbid cardiac effects which would be of greater significance than tremor control.

The role of anticonvulsant medication in preventing ''rum fits'' in alcohol withdrawal has been a subject of debate. In the first 48 hours of treatment when rum fits usually occur, dilantin, in the usual doses, has limited bioavailability. A large series of cases shows the combined use of dilantin and chlordiazepoxide to reduce seizures (71). The additive effect of alcohol and opiates in the central nervous system can be fatal. However, the withdrawal syndrome of people addicted to the two substances is completely different (74).

VIII. CONCLUSIONS

I have reviewed here examples of the interactions of ethanol with a wide variety of drugs. These and other recognized interactions make it imperative that the physician know the alcohol intake habits of his patients before prescribing and be aware of the reactions that could occur if alcohol is taken (75). It is important that the physician be adept in ways to circumvent denial by his patients of use of alcohol, as this common characteristic of alcoholics can readily be a source of diagnostic error or of unwise therapeutic advice.

The use of alcohol will continue by a large number of our patients. The physician must deal with it in a scientific, dispassionate manner in order to achieve good patient care.

REFERENCES

1. Streissguth, A. P., Jones, K. L., Smith, D. W., and Ulleland, C. N. Pattern of malformation in offspring of chronic alcoholic mothers. *Lancet* **i,** 1267–1270 (1973).
2. Seixas, F. A. Alcohol and its drug interactions. *Ann. Intern. Med.* **83,** 86–92. (1975).
3. Wallgren, H., and Barry, H. ''Actions of Alcohol, Chronic and Clinical Aspects,'' Vol. 2, p. 656. Am. Elsevier, New York. 1970.
4. Kitto, W. Antibiotics and ingestion of alcohol. *J. Am. Med. Assoc.* **193,** 411 (1965).
5. Kissen, B., and Begleiter, H. ''The Biology of Alcoholism,'' Vol. 3, p. 144. Plenum, New York, 1974.
6. Seixas, R. Alcohol, a carcinogen? *Ca* **25,** 62–65 (1975).
7. See reference 5, p. 139.
8. Fridman, V. Combined effect of alcohol and other toxic substances: Lead, mercury and others. *Alkoholizan (Belgrade)* **8,** 51–55 (1968).
9. Keller M., ed., Drug interactions with congeners, *Q. J. Stud. Alcohol* **5,** Suppl. (1970).

10. Tacker, M., Creavin, P. J., and McIsaac, W. M. Alterations in tyramine metabolism by ethanol. *Biochem. Pharmacol.* **19**, 604–607 (1970).

11. Goodman, L. S., and Gilamn, A. eds., "The Pharmacological Basis of Therapeutics," 3rd ed. Macmillan, New York, 1965.

12. Thurman, R. G., and Bunzel, H. S. Alcohol dehydrogenase in microsomal ethanol oxidation. In *Alcoholism, Clin. Exp. Res.* **1**(1), 33–38 (1977).

13. Mannering, G. J., Van Hocken, R., and Makar, A. B. Role of the intracellular distribution of hepatic catalase in the perioxidative oxidation of methanol. *Ann. N.Y. Acad. Sci.* **168**, 365 (1969).

14. Caldwell, J., and Sever, P. S. The biochemical pharmacology of abused drugs, II. Alcohol and barbiturates. *Clin. Pharmacol. Ther.* **16**, 737–749 (1974).

14a. Morgan, R., and Cagan, E. J. Methyl alcohol intoxication in B Kissin and H Begleiter. *Biol. Alcoholism* **3**, 176–189 (1970).

15. Blair, A. H., and Vallee, B. L. Some catalytic properties of human liver alcohol dehydrogenase. *Biochemistry* **5**, 2026–2034 (1966).

16. Dixon, R. L., and Rall, D. P. Enhancement of ethanol toxicity by ethacrynic acid. *Proc. Soc. Exp. Biol. Med.* **118**, 970–973 (1965).

17. Handel, K. The present practice of driving laws. *In* "Handbuch der Verkehrsmedizin" (K. Wagner, ed.), pp. 72–88. Springer-Verlag, Berlin and New York, 1968.

18. See reference 5, p. 145.

19. See reference 3, p. 170.

20. Becker, C. L., Roe, R. L., and Scott, R. A. "Curriculum on Pharmacology, Neurology and Toxicology." Medcom Press, New York, 1974.

21. Lieber, C. S., Jones, D. P., Losowsky, M. S., and Davidson, C. S., Interrelationship of uric acid and ethanol. *J. Clin. Invest.* **41**, 1863–1870 (1962).

22. Ammon, H. P. T. Alcohol and puromycin C—two mutually reinforcing factors in the production of fatty liver. *Z. Gesamte Exp. Med.* **152**, 56–61 (1970).

23. Arky, R. A. Veverbrants, E., and Abramson, E. A., Irreversible hypoglycemia; a complication of alcohol and insulin. *J. Am. Med. Assoc.* **206**, 575–578 (1968).

24. Teschke, R., Hasumura, Y., and Lieber, C. S. Hepatic microsomal ethanol oxidizing system, solubilization, isolation and characterization. *Arch. Biochem. Biophys.* **163**, 404–416 (1974).

25. Seminar discussion. *Alcoholism, Clin. Exp. Res.* **1**(1), 50 (1977).

26. Kater, R. M. H., Zieve, P., Tobon, F., Roggin, G., and Iber, F. L. Accelerated metabolism of drugs in alcoholics. *Gastroenterology* **56**, 412 (1969).

27. Gupta, R. C., and Kofold, J. Toxicological statistics for barbiturates, other sedatives and tranquillizers in Ontario. A 10-year survey. *Can. Med. Assoc. J.* **94**, 863–865 (1966).

28. See reference 11.

29. Broughton, P. M., Higgins, G., and O'Brien, J. R. P. Acute barbiturate poisoning. *Lancet* No. 270, 180–184 (1956).

30. Lieber, C. S. Liver adaptation to injury in alcoholism. *N. Engl. J. Med.* **288**, 356–362 (1973).

31. Lieber, C. S., and DiCarli, L. M. Significance and characteristics of hepatic microsomal oxidation in the liver. *Drug Metab. Dispos.* **1**, 428–440 (1900).

32. Misra, P. S., Lefebre, A., and Ishii, H. Increase of ethanol meprobamate and pentobarbital metabolism after chronic ethanol administration in man and in rats. *Am. J. Med.* **51**, 346–351 (1971).

33. Teschke, R., Hasumura, Y., and Lieber, C. S. NADPH dependent oxidation by methanol, ethanol, propanolol and butanol by hepatic microsomes. *Biochem. Biophys. Res. Commun.* **60**, 851–857 (1974).

34. Teschke, R., Hasumura, Y., and Lieber, C. S. Increased carbon tetrachloride nepatotoxicity and its mechanism after chronic ethanol consumption. *Gastroenterology* **66**, 415–422 (1974).

35. Hald, S., and Jacobsen, E. The formation of acetaldehyde in the organism after ingestion of antabuse (tetraethylthiruram disulfide) and alcohol. *Acta Pharmacol. Toxicol. (Copenhagen)* **4,** 305–310 (1948).

35a. Fox, R. Antabuse as an adjunct to psychotherapy in alcoholism. *N.Y. State J. Med.* **58,** 9 (1958).

35b. Talbott, G. D., and Gander, O. Antabuse, 1973. *Md. State Med. J.* **22,** 60–63 (1973).

35c. Ewing, J. "How to Help the Chronic Alcoholic." Cent. Alcohol Stud., Univ. of North Carolina, Chapel Hill, 1974.

35d. Bourne, P., Alford, J. A., and Bowcock, J. Z. Treatment of skid row alcoholics with disulfiram. *Q. J. Stud. Alcohol* **27,** 42, (1966).

35e. Baekeland, F., Lundwall, L., Kissin, B., Shanahan, T. Correlates of outcome in disulfiram treatment of alcoholism. *J. Nerv. Ment. Dis.* **153,** 1–9 (1971).

35f. Hayman, M. Treatment of alcoholism in private practice with a disulfiram oriented program. *Q. J. Stud. Alcohol* **26,** 460 (1965).

35g. Kimmel, M. E. Antabuse in a clinic program. *Am. J. Nurs.* **7,** 1173 (1971).

35h. Wallerstein, R. S. "Hospital Treatment of Alcoholism," Menninger Clinic Monogr. Ser., No. 11. Basic Books, New York, 1956.

36. Goldstein, M. Anagnoste, B., and Lauber, E. Inhibition of dopamine beta hydroxylase by disulfiram. *Life Sci.* **3,** 763 (1964).

37. Ewing, J. A., Rouse, B. A., Mueller, R. A., and Silver, D. Can dopamine beta hydroxylase levels predict adverse reactions to disulfuram. A research note. *Alcoholism, Clin. Exp. Res.* **2**(1), 93–94 (1978).

38. Levy, M. S., Livingstone, B. I., and Collins, D. M. A clinical comparison of disulfiram and calcium carbimide. *Am. J. Psychiatry* **123,** 1018–1022 (1967).

39. Perman, E. S. Intolerance to alcohol. *N. Engl. J. Med.* **263,** 114 (1965).

40. See reference 39.

41. Krantz, J. C., Jr. The problem of modern drug incompatibilities. *Am. J. Pharm.* **139,** 115–121 (1967).

42. Truitt, E. B., Jr., Duritz, G., Morgan, A. M., and Prouty, B. W. Disulfiram-like actions produced by hypoglycemic sulfonurea compounds. *Q. J. Stud. Alcohol* **23,** 197–207 (1962).

43. Signorelli, S. Tolerance in patients on chlorpropamide. *Ann. N.Y. Acad. Sci.* **74,** 900–903 (1959).

44. Agarnoff, D. L., and Horowitz, A. Drug interactions. *Pharmacol. Phys.* **4,** 1–7 (1970).

45. See reference 39.

46. See reference 20.

47. Kalant, H., LeBlanc, A. E., Guttman, M., and Khanna, J. M. Metabolic and pharmacologic interaction of ethanol and metronidazole in the rat. *Can. J. Physiol. Pharmacol.* **50,** 476–484 (1972).

48. See reference 41.

49. Noble, E. P., Alkana, R. L., and Parker, F. S. Ethanol-induced CNS depression and its reversal: A review. *Biomed. Res. Alcohol Abuse Probl., Conf. Proc.; Nonmed. Use Drugs Directorate, Natl. Health Welfare, Halifax, Can.* pp. 308–367 (1974).

50. Blum, K., Hamilton, M. G., and Wallace, J. E. "Alcohol and Opiates: Neurochemical and Behavioral Mechanisms." Academic Press, New York, 1977.

51. Jaffee, J. H. Narcotic analgesics. *In* "The Pharmacological Basis of Therapeutics" (L. S. Goodman and A. Gilman, eds.) 3rd ed., pp. 247–284. Macmillan, New York, 1965.

52. Myers, R. D., and Melchior, C. L. Alcohol drinking: Abnormal intake caused by tetrahydro-papapaverline in brain. *Science* **196,** 554–556 (1977).

53. See reference 5, p. 136.

54. See reference 5, p. 136.

55. Pawan, G. L. S. The effect of vitamin supplements and various sugars on the rate of alcohol metabolism in men. *Biochem. J.* **107,** 25 (1968).
56. Woods, H. F., and Alberti, K. G. Dangers of intravenous fructose. *Lancet* **ii,** 1354–1357 (1972).
57. Yu, D. T., Burch, H. B., and Phillips, M. J. Pathogenesis of fructose hepatotoxicity. *Lab. Invest.* **30,** 1, 85 (1974).
58. Bode, J. C., Zelder, O., and Rumpelt, H. J. Depletion of liver adenosine phosphate and metabolic effects of I.V. infusion of fructose and sorbitol in man and in the rate. *Eur. J. Clin. Invest.* **3,** 436–441 (1974).
59. See reference 58.
60. Seixas, F. A., ed., "Treatment of the Alcohol Withdrawal Syndrome." Natl. Counc. Alcoholism, New York, 1971.
61. What do we do about long-term sedatives? *Ann. N.Y. Acad. Sci.* **252,** 378–399 (1975).
62. Rada, R. T., Kellner, R., and Buchanan, J. G. Chlordiazepoxide and alcohol: A fatal overdose. *J. Forensic Sci.* **20**(3), 00.
63. Jarvik, M. Drugs used in the treatment of psychiatric disorders. *In* "The Pharmacological Basis of Therapeutics" (L. S. Goodman and A. Gilman, eds.), 3rd Ed. Macmillan, New York, 1965.
64. Kaim, S. C., Klett, C. J., Rothfield, B., *et al.* "Treatment of the Acute Alcohol Withdrawal State: A Comparison of Four Drugs in Treatment of the Alcohol Withdrawal Syndrome." Natl. Counc. Alcoholism, New York, 1971.
65. Milner, G. Cumulative lethal dose of alcohol in mice given amitriptyline. *J. Pharm. Sci.* **57,** 2005–2006 (1968).
66. Theobald, W., and Stenger, E. G. Reciprocal potentiation between alcohol and psychotropic drugs. *Arzneim.-Forsch.* **12,** 531–533 (1962).
67. Kline, N. S. Lithium therapy appears to have significant beneficial effects on chronic alcoholism. *Physician's Alcohol Newsl.* **8,** 1 (1970).
68. Reynolds, C. M., Merry, V., and Coppen, A. Prophylactic treatment of alcoholism by lithium carbonate, an initial report. *Alcoholism, Clin. Exp. Res.* **1**(2), 00 (1977).
69. Sellers, E. M., and Cooper, S. D. Lithium treatment of alcohol withdrawal. *Clin. Pharmacol. Ther.* **15,** 218–219 (1974).
70. Zilm, D. H., and Sellers, E. M. Propranolol effect on tremor in alcoholic withdrawal. *Ann. Intern. Med.* in press (1979).
71. Sampliner, R., and Iber, F. L. Diphenylhydantoin control of alcohol withdrawal seizures. *J. Am. Med. Assoc.* **230,** 1430–1432 (1974).
72. Antihistamines and alcohol. *Br. Med. J.* **i,** 403 (1954).
73. McIver, A. K. Drug incompatabilities. *Pharm. J.* **195,** 609–612 (1954).
74. Brecher, E. M. "Licit and Illicit Drugs," pp. 65, 251. Little, Brown, Boston, Massachusetts, 1972.
75. Hathcock, J., and Coon, J. M., "Nutrition and Drug Interrelations." Academic Press, New York, 1978.

EDITORIAL COMMENT

The interaction between alcohol and various drugs is part of the broad subject of "Nutrition and Drug Interrelations" (1). The peculiar "nutrient-drug characteristic" of alcohol appears to multiply the number and biological types of interactions, the list of which becomes long and therefore difficult to remember. Some order is imparted to the seemingly complex variety of interactions by an understanding of the MEOS (microsomal ethanol oxidizing system), the increasing activity of which is induced by chronic use of alcohol. The MEOS may be the mechanism for the metabolism of a number of drugs. Large intakes of alcohol may competitively decrease the disposal of these drugs, whereas degradation of these drugs may be accelerated in the habitual alcohol user who is "dry" at the moment.

The ingestion of alcohol after taking disulfiram (Antabuse) or some of the other drugs mentioned by Doctor Seixas results in symptoms of flushing, nausea, vomiting, chest pain, hypotension, and even convulsions. This reaction is the basis for one mode of treatment of alcoholism. There are two proposed mechanisms by which these symptoms occur: (i) an inhibition of aldehyde dehydrogenase, the enzyme which catalyzes the second step in the oxidation of alcohol, the transformation of acetaldehyde to acetic acid or acetate. The symptoms are attributed to the accumulation of acetaldehyde. Aldehyde dehydrogenase also catalyzes the conversion of 5-hydroxyindoleacetaldehyde to 5-hydroxyindolacetic acid and is competitively inhibited by accumulation of acetaldehyde. (ii) Liberation of increased amounts of serotonin with a block in its metabolism. It appears that in the case of two drugs used for treatment of diabetes, chlorpropamide and tolbutamide, aldehyde dehydrogenase is inhibited and thus the metabolism of serotonin to 5-hydroxytryptophol is favored over the production of 5-hydroxyindolacetic acid (2). The metabolism of alcohol tends to decrease the DPN over DPNH ratio thus also favoring production of 5-hydroxytryptophol. The precise mechanisms for the symptoms of the Antabuse syndrome are not known with certainty but appear to involve both acetaldehyde and altered serotonin metabolism. The Antabuse phenomenon can be frightening. The restaurant diner who becomes acutely ill during his meal could be a victim of an interaction between a drug such as chlorpropamide and his before-dinner drink, or he could be experiencing the so-called Chinese Restaurant Syndrome from excess monosodium glutamate or perhaps some form of food allergy.

Ingestion of alcohol will cause symptoms similar to the Antabuse phenomenon in persons having a carcinoid tumor, apparently through the release of serotonin products. Still another interaction between alcohol and malignant tumors is the occurrence of pain and perhaps swelling of tissues involved with lymphoma (Hodgkin's type, in particular) after the taking of alcohol. This occurs in only a small proportion of instances of Hodgkin's disease, but when the person comments to his physician that "shortly after I had this drink, I felt a pain in the side of my neck and then I discovered a lump there," the diagnosis of a lymphoma is probable. The mechanism for this reaction is not known.

The potential for interaction between drugs prescribed in therapy and socially acceptable quantities of alcohol should be kept in mind by the prescribing physician, and the patient should be cautioned concerning the limitations that may be in effect when he is taking antihistamines, sedatives, tranquilizers, or the tricyclic antidepressants. He should know that even small amounts of alcohol may make driving an automobile unsafe when combined with these drugs. A physician should describe to the patient the nature of the undesirable interactions.

A physician likewise should bear in mind the quantities of alcohol "hidden" in liquid medications that he may prescribe and also be aware of the amounts contained in widely advertised, over-the-counter "remedies" for colds and coughs, "tonics," some liquid vitamin–iron preparations, and the like. Certain of these contain 10–25% alcohol. Elixirs for respiratory relief may contain 15–20% alcohol; some of the antihistamine–decongestants, 15% or so; and terpin hydrate elixir, 42%.

REFERENCES

1. Hathcock, J., and Coon, J. M. "Nutrition and Drug Interrelations." Academic Press, New York, 1978.
2. Podginy, H., and Bressler, R. Biochemical basis of the sulfonylurea-induced antabuse syndrome. *Diabetes* **17,** 679 (1968).

17

INCORPORATING ALCOHOLIC BEVERAGES INTO THERAPEUTIC DIETS: SOME POTENTIALITIES AND PROBLEMS

K. M. West

I. Some Problems 257
    A. Inadequate Dietary Histories 257
    B. Ignorance about the Metabolic Effects of Alcohol 258
    C. Sensitivity to the Combination of Alcohol and Sulfonylurea 260
II. Should Alcohol Be Allowed in the Diets of Fat People? 260
III. Does Alcohol Stimulate Weight Gain in the Undernourished? 261
    References 262

Other authors in this volume have covered very well the present state of knowledge on the effects of alcohol consumption on several of the major disease states. I will only give attention to some practical clinical problems that arise when therapeutic diets are necessary in patients who enjoy drinking. Since diabetes is my major interest I will draw mainly on experience in this field. However, I do believe that some of this experience has more general relevance.

Among the better general references on alcoholic beverages in clinical medicine is the book of Leake and Silverman (1966). Wine and health are discussed in the book edited by Lucia (1963). Hollands (1976) has published a good review on alcohol in the diet therapy of diabetes. Alcohol and carbohydrate metabolism have been reviewed by Marks (1978).

I. SOME PROBLEMS

A. Inadequate Dietary Histories

Certain kinds of therapeutic diets may be conceived and implemented quite effectively without a detailed review of the patient's dietary history. On the other hand, there are certain types of diets that are not likely to be implemented effectively without a background of information on the patient's life style and his dietary preferences and propensities. There is now a considerable body of evidence, for example, that only a very small percentage of diabetics understand and follow appropriate diets (West, 1973). It is also clear that a major reason for this is the frequency with which the dietary prescriptions are designed without taking into account the dietary habits and preferences of the patient and those with whom he lives. I invite you to select 20 diabetics at random and ask them to tell you exactly what they have been told about the use of alcohol. One of the discouraging results of such surveys is the evidence that persons providing the written or verbal dietary instruction usually fail to ask in detail what the patient likes to drink; and when, where, why, and with whom.

A special aspect of this particular problem is that physicians and dietitians often fail to recognize that the patient is alcoholic. It is difficult to develop and implement a therapeutic diet of any kind in an alcoholic patient. But some dietary objectives can be achieved despite alcoholism provided that the dietary strategies are conceived with full knowledge that the patient is alcoholic.

In nonalcoholic patients who like to drink it is often desirable to regulate or modify patterns of consumption of alcohol, but often alcohol can be included in the dietary prescription. When a diabetic patient takes an occasional drink, the effects of blood glucose levels are usually trivial. On the other hand, when patients drink frequently in moderate amounts, it is desirable that this consumption be a specified part of the prescribed diet and not a separate unregulated and unknown addition to the diet prescription. Thus, a careful dietary history is crucial to the development of a dietary prescription that will really work.

Incidentally, there is another indication of the strong disinclination to include in the dietary history the specifics of alcohol consumption. An examination of dietary survey data reported in the scientific literature reveals that intake of calories from carbohydrate, fat, and protein are invariably recorded with care, while data on caloric intake from alcohol are usually absent. Quite frequently no explanation is provided. I was embarrassed to find that some of my own reports contained this common defect.

B. Ignorance about the Metabolic Effects of Alcohol

This problem has two aspects. There are certain aspects not known to medical science. In diabetes, for example, much has been learned about the acute effects of alcohol and its metabolites. Knowledge is quite incomplete on the indirect effects of alcohol over longer periods, however. This is, perhaps, best illustrated by reviewing the history of our experience with dietary fat. It appeared that this nutrient would be ideal for diabetes. It did not seem to raise blood glucose at all. We have learned subsequently, however, that many of the indirect effects of fat are profoundly deleterious. Insulin requirements are, in fact, at least as high on high-fat, low-carbohydrate diets; the fat sometimes induces atherosclerosis; under certain circumstances ketogenesis is enhanced. Single dosages of moderate amounts of alcohol produce little effect on the blood glucose, but I am not aware of any data on insulin requirement or blood glucose levels when moderate amounts of alcohol are substituted isocalorically for carbohydrate or fat over a period of weeks.

In some patients alcohol induces potentially deleterious elevations of blood lipids. These might be particularly disadvantageous in diabetics. One of the main difficulties here is that we do not know whether triglyceridemia itself is an independent risk factor for atherosclerosis in either diabetics or nondiabetics. Evidence is impressive, however, that risk of atherosclerosis is related to serum cholesterol levels. Particularly in diabetics, one should be sure that the patient in whom moderate alcohol is allowed is not one of those whose serum cholesterol levels are sensitive to alcohol.

The second aspect of this problem is ignorance of physicians, dietitians, and patients concerning the nutrients in alcoholic beverages and their effects. Only a very small portion of physicians and a minority of dietitians know how much carbohydrate is in the popular brands of beer, or which wines contain only trivial amounts of carbohydrate. Even more importantly, few keep this information readily available in order to provide this information to their patients. Among the several sources of information on alcoholic beverages and common mixes are handy tables in the book of Leake and Silverman (1966), the "Manual of Applied Nutrition" (1973), and in the "Mayo Clinic Diet Manual" (1971). Many other diet manuals contain this information, but many do not.

A common procedure has been to advise diabetic patients to exchange the carbohydrate calories in beers or wines for the "bread exchange" units in standard diets, while substituting alcohol calories for fat exchanges. We need to learn more about the degree to which the effects of moderate amounts of alcohol are really equivalent to equicaloric amounts of fat. At least on the surface this would appear to be an appropriate exchange. When taken in modest amounts, neither nutrient produces much immediate effect on blood glucose levels. On the other hand, the effects of fat are quite variable depending on the circumstances of ingestion. Under some conditions the presence of fat in a meal tends to delay the rate at which the stomach empties. Less well known is the very considerable insulinogenic effect of fat that prevails under certain circumstances. When given intravenously fat does not stimulate insulin secretion. But when administered orally as part of a mixed meal, fat may stimulate the gut to produce substantial amounts of insulinogenic hormone (Dobbs *et al.*, 1975). Thus, the effects of dietary fat may differ considerably in diabetics with and those without β-cell function.

An advantage of the kinds of trades suggested above is demonstrating to the patient the high caloric content of alcoholic beverages. The beer drinker is usually impressed to find that a single beer is equivalent to one bread exchange and two fat exchanges, or that a single drink of liquor is equivalent to three fat exchanges.

There are several common problems that arise in lean insulin-dependent patients who drink. In patients who drink in moderation, alcohol itself does not induce appreciable hypoglycemia. But frequently "social drinking" is associated with delay of mealtime. This may lead to hypoglycemia. The prophylaxis is, fortunately, simple. If the meal is unavoidably delayed, carbohydrate should be taken with, or in addition to, the alcoholic beverage. It should also be stressed that this carbohydrate should be taken almost continuously if the meal is substantially delayed (e.g., 10 gm each 30 minutes).

Even if a hypoglycemic episode is not attributable to the drinking, the significance of hypoglycemia is often misinterpreted in patients who have been drinking. Their odd behavior is often mistakenly attributed to drunkenness. Commonly, these people are arrested, in which case they are not usually given the carbohydrate they require. Another common phenomenon is that the patient who has a few drinks is less likely to give attention to his usual routines for management of diabetes. He may, for example, fail to take his bedtime feeding.

C. Sensitivity to the Combination of Alcohol and Sulfonylurea

A small percentage of diabetics who take sulfonylureas have a peculiar idiosyncracy to the combination of alcohol and sulfonylurea. The basis of this is not known. These patients have episodes very similar to those experienced by patients who drink when they are taking disulfuram (Antabuse). Symptoms may be moderate or severe. They include "hot flashes" and a general sense of ill-being. Blood pressure may fall. The episodes are seen most commonly with chlorpropamide therapy, but this may be attributable only to the higher levels of blood sulfonylurea that generally prevail with this particular drug. The phenomenon is also seen with tolbutamide. The major problem here is that most physicians are unfamiliar with the existence of this phenomenon. The cause of the difficulty is therefore often unrecognized. Recently, familiarity with this propensity has been demonstrated.

II. SHOULD ALCOHOL BE ALLOWED IN THE DIETS OF FAT PEOPLE?

There is a rather general consensus to the effect that alcohol should be proscribed in the therapeutic diets of the obese. It is high in calories and not at all necessary. Most of us have the impression that drinking before meals tends to stimulate appetite, and it reduces the self-discipline required in reducing food intake. It is almost heretical, therefore, even to ask the question about whether alcohol should be a part of the dietary prescriptions of fat people. I am, however, prompted to ask the question because of the lack of direct and conclusive evidence that alcohol is particularly fattening.

It may be that people who do not drink at all are thinner than people who drink a little, moderately, or a lot, but I have found little evidence to this effect. Our data in Oklahoma Indians are quite incomplete, but so far we have not observed any relationship of alcohol intake and adiposity. Indeed, it appears that those who drink a lot are not quite as fat as the others. Simple epidemiologic observations are needed in representative samples of general populations to measure systematically the relationship of alcohol intake and adiposity. Doubtless there will be some problems in determining the extent to which results are influenced by confounding variables. It seems likely, for example, that moderate drinkers will differ from those who do not drink in respect to factors other than alcohol intake. Klatsky *et al.* (1977) have recently reported on the association between adiposity and the level of alcohol consumption as obtained by a questionnaire in a very large employed population. White men who consumed 2 to 5 drinks daily were slightly heavier than those who did not drink; but in white women and in blacks (both sexes) those who drank moderately were slightly lighter than those who did not drink.

It may be kept in mind that fat has even more calories per gram than alcohol does, and yet it has not been traditional to severely restrict fat in reducing diets. Indeed, some reducing diets have emphasized reduction of carbohydrate rather than fat. Masai tribesmen are quite thin despite high fat diets. Primitive Eskimo are very lean even though their diets contain fairly generous amounts of fat. Rural Bantus who drink heavily are usually quite thin.

In the aggregate, the available evidence in animals and man suggest that fat and alcohol are probably more fattening than other major nutrients, but this evidence is by no means complete or conclusive.

In general, I encourage obese people to limit alcohol. On the other hand, I do not think present evidence warrants the dogma that alcohol is very particularly fattening.

III. DOES ALCOHOL STIMULATE WEIGHT GAIN IN THE UNDERNOURISHED?

Most of us who enjoy a drink have the impression that a drink before lunch or supper tends to stimulate the appetite. For this reason it has been common practice to prescribe a little alcohol such as wine before meals in undernourished patients with poor appetite. I am not aware of any controlled observations, however, on whether this strategy works or not, nor do I know whether any of these persons have been, by this means, started down a pathway to alcoholic ruin. I do think one should exercise caution in prescribing alcohol to patients with poor appetite. Often they are depressed, a condition that may render them particularly susceptible to developing alcoholism.

Because of the possibility of untoward effects in this group of undernourished patients with poor appetites, it is particularly desirable that controlled observations be made in man and in animals to see whether alcohol really stimulates appetite and weight gain of appreciable degree.

REFERENCES

Dobbs, R., Faloona, G., and Unger, R. H. (1975). Effects of intravenously administered glucose on glucagon and insulin secretion during fat absorption. *Metab. Clin. Exp.* **24**, 69–75.

Hollands, M. (1976). Ethanol metabolism and diabetes. *J. Prof. Health Workers Sec. Can. Diet. Assoc.* **1**, 1.

Klatsky, A. L., Friedman, G. D., Siegelaub, A. B., and Gerard, M. J. (1977). Alcohol consumption among white, black or Oriental men and women: Kaiser–Permanente multiphasic health examination data. *Am. J. Epidemiol.* **105**, 311–323.

Leake, C. D., and Silverman, M. (1966). "Alcoholic Beverages in Clinical Medicine." Yearbook Publ., Chicago, Illinois.

Lucia, S. P. (1963). "A History of Wine as Therapy." Lipincott, Philadelphia, Pennsylvania.

"Manual of Applied Nutrition" (1973). 6th ed. Johns Hopkins Press, Baltimore, Maryland.

Marks, V. (1978). Alcohol and carbohydrate metabolism. *Clin. Endocrinol. Metab.* **7**, 333–349.

"Mayo Clinic Diet Manual" (1971). 4th ed. Saunders, Philadelphia, Pennsylvania.

West, K. M. (1973). Diet therapy of diabetes: An analysis of failure. *Ann. Intern. Med.* **79**, 425–434.

18

ALCOHOL AND KETOGENESIS

F. John Service

I. Introduction 263
II. Alcoholic Ketoacidosis 264
III. Ethanol and Ketogenesis 264
    A. Dietary Studies in Humans 264
    B. Substrate Availability 266
    C. Direct Effect of Ethanol 266
    D. Effect of Chronic Ethanol Ingestion 268
    E. Effect of Acetaldehyde and Acetate 269
    F. Summary 270
    References 270
    Editorial Comment Concerning Alcohol Use in Diabetes 273

I. INTRODUCTION

The occurrence of nondiabetic ketoacidosis in chronic alcoholics was first reported in 1940 by Dillon *et al.* (1940). Not until 1971 did the second report appear (Jenkins *et al.*, 1971), and since then there have been three others (Levy *et al.*, 1973; Cooperman *et al.*, 1974; Fulop and Hoberman, 1975). The lead case in Dillon's series was a woman who experienced five episodes of ketoacidosis in less than 2 years. The course of events leading up to each admission was similar:

> She was a chronic alcoholic and bouts of alcoholism usually if not invariably preceded her admission to the hospital. She would eat little and about 2 days before admission progressively more severe vomiting would occur. Finally, there was marked dehydration, Kussmaul breathing, acetone on the breath, and more or less severe mental confusion. Clinically the picture was identical with that of diabetic coma. [From Dillon *et al.*, 1940, reprinted by permission.]

FERMENTED FOOD BEVERAGES IN NUTRITION
Copyright © 1979 by Academic Press, Inc.
All rights of reproduction in any form reserved.
ISBN 0-12-277050-1

II. ALCOHOLIC KETOACIDOSIS

The important features of this disorder are shown in Table I and they permit the creation of a clinical profile. The predominance of females observed by other investigators (Dillon *et al.*, 1940; Jenkins *et al.*, 1971; Levy *et al.*, 1973; Cooperman *et al.*, 1974) was not found by Fulop and Hoberman (1975). It remains to be shown after more experience with this disorder whether there is a predilection for one sex. Chronic alcoholism without diabetes or with very mild diabetes is always present. For several days before the acute illness there was an alcoholic debauch followed by reduced food intake, vomiting, and usually continued ethanol intake. Results of physical examination are not unlike those observed for diabetic ketoacidosis, with Kussmaul breathing, dehydration, mental obtundation, and abdominal pain and tenderness. Laboratory evaluation shows the typical changes of acute metabolic acidosis, namely low pH, p_{CO_2}, and HCO_3. Fulop and Hoberman (1975) encountered some patients with superimposed metabolic alkalosis presumed to be due to loss of hydrogen ion as a consequence of vomiting and respiratory alkalosis from delirium tremens. The large anion gap can be accounted for by the accumulation of ketoacids, chief of which is β-hydroxybutyric acid. Because of the altered (reduced) redox state of the hepatocyte which arises from the oxidation of ethanol, acetoacetate is reduced to β-hydroxybutyrate, and this results in an increased β-hydroxybutyrate/acetoacetate ratio. Because the qualitative test (nitroprusside) for serum ketones reacts only with acetoacetate, a disproportionately low positive result may be oberved in the presence of severe β-hydroxyacidemia. The altered redox state of the hepatocyte also promotes the conversion of pyruvate to lactate and, secondarily, causes hyperlactatemia, which contributes to the anion gap and the metabolic acidosis. Abnormal liver function is often observed. Modest hyperglycemia or severe hypoglycemia may be observed. Treatment requires correction of the dehydration and administration of glucose and sodium bicarbonate.

III. ETHANOL AND KETOGENESIS

A. Dietary Studies in Humans

Ethanol consumption in man is ketogenic. The addition of ethanol to a ketogenic diet resulted in increased ketonemia in nondiabetic and mildly diabetic subjects (Schlierf *et al.*, 1964). When dietary carbohydrate was replaced with ethanol or fat in chronic alcoholics, greater ketogenesis was associated with ethanol (Lefèvre and Lieber, 1967). When alcohol was the only source of calories over a period of several days, severe ketonemia (Lefèvre *et al.*, 1970) associated with elevated levels of free fatty acids (Schapiro *et al.*, 1965) was observed.

Table I. Important Findings in Patients with Alcoholic Ketoacidosis

Patient findings	Source			
	Jenkins et al. (1971)	Levy et al. (1973)	Cooperman et al. (1974)	Fulop and Hoberman (1975)
Number	3	5	6	24
Sex ratio, F/M	3/0	2/3	6/0	7/17
Age (years)	27–44	37–69	31–51	25–76
Alcoholism	3	5	6	24
Diabetes	0	0	0	6
Kussmaul respiration	3	3	1	—
Mental obtundation	2	3	0	—
pH	7.07–7.21	6.96–7.29	7.09–7.28	6.7–7.63
p_{CO_2} mm Hg	—	<10–14	12–32	11–37
HCO—(mEq/liter)	6–7	5–8	5–13	11–35
β-Hydroxybutyric acid (mmole/liter)	3.8–20.5	5.5–19.2	7.3–11.4	1.4–22.5
Lactate (mmole/liter)	1.5–5	1.9–18.3	0.3–2.6	1.0–20.6
Anion gap	5–30	15–34	9–22	—
Blood glucose (mg/dl)	57–330	25–235	75–275	5–193

B. Substrate Availability

The pathogenesis of alcoholic ketoacidosis can be considered from the points of view of availability of substrate for ketogenesis, the effect of ethanol and its metabolites on hepatic metabolism, and the effect of chronic ethanol consumption on hepatic metabolism.

Free fatty acids are the substrate for ketone body production. Ethanol administered to normal subjects after prolonged fasting has been shown to increase free fatty acids (Field *et al.*, 1963); no increase was observed after an overnight fast (Lieber *et al.*, 1962b). The relative or absolute state of starvation that precedes the development of alcoholic ketoacidosis and the effects of low plasma insulin (Jenkins *et al.*, 1971; Levy *et al.*, 1973; Cooperman *et al.*, 1974) and elevated cortisol (Levy *et al.*, 1973; Cooperman *et al.*, 1974), growth hormone (Levy *et al.*, 1973; Cooperman *et al.*, 1974), and probably glucagon promote lipolysis, increased circulating free fatty acids (Raben and Hollenberg, 1959; Renold and Cahill, 1965), and ketogenesis. When measured in alcoholic ketoacidosis (Jenkins *et al.*, 1971; Levy *et al.*, 1973; Cooperman *et al.*, 1974), free fatty acids have been elevated, often to levels greater than those observed for starvation (Cahill *et al.*, 1966) or diabetic ketoacidosis (Laurell, 1956; Bierman *et al.*, 1957; Gerich *et al.*, 1971).

C. Direct Effect of Ethanol*

The effects of ethanol and its metabolites on ketogenesis are related directly to the metabolism of ethanol and to mitochondrial damage consequent to chronic ethanol ingestion (Lieber, 1976, 1977). The oxidation of ethanol to acetaldehyde is accomplished primarily by the cytosolic enzyme alcohol dehydrogenase, which is linked to the NAD/NADH couple (Teschke *et al.*, 1976) (Fig. 1). The excess NADH resulting from ethanol oxidation generates an excess of reducing equivalents in the hepatocyte and this is responsible for a variety of metabolic disturbances. The altered redox state promotes the conversion of pyruvate to lactate, the result being an accumulation of intracellular lactate and subsequently hyperlactatemia (Lieber *et al.*, 1962a). As a result of competition between lactate and urate for renal tubular excretion, hyperuricemia often ensues (Lieber *et al.*, 1962a). The excess NADH promotes the conversion of dihydroxyacetone phosphate to α-glycerophosphate, to which fatty acids become esterified to form triglyceride (Nikkila and Ojala, 1963). Fatty acid synthesis is accelerated because of the accumulation of NADPH, which is generated in the presence of NADH by the transhydrogenation of NADP (Gordon, 1972).

The overall effect of the excess reducing equivalents is enhanced lipogenesis,

---

*Abbreviations used in this section: NAD, nicotinamide adenine dinucleotide; NADH, NAD reduced form; NADP, NAD phosphate; NADPH, NADP reduced form.

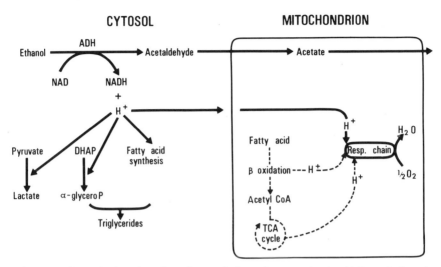

Fig. 1. Effect of ethanol on hepatic metabolism. As a result of oxidation of ethanol, reducing equivalents are generated which stimulate formation of lactate, α-glycerophosphate, and fatty acids; the latter two subsequently form triglycerides. Reducing equivalents also enter mitochondrion, where they compete for sites on respiratory chain with reducing equivalents from β-oxidation of fatty acids and from tricarboxylic acid cycle. The latter cycle is inhibited also by direct effect of reducing equivalents on several of its reaction sequences. ADH, alcohol dehydrogenase; DHAP, dihydroxy acetone phosphate; α-glycero-P, α-glycerophosphate; TCA, tricarboxylic acid cycle.

which leads to the accumulation of fat within the liver. Not all the excess hydrogen generated as a result of ethanol oxidation is disposed of in the cytosol; by various shuttle mechanisms some hydrogen can be transferred into the mitochondrion where it competes with reducing equivalents generated by β-oxidation of fatty acids and the tricarboxylic acid cycle for entry into the respiratory chain. The overall effect within the mitochondrion is inhibition of fatty acid oxidation because of impaired β-oxidation and suppression of tricarboxylic acid cycle activity. The latter is slowed also because excess reducing equivalents in the form of NADH impair several reactions—for example, those associated with isocitrate dehydrogenase, α-ketoglutarate dehydrogenase, malate dehydrogenase, and pyruvate dehydrogenase. This is compounded by the diminished availability of respiratory chain activity, on which the tricarboxylic acid cycle depends, because of the excess hydrogen within the mitochondrion. Besides the demonstration in a variety of animal hepatic preparations (Lieber and Schmid, 1961; Lieber *et al.,* 1967; Blomstrand and Kager, 1973; Blomstrand *et al.,* 1973; Ontko, 1973), it has been shown also in humans (Wolfe *et al.,* 1976) that ethanol impairs the oxidation of fatty acids and increases the splanchnic storage of fat.

D. Effect of Chronic Ethanol Ingestion

Chronic ethanol ingestion leads to a proliferation of hepatic smooth endoplasmic reticulum (Iseri *et al.,* 1964, 1966). It is there in the microsomes that what is usually a minor contributor to ethanol oxidation, namely, the microsomal ethanol-oxidizing system, becomes major and may even exceed the contribution of the alcohol dehydrogenase pathway in chronic alcoholism (Lieber and DeCarli, 1972). Other effects of the proliferation of the smooth endoplasmic reticulum, besides the increased metabolism of alcohol, are the increased degradation of other drugs (Misra *et al.,* 1971), increased synthesis of cholesterol (Lefèvre *et al.,* 1972), and increased secretion of lipoproteins (Díaz Gomez *et al.,* 1973).

Chronic ethanol consumption leads to striking structural abnormalities of mitochondria (Suoboda and Manning, 1964) which are directly attributable to ethanol and not to dietary deficiencies (Rubin and Lieber, 1968). Furthermore, a number of abnormalities of mitochondrial function have been observed in hepatocytes which lead to enhanced ketogenesis (Fig. 2). In isolated hepatic mitochondria from rats chronically treated with ethanol, a 30% increase in the activity of acylcarnitine transferase over that in control animals has been observed (Cederbaum *et al.,* 1975). Total oxidation of the fatty acids once translocated into the mitochondrion requires the activities of the β-oxidation sequence (which generates acetyl-CoA) and tricarboxylic acid cycle, both of which are linked to the respiratory chain. The rates of oxygen consumption in mitochondria

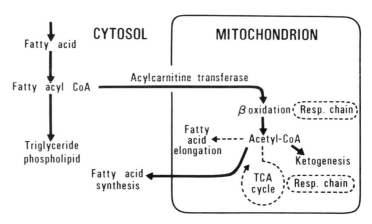

Fig. 2. Effect of chronic ethanol ingestion on hepatic metabolism. Fatty acid entry into the mitochondrion and β-oxidation are accelerated. Respiration is depressed and oxidative phosphorylation is impaired. Tricarboxylic acid cycle is slowed because of excess reducing equivalents and depressed respiration; result is depression of total fatty acid oxidation. Net effect is diversion of increased generation of acetyl-CoA into ketone body formation.

from ethanol-fed rats have been shown to be depressed 10–15% in state 4 (absence of adenosine 5′-diphosphate) and depressed 30% in state 3 (presence of adenosine 5′-diphosphate) in comparison with controls (Cederbaum *et al.,* 1975). These observations indicate that chronic ethanol consumption results not only in a defect in the respiratory chain but also in an impairment of oxidative phosphorylation.

Despite the accelerated transfer of fatty acids into the mitochondrion, total oxidation of fatty acids is depressed and ketogenesis is accelerated in ethanol-fed rats (Cederbaum *et al.,* 1975). Fatty acid oxidation has been studied in the presence of an artificial electron transfer system (formazan production) under anaerobic conditions, which eliminates the effect of the defective respiratory chain. The rate of formazan formation, which reflects the activity of the artificial electron transfer system, was higher in ethanol-fed rats than in pair-fed controls (Cederbaum *et al.,* 1975). These observations are consistent with increased β-oxidation or increased activity of the tricarboxylic acid cycle. When fluoroacetate, an inhibitor of the tricarboxylic acid cycle, was added to the system, no change in the increased formazan production was observed (Cederbaum *et al.,* 1975). Consequently, it was concluded that in contrast to the acute effect of ethanol, chronic ethanol ingestion resulted in increased β-oxidation of fatty acids. The tricarboxylic acid cycle activity, on the other hand, is suppressed in ethanol-fed rats (Cederbaum *et al.,* 1975) because of the altered redox state of the mitochondrion, just as it is after acute ethanol ingestion. CO_2 production from various ^{14}C-labeled tricarboxylic acid intermediates has been found to be decreased 15–25% in ethanol-fed rats (Lieber *et al.,* 1963). That the increased acetyl-CoA generated from the accelerated β-oxidation of fatty acids in chronic ethanol ingestion is diverted into ketogenesis is supported by the observation of increased formazan production and ketogenesis but reduced total oxidation of fatty acids (Cederbaum *et al.,* 1975).

E. Effect of Acetaldehyde and Acetate

Acetaldehyde, the immediate oxidation product of ethanol, damages the structure of the mitochondrion, and this leads to metabolic aberrations similar to those seen with chronic ethanol ingestion. Furthermore, acetaldehyde dehydrogenase, which is intramitochondrial in location, is depressed, and this results in impaired clearance and degradation of acetaldehyde. The addition of acetaldehyde to isolated hepatic mitochondria inhibited the state 3 rate of oxygen consumption and depressed CO_2 production from tricarboxylic acid cycle intermediates (Cederbaum *et al.,* 1976). The impairment of the tricarboxylic acid cycle probably was due to impaired respiratory chain activity, the altered redox state from excess NADH, and competition for NAD between acetaldehyde and various tricarboxylic acid cycle intermediates (Cederbaum *et al.*, 1976). Acetate, which is metabolized primarily in extrahepatic tissue, probably contributes little to the effects of chronic ethanol ingestion.

F. Summary

Alcoholic ketoacidosis is now recognized as a distinct clinical entity. The pathogenesis of ketone body formation in this disorder appears to be multifactorial. There is an appropriate substrate (increased fatty acids) and a hormonal (decreased insulin, increased cortisol, glucagon and growth hormone) and nutritional (starvation) milieu. Superimposed are the direct effects of ethanol metabolism and its metabolite, acetaldehyde, and the disruptive effect of chronic ethanol ingestion on mitochondrial function. The result is increased fatty acid transport to the liver, increased fatty acid transfer into the mitochondrion, increased β-oxidation, and the diversion of the increased acetyl-CoA from the tricarboxylic acid cycle to ketogenesis.

REFERENCES

Bierman, E. L., Dole, V. P., and Roberts, T. N. (1957). An abnormality of nonesterified fatty acid metabolism in diabetes mellitus. *Diabetes* **6**, 475–479.

Blomstrand, R., and Kager, L. (1973). The combustion of triolein-1-^{14}C and its inhibition by alcohol in man. *Life Sci.* **13**, 113–123.

Blomstrand, R., Kager, L., and Lantto, O. (1973). Studies on the ethanol-induced decrease of fatty acid oxidation in rat and human liver slices. *Life Sci.* **13**, 1131–1141.

Cahill, G. F., Jr., Herrera, M. G., Morgan, A. P., Soeldner, J. S., Steinke, J., Levy, P. L., Reichard, G. A., and Kipnis, D. M. (1966). Hormone-fuel interrelationships during fasting. *J. Clin. Invest.* **45**, 1751–1769.

Cederbaum, A. I., Lieber, C. S., Beattie, D. S., and Rubin, E. (1975). Effect of chronic ethanol ingestion on fatty acid oxidation by hepatic mitochondria. *J. Biol. Chem.* **250**, 5122–5129.

Cederbaum, A. I., Lieber, C. S., and Rubin, E. (1976). Effect of chronic ethanol consumption and acetaldehyde on partial reactions of oxidative phosphorylation and CO_2 production from citric acid cycle intermediates. *Arch. Biochem. Biophys.* **176**, 525–538.

Cooperman, M. T., Davidoff, F., Spark, R., and Pallotta, J. (1974). Clinical studies of alcoholic ketoacidosis. *Diabetes* **23**, 433–439.

Díaz Gómez, M. I., Castro, J. A., de Ferreyra, E. C., D'Acosta, N., and de Castro, C. R. (1973). Irreversible binding of ^{14}C from ^{14}CCl$_4$ to liver microsomal lipids and proteins from rats pretreated with compounds altering microsomal mixed-function oxygenase activity. *Toxicol. Appl. Pharmacol.* **25**, 534–541.

Dillon, E. S., Dyer, W. W., and Smelo, L. S. (1940). Ketone acidosis in nondiabetic adults. *Med. Clin. North Am.* **24**, 1813–1822.

Field, J. B., Williams, H. E., and Mortimore, G. E. (1963). Studies on the mechanism of ethanol-induced hypoglycemia. *J. Clin. Invest.* **42**, 497–506.

Fulop, M., and Hoberman, H. D. (1975). Alcoholic ketosis. *Diabetes* **24**, 785–790.

Gerich, J. E., Martin, M. M., and Recant, L. (1971). Clinical and metabolic characteristics of hyperosmolar nonketotic coma. *Diabetes* **20**, 228–238.

Gordon, E. R. (1972). Effect of an intoxicating dose of ethanol on lipid metabolism in an isolated perfused rat liver. *Biochem. Pharmacol.* **21**, 2991–3004.

Iseri, O. A., Gottlieb, L. S., and Lieber, C. S. (1964). The ultrastructure of ethanol-induced fatty liver. *Fed. Proc., Fed. Am. Soc. Exp. Biol.* **23**, 579. (Abstr.)

Iseri, O. A., Lieber, C. S., and Gottlieb, L. S. (1966). The ultrastructure of fatty liver induced by prolonged ethanol ingestion. *Am. J. Pathol.* **48**, 535–555.

Jenkins, D. W., Eckel, R. E., and Craig, J. W. (1971). Alcoholic ketoacidosis. *J. Am. Med. Assoc.* **217**, 177–183.

Laurell, S. (1956). Plasma free fatty acids in diabetic acidosis and starvation. (Letter to the editor.) *Scand. J. Clin. Lab. Invest.* **8**, 81–82.

Lefèvre, A., and Lieber, C. S. (1967). Ketogenic effect of alcohol. *Clin. Res.* **15**, 324. (Abstr.)

Lefèvre, A., Adler, H., and Lieber, C. S. (1970). Effect of ethanol on ketone metabolism. *J. Clin. Invest.* **49**, 1775–1782.

Lefèvre, A. F., DeCarli, L. M., and Lieber, C. S. (1972). Effect of ethanol on cholesterol and bile acid metabolism. *J. Lipid Res.* **13**, 48–55.

Levy, L. J., Duga, J., Girgis, M., and Gordon, E. E. (1973). Ketoacidosis associated with alcoholism in nondiabetic subjects. *Ann. Intern. Med.* **78**, 213–219.

Lieber, C. S. (1976). The metabolism of alcohol. *Sci. Am.* **234**, 25–33.

Lieber, C. S. (1977). The metabolic basis of alcohol's toxicity. *Hosp. Pract.* **12**, 73–80.

Lieber, C. S., and DeCarli, L. M. (1972). The role of the hepatic microsomal ethanol oxidizing system (MEOS) for ethanol metabolism *in vivo*. *J. Pharmacol. Exp. Ther.* **181**, 279–287.

Lieber, C. S., and Schmid, R. (1961). The effect of ethanol on fatty acid metabolism: Stimulation of hepatic fatty acid synthesis in vitro. *J. Clin. Invest.* **40**, 394–399.

Lieber, C. S., Jones, D. P., Losowsky, M. S., and Davidson, C. S. (1962a). Interrelation of uric acid and ethanol metabolism in man. *J. Clin. Invest.* **41**, 1863–1870.

Lieber, C. S., Leevy, C. M., Stein, S. W., George, W. S., Cherrick, G. R., Abelmann, W. H., and Davidson, C. S. (1962b). Effect of ethanol on plasma free fatty acids in man. *J. Lab. Clin. Med.* **59**, 826–832.

Lieber, C. S., Jones, D. P., Mendelson, J., and DeCarli, L. M. (1963). Fatty liver, hyperlipemia and hyperuricemia produced by prolonged alcohol consumption, despite adequate dietary intake. *Trans. Assoc. Am. Physicians* **76**, 289–300.

Lieber, C. S., Lefèvre, A., Spritz, N., Feinman, L., and DeCarli, L. M. (1967). Difference in hepatic metabolism of long- and medium-chain fatty acids: The role of fatty acid chain length in the production of the alcoholic fatty liver. *J. Clin. Invest.* **46**, 1451–1460.

Misra, P. S., Lefèvre, A., Ishii, H., Rubin, E., and Lieber, C. S. (1971). Increase of ethanol, meprobamate and pentobarbital metabolism after chronic ethanol administration in man and in rats. *Am. J. Med.* **51**, 346–351.

Nikkila, E. A., and Ojala, K. (1963). Role of hepatic L-α-glycerophosphate and triglyceride synthesis in production of fatty liver by ethanol. *Proc. Soc. Exp. Biol. Med.* **113**, 814–817.

Ontko, J. A. (1973). Effects of ethanol on the metabolism of free fatty acids in isolated liver cells. *J. Lipid Res.* **14**, 78–86.

Raben, M. S., and Hollenberg, C. H. (1959). Effect of growth hormone on plasma fatty acids. *J. Clin. Invest.* **38**, 484–488.

Renold, A. E., and Cahill, G. F., Jr. (1965). Metabolism of isolated adipose tissue: A summary. *In* "Handbook of Physiology. Sect. 5: Adipose Tissue" (A. E. Renold and G. F. Cahill, Jr., eds.), pp. 483–490. Am. Physiol. Soc., Washington, D.C.

Rubin, E., and Lieber, C. S. (1968). Alcohol-induced hepatic injury in nonalcoholic volunteers. *N. Engl. J. Med.* **278**, 869–876.

Schapiro, R. H., Scheig, R. L., Drummey, G. D., Mendelson, J. H., and Isselbacher, K. J. (1965).

Effect of prolonged ethanol ingestion on the transport and metabolism of lipids in man. *N. Engl. J. Med.* **272**, 610–615.

Schlierf, G., Gunning, B., Uzawa, H., and Kinsell, L. W. (1964). The effects of calorically equivalent amounts of ethanol and dry wine on plasma lipids, ketones and blood sugar in diabetic and nondiabetic subjects. *Am. J. Clin. Nutr.* **15**, 85–89.

Suoboda, D. J., and Manning, R. T. (1964). Chronic alcoholism with fatty metamorphosis of the liver: Mitochondrial alterations in hepatic cells. *Am. J. Pathol.* **44**, 645–662.

Teschke, R., Hasumura, Y., and Lieber, C. S. (1976). Hepatic ethanol metabolism: Respective roles of alcohol dehydrogenase, the microsomal ethanol-oxidizing system, and catalase. *Arch. Biochem. Biophys.* **175**, 635–643.

Wolfe, B. M., Havel, J. R., Marliss, E. B., Kane, J. P., Seymour, J., and Ahuja, S. P. (1976). Effects of a 3-day fast and of ethanol on splanchnic metabolism of FFA, amino acids, and carbohydrates in healthy young men. *J. Clin. Invest.* **57**, 329–340.

EDITORIAL COMMENT CONCERNING ALCOHOL
USE IN DIABETES

The physician often must advise a diabetic patient concerning the use of alcoholic beverages. The preceding paper by Doctor Service is devoted to the uncommon clinical entity of alcoholic ketoacidosis. The insight that it gives into the mechanisms by which alcohol favors ketogenesis will help the physician advise his diabetic patient who asks whether he can use alcoholic beverages.

Diabetes beginning in childhood or young adult life is commonly characterized by complete or nearly complete insulin deficiency and by susceptibility to ketosis. Some idea of the risk of developing significant ketosis after alcohol ingestion could be obtained by testing the urine for ketones from time to time after the use of alcohol. If ketonuria is found more consistently and in larger amounts after alcohol use then the diabetic should be advised to be cautious in the use of alcohol. Diabetes beginning after middle age is usually characterized by a resistance to ketosis and a common finding of obesity. Such individuals would be less likely to exhibit ketonuria after ingestion of alcohol, but this could be verified for each patient by trial. In diabetes of adult-onset variety the regular use of alcohol might contribute to obesity and the obesity in turn could aggravate hyperglycemia and glycosuria.

Both juvenile and adult onset diabetes are associated with accelerated atherosclerosis manifested as coronary artery disease or occlusive vascular disease affecting the lower extremities. Since hypertriglyceridemia may be induced by the use of alcohol and since hypertriglyceridemia does appear to contribute to atherogenesis, one could question whether alcohol should not be forbidden to the diabetic. If this is an issue, then comparison of serum triglyceride levels after several weeks of abstinence might be compared with measurements done after several weeks of a representative consumption of alcohol. On the other hand, the use of alcohol may have effects on atherogenesis which go beyond variations in triglyceride levels. It has been reported that the use of alcohol in moderation by nondiabetic individuals may increase high density lipoprotein cholesterol, a class of lipoproteins which tend to protect against coronary artery disease [Hulley, S. B., Cohen, R., and Widdowson, G. Plasma high-density lipoprotein cholesterol level. *J. Am. Med. Assoc.* **238**, 2269–2271 (1977)] and may decrease the frequency of clinical instances of coronary artery disease [Yano, K., Rhoads, G. G., and Kagan, A. Coffee, alcohol and risk of coronary heart disease among Japanese men living in Hawaii. *N. Engl. J. Med.* **297**, 405–409 (1977)]. On the basis of these reports, one might conclude that the use of alcohol probably may be permitted in diabetics with minimal concern that vascular disease will thereby be accelerated.

Alcohol also has the property of blocking gluconeogenesis and thus impairs one of the homeostatic protections against hypoglycemia. As long as hepatic glycogen stores have not been depleted by fasting, impairment of gluconeogenesis should not result in hypoglycemia. The diabetic who has taken an excessive dose of insulin will experience hypoglycemia, but ordinarily under these circumstances the liver will be well-stocked with glycogen. Hence, in the diabetic who eats with reasonable regularity, alcohol should not render him particularly vulnerable to hypoglycemic episodes. The immoderate use of alcohol with impairment of judgment may lead to delayed meals or missed injections of insulin and thus create major problems in diabetes control. Diabetics are commonly warned that if they have had even one or two drinks, have the odor of alcohol on their breaths, and then experience a hypoglycemic episode they may be falsely accused of drunkenness and the hypoglycemia may therefore not be identified and treated.

Thus alcohol may be expected to impair some of the homeostatic mechanisms which protect against hypoglycemia, on the one hand, and ketosis, on the other. If the diabetic is reasonably regular and consistent in his eating and if insulin is used in appropriate amounts alcohol in moderate quantities likely will produce no perceptible effects on glycosuria, glycemia, or ketonuria. Nevertheless, alcohol does have the potential for impairing homeostatic metabolic mechanisms and may be looked upon as possibly having a destabilizing influence on the character of the diabetes.

The diabetic often asks whether alcohol should be figured into the diet and if so whether it should be exchanged for carbohydrate or for fat. A common practice is to calculate it into the diet as a form of fat, perhaps since it is ketogenic. To give alcohol the status of a fat or a fruit exchange, however, would be to suggest to the diabetic patient that alcohol is nutritionally identical and inappropriately large amounts of alcohol might therefore be consumed with the rationalization that, after all, it is a recognized food exchange. In practice it is probably best to recommend that the diabetic consume a conventional diet devised in terms of the American Diabetes Association exchanges and that alcohol be used only in moderation and then in addition to the prescribed diet. If weight control is a problem the use of alcohol will need to be limited and if the patient, because of obesity, should require more severe calorie restriction (e.g., calorie intake less than perhaps 1200 calories) the use of alcohol probably should be forbidden until a satisfactory weight is achieved and a weight maintaining diet is to be followed.

V

Effects of Misuse and Excess

19

EFFECTS OF ALCOHOL ON GASTROINTESTINAL AND PANCREATIC FUNCTION IN ALCOHOLICS

Esteban Mezey and Charles H. Halsted

I. Introduction	277
II. Absorption and Metabolism of Ethanol by the Gastrointestinal Tract	278
A. Absorption	278
B. Metabolism	279
III. Alcohol and the Stomach	281
A. Effect of Alcohol on the Gastric Mucosa	281
B. Ethanol and Gastric Motility	283
C. Ethanol and Gastric Secretion	283
IV. Alcoholism and the Small Intestine	284
A. Morphology	284
B. Intestinal Metabolism	284
C. Intestinal Enzymes	285
D. Alcoholism and Intestinal Absorption	286
V. Alcoholism and the Pancreas	291
A. Effects of Alcohol on the Pancreas	291
B. Malnutrition and Pancreatic Function	294
C. Alcoholism and Pancreatic Carcinoma	296
VI. Summary of Research Needs	296
References	297

I. INTRODUCTION

Chronic alcoholism is a frequent cause of gastrointestinal and pancreatic dysfunction and the principal cause of malnutrition in countries with adequate food supplies. Although decreased dietary intake is common in alcoholics, maldigestion and malabsorption of nutrients may play a significant role in the malnutrition of alcoholics. In addition, alcohol-related diseases of the gastrointestinal tract and pancreas often are the cause of morbidity and mortality in alcoholic patients. Most of the changes in the function of the gastrointestinal tract and pancreas have been demonstrated to occur after the ingestion or administration of excessive amounts of ethanol. Furthermore, in most cases the ethanol ingested was in the form of distilled spirits or was administered as ethanol in high concentrations. There is a paucity of information on the effects of moderate ingestion of ethanol and of its ingestion as fermented beverages such as beer or wine.

II. ABSORPTION AND METABOLISM OF ETHANOL BY THE GASTROINTESTINAL TRACT

A. Absorption

Ethanol is absorbed rapidly by the gastrointestinal tract. The absorption starts in the stomach and continues in the upper small intestine. During active drinking the small intestine is exposed to ethanol directly via intestinal transit and by way of diffusion from the blood stream. The high concentration of ethanol reached in the jejunum immediately after ingestion of a dose of ethanol (Fig. 1) decreases rapidly, reaching levels that are in equilibrium with the vascular space by 120 minutes after ingestion (Halsted *et al.*, 1973a). The absorbed ethanol is distrib-

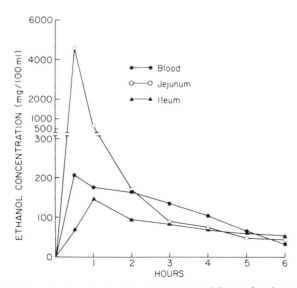

Fig. 1. Distribution of ethanol in the blood, jejunum, and ileum after the acute oral ingestion of ethanol, 0.8 gm/kg body weight, in one patient. (From Halsted *et al.*, 1973a, reproduced by permission.)

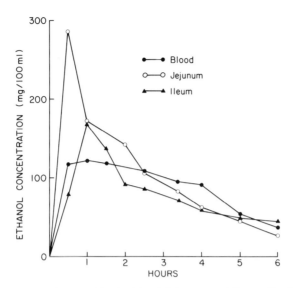

Fig. 2. Distribution of ethanol in the blood, jejunum, and ileum after the intravenous administration of ethanol, 0.8 gm/kg body weight, in one patient. (From Halsted *et al.*, 1973a, reproduced by permission.)

uted throughout the body water and appears in the urine, pulmonary alveolar air, spinal fluid, and the remainder of the intestine. The appearance and increase of ethanol in the ileum parallel the levels in the vascular space, suggesting that ethanol enters the ileum from the vascular space rather than by traveling down the length of the intestine. The rates of absorption of alcohol are slower when it is ingested with meals, and in the form of beer and wine, than when ingested in the form of spirits (Leake and Silverman, 1966).

Ethanol administered by the intravenous route (Fig. 2) also is distributed throughout the gastrointestinal tract at concentrations paralleling those found in the vascular space (Halsted *et al.*, 1973a).

B. Metabolism

Alcohol dehydrogenase, an enzyme present in the supernatant fraction of tissue homogenates, which is the principal enzyme catalyzing the oxidation of ethanol in the liver (Isselbacher and Greenberger, 1964), has been found to be distributed in a number of tissues including the intestine in both the rat (Spencer *et al.*, 1964; Carter and Isselbacher, 1971; Mistilis and Garske, 1969) and man (Spencer *et al.*, 1964). The alcohol dehydrogenase activities obtained in various rat tissues in our laboratory are shown in Table I. In the gastrointestinal tract the activity of alcohol dehydrogenase was found to be highest in the mucosa of the stomach and upper jejunum and lower in the ileum. Rat gastric alcohol dehy-

Table I. Tissue Distribution of Alcohol Dehydrogenase Activity in the Rat

Tissue	Number of animals	Alcohol dehydrogenase[a] (μmole/mg protein/hour)
Liver	24	1.32 ± 0.243
Gastric mucosa	8	6.37 ± 1.632
Intestinal mucosa		
Upper third	8	0.282 ± 0.197
Middle third	8	0.264 ± 0.156
Distal third	8	0.135 ± 0.049
Retina	6	0.256 ± 0.106
Brain	8	0.275 ± 0.184
Kidney	8	0.112 ± 0.074
Testis	8	0.068 ± 0.037

[a] Alcohol dehydrogenase activity was determined as described by Mezey *et al.* (1968). Ethanol concentrations used in the reaction mixtures to obtain maximum activities were 18 mM for liver, retina, kidney, and testis; 108 mM for intestinal mucosa and brain; and 900 mM for gastric mucosa. The values are expressed as means ± SD.

drogenase differs from the liver enzyme in that it has a much higher K_m for ethanol, and a different electrophoretic mobility. The K_m of gastric alcohol dehydrogenase is estimated to be 400 mM as compared with a K_m of 1.0 mM for liver alcohol dehydrogenase (Cederbaum *et al.,* 1975). The enzyme activities in the liver and the stomach, measured at a substrate concentration of 18 mM, were 1.32 ± 0.243 and 0.290 ± 0.126 (SD) μmoles/mg protein/hour, respectively. However, at a substrate concentration of 900 mM the activities in the liver and stomach were 0.471 ± 0.167 and 6.37 ± 1.632 μmoles/mg protein/hour, respectively. Since concentrations of 400 mM (1840 mg/100 ml) or more are commonly reached in the stomach immediately after the ingestion of alcoholic beverages, gastric alcohol dehydrogenase may play a role in the initial oxidation of alcohol. The activity of alcohol dehydrogenase in the upper intestine was about one-fifth of the activity found in the liver. Other investigators have found similar activities of alcohol dehydrogenase in the upper small intestine of the rat (Spencer *et al.,* 1964; Carter and Isselbacher, 1971; Mistilis and Garske, 1969); however, they obtained lower comparative liver alcohol dehydrogenase activities, so that in their reports the activity of alcohol dehydrogenase in the intestine calculates to between one-half and one and a half times that found in the liver. The discrepancies in the activities of rat liver alcohol dehydrogenase may be related to differences in assay methodology. Intestinal alcohol dehydrogenase has been found to have at least two isoenzymes on agar gel electrophoresis in the rhesus monkey (von Wartburg and Papenberg, 1966) but has not been characterized further. Krebs and Perkins (1970) have suggested that intestinal alcohol dehydrogenase may be of bacterial origin. However, Carter and Isselbacher (1971) showed that ethanol is metabolized to carbon dioxide by rat stomach and small intestinal slices. The rate of metabolism was about 68% of that found by liver slices when compared on a wet-weight basis and was not different in intestinal slices obtained from germ-free animals.

III. ALCOHOL AND THE STOMACH

Clinically, it is generally appreciated that chronic alcoholics have a higher than usual incidence of upper gastrointestinal bleeding, associated with gastritis and often with peptic ulceration of the stomach or duodenum. In a retrospective survey of upper gastrointestinal bleeding at the University of California, Davis, and the Martinez Veterans Administration Hospital, the leading causes were gastritis (25%), peptic ulcer (24%), and esophageal varices (19%) as compared to percentages of 5, 59, and 0% in nonalcoholics (Belber, 1978). Such clinical observations raise questions about the specific effect of alcohol on the gastric mucosa and on gastric secretory and motility functions. These physiologic effects will be reviewed separately.

A. Effect of Alcohol on the Gastric Mucosa

1. Acute Effects of Alcohol

The first scientific recording of the effect of alcohol on the gastric mucosa comes from the descriptions of Beaumont (1833), who observed "apthous patches," i.e., erythematous lesions, in the stomach wall of his gastronomy patient following alcoholic binges. A century later, Palmer, another U.S. Army physician, performed gastroscopy and gastric biopsy on 34 soldiers within 6 hours of excessive drinking and found acute gastritis (erythema and erosion) with glandular necrobiosis, edema, and leukocytosis in 30 of them (Palmer, 1954). Prompted by variable observations of biopsy, surgically and autopsy-obtained mucosa in alcoholic patients, Williams (1956) administered graded doses of ethanol to guinea pigs by stomach tube. The gastric mucosa obtained 24 hours later showed multiple hemorrhages and erosions in animals which had received more than 5 ml of 40% ethanol, or 5 ml of conventional Scotch whiskey. More recent studies (Dinoso *et al.*, 1970, 1976; Eastwood and Kirchner, 1974) have confirmed the acute damaging effects of ethanol on the gastric mucosa of experimental animals. These studies showed that (a) the effect of ethanol on Heidenhain pouches of dogs (depletion of mucin granules followed by edema and hemorrhagic congestion of lamina propria) is aggravated by the presence of acid (Dinoso *et al.*, 1970), and (b) ethanol in concentration greater than 25% is damaging to gastric mucosa, even in the absence of acid (Eastwood and Kirchner, 1974). The electron microscopic changes (disruption of apical membrane and widening of intracellular spaces) is dependent on ethanol concentration and is associated with protein, potassium, and sodium influx into the stomach and efflux of hydrogen ions from the stomach (Dinoso *et al.,* 1976). Ethanol is one of a variety of agents that break the gastric mucosal barrier to back diffusion of acid, potentially resulting in local histamine release, increased capillary permeability, interstitial hemorrhage, and necrosis (Davenport, 1967, 1969). The acute effect of ethanol on the human gastric mucosal barrier was studied by Geall *et al.* (1970) who showed a marked suppression of potential differences between intragastric contents and the systemic circulation (i.e., back diffusion of H^+ ions) following ingestion of 25 and 50 ml of whiskey.

2. Chronic Effects of Alcohol

Dinoso *et al.* (1972) performed gastric biopsies and gastric analyses on 70 chronic alcoholic patients with a mean duration of 13 years of alcoholism, 4–8 weeks after cessation of alcohol intake. Chronic atrophic gastritis was found in 24% of specimens obtained from the body of the stomach and in 66% of specimens of antral mucosa compared to a 4% incidence of atrophic gastritis in a control group of similar-aged subjects. In several patients, serial biopsies over a 6- to 9-month period showed recovery of atrophic gastritis to a normal mucosa. In the entire group, the maximum stimulated acid output (MAO) was no different than in the control group, while the lowest acid outputs were found in the patients with atrophic gastritis.

Gastric mucosal morphology has been correlated with blood and protein loss during the ingestion of ethanol. Dinoso *et al.* (1973) studied fecal blood loss in 18 chronic alcoholics—six with normal gastric mucosa, six with superficial gastritis, and six with atrophic gastritis—by calculating the excretion in the stools of radioactivity following intravenous infusion of ^{51}Cr-labeled autologous red cells, before and after ingestion of 200 ml of 40% ethanol. The ethanol-stimulated mean fecal blood loss rose to 1.7 ml per day in the atrophic gastritis group, while there were no changes in the other groups. These studies, demonstrating susceptibility of alcoholics with atrophic gastritis to ethanol-induced bleeding, contrasted with a previous observation which showed no effect of alcohol on gastric blood loss in normal volunteers (Bouchier and Williams, 1969). The effect of ethanol ingestion on gastrointestinal protein loss was studied by the method of clearance of intravenously administered ^{51}Cr-labeled albumin (Chowdhury *et al.,* 1977). Patients with both superficial and atrophic gastritis had accelerated protein loss into the feces following the ingestion of 200 ml of 40% ethanol. Studies in which gastric protein loss was quantitated with a stomach tube suggested that most of the gastrointestinal protein loss in drinking patients with atrophic gastritis could be accounted for by plasma leakage into the stomach. These studies suggest that accelerated gastrointestinal catabolism of protein may be a significant factor in the overall nutrition of alcoholics.

B. Ethanol and Gastric Motility

The nausea and vomiting that are often encountered in acute intoxication could be related to a retarding effect of ethanol on gastric emptying. Whether or not ethanol delays gastric emptying may depend on the presence of food in the stomach. In a study of normal subjects using a phenol red marker, Cooke (1970) found that the ingestion of 6% ethanol in 250 ml of water had no effect on gastric emptying time as compared to water alone. Barboriak and Meade (1970) studied the effect of 4 ounces of whiskey on gastric emptying of a high fat (60 gm) meal of buttered toast and eggs using a ^{51}Cr marker and a scintillation scanning technique. In seven subjects, the prior ingestion of whiskey prolonged the half-time of gastric emptying by a mean of 99 minutes. The mechanism for the effect of ethanol on gastric emptying is uncertain and may involve one or several gastrointestinal hormones (e.g., gastrin, cholecystokinin).

C. Ethanol and Gastric Secretion

In spite of the past practice of using ethanol as a test meal to measure gastric secretion, several recent studies indicate that intragastric ethanol has no stimulatory effect on acid secretion whatsoever. Cooke (1970) found no increase in gastric acidity following a test meal of 6% ethanol to healthy subjects. Subsequent studies using oral doses of 8, 12, and 16% ethanol in volumes of 350 ml had no stimulatory effect when compared to water alone (Cooke, 1972). These studies contrast with those of Hirschowitz *et al.* (1956), who administered 4, 8, and 16% ethanol intravenously in 250 ml of saline over a 30-minute period and found consistent increases in gastric acidity in the following 1-hour period. Cooke (1972) and Cooke and Grossman (1968) were unable to find any direct stimulatory effect of intragastric ethanol on the serum gastrin level. Thus, if ethanol does stimulate gastric acid output, the effect would seem to occur only after ethanol reaches an undefined level in the vascular circulation. Whether or not systemic ethanol directly affects the parietal cell and/or gastric releasing cells in the antral mucosa is yet to be determined.

In summary, ethanol has an acute injurious effect on the gastric mucosa and is rightly classified as a breaker of the gastric mucosal barrier. There is no good evidence that repeated alcohol ingestion leads to hyperacidity; in fact, the opposite finding of decreased acid output in association with mucosal atrophy seems to be the rule. Atrophic gastritis in chronic alcoholism predisposes a higher risk for gastric blood and protein loss. Peptic ulceration occurring in chronic alcoholic patients would seem to be the result of some other factor than ethanol-induced acid secretion.

IV. ALCOHOLISM AND THE SMALL INTESTINE

The factors that affect the small intestine in alcoholism include both acute and chronic effects of ethanol and the effects of attendant nutritional deficiency. As indicated in Section II,A, during active drinking, the upper jejunum is frequently exposed to ethanol concentration of 2% or more. The major attendant nutritional deficiency in alcoholism is that of folate, a vitamin essential for DNA metabolism and cellular replication.

A. Morphology

Small intestinal biopsies of chronic alcoholics, taken soon after a period of binge drinking, have usually shown normal mucosa on light microscopy (Mezey *et al.*, 1970; Halsted *et al.*, 1971). In two reports, however, intestinal biopsies of several folate deficient alcoholics showed megalocytic changes of the surface epithelium, i.e., enlargement of epithelial cell nuclei and shortening and widening of the columnar cell (Bianchi *et al.*, 1970; Hermos *et al.*, 1972). These changes, identical to those shown in pernicious anemia (Faroozan and Trier, 1967), probably reflect the profound folate deficiency associated with alcoholism in these patients. Whereas there are no clear-cut clinical reports of altered gut morphology in folate deficiency in the absence of alcoholism, both Klipstein *et al.* (1973) and Howard *et al.* (1974) have demonstrated megalocytic changes in the epithelial cells of rats made folate deficient by diet. Acute hemorrhagic erosions were produced in jejunal villi of rats given intragastric ethanol in a dose of 3 gm/kg body weight, the severity of the histologic response correlating with the concentration of ethanol, which varied from 5 to 50 gm/100 ml (Baraona *et al.*, 1974). Similar acute jejunal histologic changes were noted by Hoyumpa *et al.* (1975a) in rats after the intragastric administration of ethanol (7.5 gm/kg). Chronic alcohol feeding to rats for 9 to 12 months and to human volunteers for 2 to 3 months resulted in ultrastructural changes in intestinal epithelial cells similar to those produced in hepatocytes—distorted intestinal epithelial cell mitochondria with dilated endoplasmic reticulum. In the rats, slight villus shortening was associated with decreased activities of mucosal surface disaccharidase enzymes and increased activity of the crypt enzyme thymidine kinase (Rubin *et al.*, 1972).

B. Intestinal Metabolism

1. Glucose

Rosensweig *et al.* (1971) demonstrated that jejunal glycolytic enzymes (e.g., fructose-1-phosphate aldolase and hexokinase) can be induced by feeding specific substrates to rat and man. Additional studies showed induction of jejunal glycolysis by pharmacologic doses of folic acid (15 mg per day for seven days) (Rosensweig *et al.*, 1969). Recently the same group has demonstrated that folic acid induction of jejunal glycolysis in man can be prevented by the concommitant oral administration of ethanol, 60 ml per day (Greene *et al.*, 1975). While these studies provide some insight into intestinal folate–ethanol interrelationships, their clinical significance remains unclear, since the doses of folic acid employed exceed the usual dietary component manyfold, and since the role of jejunal glycolysis in small intestine function remains to be defined.

2. Lipids

The acute administration of ethanol has been shown to result in increases in triglyceride (Carter *et al.*, 1971) and cholesterol synthesis by intestinal slices (Middleton *et al.*, 1971), triglyceride content of the small intestinal mucosa, and lymphatic output of triglycerides, cholesterol, and phospholipids (Mistilis and Ockner, 1972). The increased triglyceride synthesis could be partially suppressed by pyrazole, while the increased cholesterol synthesis (Carter *et al.*, 1971) was only demonstrable when ethanol remained in the intestinal lumen (Middleton *et al.*, 1971), suggesting that the effect of ethanol may be mediated by its metabolism in the intestinal mucosa. It has also been suggested that the observed increases in intestinal lipid synthesis and lipid output by the lymph may contribute to the hyperlipemia and fatty infiltration of the liver induced by ethanol (Mistilis and Ockner, 1972).

C. Intestinal Enzymes

Several systems have been studied, involving digestion, secretion, and absorption. Studies involving alcohol and the metabolism of glucose and lipids have been described above.

1. Disaccharidases

In an intriguing study of Czechs with functional diarrhea, Madzarovora-Nohejlova (1971) identified 10 Pilsen beer drinkers in whom jejunal sucrase and maltase activities were decreased, though lactase activity was normal. This study was interpreted as showing "exhaustion" of jejunal disaccharidase by exposure to the high maltose content of Pilsen beer. More recently, Perlow *et al.* (1977) performed lactose tolerance tests (1 gm/kg body weight) and measurements of jejunal disaccharidase activities on 21 black and 20 Caucasian alcoholic patients, all of whom had been drinking heavily for at least 1 month, and none of whom showed evidence of clinical malnutrition. An age and race-matched nonalcoholic control group was studied for comparison. Among blacks, lactase deficiency, correlating with lactose intolerance, was present in half the controls but universally among the alcoholics. The difference in lactase activities between nonalcoholic and alcoholic Caucasian patients was not significant. In both racial groups, about a third of the alcoholics had subnormal jejunal sucrase activity. Lactase and sucrase activities rose in five alcoholics of both races who were rebiopsied after 3 weeks of abstinence. In all instances, intestinal morphology was normal. These studies showed increased susceptibility of black alcoholics to milk-induced diarrhea and also imply that dietary sucrose may be a cause for diarrhea in as many as one-third of all alcoholics. Of additional importance, these data suggest a particular susceptibility of brush border enzymes to damage by alcohol, even though electron microscopic studies have not shown any specific effect on the microvillus membrane (Rubin *et al.*, 1972).

2. Enzyme Systems Involved in Intestinal Transport

Greene *et al.* (1971) showed that methanol and ethanol stimulate the activity of adenyl cyclase in homogenates of rat and human jejunal mucosa. These findings may be relevant to intestinal secretion, which is mediated by the cyclic AMP system. However, the effects on adenyl cyclase could only be demonstrated at high ethanol concentratins (11–33%), which are likely to be reached only transiently in the jejunum during active drinking (Halsted *et al.*, 1973a).

An attractive hypothesis that ethanol impairs sodium and water transport by inhibition of mucosal Na^+, K^+-dependent ATPase, the "sodium pump" of the lateral basement membrane of the absorbing epithelial cell, has been indirectly supported. Two studies showed that acute *in vitro* exposure to ethanol diminishes jejunal ATP content (Carter and Isselbacher, 1973; Krasner *et al.*, 1976a). While brain Na^+, K^+-ATPase is inhibited *in vitro* by 2% ethanol (Israel *et al.*, 1965), Krasner *et al.* (1976a) were unable to demonstrate diminished ATPase in whole mucosal homogenates of jejunal mucosa of guinea pigs fed ethanol, 2.5 gm/kg per day, for 2 weeks. On the other hand, the findings of Hoyumpa *et al.* (1975a), that both ouabain and 2.5% ethanol block the exit step of [^{35}S]thiamine from everted gut sacs of rat intestine, suggest inhibition by ethanol of intestinal Na^+, K^+-ATPase. A recent report that Na^+, K^+-ATPase in a purified basolateral membrane preparation of jejunal and ileal mucosa is suppressed by 0.5 M ethanol (Hoyumpa *et al.*, 1976), supports the hypothesis that inhibition of this enzyme is the mechanism by which ethanol blocks transport of water, sodium, and certain water soluble vitamins.

D. Alcoholism and Intestinal Absorption

In considering the multitude of studies of alcoholism on intestinal absorption, it is useful to distinguish studies of acute and chronic effects of ethanol from studies that suggest that the inhibitory effects of alcoholism on intestinal absorption are dependent on associated nutritional factors. As described below, the steatorrhea of chronic alcoholism is probably related mostly to inhibitory effects of alcohol and associated protein deficiency on pancreatic exocrine function. This discussion will deal with processes related to intestinal transport of water-soluble substances, including electrolytes, amino acids, and the water soluble vitamins B_{12}, folic acid, and thiamine.

The subjects utilized for clinical study of alcoholism on absorption have mainly been derelicts studied after drinking binges. About one-third of binge-drinking alcoholics give histories of diarrhea while drinking (Small *et al.*, 1959). Diarrhea may be caused in part by altered small intestinal motility. Robles *et al.* (1974) showed that alcohol acutely suppressed impeding contractions (type I) in the jejunum and enhanced propulsive contractions (type III) in the ileum of volunteers fed alcohol either acutely or chronically. Diarrhea in alcoholism is also partially the result of impaired jejunal fluid transport. The techniques of measuring fluid transport include triple lumen tube perfusion in human subjects and *in vivo* loop perfusion in experimental animals.

1. Water and Sodium Absorption

Jejunal perfusion studies have shown that water and sodium transport are decreased in binge-drinking alcoholics (Halsted *et al.*, 1971, 1973b; Krasner *et al.*, 1976b). Recent studies (Mekhjian and May, 1977) showed that decreased absorption or net secretion of water and sodium follows the administration to normal volunteers of a folate-depleted or folate-supplemented diet together with ethanol as 36% of calories. In a prospective study of four patients on various dietary regimens, Halsted *et al.* (1973b) induced net jejunal water and sodium secretion in two alcoholics given ethanol and a folate deficient diet, in a sober patient made folate deficient by diet, and in a patient fed ethanol 300 gm daily for 3 weeks with a nutritious, folate-repleted diet. More recently, Goetsch and Klipstein (1977) showed that dietary induction of folate deficiency in the rat is followed by net secretion of water and sodium into perfused jejunal loops. Thus, water and electrolyte malabsorption, a frequent manifestation of binge drinking, may reflect the combined effects of chronic ethanol ingestion and folate deficiency.

2. D-Xylose

This poorly metabolized pentose is used clinically to test mucosal transport. Confusing and sometimes conflicting results have been obtained from D-xylose testing of alcoholics. In the studies mentioned above (Halsted *et al.*, 1971; Krasner *et al.*, 1976b; Mezey *et al.*, 1970), D-xylose malabsorption was found in about one-third of all patients on admission to hospital. Mezey (1975) found D-xylose malabsorption in 10 recently admitted alcoholics, which became normal in eight patients after 2 weeks of normal diet in spite of continued ingestion of alcohol. In the same study, D-xylose malabsorption was induced by an acute dose

of ethanol, 0.8 gm/kg, in seven chronic alcoholics. Halsted *et al.* (1973b) induced D-xylose malabsorption in two alcoholics who drank ethanol during dietary induction of folate deficiency, but not in a sober patient on the same diet.
Lindenbaum and Lieber (1975) found that the chronic administration of ethanol
for 13–37 days to six normal volunteers enhanced D-xylose absorption. The
effect could not be explained by changes in tissue disappearance or urinary
excretion after intravenous D-xylose administration. Thus, D-xylose malabsorption is frequent in recently drinking alcoholics, and may require the combined
factors of folate deprivation and an acute ethanol load, whereas chronic ethanol
administration in nutritionally adequate patients actually enhances D-xylose
absorption.

3. Thiamine Absorption

Thiamine absorption has been studied both acutely and chronically in normal
volunteers and in chronic alcoholic patients. Measuring plasma levels and urine
excretion of radioactivity following oral doses of [^{35}S]thiamine (5 mg), Thomson
et al. (1970) found significant intestinal malabsorption in 34 chronic alcoholic
patients. In 12 patients with fatty liver and a variety of vitamin deficiencies,
thiamine malabsorption was corrected by a nutritious diet for 6 weeks. The
authors suggested that both nutritional deficiency and elevated portal venous
pressure contributed to initial thiamine malabsorption. In addition, the direct
effect of ethanol on thiamine absorption was shown by suppression of plasma and
urine radioactivity by the oral administration of ethanol (1.5 gm/kg) in three of
nine normal subjects given 5 mg oral [^{35}S]thiamine.

Recent studies from Hoyumpa *et al.* (1975a,b) and Howard *et al.* (1974) have
partially clarified the mechanism of thiamine transport and have provided two
possible mechanisms for the inhibitory effect of alcoholism. Using *in vivo* jejunal loops and everted gut sacs of rat intestine, Hoyumpa *et al.* (1975a,b)
showed that net transport of [^{35}S]thiamine is active, i.e., shows saturation kinetics, at low concentration (0.06 to 2.0 μM), but is passive at higher concentrations. Using the everted gut sac technique, they demonstrated inhibition of the
exit phase of [^{35}S]thiamine from tissue to serosal fluid in rats that had 1 hour
previously been given intragastric ethanol, 7.5 gm/kg body weight. Inhibition
only occurred at the lower, actively transported concentration of thiamine, and
correlated with histological evidence of acute jejunal mucosal damage. Similar
effects on the exit phase of thiamine transport were obtained with ouabain, an
inhibitor of Na^+, K^+-ATPase. These data, supported by recent studies using
basolateral membranes (Hoyumpa *et al.* (1976), suggest that ethanol specifically
impairs the Na^+, K^+-ATPase or "sodium pump" mechanism, which may be
essential for thiamine transport. Howard *et al.* (1974) made rats folate deficient
by diet over a 10-month period, and then studied [^{35}S]thiamine transport using
the *in vivo* intestinal loop perfusion method. Compared to pair-fed controls, the

folate-depleted rats showed significantly less absorption from a low (0.5 μM), but not high (17.5 μM) concentration of thiamine. These data correlated with histologic changes of megalocytosis in the villus epithelium. Thus, folate deficiency was also shown to impair the active transport mechanism of thiamine absorption. In summary, these cumulative studies have shown depressed thiamine absorption in chronic alcoholism, which can be ascribed both to a toxic effect of ethanol and to nutritional folate deprivation.

4. Folate

Folate deficiency occurs in the majority of binge-drinking alcoholics and is the most common cause of anemia in this population group (Herbert *et al.*, 1963; Hines, 1975). In addition to dietary inadequacy and possible defective hepatic folate storage, intestinal malabsorption may be an important cause of folate deficiency in alcoholism. Three studies have been reported from our laboratory which deal with the intestinal absorption of simple folic acid, or pteroylmonoglutamate. In our initial study (Halsted *et al.*, 1967), we demonstrated lower levels of serum radioactivity in a group of 23 recently drinking alcoholics, compared to a matched control group, after the oral administration of [^{3}H]-pteroylmonoglutamate (^{3}H-PG-1), 1.5 μg/kg, and an intravenous tissue saturating dose of nonradioactive folic acid. There was no consistent change in absorption of ^{3}H-PG-1 in normal volunteers studied before and after acute alcoholic intoxication. Subsequently, the intestinal perfusion method was used to measure the jejunal uptake (luminal disappearance) of ^{3}H-PG-1 in 11 alcoholics studied within 48 hours of admission to the hospital (Halsted *et al.*, 1971). These studies used an isotonic saline solution containing glucose, 16.7 mM, and ^{3}H-PG-1 in a concentration of 25 μg/liter. Eight patients, all with clinical or laboratory evidence of malnutrition, including five with low serum folate levels, had lower jejunal uptake of ^{3}H-PG-1 as compared to three others, all of whom were well nourished. On follow-up, the jejunal uptake of ^{3}H-PG-1 rose to a normal level after two weeks of hospital diet, despite continued ethanol ingestion in five of the patients. The subsequent administration of ethanol, 256 gm per day, for two weeks with a hospital diet, failed to suppress the jejunal uptake of ^{3}H-PG-1 in seven of nine well-nourished alcoholics. These studies suggested that lack of a nutritional factor or factors resulted in malabsorption of folic acid. Subsequently, a prospective study was performed (Halsted *et al.*, 1973b) in which the jejunal uptake of ^{3}H-PG-1 was studied in four patients, all recovered and well-nourished at the start. Two patients were administered a folate-deficient diet, to which was added ethanol, 256 gm per day, in divided doses; one patient remained sober but ate the folate-deficient diet; and the fourth patient ate a nutritious hospital diet but in addition drank ethanol, 300 gm per day for three weeks. The time for development of folate deficiency by diet (low serum folate level with megaloblastic bone marrow) was 7 weeks in the patients who drank alcohol but 9 weeks in the patient who remained sober. No change in serum folate level was found in the patient who drank excessive alcohol with a regular diet. Compared to the initial values, the jejunal uptake of ^{3}H-PG-1, glucose, and sodium decreased after induction of folate deficiency in the two patients who drank alcohol. These values were corrected by the administration of folic acid, 5 mg by mouth per day for 2 weeks, in spite of continued ingestion of alcohol. Dietary induction of folate deficiency in the sober patient and administration of ethanol with a nutritious diet to the fourth patient resulted in net water secretion into the jejunum but no change in folic acid uptake. These studies suggested that the combination of dietary folate deficiency and alcohol ingestion results in impaired intestinal absorption of ^{3}H-PG-1, glucose, and sodium, while either factor alone, while inducing net water and sodium secretion into the jejunum, is insufficient to decrease folic acid absorption. Protein deficiency, also frequent in alcoholics, was not found to impair folic acid absorption *in vivo* in rats fed diets devoid of protein for 2 weeks (Halsted *et al.*, 1974). Still to be determined are the effects of alcohol ingestion and associated malnutrition on the digestion and absorption of pteroylpolyglutamates. Absorption of these conjugated folates requires prior hydrolysis to pteroylmonoglutamates (Halsted *et al.*, 1975). Thus, malabsorption of conjugated folate could result from inhibition of the enzymatic hydrolysis step as well as impairment of transport of PG-1.

5. Vitamin B$_{12}$

Absorption of vitamin B$_{12}$ was measured by the Schilling test (48-hour urine collection after administration of [^{57}Co]vitamin B$_{12}$, 0.75 μg, with intrinsic factor and flushing doses of parenteral vitamin B$_{12}$) in eight nutritionally repleted chronic alcoholic patients who were studied before and after drinking ethanol for 11–38 days at doses of 173–253 gm per day (Lindenbaum and Lieber, 1975). Supplemental vitamins were given daily. Decreased vitamin B$_{12}$ absorption was induced in six subjects who received the highest daily doses of ethanol. These results were not corrected by administration of pancreatic extract, shown by others to correct vitamin B$_{12}$ malabsorption in alcoholics with pancreatic insufficiency (Toskes and Deren, 1973). The effect of ethanol appeared to be specific for vitamin B$_{12}$ absorption since, paradoxically, enhanced absorption of D-xylose was observed in six of the patients following chronic alcohol administration.

6. Amino Acids

Few studies have been made of the effect of alcoholism on amino acid absorption. Using the jejunal perfusion method, Israel *et al.* (1969) found significant suppression of uptake of L-methionine when the perfusion solution contained 2% ethanol. Additional studies showed 50% inhibition of *in vivo* absorption of L-phenylalanine, but not of D-phenylalanine in rats given intragastric ethanol, 2.5 gm/kg body weight (Israel *et al.*, 1968). Another study using everted gut sacs demonstrated inhibition by 3% ethanol of transport of six different essential L-amino acids and of D-glucose (Chang *et al.*, 1967). Additional studies of the effect of alcoholism on protein nutrition include the recent observation of a protein losing gastroenteropathy in chronic alcoholics, cited above (Perlow *et al.*, 1977).

In summary, intestinal morphology, enzymology, and intestinal absorption in alcoholism are affected by the three factors of acute alcohol ingestion, chronic exposure to alcohol, and malnutrition, specifically folate deficiency. Whereas the studies cited have shown malabsorption of many different nutrients in chronic alcoholism, the relative effect of intestinal malabsorption on the poor nutrition of binge-drinking alcoholics remains to be established.

V. ALCOHOLISM AND THE PANCREAS

Alcoholism is one of the principal causes of pancreatitis. Most of the pancreatitis in alcoholism falls in the category of chronic relapsing pancreatitis, in which acute episodes of abdominal pain develop from a gland which has irreversible changes of the ducts and parenchyma associated with chronic inflammation and fibrosis (Strum and Spiro, 1971). The development of the first episode of abdominal pain usually occurs after 5 to 10 years of alcohol intake. The occurrence of single episodes of acute pancreatitis after alcohol intake, such as is observed to occur in association with gallstones, has been documented only occasionally (Thomson *et al.*, 1975). Also, there is no evidence that episodes of acute pancreatitis precede the development of chronic pancreatitis. In one study the mean age of onset of acute pancreatitis was 50.7 years, which is 13 years later than the mean age of onset of 37.7 years, found for chronic pancreatitis (Sarles *et al.*, 1965). Chronic pancreatitis can also develop insidiously in the absence of episodes of abdominal pain. In one study, autopsy evidence of pancreatitis was found in 12.6% of alcoholic patients in whom the disease was not suspected during life (Czernobilsky and Mikat, 1964). Both a toxic effect of ethanol and malnutrition have been considered as causes of chronic pancreatitis in alcoholism.

A. Effects of Alcohol on the Pancreas

Various studies suggest that there is a relationship between the amounts of alcohol consumed and the prevalence of chronic pancreatitis. In an international survey of chronic pancreatitis based on autopsy material, a daily consumption of alcohol of the order of 150 gm was found (Sarles, 1973). In Marseilles, patients with chronic calcifying pancreatitis consumed an average of 175 gm per day as compared with 74 gm per day for control subjects (Sarles *et al.*, 1965). Of the 55 patients with chronic calcifying pancreatitis studied, only one abstained from alcohol, while 5 ingested less than 50 gm per day. In a recent study from Portugal (Bordalo *et al.*, 1974), pancreatic insufficiency was observed only in subjects ingesting more than 160 gm of alcohol per day.

It has been suggested that alcohol may cause pancreatitis by a dual effect: the stimulation of pancreatic secretion with concomitant obstruction to pancreatic flow. The increased pancreatic duct pressure from such a combination of events would lead to pancreatic tissue edema, inflammation, and necrosis, and eventually parenchymal atrophy and fibrosis. One of the principal difficulties in accepting this theory is that there is no evidence that acute pancreatic lesions precede the development of chronic pancreatitis. The known effects of the acute and chronic administration of alcohol on the pancreas in man and experimental animals are complex, and they do not yet provide a clear picture of the pathogenesis of chronic pancreatitis in alcoholism.

1. Acute Effects of Alcohol

Alcohol distributes in body water and appears in pancreatic secretions in concentrations similar to those achieved in the peripheral blood (Beck *et al.*, 1974). A number of early studies showed that alcohol administered either orally or intravenously stimulated pancreatic secretion in dogs (Walton *et al.*, 1960) or man (Lowenfels *et al.*, 1968). The increases in pancreatic secretion were attributed to an increase in endogenous secretin released when hydrochloric acid secretion, enhanced by alcohol, reached the duodenum. Recently, in the human, the administration of 60 ml of 86 proof vodka was shown to result in an increase in plasma secretin (Straus *et al.*, 1975). Also possible is that the observed increases in pancreatic secretion were mediated by gastrin since perfusion of a denervated gastric pouch in the dog with ethanol increased pancreatic secretion despite the diversion of gastric secretion (Schapiro *et al.*, 1968). Increases in serum gastrin have been demonstrated after the oral or intravenous administration of alcohol in a dose of 32 gm (Becker *et al.*, 1974), and gastrin is known to have a stimulatory effect on the pancreas (Wormsley, 1966). Administration of a large dose of alcohol (1 gm/kg) intravenously in one study in rabbits resulted in a decrease in the pancreatic output of volume and bicarbonate but in no changes in the output of protein. Also, addition of a large amount of alcohol (3% w/v) to fluid bathing the pancreas *in vitro* resulted in decreased pancreatic secretion of volume, bicarbonate, and protein (Solomon *et al.*, 1974).

Alcohol, in contrast to its usual effect of stimulating basal pancreatic output, has an inhibitory effect on the secretin and cholecystokinin-stimulated output of the pancreas. Davis and Pirola (1966) were the first to demonstrate that the intraduodenal or intravenous administration of ethanol in a dose of 0.6 gm/kg resulted in decreases in the pancreatic output of volume stimulated by secretin and cholecystokinin. Subsequently, Mott *et al.* (1972) confirmed the above finding using the same dose of ethanol by the intravenous route, and demonstrated, in addition, inhibition in the output of bicarbonate and enzymes. Furthermore, they showed that the maximum inhibition of stimulated pancreatic output coincided with peak levels of alcohol in the blood of the order of 200 mg/100 ml. The effect of ethanol in decreasing stimulated pancreatic output was initially attributed to alcohol-induced spasm of the sphincter of Oddi, inferred from the observation that alcohol administration increased bile duct pressures (Pirola and Davis, 1968). However, this mechanism was excluded by the demonstration of similar decreases in pancreatic stimulated output in dogs with cannulae in their pancreatic ducts after the intravenous administration of alcohol (Bayer *et al.*, 1972; Tiscornia *et al.*, 1974). In addition, the increased pressure of the sphincter of Oddi, following alcohol ingestion, was found to be too small to explain the decrease in pancreatic secretion (Capitaine and Sarles, 1971). The decrease in stimulated pancreatic output produced by alcohol appears to be mediated by cholinergic nerves since it is blocked by atropine and pentolinium (Tiscornia *et al.*, 1975).

2. Chronic Effects of Alcohol

A paradoxical increase in secretin-stimulated pancreatic volume output was found by Dreiling *et al.* (1973) in 17 patients during the early stages of alcoholism which, however, declined in 11 patients who continued alcohol intake. The initial increase in pancreatic output was ascribed by the authors to an increase in pancreatic mass due to regeneration in association with early inflammation of the pancreas. This early stage of injury precedes the development of chronic pancreatitis in which the parenchyma is replaced by fibrosis. Similar increases in secretin and pancreozymin-stimulated pancreatic output of volume, bicarbonate, and protein were found in dogs after the administration of ethanol for a period of 6 weeks to 3 months following which there was a decrease toward baseline levels (Tiscornia *et al.*, 1974). Recent studies by the same authors did not show any change in secretin and cholecystokinin-stimulated pancreatic output after ethanol feeding of dogs from 6 weeks to 24 months, although an increase in secretin-stimulated volume and bicarbonate output was found after 24 months of ethanol feeding (Sarles *et al.*, 1977).

The administration of an acute intravenous dose of ethanol resulted in an increase in the secretin- and cholecystokinin-stimulated pancreatic output of protein in dogs after ethanol feeding for 12 months (Tiscornia *et al.*, 1974). A similar increase in protein output of the pancreas was found in rats after ethanol feeding for 13 months (Cavarzan *et al.*, 1975). Rats fed ethanol for 30 months developed pancreatic lesions, consisting of acinar cell atrophy, duct proliferation, chronic inflammation, and fibrosis, which are indistinguishable from those observed in chronic pancreatitis in man. An increase in the concentration of protein in the pancreatic juice and precipitation of protein in the pancreatic ducts are also seen. It is suggested that an increased secretion of protein with resultant precipitation leading to obstruction may be a mechanism for the development of chronic pancreatitis (Sarles *et al.,* 1971a). The mechanism for the increased pancreatic secretion of protein after chronic alcohol consumption remains to be determined. Chronic ethanol feeding in rats has been found to result in a decreased synthesis of pancreatic protein and phospholipids. A decrease in the incorporation of DL-[1-^{14}C]leucine into pancreatic protein was demonstrated in both *in vivo* and *in vitro* experiments after 48 weeks of ethanol feeding (Sardesai and Orten, 1968); while a decrease incorporation of ^{32}P into phospholipids was demonstrated after 65 days of ethanol feeding (Orrego-Matte *et al.,* 1969).

B. Malnutrition and Pancreatic Function

Chronic alcoholism is a frequent cause of malnutrition and vitamin deficiencies. Pancreatic dysfunction has been well documented in malnourished populations. Chronic pancreatitis, indistinguishable from the pancreatitis seen in the chronic alcoholic, is a common finding in populations with a low protein intake and without a significant alcohol intake (Sarles, 1974). The pancreatic dysfunction in infants with kwashiorkor (Gomez *et al.,* 1954) and in some adults with malnutrition (Tandon *et al.,* 1969) is frequently reversible to normal on protein repletion. Rats fed a protein-free diet show marked decreases in bicarbonate, trypsin, and volume of pancreatic juice produced after secretin stimulation, all of which are reversed to normal on protein refeeding (Lemire and Iber, 1967).

Pancreatic function was evaluated by means of the secretin-stimulation test in 32 chronic alcoholics 1–4 days after admission to the hospital and discontinuation of alcohol intake. All these patients had been ingesting alcohol in excess of 200 gm per day, and dietary intake had been poor (less than 20 gm of protein per day) for a period of at least 10 days prior to admission in 25 (78%) of them. Decreases in the output of volume, bicarbonate, and amylase were found in 14 (44%) of the patients. In most of the patients pancreatic function returned to normal after institution of a normal diet for 2 weeks whether or not alcohol ingestion had been continued in a dose of 134 gm per day. These studies demonstrated that subclinical malnutrition may play a role in the pathogenesis of pancreatic dysfunction (Mezey *et al.,* 1970). Later studies, in which the pancreas was stimulated with secretin and cholecystokinin, confirmed the findings of decreased output of bicarbonate and pancreatic enzymes; in those patients, however, volume output was found to be normal. Further studies (Mezey and Potter, 1976) demonstrated that the transient pancreatic dysfunction frequently observed in the actively drinking alcoholic patients was caused by protein deficiency. Five patients with decreases in the pancreatic output of bicarbonate and enzymes after secretin and cholecystokinin stimulation were placed on 250 gm of alcohol a day and given diets of different protein content. Initial administration of a low protein (25 gm) 1800 calorie diet for a period of 10 days resulted in no improvement of pancreatic function. By contrast, institution of a normal protein (100 gm) 2600 calorie diet for 10 days resulted in a return to normal in the output of bicarbonate, amylase, lipase, and chymotrypsin, while readministration of the low protein diet for another 10 days resulted in a decreased output of amylase and chymotrypsin. The outputs of volume, trypsin, and protein remained unchanged throughout the study. Mild steatorrhea was observed in the two patients with the lowest lipase output at admission, with recovery to normal following institution of the normal protein diet. The presence of steatorrhea in chronic alcoholic patients correlated best with a low lipase output following secretin and cholecystokinin stimulation (Fig. 3). In support of this mechanism for steatorrhea are previous studies in alcoholic patients demonstrating abnormal hydrolysis of triglycerides in association with decreased lipase concentration, but normal intraluminal bile salt concentration, with improvement toward normal following abstinence (Roggin *et al.*, 1972).

Adaptive increases in the concentrations of amylase chymotrypsin and lipase in pancreatic homogenates have been demonstrated in rats fed high carbohydrate (Grossman *et al.*, 1943), high protein (Abdeljlil and Desnuelle, 1964), and high fat (Bŭcko and Kopec, 1968) diets, respectively. Although dietary adaptation may have contributed to the increased pancreatic output of amylase and chymotrypsin following the administration of the normal diet, it is an unlikely mechanism for the overall recovery of pancreatic secretion, in particular, for the increases in the output of bicarbonate and lipase. Some of the changes in pancreatic enzyme output may have resulted from the interaction of alcohol with the protein content of the diet, since chronic alcohol administration has been shown to decrease the pancreatic content of enzymes in rats fed a low protein diet, while it increases them in rats fed a high protein and high lipid diet (Sarles *et al.*, 1971b). The significance of pancreatic dysfunction in the chronic alcoholic, as related to the development of chronic pancreatitis, remains unknown.

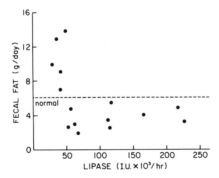

Fig. 3. Relationship between fecal fat excretion and pancreatic output of lipase following cholecystokinin–pancreozymin–secretin stimulation. (From Mezey, 1975, reproduced by permission.)

C. Alcoholism and Pancreatic Carcinoma

Alcoholism has been reported to be associated with an increased prevalence of cancer of the mouth, pharynx, and esophagus (Kamionkowski and Fleshler, 1965), but not of the stomach or intestine. In a preliminary retrospective review of 83 patients with carcinoma of the pancreas, 75% of the patients were found to have a history of chronic alcoholism (Burch and Ansari, 1968). This was much higher than the 14% prevalence of chronic alcoholism found in a matched control group of patients. The mean duration of chronic alcoholism in the patients with carcinoma of the pancreas was 15 years. Only 5.5% of the carcinoma of the pancreas was associated with chronic inflammation of the pancreas, and a history of episodes of pancreatitis was obtained in only 2.4% of the patients, suggesting that pancreatitis does not precede the development of carcinoma of the pancreas. In a recent review of the causes of death of 874 white chronic alcoholic patients admitted to a mental institution, the incidence of death from carcinoma of the pancreas was found to be lower than expected on the basis of reported United States age corrected death rates for carcinoma of the pancreas (Monson and Lyon, 1975).

VI. SUMMARY OF RESEARCH NEEDS

Areas which require further study are listed below.

1. Investigation of the physiologic role of alcohol dehydrogenase activities in the stomach and various parts of the intestine, and of the role that alcohol ingestion may have in interfering with those physiologic functions.

2. Sources of endogenous ethanol production in the gastrointestinal tract, and of its metabolic fate and effects.

3. Study of the effects of moderate drinking and of smaller doses of ethanol on gastrointestinal and pancreatic function.

4. Further differentiation of the effects of acute alcoholism from those of chronic alcoholism on intestinal and pancreatic function and morphology at the clinical and experimental level. Special consideration has to be given to (a) the use of alcohol concentrations found in the gut during active drinking and (b) differentiation of the effects of alcohol from those associated with nutritional deficiency.

5. Specific areas which require definition in view of the association of alcoholism and malnutrition are (a) studies of protein nutrition in alcoholism, in particular of peptide absorption and protein loss from the gut; and (b) a more precise definition of the role of malabsorption in the deficiency of water soluble vitamins such as folate and thiamine, in light of new concepts of mechanism of intestinal transport.

6. More comprehensive studies of biochemical changes produced in the intestine and pancreas by chronic ethanol ingestion and/or nutrient deficiencies. Correlation of changes in biochemical parameters with changes in function and morphology is necessary to unravel the pathogenesis of intestinal and pancreatic dysfunction. For example, studies of the effects of alcohol and nutrient deficiencies on protein and enzyme synthesis and degradation in the pancreas would provide a better understanding of the secretory and morphologic changes found in the pancreas of man and experimental animals after long-term aocohol ingestion.

7. Further investigation of the effects of alcohol and nutrient deficiencies on the secretion and action of hormones which regulate pancreatic secretion.

8. More investigation of the relationship, if any, between acute and chronic pancreatitis.

9. More precise information on the amounts of alcohol intake necessary for the development of chronic pancreatitis in the patient and in the experimental animal.

REFERENCES

Abdeljlil, A. B., and Desnuelle, P. (1964). Sur l'adaptation des enzymes exocrines du pancréas à la composition du régime. *Biochim. Biophys. Acta* **81**, 136–149.

Baraona, E., Pirola, R. C., and Lieber, C. S. (1974). Small intestinal damage and changes in cell population produced by ethanol ingestion in the rat. *Gastroenterology* **66**, 226–234.

Barboriak, J. J., and Meade, R. C. (1970). Effect of alcohol on gastric emptying in man. *Am. J. Clin. Nutr.* **23**, 1151–1153.

Bayer, M., Rudick, J., Lieber, C. S., and Janowitz, H. D. (1972). Inhibitory effect of ethanol on canine exocrine pancreatic secretion. *Gastroenterology* **63**, 619–626.

Beaumont, W. (1833). "Experiments and Observation on the Gastric Juice and the Physiology of Digestion," p. 280. J. P. Allen, Plattsburgh, New York.

Beck, I. T., Paloschi, G. B., Dinda, P. K., and Beck, M. (1974). Effect of intragastric administration of alcohol on the ethanol concentrations and osmolality of pancreatic juice, bile, and portal and peripheral blood. *Gastroenterology* **67**, 484–489.

Becker, H. D., Reeder, D. D., and Thompson, J. C. (1974). Gastrin release by ethanol in man and in dogs. *Ann. Surg.* **179**, 906–909.

Belber, J. P. (1978). Gastroscopy and Duodenoscopy. *In* "Gastrointestinal Disease" (M. H. Sleisenger and J. S. Fortran, eds.), 2nd ed., pp. 691–713. Saunders, Philadelphia, Pennsylvania.

Bianchi, A., Chipman, D. W., Dreskin, A., and Rosensweig, N. (1970). Nutritional folic acid deficiency with megaloblastic change in the small bowel epithelium. *N. Engl. J. Med.* **282**, 859–861.

Bordalo, O., Batista, A., Noronha, M., Lamy, J., and Dreiling, D. A. (1974). Effects of ethanol on liver morphology and pancreatic function in chronic alcoholism. *Mt. Sinai J. Med., N.Y.* **41**, 722–731.

Bouchier, I. A. D., and Williams, H. S. (1969). Determination of faecal blood-loss after combined alcohol and sodium-acetylsalicylate intake. *Lancet* **i**, 178–180.

Bučko, A., and Kopec, Z. (1968). Adaptation of enzymes activity of the rat pancreas on altered food intake. *Nutr. Dieta* **10**, 276–287.

Burch, G. E., and Ansari, A. (1968). Chronic alcoholism and carcinoma of the pancreas. *Arch. Intern. Med.* **122**, 273–275.

Capitaine, Y., and Sarles, H. (1971). Action de l'ethanol sur le tonus du sphincter d'Oddi chez l'homme. *Biol. Gastro-Enterol.* **3**, 231–236.

Carter, E. A., and Isselbacher, K. J. (1971). The metabolism of ethanol to carbon dioxide by stomach and small intestinal slices. *Proc. Soc. Exp. Biol. Med.* **138**, 817–819.

Carter, E. A., and Isselbacher, K. J. (1973). Effect of ethanol on intestinal adenosine triphosphate (ATP) content. *Proc. Soc. Exp. Biol. Med.* **142**, 1171–1173.

Carter, E. A., Drummey, G. D., and Isselbacher, K. J. (1971). Ethanol stimulates triglyceride synthesis by the intestine. *Science* **174**, 1245–1247.

Cavarzan, A., Texeira, A. S., Sarles, H., Palasciano, G., and Tiscornia, O. (1975). Action of intragastric ethanol on the pancreatic secretion of conscious rats. *Digestion* **13**, 145–152.

Cederbaum, A. I., Pietrusko, R., Hempel, J., Becker, F. F., and Rubin, E. (1975). Characterization of a nonhepatic alcohol dehydrogenase from rat hepatocellular carcinoma and stomach. *Arch. Biochem. Biophys.* **171**, 348–360.

Chang, T., Lewis, J., and Glazko, A. J. (1967). Effect of ethanol and other alcohols on the transport of amino acids and glucose by everted sacs of rat small intestine. *Biochim. Biophys. Acta* **135**, 1000–1007.

Chowdhury, A. R., Malmud, L. S., and Dinoso, V. P., Jr. (1977). Gastrointestinal plasma protein loss during ethanol ingestion. *Gastroenterology* **72**, 37–40.

Cooke, A. R. (1970). The simultaneous emptying and absorption of ethanol from the human stomach. *Am. J. Dig. Dis.* **15**, 449–454.

Cooke, A. R. (1972). Ethanol and gastric function. *Gastroenterology* **62**, 501–502.

Cooke, A. R., and Grossman, M. I. (1968). Comparison of stimulants of antral release of gastrin. *Am. J. Physiol.* **215**, 314–317.

Czernobilsky, B., and Mikat, K. W. (1964). The diagnostic significance of interstitial pancreatitis found at autopsy. *Am. J. Clin. Pathol.* **41**, 33–43.

Davenport, H. W. (1967). Ethanol damage to canine oxyntic glandular mucosa. *Proc. Soc. Exp. Biol. Med.* **126**, 657–662.

Davenport, H. W. (1969). Gastric mucosal hemorrhage in dogs. Effects of acid, aspirin, and alcohol. *Gastroenterology* **56**, 439–449.

Davis, A. E., and Pirola, R. C. (1966). The effects of ethyl alcohol on pancreatic exocrine function. *Med. J. Aust.* **2**, 757–760.

Dinoso, V. P., Jr., Chey, W. Y., Siplet, H., and Lorber, S. H. (1970). Effects of ethanol on the gastric mucosa of the Heidenhain pouch of dogs. *Am. J. Dig. Dis.* **15**, 809–817.

Dinoso, V. P., Jr., Chey, W. Y., Braverman, S. P., Rosen, A. P., Ottenberg, D., and Lorber, S. H. (1972). Gastric secretion and gastric mucosal morphology in chronic alcoholics. *Arch. Intern. Med.* **130**, 715–719.

Dinoso, V. P., Jr., Meshkinpour, H., and Lorber, S. H. (1973). Gastric mucosal morphology and faecal blood loss during ethanol ingestion. *Gut* **14**, 289–292.

Dinoso, V. P., Jr., Ming, S., and McNiff, J. (1976). Ultrastructural changes of the canine gastric mucosa after topical application of graded concentrations of ethanol. *Am. J. Dig. Dis.* **21**, 626–632.

Dreiling, D. A., Greenstein, A., and Bordalo, O. (1973). Newer concepts of pancreatic secretory patterns. Pancreatic secretory mass and pancreatic secretory capacity: Pancreatic hypersecretion. *M. Sinai J. Med., N.Y.* **40**, 666–676.

Eastwood, G. L., and Kirchner, J. P. (1974). Changes in the fine structure of mouse gastric epithelium produced by ethanol and urea. *Gastroenterology* **67**, 71–84.

Faroozan, P., and Trier, J. S. (1967). Mucosa of the small intestine in pernicious anemia. *N. Engl. J. Med.* **277**, 553–558.

Geall, M. G., Phillips, S. F., and Summerskill, W. H. J. (1970). Profile of gastric potential difference in man. Effects of aspirin, alcohol, bile, and endogenous acid. *Gastroenterology* **58**, 437–443.

Goetsch, C. A., and Klipstein, F. A. (1977). Effect of folate deficiency of the intestinal mucosa on jejunal transport in the rat. *J. Lab. Clin. Med.* **89**, 1002–1008.

Gomez, F., Galvan, R. R., Cravioto, J., and Frenk, S. (1954). Estudio sobre el niño desnutrido. XI. Actividad enzimática del contenido duodenal en niños con desnutricion de tercer grado. *Pediatrics* **13**, 544–548.

Greene, H. L., Herman, R. H., and Kraemer, S. (1971). Stimulation of jejunal adenyl cyclase by ethanol. *J. Lab. Clin. Med.* **78**, 336–342.

Greene, H. L., Stifel, F. B., Herman, R. H., Herman, Y. F., and Rosensweig, N. S. (1975). Ethanol-induced inhibition of human intestinal enzyme activities: Reversal by folic acid. *Gastroenterology* **67**, 434–440.

Grossman, M. I., Greengard, H., and Ivy, A. C. (1943). The effect of dietary composition on pancreatic enzymes. *Am. J. Physiol.* **138**, 676–682.

Halsted, C. H., Griggs, R. C., and Harris, J. W. (1967). The effect of alcoholism on the absorption of folic acid (H^3-PGA) evaluated by plasma levels and urine excretion. *J. Lab. Clin. Med.* **69**, 116–131.

Halsted, C. H., Robles, E. A., and Mezey, E. (1971). Decreased jejunal uptake of labeled folic acid (^{3}H-PGA) in alcoholic patients: Roles of alcohol and nutrition. *N. Engl. J. Med.* **285**, 701–706.

Halsted, C. H., Robles, E. A., and Mezey, E. (1973a). Distribution of ethanol in the human gastrointestinal tract. *Am. J. Clin. Nutr.* **26**, 831–834.

Halsted, C. H., Robles, E. A., and Mezey, E. (1973b). Intestinal malabsorption in folate deficient alcoholics. *Gastroenterology* **64**, 526–532.

Halsted, C. R., Bhanthumnavin, K., and Mezey, E. (1974). Jejunal uptake of tritiated folic acid in the rat studied by *in vivo* perfusion. *J. Nutr.* **104**, 1674–1680.

Halsted, C. H., Baugh, C. H., and Butterworth, C. E., Jr. (1975). Jejunal perfusion of simple and conjugated folates in man. *Gastroenterology* **68**, 261–269.

Herbert, V., Zalusky, R., and Davidson, C. S. (1963). Correlation of folate deficiency with alcoholism and associated macrocytosis, anemia, and liver disease. *Ann. Intern. Med.* **58**, 977–988.

Hermos, J. A., Adams, W. H., Lin, Y. K., and Trier, J. (1972). Mucosa of the small intestine in folate deficient alcoholics. *Ann. Intern. Med.* **76**, 957–965.

Hines, J. D. (1975). Hematological abnormalities involving vitamin B_6 and folate metabolism in alcoholic subjects. *Ann. N.Y. Acad. Sci.* **252**, 316–327.

Hirschowitz, B. I., Pollard, M. H., Hartwell, S. W., Jr., and London, J. (1956). The action of ethyl alcohol on gastric acid secretion. *Gastroenterology* **30**, 244–253.

Howard, L., Wagner, C., and Schenker, S. (1974). Malabsorption of thiamine in folate deficient rats. *J. Nutr.* **104**, 1024–1032.

Hoyumpa, A. M., Jr., Breen, K. J., Schenker, S., and Wilson, F. A. (1975a). Thiamine transport across the rat intestine. Effect of ethanol. *J. Lab. Clin. Med.* **86**, 803–816.

Hoyumpa, A. M., Jr., Middleton, H. M., III, Wilson, F. A., and Schenker, S. (1975b). Thiamine transport across the rat intestine. *Gastroenterology* **68**, 1218–1227.

Hoyumpa, A. M., Jr., Nichols, S., Wilson, F. A., and Schenker, S. (1976). Possible mechanism of inhibition of intestinal thiamine transport by ethanol. *Clin. Res.* **24**, 565A.

Israel, Y., Kalant, H., and Laufer, I. (1965). Effect of ethanol on Na, K, Mg-stimulated microsomal ATPase activity. *Biochem. Pharmacol.* **14**, 1803–1914.

Israel, Y., Salazar, I., and Rosenmann, E. (1968). Inhibitory effects of alcohol on intestinal amino acid transport *in vivo* and *in vitro*. *J. Nutr.* **96**, 499–504.

Israel, Y., Valenzuela, J. E., Salazar, I., and Ugarte, G. (1969). Alcohol and amino acid transport in the human small intestine. *J. Nutr.* **98**, 222–224.

Isselbacher, K. J., and Greenberger, N. J. (1964). Metabolic effects of alcohol on the liver. *N. Engl. J. Med.* **270**, 351–356, 402–410.

Kamionkowski, M. P., and Fleshler, B. (1965). The role of alcoholic intake in esophageal carcinoma. *Am. J. Med. Sci.* **249**, 696–700.

Klipstein, F. A., Lipton, S. D., and Schenk, E. A. (1973). Folate deficiency of the intestinal mucosa. *Am. J. Clin. Nutr.* **26**, 728–737.

Krasner, N., Carmichael, H. A., Russell, R. I., Thompson, G. G., and Cochran, K. M. (1976a). Alcohol and absorption from the small intestine. Effect of ethanol on ATP and ATPase activity in guinea pig jejunum. *Gut* **17**, 249–251.

Krasner, N., Cochran, K. M., Russell, R. I., Carmichael, H. A., and Thompson, G. G. (1976b). Alcohol and absorption from the small intestine. Impairment of absorption from the small intestine in alcoholics. *Gut* **17**, 245–248.

Krebs, H., and Perkins, J. R. (1970). The physiological role of liver alcohol dehydrogenase. *Biochem. J.* **118**, 635–644.

Leake, C. D., and Silverman, M. (1966). "Alcoholic Beverages in Clinical Medicine," p. 48. Year Book Medical Publ., Chicago, Illinois.

Lemire, S., and Iber, F. L. (1967). Pancreatic secretion in rats with protein malnutrition. *Johns Hopkins Med. J.* **120**, 21–25.

Lindenbaum, J., and Lieber, C. S. (1975). Effect of chronic ethanol administration on intestinal absorption in man in the absence of nutritional deficiency. *Ann. N.Y. Acad. Sci.* **252**, 228–234.

Lowenfels, A. B., Masih, B., Lee, T. C. Y., and Rohman, M. (1968). Effect of intravenous alcohol on the pancreas. *Arch. Surg.* **96**, 440–441.

Madzarovora-Nohejlova, J. (1971). Intestinal disaccharidase activity in adults and chronic Pilsen beer drinkers. *Biol. Gastro-Enterol.* **4**, 325–332.

Mekhjian, H. S., and May, E. S. (1977). Acute and chronic effects of ethanol on fluid transport in the human small intestine. *Gastroenterology* **72**, 1280–1286.

Mezey, E. (1975). Intestinal function in chronic alcoholism. *Ann. N.Y. Acad. Sci.* **252**, 215–227.

Mezey, E., and Potter, J. J. (1976). Changes in exocrine pancreatic function produced by altered dietary protein intake in drinking alcoholics. *Johns Hopkins Med. J.* **138**, 7–12.

Mezey, E., Cherrick, G. R., and Holt, P. R. (1968). Biliary excretion of alcohol dehydrogenase. *J. Lab. Clin. Med.* **71**, 798–806.

Mezey, E., Jow, E., Slavin, R. E., and Tobon, F. (1970). Pancreatic function and intestinal absorption in chronic alcoholism. *Gastroenterology* **59**, 657–664.

Middleton, W. R. J., Carter, E. A., Drummey, G. D., and Isselbacher, K. J. (1971). Effect of oral ethanol administration on intestinal cholesterogenesis in the rat. *Gastroenterology* **60**, 880–887.

Mistilis, S. P., and Garske, A. (1969). Induction of alcohol dehydrogenase in liver and gastrointestinal tract. *Aust. Ann. Med.* **18**, 227–231.

Mistilis, S. P., and Ockner, R. K. (1972). Effects of ethanol on endogenous lipid and lipoprotein metabolism in small intestine. *J. Lab. Clin. Med.* **80**, 34–46.

Monson, R. R., and Lyon, J. L. (1975). Proportional mortality among alcoholics. *Cancer (Philadelphia)* **36**, 1077–1079.

Mott, C., Sarles, H., Tiscornia, O., and Gullo, L. (1972). Inhibitory action of alcohol on human exocrine pancreatic secretion. *Am. J. Dig. Dis.* **17**, 902–910.

Orrego-Matte, H., Navia, E., Feres, A., and Costamaillere, L. (1969). Ethanol ingestion and incorporation of ^{32}P into phospholipids on pancreas in the rat. *Gastroenterology* **56**, 280–285.

Palmer, E. D. (1954). Gastritis: A re-evaluation. *Medicine (Baltimore)* **33**, 199–290.

Perlow, W., Baraona, E., and Lieber, C. S. (1977). Symptomatic intestinal disaccharidase deficiency in alcoholics. *Gastroenterology* **77**, 680–684.

Pirola, R. C., and Davis, A. E. (1968). Effects of ethyl alcohol on sphincter resistance at the choledocho-duodenal junction. *Gut* **9**, 557–560.

Robles, E. A., Mezey, E., Halsted, C. H., and Schuster, M. M. (1974). Effect of ethanol on motility of the small intestine. *Johns Hopkins Med. J.* **135**, 17–24.

Roggin, G. M., Iber, F. L., and Linscheer, W. G. (1972). Intraluminal fat digestion in the chronic alcoholic. *Gut* **13**, 107–111.

Rosensweig, N. S., Herman, R. H., Stifel, F. B., and Herman, Y. F. (1969). Regulation of human jejunal glycolytic enzymes by oral folic acid. *J. Clin. Invest.* **48**, 2038–2045.

Rosensweig, N. S., Herman, R. H., and Stifel, F. B. (1971). Dietary regulation of small intestinal enzyme activity in man. *Am. J. Clin. Nutr.* **24**, 65–69.

Rubin, E., Rybak, B. J., Lindenbaum, J., Gerson, C. D., Walker, G., and Lieber, C. S. (1972). Ultrastructural changes in the small intestine induced by ethanol. *Gastroenterology* **63**, 801–814.

Sardesai, V. M., and Orten, J. M. (1968). Effects of prolonged alcohol consumption in rats on pancreatic protein synthesis. *J. Nutr.* **96**, 241–246.

Sarles, H. (1973). An international survey on nutrition and pancreatitis. *Digestion* **9**, 389–403.

Sarles, H. (1974). Chronic calcifying pancreatitis. Chronic alcoholic pancreatitis. *Gastroenterology* **66**, 604–616.

Sarles, H., Sarles, J. C., Camatte, R., Muratore, R., Gaini, M., Guien, C., Pastor, J., and Le Roy, F. (1965). Observations on 205 confirmed cases of acute pancreatitis, recurring pancreatitis, and chronic pancreatitis. *Gut* **6**, 545–559.

Sarles, H., Lebreuil, G., Tasso, F., Figarella, C., Clemente, F., Devaux, M. A., Fagonde, B., and Payan, H. (1971a). A comparison of alcoholic pancreatitis in rat and man. *Gut* **12**, 377–388.

Sarles, H., Figarella, C., and Clemente, F. (1971b). The interaction of ethanol, dietary lipids, and proteins on the rat pancreas. I. Pancreatic enzymes. *Digestion* **4**, 13–22.

Sarles, H., Tiscornia, O., and Palasciano, G. (1977). Chronic alcoholism and canine exocrine pancreas secretion. A long-term follow-up study. *Gastroenterology* **72**, 238–243.

Schapiro, H., Wruble, L. D., Estes, J. W., and Britt, L. G. (1968). Pancreatic secretion stimulated by the action of alcohol on the gastric antrum. *Am. J. Dig. Dis.* **13**, 536–539.

Small, M., Longarini, A., and Zamcheck, N. (1959). Disturbances of digestive physiology following acute drinking episodes in "Skid-row" alcoholics. *Am. J. Med.* **27**, 575–585.

Solomon, N., Solomon, T. E., Jacobson, E. D., and Shanbour, L. L. (1974). Direct effects of alcohol on *in vivo* and *in vitro* exocrine pancreatic function and metabolism. *Am. J. Dig. Dis.* **19**, 253–260.

Spencer, R. P., Brody, K. R., and Lutters, B. M. (1964). Some effects of ethanol on the gastrointestinal tract. *Am. J. Dig. Dis.* **9**, 599–604.

Straus, E., Urbach, H. J., and Yalow, R. S. (1975). Alcohol-stimulated secretion of immunoreactive secretin. *N. Engl. J. Med.* **293**, 1031–1032.

Strum, W. B., and Spiro, H. M. (1971). Chronic pancreatitis. *Ann. Intern. Med.* **74**, 264–277.

Tandon, B. N., George, P. K., Sama, S. K., Ramachandran, K., and Gandhi, P. C. (1969). Exocrine pancreatic function in protein calorie malnutrition disease of adults. *Am. J. Clin. Nutr.* **22**, 1476–1482.

Thomson, A. L., Baker, H., and Leevy, C. M. (1970). Patterns of ^{35}S-thiamine hydrochloride absorption in the malnourished alcoholic patient. *J. Lab. Clin. Med.* **76**, 34–45.

Thomson, W. O., Imrie, C. W., and Joffe, S. N. (1975). Alcohol-associated pancreatitis in a 15-year-old. *Lancet* **ii**, 1256.

Tiscornia, O. M., Palasciano, G., and Sarles, H. (1974). Effects of chronic ethanol administration on canine exocrine pancreatic secretion. *Digestion* **11**, 172–182.

Tiscornia, O. M., Palasciano, G., and Sarles, H. (1975). Atropine and exocrine pancreatic secretion in alcohol-fed dogs. *Am. J. Gastroenterol.* **63,** 33–36.

Toskes, P. P., and Deren, J. J. (1973). Vitamin B_{12} absorption and malabsorption. *Gastroenterology* **65,** 662–683.

von Wartburg, J. P., and Papenberg, J. (1966). Alcohol dehydrogenase and ethanol metabolism. *Psychosom. Med.* **28** (II), 405–413.

Walton, B., Schapiro, H., and Woodward, E. R. (1960). The effect of alcohol on pancreatic secretion. *Surg. Forum* **11,** 365–366.

Williams, A. W. (1956). Effects of alcohol on gastric mucosa. *Br. Med. J.* **i,** 256–259.

Wormsley, K. G. (1966). Effect of gastrin-like pentapeptide on stomach and pancreas. *Lancet* **i,** 993–996.

20

LIVER ABNORMALITIES IN ALCOHOLISM: ALCOHOL CONSUMPTION AND NUTRITION

Esteban Mezey and Patricia B. Santora

	I.	Introduction	303
	II.	Alcohol Consumption and Beverage Choice	304
		A. Patients Studied	304
		B. Data on Consumption and Beverage Choice	304
		C. Laboratory Abnormalities	305
		D. Liver Histology	307
		E. Comments	307
	III.	Malnutrition and Liver Disease	311
		A. Liver Disease in Malnourished Populations	311
		B. Experimental Production of Liver Disease with Deficient Diets	311
		C. Effect of Alcohol on Absorption and Metabolism of Nutrients	312
	IV.	Summary of Research Needs	313
		References	314
		Editorial Comment	316

I. INTRODUCTION

Chronic alcoholism is associated with an enhanced risk of liver disease. Lelbach (1974) demonstrated a high degree of correlation between the length of time and amount of abuse of alcoholic beverages and the incidence of cirrhosis; however, less than 20% of chronic alcoholics eventually develop cirrhosis, suggesting that factors other than alcohol—either dietary, toxic, or genetic—may also contribute to liver disease in alcoholism. There is a lack of quantitative data on the daily consumption of alcohol in relation to early manifestations of alcoholic liver disease, and it remains to be determined whether or not the type of beverage ingested (spirits, wine, beer, or combinations thereof) affects the incidence of liver abnormalities. Past attempts to correlate alcohol consumption by beverage choice with pathological complications of alcoholism, including cirrhosis, have been inconsistent and often contradictory. It had been suggested that the mortality rate from cirrhosis was directly related to the consumption of alcohol in the form of wine (Schmidt and Bronetto, 1962), spirits (Wallgren, 1960), and wine and spirits (Terris, 1967). In another study, the prevalence of abnormal liver tests and types of histological hepatic change were similar in beer and spirits drinkers and related to only the amount and duration of alcohol consumption (Lelbach, 1968).

The first part of this chapter provides data on consumption estimates and beverage choice in a group of noncirrhotic alcoholic patients in relation to biochemical and histological liver abnormalities, while the latter part of the chapter deals with the possible role of malnutrition as a factor in the pathogenesis of liver disease in the alcoholic.

II. ALCOHOL CONSUMPTION AND BEVERAGE CHOICE

A. Patients Studied

One hundred and sixteen male chronic alcoholic patients ranging in age from 23 to 63 years (mean, 37.9 years) were studied. All the patients had been admitted to the Alcoholism Metabolic Unit at Baltimore City Hospitals on a voluntary basis to seek help for their drinking problem. Those alcoholics showing overt clinical evidence of cirrhosis were not admitted to the Metabolic Unit and hence not included in this study. The majority of the study group were from the lower-middle socioeconomic class.

History of alcohol intake was taken independently by two observers following admission and verified for consistency. The daily alcohol consumption for 2–3 weeks prior to admission was converted into total grams of ethanol, based on an average ethanol content of 43% spirits, 12% wine, and 4.5% beer. The average recorded alcohol consumption was calculated for each patient. When 75% of the alcohol consumption was of one beverage, the patient was recorded as drinking primarily that beverage.

B. Data on Consumption and Beverage Choice

The duration of alcohol consumption in 116 patients was less than 10 years in 13 patients (11.2%); 10–14 years in 19 (16.4%); 15–20 years in 37 (31.9%); and over 20 years in 47 (40.5%). Of these 60 (52%) consumed primarily spirits, 15 (13%) wine, and nine (8%) beer, whereas 32 (27%) drank combinations of the

Table I. Beverage Choice and Consumption of 116 Alcoholic Patients

| | | | Recent mean daily ethanol intake | |
Beverage	Number of patients	Mean age	gm	gm/kg
Spirits	60 (52%)	40.6	348.0	5.05
Wine	15 (13%)	40.5	267.1	4.14
Beer	9 (8%)	33.8	256.8	4.01
Combinations	32 (27%)	36.7	372.1	5.41
Total	116 (100%)	37.9	311.0	4.65

three beverages (Table I). The mean duration of alcohol intake among spirits alcoholics was 19.5 years. This group had the highest recent mean daily intake of ethanol (348.0 gm). Those drinking wine had the longest mean duration of alcohol intake (21.6 years). The beer alcoholics had the shortest mean duration of alcohol intake (15.1 years) and the lowest recent mean daily intake of ethanol (256.8 gm). The wine alcoholics were older than the beer alcoholics by a mean age of 6.7 years. The mean duration of alcohol intake among those drinking combinations of beverages was 17.6 years. Of these 18 drank spirits and beer, six spirits and wine, six spirits, wine, and beer, and two beer and wine. Among these patients the highest recent mean intake of 454.4 gm per day (7.14 gm/kg) was found in the six patients who drank spirits and wine.

Fifty-eight patients (50%) had previous hospital admissions for alcoholism. Most of these admissions were to state hospitals. Beverage choice was not different among the alcoholics with previous hospitalizations as compared with those without previous hospitalizations. Alcoholics with previous hospitalizations for alcoholism, however, had a longer mean duration of alcohol intake by 1 year and a higher recent mean intake by 13.1 gm per day.

C. Laboratory Abnormalities

The range of abnormal laboratory tests present at admission is shown in Table II. In general the changes were moderate. Thirty-eight (33%) of the 116 patients were anemic (hemoglobin concentration less than 14 gm/100 ml). Serum bilirubin was elevated in 30 (26%); of these 26 (87%) were 1–2 mg% and four (13%) were over 2 mg%. SGOT was abnormal in 73 (63%); of these 58 (79%) were 40–100 IU and 15 (21%) were over 100 IU. Elevations in alkaline phosphatase were found in 46 (40%); of these 38 (83%) were 12–20 King–Armstrong (KA) units and eight (17%) were over 20 KA units. Serum albumin was only mildly depressed to levels between 3.0 and 3.4 gm/100 ml in 10 (9%) of the patients. In addition the bromsulphalein retention test was done in 76 patients and found

Table II. Abnormal Laboratory Determinations in the Study Group of 116 Alcoholic Patients

Beverage	Number of patients	Number and percentage of patients with abnormal values and range of abnormal values				
		Hemoglobin[a] (gm/100 ml)	Serum bilirubin[b] (total mg/100 ml)	SGOT[c] (IU/liter)	Alkaline phosphatase[d] (KA units/100 ml)	
Spirits	60 (52%)	21 (35%) 8.9–13.9	14 (23%) 1.1–3.3	39 (65%) 46–240	20 (33%) 12.2–35.5	
Wine	15 (13%)	10 (67%)[e] 10.0–13.9	3 (20%) 1.2–1.4	8 (53%) 48–147	10 (67%) 12.7–21.4	
Beer	9 (8%)	1 (11%) 11.0	4 (44%) 1.1–1.8	5 (56%) 41–390	5 (56%) 12.4–45.0	
Combinations	32 (27%)	6 (19%) 10.2–13.8	9 (28%) 1.1–2.1	21 (66%) 44–335	11 (34%) 12.9–36.0	

[a] Hemoglobin, range of normal values: 14.0–17.0.
[b] Serum bilirubin, range of normal values: < 1.0.
[c] SGOT, serum glutamic oxaloacetic transaminase (IU); range of normal values: <40.
[d] Alkaline phosphatase (King–Armstrong units); range of normal values: 4–12.
[e] Significantly different from beer and combinations alcoholics at $p < 0.05$.

to be abnormal in 34 (45%); of these 28 (82%) were 4–15% retention and six (18%) were over 15% retention.

The incidence of anemia was highest (67%) in the patients ingesting wine and lowest in those ingesting beer (11%). Serum folate levels were obtained in 22 of the anemic patients and found to be low (less than 5 ng/ml) in 10 (45%). In addition, serum folate was measured in 29 of the nonanemic patients and found to be low in six (21%). In the wine alcoholics with anemia, folate levels were low in three of the five patients in which it was measured. The sole beer alcoholic with anemia had a normal serum folate level. Although the number of patients ingesting wine and beer is small compared to spirits and combinations drinkers, no particular beverage was seen to dominate in the incidence of abnormal laboratory determinations pertaining to the liver.

D. Liver Histology

The data in this section are based on findings in 55 patients on whom liver biopsy was performed. The indications for liver biopsy were hepatomegaly and/or abnormal liver function tests. Five patients had normal liver biopsies (Table III). There were no instances of cirrhosis. Fatty infiltration of the liver of various degrees was the principal finding in 31 (56%) of the patients. Alcoholic hepatitis was found in five patients all of whom ingested in excess of 200 gm of ethanol daily; four of these patients ingested spirits alone and one patient ingested a combination of wine and beer (Table IV). Nonspecific changes on liver biopsy, consisting of either occasional areas of focal liver cell necrosis or portal inflammation, were found in 14 patients whose daily consumption ranged from 80 gm of ethanol to levels above 300 gm, with nine cases (64%) found in patients consuming in excess of 200 gm; all of these patients ingested beverages other than beer. Various degrees of fibrosis were found in 19 (35%) of the liver biopsies and the recent daily consumption of these patients ranged from 67 gm of ethanol to levels above 300 gm. Four of the five patients with alcoholic hepatitis had elevated serum transaminase levels, while all those with normal liver biopsies had normal bilirubin levels and bromsulphalein retention tests.

E. Comments

This study demonstrates a significant incidence of laboratory abnormalities of liver function and histology in chronic alcoholic patients who had no history or obvious clinical evidence of cirrhosis. With the exception of three patients, all the patients in this study consumed in excess of 80 gm of ethanol a day, with 96 (83%) of them consuming in excess of 160 gm daily. Fatty infiltration of the liver was the principal abnormality found on liver biopsy. This is not unexpected, since fatty infiltration of the liver is readily produced in man (Rubin and Lieber,

Table III. Consumption Estimates Related to Liver Histology

Daily ethanol intake range (gm)	Number of patients	Number of patients with liver biopsy	Liver histology						
			Normal	Fatty infiltration	Fatty infiltration with fibrosis	Hepatitis	Hepatitis with fibrosis	Nonspecific changes	Nonspecific changes with fibrosis
<80	3	2	0	0	2	0	0	0	0
80–100	6	3	0	0	1	0	0	1	1
100–160	11	6	1	3	1	0	0	0	1
160–200	13	8	0	5	1	0	0	0	2
200–300	29	14	1	6	2	1	2	2	0
300+	54	22	3	9	1	0	2	4	3
Total	116	55 (47%)	5 (9%)	23 (42%)	8 (14%)	1 (2%)	4 (7%)	7 (13%)	7 (13%)

Table IV. Beverage Choice Related to Liver Histology

Beverage	Number of patients	Number of patients with liver biopsy	Liver histology						
			Normal	Fatty infiltration	Fatty infiltration with fibrosis	Hepatitis	Hepatitis with fibrosis	Nonspecific changes	Nonspecific changes with fibrosis
Spirits	60	31	3	12	3	1	3	4	5
Wine	15	7	1	1	1	0	0	2	2
Beer	9	4	0	2	2	0	0	0	0
Combinations	32	13	1	8	2	0	1	1	0
Total	116	55 (47%)	5 (9%)	23 (42%)	8 (14%)	1 (2%)	4 (7%)	7 (13%)	7 (13%)

1968) and animals (Lieber *et al.*, 1965) by the administration of ethanol. Its occurrence, however, is not limited to alcohol abusers and is a reversible condition if alcohol consumption is reduced. Surprisingly, the second most common finding was that of nonspecific abnormalities of liver histology present in 26% of the biopsies. The finding of nonspecific changes such as portal inflammation and/or areas of focal necrosis not fitting the classical types of alcoholic liver disease has not been emphasized previously. There is no evidence that alcohol intake is responsible for these nonspecific changes since these changes of liver histology are also found in nonalcoholic patients with various diseases. It was, however, a finding in 19% of 130 alcoholic patients in another study (Ugarte *et al.*, 1970). Alcoholic hepatitis was found in 9% of the patients biopsied, all of whom had been consuming more than 200 gm of ethanol daily. This is similar to the 12% incidence found in a recent study of 103 patients in Portugal who had biopsies after admission for detoxification (Bordalo *et al.*, 1974). The incidence of various degrees of hepatic fibrosis was 35% which is one-half the 67% incidence of fibrosis found by Bordalo *et al.* (1974) in the above study which also included two cases of cirrhosis. The significance, if any, of the nonspecific changes of liver histology remains to be determined. A recent study shows increases in the hepatic enzyme collagen proline hydroxylase, which is associated with hepatic collagen synthesis in these cases (Mezey *et al.*, 1976). Although the number of patients consuming wine and beer is small compared to the spirits and combinations drinkers, none of the laboratory abnormalities of liver function or of liver histology was correlated with a particular type of beverage. This suggests that the quantity of alcohol ingested, regardless of the type of beverage in which it is contained, is the primary cause of the liver abnormalities.

Of the 96 patients consuming more than 160 gm daily, 44 who had hepatomegaly and/or abnormal liver function tests were biopsied. The biopsies revealed abnormal histology in 40. This would suggest that only 41.6% of patients ingesting amounts greater than 160 gm per day developed liver injury and hence that factors besides alcohol may contribute to liver disease. The presence of liver injury, however, is not excluded by the absence of hepatomegaly and/or abnormal liver function tests.

The high incidence of spirits alcoholics in this study is similar to that found in other surveys of alcoholic patients in the United States (Patek *et al.*, 1975; Feldman *et al.*, 1974; Terry *et al.*, 1957) and contrasts with surveys from Europe (Madden and Jones, 1972; Devrient and Lolli, 1962; Lolli *et al.*, 1958) in which beer and/or wine were the principal beverages consumed. Different from most studies conducted in the United States and other countries was the low use of wine or beer as the sole beverage and the high use of beverage combinations. It appears that a poor economical situation associated with a high craving for alcohol leads many patients to ingest alcohol in any form obtainable.

While it is clear that excessive consumption of alcohol in any beverage form can lead to liver dysfunction, most of the cases of liver dysfunction in the United States appear to be induced by consumption of spirits, which is largely a reflection of beverage choice. Furthermore, although a high proportion of the alcohol consumed in the United States is in the form of beer, a relatively small proportion of the cases of liver dysfunction appears to occur in beer drinkers.

III. MALNUTRITION AND LIVER DISEASE

Alcoholism is a frequent cause of both malnutrition and vitamin deficiencies. Nutrient deficiencies in turn are responsible for a variety of symptoms and complications of chronic alcoholism and may contribute to the development of liver disease.

A. Liver Disease in Malnourished Populations

One argument in support of a role of malnutrition in the pathogenesis of alcoholic liver disease has been the increased incidence of liver disease in malnourished populations. Fatty infiltration of the liver and fibrosis occur in children with kwashiorkor (Cook and Hutt, 1967), and cirrhosis is more common in some malnourished populations in underdeveloped countries (Borhanmanesh *et al.*, 1971; Gillman and Gilbert, 1974). However, the fatty infiltration of kwashiorkor and that produced by a protein-deficient diet in the rhesus monkey (Deo and Ramalingaswami, 1960) does not progress to cirrhosis, and in the adult malnourished populations with cirrhosis there is no direct evidence implicating any specific nutrient. Furthermore, a toxin such as aflatoxin or a virus may be the causative agent and malnutrition may be important only in aggravating the disease. Also of interest is that undernourished civilians, following World War II, did not have histological evidence of hepatocellular damage (Sherlock and Walshe, 1948).

B. Experimental Production of Liver Disease with Deficient Diets

Fatty infiltration of the liver and cirrhosis has been produced in rats by feeding diets deficient in lipotropes. Furthermore, alcohol has been shown to increase the requirements of choline in the rat (Klatskin *et al.*, 1954). Primates and man are less susceptible to lipotrope deficiency than are rats, and hence the induction of liver damage by those means in rats may not be relevant to man. The species difference in the susceptibility to choline deficiency may be explained by differences in the rate of degradation of choline by the hepatic enzyme choline oxidase. The rat has a high choline oxidase activity while man has a very low activity (Sidransky and Farber, 1960). Also, ultrastructural changes produced by ethanol in man, baboons, and rats differ from those associated with choline and protein deficiencies (Lieber *et al.,* 1975).

C. Effect of Alcohol on Absorption and Metabolism of Nutrients

Alcohol can contribute to the production of specific nutrient deficiencies by decreasing the absorption and altering the metabolism of nutrients. In addition, ethanol has been shown to increase the release of water-soluble, but not fat-soluble, vitamins from the perfused liver (Sorrell *et al.,* 1974). The nutrient deficiencies that accompany the ingestion of alcohol can, in turn, contribute to its damaging effect on the liver. The beverage form in which alcohol is ingested may also be an important determinant in the development of nutrient deficiency. The vitamin content, with the exception of thiamine, is higher in beer than in wine or spirits (Leake and Silverman, 1966). Beer is particularly rich in folic acid (Herbert, 1963).

The effects of alcohol on small intestinal absorption and pancreatic function are discussed in Chapter 19. As regards the effect of ethanol on nutrient metabolism, ethanol has been shown to interfere with the conversion of vitamins such as folic acid and pyridoxine to their active coenzyme forms. Folic acid, pyridoxine, and vitamin B_{12} are necessary for cell replication. Folic acid is the most common vitamin deficiency in patients with alcoholic liver disease. Tissue folate stored in the healthy individual lasts about 3 months without any folate intake, and megaloblastic anemia does not develop until 4–5 months after cessation of folate ingestion. By contrast, in the alcoholic subject, megaloblastic anemia develops as early as 1.5–3 months after institution of a folate-deficient diet (Eichner *et al.,* 1971). Also, ethanol has been shown to suppress the hematologic response of anemic, folate-deficient patients to folic acid (Sullivan and Herbert, 1964). Decreased absorption, decreased hepatic uptake and storage, or decreased conversion of folate to 5-methyltetrahydrofolic acid, the form active in the synthesis of deoxyribonucleic acid, may explain the above observations. The acute administration of ethanol results in a fall in serum folate levels in alcoholic patients and normal subjects, suggesting that ethanol interferes with the formation or release of 5-methyltetrahydrofolic acid (Paine *et al.,* 1973). Recent studies suggest that ethanol inhibits the biliary excretion and normal enterohepatic circulation of 5-methyltetrahydrofolic acid (Hillman *et al.,* 1977).

Decreases in plasma pyridoxal-5-phosphate (PLP), the active form of pyridoxine, have been demonstrated in alcoholic patients with (Mitchell *et al.,* 1976) and without liver dysfunction (Lumeng and Li, 1974). In alcoholic patients, after the ingestion of alcohol for 2 weeks, the administration of pyridoxine failed to correct serum PLP levels to normal, while the administration of PLP resulted in prompt restoration of the levels to normal and disappearance of sideroblastic alterations in the bone marrow. These studies suggested that ethanol interfered with the conversion of pyridoxine to PLP (Hines and Cowan, 1970). More recent studies show that the effect of alcohol is mediated by acetaldehyde which accelerates the degradation of PLP (Lumeng and Li, 1974). The role of acetaldehyde is supported by experiments showing that inhibition of alcohol dehydrogenase with 4-methylpyrazole abolishes the effect of ethanol in diminishing the hepatic content and rate of release of PLP from the perfused rat liver (Veitch *et al.,* 1975). Pyridoxine deficiency has been shown to enhance the cytotoxicity of acetaldehyde to liver cells *in vitro* (Kakumu *et al.,* 1976). Feeding of pyridoxine-deficient diets resulted in mild fatty liver in rats (French and Castagna, 1967) and in cirrhosis in monkeys (Wizgird *et al.,* 1965).

Repair of liver injury induced by alcohol may be associated with an increased requirement of some nutrients. The production of liver dysfunction with chronic alcohol feeding despite a normal diet in experimental subjects was associated with the development of a low serum folate level and decreased deoxyribonucleic acid synthesis which could be corrected to normal only by the administration of increased amounts of folate (Leevy, 1967). However, whether or not the requirements of most nutrients change following alcohol ingestion in man remains to be determined.

IV. SUMMARY OF RESEARCH NEEDS

A prospective study relating alcohol consumption, beverage choice, and nutritional state with liver disease is needed since the information obtained from history of past experience is often unreliable. A prospective study should include the following.

1. Evaluation of the consumption of moderate and small amounts of alcohol on the liver. Most studies have dealt with the development of cirrhosis in individuals ingesting more than 150 gm of ethanol a day.

2. Detection of the development of early stages of liver dysfunction by sensitive techniques.

3. Laboratory measurements of the development of nutritional deficiencies to correlate with information on alcohol and dietary intake.

4. Investigation of changes in the requirements of nutrients during alcohol consumption and the development of liver dysfunction.

Clearly, prospective studies as just proposed are difficult to conduct and must be accompanied by efforts to reduce excessive consumption of alcoholic beverages by patients in the study. Nevertheless, failures in prevention and treatment in any large series would probably be sufficiently frequent to yield a significant spread in consumption groups.

ACKNOWLEDGMENTS

This work was supported in part by the United States Public Health Service, Grant Number AA00626, and by a grant from the United States Brewers Association, Inc.

REFERENCES

Bordalo, O., Batista, A., Noronha, M., Lamy, J., and Dreiling, D. A. (1974). Effects of ethanol on liver morphology and pancreatic function in chronic alcoholism. *Mt. Sinai J. Med., N.Y.* **41,** 722–731.

Borhanmanesh, F., Ghavami, A., Dutz, W., and Bagheri, S. (1971). Cirrhosis of the liver in Iran: A prospective study of 66 cases. *J. Chron. Dis.* **23,** 891–905.

Cook, G. C., and Hutt, M. S. R. (1967). The liver after Kwashiorkor. *Br. Med. J.* **iii,** 454–457.

Deo, M. G., and Ramalingaswami, V. (1960). Production of periportal fatty infiltration of the liver in the Rhesus monkey by a protein deficient diet. *Lab. Invest.* **9,** 319–329.

Devrient, P., and Lolli, G. (1962). Choice of alcoholic beverage among 240 alcoholics in Switzerland. *Q. J. Stud. Alcohol* **23,** 459–467.

Eichner, E. R., Pierce, H. L., and Hillman, R. S. (1971). Folate balance in dietary-induced megaloblastic anemia. *N. Engl. J. Med.* **284,** 933–938.

Feldman, J., Su, W. H., Kaley, M. M., and Kissin, B. (1974). Skid row and inner city alcoholics: A comparison of drinking patterns and medical problems. *Q. J. Stud. Alcohol* **35,** 565–576.

French, S. W., and Castagna, J. (1967). Some effects of chronic ethanol feeding on vitamin B6 deficiency in the rat. *Lab. Invest.* **16,** 526–531.

Gillman, J., and Gilbert, C. (1974). Aspects of nutritional liver disease—human and experimental. *Ann. N.Y. Acad. Sci.* **57,** 737–749.

Herbert, V. (1963). A palatable diet for producing experimental folate deficiency in man. *Am. J. Clin. Nutr.* **12,** 17–20.

Hillman, R. S., McGuffin, R., and Campbell, C. (1977). Alcohol interference with the folate enterohepatic cycle (abstract). *Clin. Res.* **25,** 518A.

Hines, J. D., and Cowan, D. H. (1970). Studies on the pathogenesis of alcohol-induced sideroblastic bone-marrow abnormalities. *N. Engl. J. Med.* **283,** 441–446.

Kakumu, S., Levinson, R., Baker, H., and Leevy, C. M. (1976). Enhanced acetaldehyde toxicity in vitamin B6 deficiency (abstract). *Gastroenterology* **71,** 914.

Klatskin, G., Krehl, W. A., and Conn, H. O. (1954). The effect of alcohol on choline requirement: I. Changes in the rat's liver following prolonged ingestion of alcohol. *J. Exp. Med.* **100,** 605–614.

Leake, C. D., and Silverman, M. (1966). "Alcoholic Beverages in Clinical Medicine," pp. 14–47. Yearbook Publ., Chicago, Illinois.

Leevy, C. M. (1967). Clinical diagnosis, evaluation and treatment of liver disease in alcoholics. *Fed. Proc., Fed. Am. Soc. Exp. Biol.* **26,** 1474–1481.

Lelbach, W. K. (1968). Liver damage from different alcoholic drinks. *Ger. Med. Mon.* **13,** 31–39.

Lelbach, W. K. (1974). Organic pathology related to volume and pattern of alcohol use. *In* "Research Advances in Alcohol and Drug Problems" (R. J. Gibbins, Y. Israel, H. Kalant, R. E. Popham, W. Schmidt, and R. G. Smart, eds.), Vol. 1, pp. 93–198. Wiley, New York.

Lieber, C. S., Jones, D. P., and DeCarli, L. M. (1965). Effects of prolonged ethanol intake: Production of fatty liver despite adequate diets. *J. Clin. Invest.* **44,** 1009–1021.

Lieber, C. S., DeCarli, L. M., and Rubin, E. (1975). Sequential production of fatty liver, hepatitis and cirrhosis in sub-human primates fed ethanol with adequate diets. *Proc. Natl. Acad. Sci. U.S.A.* **72,** 437–441.

Lolli, G., Golder, G. M., Serianni, E., Bonfiglio, G., and Balboni, C. (1958). Choice of alcoholic beverage among 178 alcoholics in Italy. *Q. J. Stud. Alcohol* **19**, 303–308.

Lumeng, L., and Li, T. K. (1974). Vitamin B6 metabolism in chronic alcohol abuse. Pyridoxal phosphate levels in plasma and the effects of acetaldehyde on pyridoxal phosphate synthesis and degradation in human erythrocytes. *J. Clin. Invest.* **53**, 693–704.

Madden, J. S., and Jones, D. (1972). Bout and continuous drinking in alcoholism. *Br. J. Addict.* **67**, 245–250.

Mezey, E., Potter, J. J., and Maddrey, W. C. (1976). Hepatic collagen proline hydroxylase activity in alcoholic liver disease. *Clin. Chim. Acta* **68**, 313–320.

Mitchell, D., Wagner, C., Stone, W. J., Wilkinson, G. R., and Schenker, S. (1976). Abnormal regulation of plasma pyridoxal 5′-phosphate in patients with liver disease. *Gastroenterology* **71**, 1043–1049.

Paine, C. J., Eichner, E. R., and Dickson, V. (1973). Concordance of radioassay and microbiologic assay in the study of the ethanol-induced fall in serum folate level. *Am. J. Med. Sci.* **266**, 135–138.

Patek, A. J., Toth, I. G., Saunders, M. G., Castro, G. A. M., and Engel, J. J. (1975). Alcohol and dietary factors in cirrhosis: An epidemiological study of 304 alcoholic patients. *Arch. Intern. Med.* **135**, 1053–1057.

Rubin, E., and Lieber, C. S. (1968). Alcohol-induced hepatic injury in nonalcoholic volunteers. *N. Engl. J. Med.* **278**, 869–876.

Schmidt, W., and Bronetto, J. (1962). Death from liver cirrhosis and specific beverage consumption: An ecological study. *Am. J. Public Health* **52**, 1473–1482.

Sherlock, S., and Walshe, V. (1948). Effect of undernutrition in man on hepatic structure and function. *Nature (London)* **161**, 604.

Sidransky, H., and Farber, E. (1960). Liver choline oxidase activity in man and in several species of animals. *Arch. Biochem. Biophys.* **87**, 129–133.

Sorrell, M. F., Baker, H., Barak, A. J., and Frank, O. (1974). Release by ethanol of vitamins into rat liver perfusates. *Am. J. Clin. Nutr.* **27**, 743–745.

Sullivan, L. W., and Herbert, V. (1964). Suppression of hematopoiesis by ethanol. *J. Clin. Invest.* **43**, 2048–2062.

Terris, M. (1967). Epidemiology of cirrhosis of the liver: National mortality data. *Am. J. Public Health* **57**, 2076–2088.

Terry, J., Lolli, G., and Golder, G. M. (1957). Choice of alcoholic beverage among 531 alcoholics in California. *Q. J. Stud. Alcohol* **18**, 417–428.

Ugarte, G., Iturriaga, H., and Insunza, I. (1970). Some effects of ethanol on normal and pathologic livers. *Prog. Liver Dis.* **3**, 355–370.

Veitch, R. L., Lumeng, L., and Li, T. K. (1975). Vitamin B6 metabolism in chronic alcohol abuse. The effect of ethanol oxidation on hepatic pyridoxal 5′ phosphate metabolism. *J. Clin. Invest.* **55**, 1026–1032.

Wallgren, H. (1960). Alcoholism and alcohol consumption. *Alkoholpolitik* **23**, 177–179.

Wizgird, J. P., Greenberg, L. D., and Moon, H. D. (1965). Hepatic lesions in pyridoxine-deficient monkeys. *Arch. Pathol.* **79**, 317–323.

EDITORIAL COMMENT

In the voluminous literature on alcohol consumption in the United States few solid data have been presented on the differences in disease phenomena following the use of different types of alcoholic beverages. Perhaps there are no significant differences, but this will not be established until good data are in hand. There are really two questions. (1) Are there different effects from consumption of the same quantities of ethanol in different beverage forms? (2) Which beverages, in fact, are responsible for the various adverse effects of alcohol abuse? This paper has presented some evidence on the latter point.

21

EFFECTS OF ALCOHOL ON THE CARDIOVASCULAR SYSTEM

Arthur L. Klatsky

I. Introduction and Historical Review 317
II. Effects of Alcohol on Cardiovascular Physiology,
   Biochemistry, and Structure 319
   A. Heart Rate, Blood Pressure, Cardiac Output, and
      Vascular Resistance 319
   B. Myocardial Function 319
   C. Coronary Circulation 321
   D. Myocardial Biochemistry 321
   E. Myocardial Structure 322
III. Alcohol and Cardiovascular Disease 323
   A. Partially Defined Syndromes 323
   B. Alcoholic Heart Disease 327
   C. Relation to Hypertension 330
   D. Relation to Coronary Disease 332
IV. Summary and Research Needs 335
   A. Alcoholic Heart Disease 335
   B. Hypertension 335
   C. Coronary Heart Disease 335
   D. General Approach 335
   References 336

I. INTRODUCTION AND HISTORICAL REVIEW

The effects of ethyl alcohol on the cardiovascular system and especially on the heart itself have excited the interest of clinicians and investigators for well over a century. While much has been learned, the areas of solid knowledge are few, and there are a number of apparent paradoxes. The physiologic, clinical, and epidemiologic evidence cannot at this time be integrated into definitive general concepts. It is clear that past attempts to generalize have had the effect of slowing progress in understanding this subject. Imprecise and implicitly judgmental terms such as "alcoholic," "alcohol abuse," "problem drinkers," and (with reference to amount of drinking) "excessive," "heavy," "moderate," and "light" should be avoided, and, when references using these terms are cited, the words will therefore be placed in quotation marks.

A number of famous physicians of the nineteenth century commented about an apparent relation between chronic use of larger than average amounts of alcohol and heart disease. Among the first was the German neurologist, Nikolaus Friedreich (1861), who described cardiac hypertrophy associated with "alcoholism." The English physician W. H. Walsche (1873) described a condition which he referred to as "patchy cirrhosis of the heart" in "chronic alcoholics." The German pathologist Otto Böllinger (1884) described cardiac dilatation and hypertrophy, which he considered common among Bavarian beer drinkers. This became known as the "Münchener bierherz." It was estimated that the average yearly consumption of beer in Munich at the time was 432 liters, compared with 82 liters in other parts of Germany. Strümpell (1890), Osler (1899), and Steell (1893), among others, made similar comments in the last decade of the nineteenth century. The epidemic of arsenic-beer drinkers' disease occurred in Manchester, England in 1900. Before this event, Steell (1893), in a report of 25 cases, stated "not only do I recognize alcoholism as one of the causes of muscle failure of the heart but I find it a comparatively common one." Following the arsenic-beer episode, Steell (1906) wrote in his textbook that "in the production of the combined affection of the peripheral nerves and the heart met with in beer drinkers, arsenic has been shown to play a conspicuous part." In another great textbook, "The Study of the Pulse" MacKenzie (1902) described cases of heart muscle failure attributed to chronic use of alcohol and first used the term "alcoholic heart disease." During the early part of the twentieth century, there was general doubt that alcohol had a direct role in producing cardiac muscle disease, although Vaquez (1921) took a strong view in favor of such a relation and described in detail an alcohol-induced syndrome which, in our time, would be called a chronic low-output cardiomyopathy.

After the detailed descriptions of cardiovascular beriberi by Aalsmeer and Wenckebach (1929) and Keefer (1930), the concept of "beriberi heart disease" dominated thinking about the effects of alcohol on the heart for several decades. From the mid-1950s to the present time, increasing interest has been evident in possible direct effects of alcohol upon the heart, separate from or in addition to deficiency states. During this period independent interest has arisen, largely through epidemiologic studies, in possible relations of chronic alcohol use to hypertension and coronary heart disease.

II. EFFECTS OF ALCOHOL ON CARDIOVASCULAR PHYSIOLOGY, BIOCHEMISTRY, AND STRUCTURE

A. Heart Rate, Blood Pressure, Cardiac Output, and Vascular Resistance

Most investigators report that, in healthy humans, "low" or "moderate" alcohol doses produce slight increase in heart rate and cardiac output with some increase in blood pressure (systolic more than diastolic). Peripheral resistance changes little; cutaneous resistance decreases, but this is compensated by visceral vasoconstriction. Data of this sort have been reported by Grollman (1930), Davison (1949), and Eliaser and Giansiracusa (1956), among others, with alcohol doses of approximately 30–75 ml. The extent to which these changes are direct and the degree to which they represent indirect nervous system regulation of the circulation in humans have not been elucidated. It has long been known (Eliaser and Giansiracusa, 1956) that doses of alcohol sufficient to produce central nervous system respiratory depression also produce hypotension, bradycardia, and, ultimately, asystole. Nervous system mechanisms have been presumed to predominate in these effects.

B. Myocardial Function

1. General Summary

The myocardial effects of alcohol have been explored in a number of studies of humans (normal and diseased), intact animals, and isolated diseased animal heart muscle preparations. The results vary with dose, route, and time course of ethanol administration, parameters measured, and pathologic state of the subjects. Most studies suggest that, in sufficient doses, alcohol decreases myocardial contractility. The dose required may be lower if there is clinical evidence of heart muscle disease, *or* if the subject has ingested substantial amounts of alcohol for a long time.

2. Acute Studies

Studies of isolated heart muscle fibers (Spann *et al.*, 1968; Gimeno *et al.*, 1962) suggest that acute exposure to alcohol is associated with depression of myocardial contractility. Animal studies of intact anesthetized (Regan *et al.*, 1966; Webb *et al.*, 1966; Mendoza *et al.*, 1971) or conscious (Horwitz and Atkins, 1974) dogs also provide evidence of decreased myocardial contractility at blood alcohol levels of 100–200 mg per 100 ml. Development of compensatory mechanisms was demonstrated by the studies of Wong (1973) which showed a greater depressant effect in animals also given autonomic blockade by propranolol and atropine. On the other hand, Horwitz and Atkins (1974) did not find an autonomic blockade of the depressant effect of ethanol on the hearts of conscious dogs.

In normal humans, depression of left ventricular function as measured by direct techniques has been found by several investigators (Mendoza *et al.*, 1971; Newman and Valicenti, 1971) at blood alcohol levels of 75–250 mg per 100 ml. The work of Ahmed *et al.* (1973), using systolic time intervals, showed myocardial depression in normal volunteers, to whom was administered an 81-ml dose (6 oz. of Scotch whiskey) of alcohol, which resulted in blood levels of 75–110 mg per 100 ml. Similarly, a single dose of ethyl alcohol given to healthy volunteers (Delgado *et al.*, 1975) with resultant blood alcohol levels of 75–138 mg per 100 ml demonstrated a slight but significant decrease in left ventricular ejection fraction by echocardiography. In the last two studies cited (Ahmed *et al.*, 1973; Delgado *et al.*, 1975) the decreased myocardial *contractility* was *not* associated with decreased *performance* of the heart as a pump, which led the authors to suggest that compensatory mechanisms came into play to preserve overall circulatory function in normals. The work of Delgado *et al.* (1975), as well as the earlier work of Juchems and Klobe (1969), indicated poor correlation of the effects observed with dose levels within the ranges studied, which also suggested the development of compensatory mechanisms. Furthermore, studies of maximal or near maximal cardiac exercise performance of normal humans by Riff *et al.* (1969) and Blomqvist and co-workers (1970) showed little effect of blood alcohol levels of 85 mg per 100 ml to 200 mg per 100 ml. The acute myocardial depressant effects of alcohol may be stronger and evident at lower alcohol dose levels in persons with preexisting heart disease not related to drinking. For example, Conway (1968) presented such data for patients with coronary artery disease and concluded that ''3–4 whiskeys'' had a profound myocardial depressant effect. Gould *et al.* (1971, 1972) reported similar results with 60 ml of alcohol administered to volunteers with various types of heart disease. In a fascinating study of anesthetized dogs administered a standard nonpenetrating blow to the chest, Liedtke and DeMuth (1975) demonstrated that the prior administration of alcohol greatly increased mortality and decreased the mechanical performance of the traumatized heart.

3. Chronic Studies

Burch *et al.* (1971) reported deleterious effects of alcohol on cardiac function in well-nourished mice fed large amounts of alcohol for as short a time as 7–10 weeks. One study (Maines and Aldinger, 1967) of chronic ethanol administration (25% by volume) in rats showed a decrease in ventricular force, but another (Lochner *et al.*, 1969) showed no definite abnormality of function in the same species of rats fed less ethanol (15% by volume). In a study of dogs fed approximately one third their caloric intake in the form of alcohol (Pachinger *et al.*, 1973), no functional and cardiovascular abnormalities were observed. However,

Regan *et al.* (1974) found definite hemodynamic impairment in dogs fed a like proportion of their calories as alcohol for an 18-month period. Thus, it seems probable that *dose* and *duration* of chronic alcohol use both are important in production of functional cardiac abnormalities in animals. No experiment has yet produced frank congestive heart failure in well-nourished animals.

There is convincing evidence of functional myocardial abnormalities in humans with a long history of substantial alcohol intake, and *no* evidence of clinical cardiac disease. The physiologic studies of Gould *et al.* (1969), Regan *et al.* (1969), and Limas and co-workers (1974) all agree on this matter. Noninvasive measurements have also confirmed these findings (Spodick *et al.,* 1972) and some (Wu *et al.,* 1976) have suggested that men are more susceptible than women to such effects. This phenomenon has been called "preclinical" cardiomyopathy.

C. Coronary Circulation

The substantial body of evidence that ethanol has deleterious effects on myocardial function, perhaps even more pronounced in the presence of heart disease, does not seem to be parallelled by a similar degree of consistency in known effects on the coronary circulation. In fact, the work by Mendoza *et al.* (1971) already cited as showing acute impairment by alcohol of myocardial contractility in humans and dogs showed concomitant increase in "effective coronary blood flow" in both humans and dogs. Similarly, the studies by Regan and co-workers (1969) which showed deleterious effects of alcoholism on myocardial function (and metabolism) also found apparent increase in coronary flow and narrowing of myocardial arteriovenous oxygen difference. Earlier, Ganz (1963) also reported apparent increase in coronary flow. Other work by Regan and colleagues (1966), as well as work by Webb and Degerli (1965), showed apparent acute *decrease* in coronary flow with administration of alcohol. Among those reporting *no effect* on coronary flow are Schmitthenner *et al.* (1958) and Wendt *et al.* (1962).

D. Myocardial Biochemistry

Little fundamental knowledge exists concerning possible metabolic defects or abnormalities in myocardial cells which could be related to functional effects or to clinical syndromes. Although it is now believed that many of the consequences of alcoholic toxicity to liver cells may be attributable to an increased concentration of reduced nicotinamide adenine dinucleotide (NADH) in hepatic cells (Isselbacher, 1977), this is unlikely to be the case in heart cells, as these lack the enzyme alcohol dehydrogenase. It is not known whether the effects on the heart in intact animals or humans are due to alcohol itself, some metabolite of alcohol (such as acetaldehyde), to associated metabolic consequences of alcohol use (e.g., hypomagnesemia, acidosis, higher catechol levels), or to some combination of these.

Animal experiments in isolated hearts and heart homogenates (Lochner *et al.*, 1969; Kikuchi and Kako, 1970) seem to indicate that alcohol inhibits fatty acid oxidation and causes increased incorporation of fatty acids into triglycerides. The work of Regan's group (Regan *et al.*, 1966, 1969) and of Wendt *et al.* (1966) demonstrated "leakage" of potassium, phosphate, and SGOT from myocardial cells for 4 to 5 hours following oral alcohol administration at blood alcohol concentrations of 200 mg per 100 ml. Both groups also demonstrated decreased uptake of free fatty acids by the heart, consistent with the possibility of inhibition of fatty acid oxidation by alcohol. Wendt and co-workers (1966) also demonstrated leakage of intramitochondrial enzymes (isocitric dehydrogenase and malic dehydrogenase) after alcohol ingestion, even in patients without clinical evidence of heart disease and with no substantial mechanical impairment of function.

In a recent summary of the subject of ethyl alcohol and myocardial metabolism, Whereat and Perloff (1973) state that knowledge is "vanishingly small" and that the possible mechanisms relating chronic use of alcohol to heart muscle disease are, at this time, theoretical.

E. Myocardial Structure

As previously stated, cardiac hypertrophy and fibrosis were described as early as 1861 (Friedreich, 1861) in chronic regular users of alcohol. The gross and microscopic pathology has been very well summarized by Burch *et al.* (1966), as well as by Alexander (1966a,b). Dilatation of all cardiac chambers, pale, flabby myocardium, and patchy endocardial thickening with mural thrombi are seen grossly. Microscopically, variation in muscle fiber size, vacuolization, edema, fatty change, focal inflammatory change, and patchy or diffuse fibrosis are seen. As Burch *et al.* (1966) point out, the findings are nonspecific and "common to other types of cardiomyopathies." The prevalence of pathologic abnormalities in the hearts of chronic alcohol users was shown by Schenk and Cohen (1970), who studied hearts of 97 patients with a history of "excessive alcohol intake." Twenty of these patients had clinical evidence of heart disease during life, but a substantial majority of those with no heart disease during life had enlarged hearts and nonspecific myocardial fibrosis microscopically. These findings were much less prevalent in controls not known to be "excessive" alcohol users.

Ultrastructural abnormalities have been extensively studied in autopsied hearts of persons thought to have alcohol-related heart disease. Severe widespread mitochondrial damage, swelling of sarcoplasmic reticulum, myofibrillar disruption, and disturbance of intercalated disc structure have been noted, as well as accumulation of triglycerides (Burch *et al.,* 1966) and glycoproteins (Alexander, 1966b). These findings also are not pathognomonic, according to Mitchell and Cohen (1970). In humans, it is difficult to rule out or control possible nutritional factors in these findings, but generally similar ultrastructural abnormalities have been reported by Burch *et al.* (1971) and Segel *et al.* (1975) in well-nourished animals fed large amounts of ethanol.

In the work by Burch *et al.,* the animals allowed a balanced laboratory chow *ad libitum* showed a net gain in weight during the experiment. In the experiments of Segel *et al.,* the animals were fed 5, 10, or 25% ethanol solution with their regular chow diet. Some of the animals reported by Segel and colleagues were also fed a multivitamin supplement. This supplement did protect somewhat against the ethanol-induced decline in ATPase activity, but showed no other evidence of protection against ethanol-induced cardiac toxicity. It appears unlikely from the evidence that the results were due to nutrient dilution by the alcohol consumed.

In a study by Bulloch and co-workers (1972) of living "alcoholic" patients, myocardial biopsies were taken. The only unequivocal ultrastructural abnormalities in these patients, who had clinical evidence of heart disease, was dilatation of the sarcoplasmic reticulum and intercalated discs. The advanced ultrastructural abnormalities found in autopsied patients are not specific to alcohol toxicity. It is possible that, to some extent, they are sequellae of advanced recurrent heart failure or are secondary to other coexistent factors.

III. ALCOHOL AND CARDIOVASCULAR DISEASE

A. Partially Defined Syndromes

There are three fairly definite clinical cardiovascular syndromes in which alcohol has been thought to play a role. In this chapter, these are called "arsenic-beer drinkers' disease," "cobalt-beer drinkers' disease," and "cardiovascular beriberi." In all three entities, the role of alcohol is either partial, indirect, or both of these. This fact may provide partial insight into the subject and provides a clue pointing to proper lines of investigation into even less well-understood alcohol-related cardiovascular diseases.

1. Arsenic-Beer Drinkers' Disease

In 1900 an epidemic of over 6000 cases (over 70 deaths) occurred in and near Manchester, England, which proved to be due to a contamination of beer by small amounts of arsenic. The disorder produced signs and symptoms of the skin, nervous system, gastrointestinal tract, but especially prominent were cardiovascular manifestations. For example, in a beautifully written clinical description in

The Lancet, Dr. Ernest Reynolds (1901) made the following statements: (1) "cases were associated with so much heart failure and so little pigmentation that they were diagnosed as beri-beri. . . . " (2) "so great has been the cardiac muscle failure that . . . undoubtedly the principal cause of death has been cardiac failure. . . . " (3) "at the post-mortem examinations the only prominent signs were the interstitial nephritis and the dilated flabby heart. . . . "

There were a number of lively entries in *The Lancet* over the next several years (Royal Commission on Arsenical Poisoning, 1901a,b; Tunnicliffe and Rosenheim, 1901; Gowers, 1901). There were allusions to possible earlier outbreaks, including one in France 12 years earlier when arsenic had been placed into *wine*. It was determined that the probable source of the arsenic was contaminated sulfuric acid used to treat cane sugar. The result in the affected beer was arsenic content of 2–4 parts per million (approximately 0.2–0.4 mg/liter). Among the more important references made were to (1) the "peculiar idiosyncrasy which some people seemed to have" (Royal Commission on Arsenical Poisoning, 1901a), i.e., many persons seemed to become ill who drank less beer than others not affected. (2) "the amount of arsenic . . . actually consumed by the patients was not sufficient to explain the poisoning" (Tunnicliffe and Rosenheim, 1901). Sir William Gowers (1901) pointed out, for example, that he prescribed ten times the amount of arsenic involved over long periods of time for treatment of epilepsy, with no ensuing evidence of toxicity. The Royal Commission (1903) suggested that "alcohol predisposed people to arsenic poisoning." As best as one can determine, no one suggested the converse possibility.

2. Cobalt-Beer Drinkers' Disease

Sixty-five years later, the condition called "cobalt-beer drinkers' disease" was recognized. It has certain similarities to arsenic-beer drinkers' disease. In the mid-1960s, reports of epidemic outbreaks of an acute congestive cardiomyopathy appeared from Omaha (McDermott *et al.,* 1966; Sullivan *et al.,* 1969), Quebec (various authors, 1967), and Minneapolis (Alexander, 1968), in North America, and in Leuven, Belgium (Kesteloot *et al.,* 1968). The disease developed suddenly in persons who generally had been users of substantial amounts of beer. The North American patients suffered a high mortality rate, but those who recovered did well despite return, by many persons, to their previous beer use habits. The Belgian cases were less acute in onset, had lower mortality, and occurred over a longer period of time.

The explanation for this phenomenon proved to be the addition of small amounts of cobalt chloride by certain breweries to improve the foaming qualities of the beer. The foaming properties had been depressed by the widespread use of detergents in taverns when washing glassware. This etiology was tracked down largely by the Quebec investigators (Morin and Daniel, 1967) and the condition has, quite justly, been widely known as "Quebec-beer drinkers' cardiomyopathy." The largest brewery in Quebec had added cobalt to all of its beer, not just the beer destined for use as draught beer. The removal of the cobalt additive ended the epidemic in all locations.

In Belgium, where the concentration of added cobalt was less, the illness was much less acute and other manifestations of chronic cobalt use, such as increased red blood cell mass (polycythemia) and thyroid enlargement (goiter), were more prevalent.

However, even in Quebec, where the cobalt doses were probably greatest, 12 liters of the contaminated beer would have provided only about 8 mg of cobalt, less than 20% of the dose commonly employed as a hematinic. The hematinic use has not been implicated as a cause of heart muscle disease, whereas the first cases of this dramatic heart condition occurred 4–8 weeks after the addition of the chemical to beer.

Thus, it was fairly conclusively established, that *both* cobalt ingestion and substantial alcohol consumption were necessary to produce this illness. It also is almost certain that most persons appropriately exposed did *not* develop the disorder. The Belgian investigators suggested (Kesteloot *et al.*, 1968) that relative protein deficiency, due to anorexia, predisposed beer drinkers to the development of disease; they suggested that chelation of cobalt with sulfhydryl groups in some amino acids might protect against disease. There was much speculation, but the biochemical mechanisms were not defined. Alexander (1969) summed up both the arsenic and cobalt episodes: "This is the second known metal induced cardiotoxic syndrome produced by contaminated beer."

The parallels between the arsenic and cobalt beer episodes are so striking that the possibility that other metals or chemicals act synergistically with alcohol to produce disease has been and should be considered. A number of drugs are cardiotoxic, and many viruses are cardiotropic. Selenium, copper, and iron have been considered as cofactors. There is some evidence also that the heart may be involved in hemochromatosis (iron-storage disease) and that high alcohol use is prevalent among patients with hemochromatosis. Deficiency of zinc, magnesium, protein, and of various vitamins has been considered as a cofactor in alcohol cardiotoxicity. Thiamine deficiency is the only one proved to cause cardiovascular malfunction.

3. Cardiovascular Beriberi

For decades the concept of cardiovascular beriberi dominated thinking about the effects of alcohol on the cardiovascular system. Although descriptions of the condition were extant for a long time, the classical detailed description by Aalsmeer and Wenckebach (1929) clearly defined a clinical picture of high-output heart failure in polished-rice eaters in Java. The heart failure was primarily right ventricular, secondary to decreased peripheral resistance in these patients. Because it seemed that a deficiency state was the cause, it was assumed that heart failure in users of large amounts of alcohol in the West was due to associated nutritional deficiencies. Although some cases in North America and Europe fitted the clinical pattern of beriberi due to rice (Keefer, 1930) and were, without doubt, due to thiamine deficiency, it was soon evident that many other cases were atypical. In an excellent description of patients with cardiovascular disease secondary to presumed nutritional deficiency states, Weiss and Wilkins (1937) noted the following: (1) many with heart failure had good nutritional state; (2) many responded poorly, if at all, to thiamine; and (3) some had a low-output state, with *left* ventricular failure prominent. Over the next 25 years or so, however, the general view prevailed (Blankenhorn, 1945; Jones, 1959; Smith and Furth, 1943; Blankenhorn *et al.,* 1946) that thiamine deficiency was the primary culprit, that the cases unresponsive to replacement of the vitamin were due to chronicity of the condition, which ultimately becomes irreversible. Blankenhorn (1945; Blankenhorn *et al.,* 1946) pointed out that this "atypical" situation was actually the rule in alcohol-related heart disease. Jones (1959), while expressing the view that alcohol was a "contributing rather than a primary" cause of heart failure, expounded the concept that thiamine deficiency might be a contributory factor to other causes of heart failure. Blacket and Palmer (1960) stated this view, which remains current: "It (beri-beri) responds *completely* to thiamine, but merges imperceptibly into *another disease* called alcoholic cardiomyopathy, which doesn't respond to thiamine."

Blacket and Palmer (1960) and Kozam and co-workers (1972) studied a few cases of high-output cardiovascular beriberi by modern physiologic techniques. These showed remarkably high cardiac outputs at rest, among the highest ever measured. The basic cardiovascular defect in acute thiamine deficiency is dilatation of peripheral small vessels, with creation of a large arteriovenous shunt. Some cases of remarkably complete response of this situation to thiamine, with return to normal within 1–2 weeks, have been documented (Blacket and Palmer, 1960). The clinical observations of Aalsmeer and Wenckebach (1929) are thus confirmed.

It is now apparent that many of the earlier cases called "beriberi heart disease" or "cardiovascular beriberi" would now be classified as "alcoholic heart disease." A possible role of thiamine deficiency in at least some cases of chronic myocardial disease not responsive to vitamin B-1 has never been proved, but cannot be disproved either. Although the concepts of "relative" or "partial" beriberi as clinical cardiac disorders are not currently popular, the possible role of thiamine deficiency or relative thiamine deficiency in the myocardial disease associated with chronic alcohol use has never been, and may never be, fully defined.

B. Alcoholic Heart Disease

1. Evidence for Entity

We shall now consider a presumed disorder (or group of disorders) due to possible direct toxic effect of alcohol on the heart without known synergism with other chemicals or deficiency states and shall outline the circumstantial evidence for the existence of alcoholic heart disease, also frequently called "alcoholic cardiomyopathy."

First, there is the sheer volume of observations, many by excellent clinicians of the nineteenth and twentieth centuries. Many series of cases in various types of practices and populations have been reported (Alexander, 1966a,b; Böllinger, 1884; Evans, 1961; MacKenzie, 1902; Ferrans, 1966; Steell, 1893; Vaquez, 1921; Brigden and Robinson, 1964; Pintar *et al.*, 1965; Burch *et al.*, 1966; Gould *et al.*, 1969). While this type of evidence is, in the broad sense of the term "anecdotal" and does not satisfy statisticians, it cannot be dismissed or brushed aside, in the author's opinion. If the association reflects biased observation (one sees what one looks for), it seems probable that a substantial majority of currently practicing clinical cardiologists have this particular bias.

Further evidence is supplied by the fact that long-standing use of substantial amounts of alcohol has been found in a large proportion of persons in almost all series of patients with unexplained heart muscle disease. This group of patients, which is surely diverse, represents those who have evidence of heart muscle disorder without evidence of a specific cause, such as congenital abnormality, valvular disease, hypertension, or coronary artery disease. This situation, which is generally called "primary myocardial disease," comprises a variable percentage (usually 2–3%) of patients hospitalized for severe heart disease. The autopsy percentage is similar. The proportion of such patients who "admit to heavy drinking" (in the usual terminology) is often >50% (Alexander, 1966a; Hamby, 1970; Sanders, 1963; Shugoll *et al.*, 1972). However, this finding is not universal and Goodwin (1972) states that "alcohol is certainly not the cause of congestive cardiomyopathy in the majority of patients." Moreover, such data generally have been presented without control comparison, and many of the series have been reported from institutions in which large proportions of patients are regular drinkers. Alexander's (1966a) report makes the best attempt at control comparison for drinking in primary myocardial disease. He reported that 80% of patients hospitalized at the Minneapolis Veterans Administration hospital (exclusive of beriberi) were defined as "heavy drinkers," versus 28% of patients with other diagnoses who were admitted to the same medical service (Alexander, 1966a). Even if this type of control data is not ideal, the difference is impressive.

Another line of evidence consists of well-documented case reports. For example, Regan and colleagues (1969) reported a patient who developed congestive failure in a 4-month period while consuming 12–16 oz. (390 – 518 ml) of Scotch whiskey daily; followed by subsidence of clinical abnormality after cessation of drinking. Schwartz and co-workers (1975) presented the history of a patient with remarkable regression of severe advanced disease with prolonged abstinence. The most impressive study along this line is a report by Demakis *et al.* (1974) documenting greater frequency of *regression* of abnormality in primary myocardial disease among patients who abstain than among those who continue to drink.

Another piece of circumstantial evidence is the existence of acute and chronic peripheral *skeletal* muscular syndromes related to alcohol (Perkoff *et al.*, 1967). The acute syndrome is uncommon but dramatic clinically. Curiously, it has only rarely been associated with evident heart disease. The chronic syndrome is probably relatively common and consists of weakness of proximal girdle muscles. Data concerning relation to heart disease are not available. Skeletal muscle biopsies reveal evidence of myositis with intracellular edema and mitochondrial and myofibrillar damage similar to those seen in "alcoholic cardiomyopathy."

Perhaps the strongest circumstantial evidence for alcoholic heart disease is the evidence already cited in detail: (1) autopsy abnormalities in 90% of "alcoholics" with no clinical evidence of heart disease; (2) acute and chronic interference with myocardial function and metabolism by alcohol; and (3) substructural abnormalities in animal and human myocardial cells related to alcohol ingestion. These facts comprise the rapidly growing case for "preclinical cardiomyopathy," which presumably progresses to alcoholic heart disease.

2. Diagnosis

The entity called alcoholic heart disease is based on circumstantial evidence, and the diagnosis is always presumptive. This fact, plus the known difficulties of obtaining accurate drinking histories from patients, leaves clinicians with amorphous diagnostic tools:

1. *A compatible drinking history.* No one knows the minimum dose of alcohol or duration of use required. It has been estimated (Parker, 1974) that 1% ± of "heavy" drinkers of 10+ years' duration will develop alcoholic heart disease, but this seems little more than a guess. Obviously a strong individual susceptibility variable is present.

2. *The exclusion of other causes of heart disease.* This is clearly necessary but is bothersome intellectually, because there is no logical reason to think that a person might not have *both* coronary and alcoholic heart disease, for example. Existence of another type of heart disease might actually predispose to alcoholic heart disease.

3. *The susceptibility of males.* Risk factors are poorly defined, but some believe males are more susceptible. Association with race and type of beverage is

probably primarily a function of location of the study and the socioeconomic group involved.

4. *Acute worsening with substantial drinking.* In actual clinical situations, direct relation of heart disease to amount of recent alcohol use may be common, but data are hard to come by.

5. *Consistent pathologic evidence.* The pathologic changes described by both light and electron microscopy have so far not proved specific or diagnostic.

3. Clinical Features

Early symptoms or signs of alcoholic heart disease are nonspecific rhythm disturbances and ECG variations, often minor. Vague and nonspecific as sinus tachycardia, premature beats, and repolarization abnormalities (ST- to T-wave changes) on the ECG may be, these are the signs at the stage when reversibility is maximal. Evans (1959) in a classic article, described T-wave configurations he considered relatively specific for early alcoholic heart disease, but most other clinicians, including the author, have not frequently observed these "characteristic" findings.

Experienced clinicians have long recognized an acute alcohol load as one of the causes of paroxysmal atrial fibrillation. Recently, Ettinger *et al.* (1976) have reported a series of such patients with various rhythm disturbances and have given the manifestation the colorful name of "The Holiday Heart Syndrome." The likelihood that persons with these early relatively minor signs will progress to chronic myocardial weakness is completely unknown, but it is likely that they represent a high-risk group.

The late picture of low-output congestive cardiomyopathy, with conduction abnormalities on the ECG, chronic arrhythmia, and high incidence of embolic complications, is the entity usually described as alcoholic heart disease. To some extent, even this clinical picture may regress, but many seem to progress inexorably despite abstinence and optional therapy. Most do not come to medical attention until the late stage has been reached. While the onset is generally insidious, the author has been impressed by a number of patients who seemed to have an apparently acute onset of severe congestive heart failure. For further details of the clinical picture of the late-stage alcoholic heart disease, which is indistinguishable from that of other forms of primary myocardial disease, the reader is referred to the excellent clinical descriptions cited (Alexander, 1966a; Evans, 1961; Ferrans, 1966; Hamby, 1970; Sanders, 1963; Brigden and Robinson, 1964; Pintar *et al.,* 1965; Burch *et al.,* 1966).

A final clinical point of interest is that coexistence with alcoholic cirrhosis of the liver has consistently been found to be uncommon. This may in part be due to statistical artifact (i.e., the existence of one major illness tends to preclude the development of another). However, it also seems likely that there is a marked individual susceptibility variation for both complications of alcohol use, and that neither is highly prevalent among regular alcohol users.

C. Relation to Hypertension

A report from the Los Angeles Heart Study (Clark *et al.*, 1967) presented data indicating that regular users of alcohol had slightly higher mean systolic and diastolic pressures. The definition of regular use of alcohol was three times per week and the differences were slight. D'Alonzo and Pell (1968) then reported that "problem drinkers" (almost all male) had a prevalence of "hypertension" (defined as ≥ 160/95) 2.3 times greater than controls (a combination of nondrinkers and "moderate" drinkers). A report from the Framingham heart study (Kannell and Sorlie, 1974), using the same definition of hypertension, indicated that the condition was twice as prevalent in persons who took 60 oz. or more (1774 ml +) of alcohol monthly than in persons who took 10–29 oz. (296 – 858 ml) per month, who had, in turn, a slightly lower prevalence of hypertension than nondrinkers. A report from Sweden based upon 70 twin pairs, discordant for alcohol use (Myrhed, 1974), showed a significant association between long-standing alcohol use and elevated pressures.

A recent report (Klatsky *et al.*, 1977b) showed a statistically strong association between blood pressure and known drinking habits of 83,947 men and women of three races (83.5% white) (Fig. 1). Using Kaiser-Permanente multiphasic health checkup questionnaire responses, persons were classified as nondrinkers or according to usual daily number of drinks: two or fewer per day; three to five per day; or six or more per day. Compared with the nondrinkers, blood pressures of men taking two or fewer drinks per day were similar. Women who took two or fewer drinks per day had slightly lower pressures. Men and women who took three or more drinks per day had higher systolic and diastolic pressures and substantially higher prevalence of blood pressures (≥160/95 mm Hg). The prevalence of hypertension, thus defined, was doubled in whites and increased by approximately 50% in blacks, when those taking six or more drinks per day were compared with nondrinkers. The associations of blood pressure and drinking were independent of age, sex, race, smoking, coffee use, former "heavy" drinking, or adiposity.

The apparent elevations of blood pressure in the Kaiser-Permanente study (Klatsky *et al.*, 1977b) were not progressive beyond 6–8 drinks per day in whites (i.e., persons taking nine or more drinks per day did *not* have higher pressures than those taking 6–8 drinks per day). In blacks, the progression of elevated pressures was not seen with use above 3–5 drinks per day. The extent to which pressure elevations might be due to immediate effect of alcohol ingested shortly before the examination could not be assessed. In any case, other studies were not

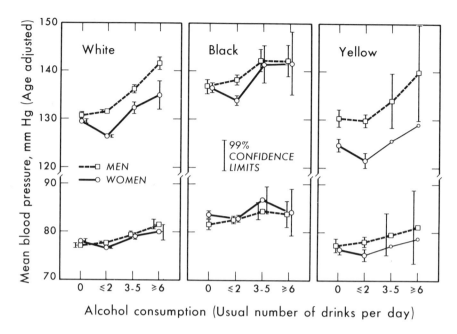

Fig. 1. Mean systolic blood pressures (upper half of figure) and mean diastolic blood pressures (lower half of figure) of white, black, or yellow men and women with known drinking habits. Small circles represent data based on less than 30 persons.

unanimous in showing acute elevation of blood pressure in response to alcohol. The possibility of an independent third factor (e.g., psychosocial "stress") resulting in both increased alcohol use and blood pressure elevation was mentioned. Although causality was not proved, it was felt that the data, together with the previous findings of others, "strongly suggest that regular use of three or more drinks of alcohol per day is a risk factor for hypertension" (Klatsky *et al.*, 1977b).

Two separate recent studies from Glasgow, Scotland, reported by Ramsay (1977) and Beevers (1977), respectively, suggest that liver function is disturbed in a disproportionate number of hypertensives. Both investigators expressed the opinion that alcohol use was involved in both the hypertension and the liver abnormalities. In an editorial comment in the issue of *The Lancet* (Editorial, 1977) which presented these articles, it was stated that, "it is becoming apparent that there is a link between alcohol ingestion and hypertension which may be of both theoretical and practical interest."

D. Relation to Coronary Disease

1. Angina Pectoris

The history of this subject is long. Heberden's classic description of angina (Heberden, 1786) reported that "wine and spiritous liquors—afford considerable relief." Thus, over the years, alcohol was widely presumed (Levine, 1951; White, 1931) to be a coronary vasodilator. However, a report by Russek *et al.* (1950), using exercise ECG studies, indicated that benefit probably came from CNS depression and not from improved coronary flow. A report by Orlando *et al.* (1976), also using exercise studies of angina patients, indicated that drinking 65–320 ml of ethanol progressively *decreased* exercise tolerance.

Attempts to study coronary blood flow in relation to acutely administered alcohol have shown conflicting results, perhaps because of differences in alcohol dose, species, and experimental conditions. One can find evidence to support increased coronary flow, decreased flow, or no effect. Thus, the relation of alcohol to the usual symptom of coronary disease, angina pectoris, is unclear: clinically, experimentally, and epidemiologically.

2. Relation to Major Coronary Events (Infarction and Sudden Death)

In the first half of this century there were a number of reports (Cabot, 1904; Hultgen, 1910; Leary, 1931; Wilens, 1947) of an apparent negative association between chronic alcohol use and atherosclerotic disease, including coronary heart disease. The concept was introduced (Ruebner *et al.*, 1961) that a statistical artifact might be involved, i.e., that the premature deaths of many regular users of substantial amounts of alcohol might preclude the development of atherosclerotic vascular disease. Reports of a negative relation between cirrhosis of the liver (Parrish and Eberly, 1961) and coronary artery disease were likewise explained as a statistical fallacy inherent in autopsy series.

Epidemiologic studies of major coronary events have not agreed with respect to a relation to habitual alcohol use, but only one (Wilhelmsen *et al.*, 1973) has shown a positive association with coronary events when control for cigarette smoking was used. The work of Paul *et al.* (1963) and the Framingham study (U.S. National Heart Institute, 1966) found no positive association, even *without* control for cigarette use. The Framingham data did show that there was a slight, but not statistically significant, negative relation between use of 30 oz. (885 ml) of alcohol monthly and coronary events. Three recent reports (Stason *et al.*, 1976; Klatsky *et al.*, 1974; Rhoads *et al.*, 1976) have indicated a negative relation between regular alcohol use and nonfatal myocardial infarction. One of these (Rhoads *et al.*, 1976) was a presentation of data from the Honolulu Heart Study. Stason *et al.* (1976) reported data from the Boston Collaborative Drug Surveillance Program which indicated no overall association; but there was evidence of a lower infarction rate in persons consuming six or more drinks daily. Several potential confounders, including smoking, were controlled for in the analysis.

Other recent epidemiologic information pertinent to this subject has been published (Klatsky *et al.*, 1974, 1976) (Fig. 2) from the Kaiser-Permanente myocardial infarction and sudden death studies, conceived and directed by Gary Friedman. In these studies (Friedman *et al.*, 1974, 1975), the subjects were men and women hospitalized for a first acute myocardial infarction and men who were victims of sudden cardiac death. By computer matching of 250,000+ stored checkup records, it was possible to select not only an age/sex/race-matched control for each myocardial infarction or sudden cardiac death patient, but a second control, which was called a "risk control," matched for presence or absence of seven coronary risk factors, including cigarette smoking and blood pressure. The use of the two control groups enabled ready determination of whether any checkup item predictive of coronary events was associated with, or independent of, the established risk factors. Nondrinkers were at a statistically significant greater risk of myocardial infarction than users of alcohol. This negative association was apparently independent of age, sex, prior coronary disease,

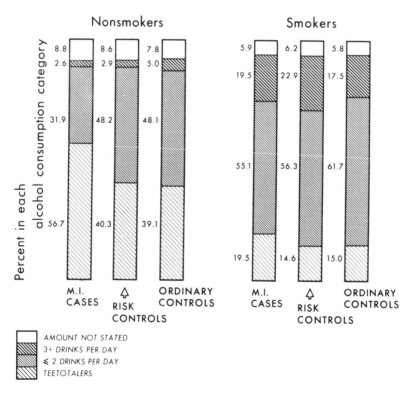

Fig. 2. The relation of alcohol consumption to cigarette smoking in myocardial infarction (M.I.) patients and both control groups. The nonsmokers include 113 myocardial infarction patients, 129 risk controls, and 179 ordinary controls. The established smokers include 185 myocardial infarction patients, 192 risk controls, and 120 ordinary controls.

risk-factor-related disease, and, of course, the matched risk factors, themselves. Explanations considered included indirect association of drinking habits with ethnic origin, psychological traits, or other unknown risk factors for myocardial infarction, as well as a possible protective effect of alcohol. It was concluded that "abstinence from alcohol may be a new risk factor for myocardial infarction" (Klatsky *et al.*, 1976).

In the Kaiser–Permanente studies, the percentage of nondrinkers was only slightly higher among the sudden cardiac death victims than among the matched risk controls. It was not felt that this represented failure of confirmation of the myocardial infarction study result, as the mechanisms of sudden cardiac death are probably different, to a substantial extent, from those involved in acute myocardial infarction. The sudden death study results were interpreted as proving that the greater risk of hospitalization for myocardial infarction among nondrinkers was *not* due to spurious selection of cases because of a higher prehospitalization death rate among users of alcohol.

Further support for a possible negative relation between alcohol use and the risk of nonfatal myocardial infarction and death from coronary heart disease is provided by the work of Yano *et al.* (1977) in a prospective epidemiologic study of a large cohort of Japanese men living in Hawaii. Drinkers of "moderate" amounts of alcohol (mostly beer) were at significantly less risk than abstainers, with control for smoking and other risk factors.

Because use of alcohol is related to two well-established risk factors for major coronary events, cigarette smoking (Klatsky *et al.*, 1976, 1977a) (Fig. 2) and hypertension (Klatsky *et al.*, 1977b) (Fig. 1), it is apparently necessary to control carefully for these traits to uncover the negative alcohol use–infarction relation. Perhaps the failure of some studies to find this association represents cancellation of increased risk related to cigarette use and hypertension by some direct or indirect protective effect of alcohol. One new clue to a possible protective effect lies in the fact that use of alcohol is related inversely to low-density lipoprotein and directly to high-density lipoprotein, as reported by Rhoads and co-workers (1976) and recently by Castelli *et al.* (1977). α-Cholesterol, believed by some to confer protection against coronary events, is thus higher in alcohol users. Further reports of evidence along these lines would be of extraordinary interest.

IV. SUMMARY AND RESEARCH NEEDS

A. Alcoholic Heart Disease

1. The circumstantial case for such a condition is very strong. Epidemiologic studies to determine prevalence among regular users of alcohol are needed. Such studies might also elucidate the question of the maximal "safe" dose of chronic alcohol use for this and other conditions.

2. Continuation of current active research seems likely to further the basic

mechanism(s) of cardiotoxicity of alcohol. Animal experiments should ultimately be expanded to include study of the role of cofactors (trace elements, deficiency states, preceding viral myocardial injury, etc.) Such work might help to explain the apparent marked individual human variation in susceptibility.

3. Further search for specific diagnostic findings is warranted, but, unfortunately, success does not seem imminent. A specific diagnostic tool would enable physicians to assess the possible *coexistence* of alcoholic heart disease and other types of cardiac disorder. Such coexistence seems likely.

B. Hypertension

1. Additional confirmation of the association with substantial regular alcohol use is desirable, with epidemiologic attempts to determine more precisely the "threshold" dose.

2. There is need to study possible indirect explanations of the association, especially chronic "stress."

3. Investigations to find mechanisms for chronic effects of alcohol on blood pressure should be undertaken. Possible effects of ethanol on the volume-regulating mechanisms (e.g., the renin–angiotensin system) would be of great interest.

C. Coronary Heart Disease

Further confirmation of the reported negative association of alcohol use with myocardial infarction is necessary before this can be accepted. Investigators in this area will need to control very carefully for established risk factors associated with alcohol use (cigarette smoking, hypertension). Further work on possible protective effects of alcohol against major coronary events should proceed (e.g., higher α-cholesterol in drinkers).

D. General Approach

Generalizations should be avoided, as the history of this subject clearly points out. The possibility that there is a "safe" or even beneficial alcohol dose is strong. The effects of alcohol on the myocardium, blood pressure, coronary disease, and other cardiovascular conditions should be considered as *independent* subjects.

ACKNOWLEDGMENTS

Portions of this article consist of research supported by grants from the United States Brewers Association, Inc., the Myocardial Infarction Branch, National Heart and Lung Institute, National Institutes of Health, the Community Service Program of the Kaiser Foundation Hospitals, and the National Center for Health Services Research and Development.

The author wishes to acknowledge the technical assistance of Della Mundy and Glenda Danneil.

REFERENCES

Aalsmeer, W. C., and Wenckebach, K. F. (1929). Herz und Kreislauf bei der Beri-Beri Krankheit. *Wien. Arch. Inn. Med.* **16**, 193–272.

Ahmed, S. S., Levinson, G. E., and Regan, T. J. (1973). Depression of myocardial contractility with low doses of ethanol in normal man. *Circulation* **48**, 378–385.

Alexander, C. S. (1966a). Idiopathic heart disease. I. Analysis of 100 cases with special reference to chronic alcoholism. *Am. J. Med.* **41**, 213–228.

Alexander, C. S. (1966b). Idiopathic heart disease. II. Electron microscopic examination of myocardial biopsy specimens in alcoholic heart disease. *Am. J. Med.* **41**, 229–234.

Alexander, C. S. (1968). The syndrome of cobalt-beer cardiomyopathy including ultrastructural changes on biopsy (abstract). *J. Lab. Clin. Med.* **72**, 850.

Alexander, C. S. (1969). Cobalt and the heart. *Ann. Intern. Med.* **70**, 411–413.

Beevers, D. G. (1977). Alcohol and hypertension. *Lancet* **ii**, 114–115.

Blacket, R. B., and Palmer, A. J. (1960). Haemodynamic studies in high output beri-beri. *Br. Heart J.* **22**, 483–501.

Blankenhorn, M. A. (1945). The diagnosis of beriberi heart disease. *Ann. Intern. Med.* **23**, 398–404.

Blankenhorn, M. A., Vilter, C. F., Scheinker, I. M., and Austin, R. S. (1946). Occidental beriberi heart disease. *J. Am. Med. Assoc.* **131**, 717–727.

Blomqvist, G., Saltin, B., and Mitchell, J. H. (1970). Acute effects of ethanol ingestion on the response to submaximal and maximal exercise in man. *Circulation* **42**, 463–470.

Böllinger, O. (1884). Ueber die Haussigkeit und Ursachen der idiopathischen Herzhypertrophie in München. *Dtsch. Med. Wochenschr. (Stuttgart)* **10**, 180.

Brigden, W., and Robinson, J. (1964). Alcoholic heart disease. *Br. Med. J.* **2**, 1283–1289.

Bulloch, R. T., Pearce, M. B., Murphy, M. L., Jenkins, B. J., and Davis, J. L. (1972). Myocardial lesions in idiopathic and alcoholic cardiomyopathy: Study by ventricular septal biopsy. *Am. J. Cardiol.* **29**, 15–25.

Burch, G. E., Phillips, J. H., Jr., and Ferrans, V. J. (1966). Alcoholic cardiomyopathy. *Am. J. Med. Sci.* **252**, 123/89–138/104.

Burch, G. E., Colcolough, H. L., Harb, J. M., and Tsui, C. Y. (1971). The effects of ingestion of ethyl alcohol, wine and beer on the myocardium of mice. *Am. J. Cardiol.* **27**, 522–528.

Cabot, R. C. (1904). The relation of alcohol to arteriosclerosis. *J. Am. Med. Assoc.* **43**, 774–775.

Castelli, W. P., Gordon, T., Hjortland, M. C., Kagan, A., Doyle, J. T., Hames, C. G., Hulley, S. B., and Zukel, W. J. (1977). Alcohol and blood lipids: The cooperative lipoprotein phenotyping study. *Lancet* **ii**, 153–155.

Clark, V. A., Chapman, J. M., and Coulson, A. H. (1967). Effects of various factors on systolic and diastolic blood pressure in the Los Angeles Heart Study. *J. Chron. Dis.* **20**, 571–581.

Conway, N. (1968). Haemodynamic effects of ethyl alcohol in patients with coronary heart disease. *Br. Heart J.* **30**, 638–644.

D'Alonzo, C. A., and Pell, S. (1968). Cardiovascular disease among problem drinkers. *J. Occup. Med.* **10**, 344–350.

Davison, F. R. (1949). "Handbook of Materia Medica, Toxicology and Pharmacology," pp. 401–409. Mosby, St. Louis, Missouri.

Delgado, C. E., Fortuin, N. J., and Ross, R. S. (1975). Acute effects of low doses of alcohol on left ventricular function by echocardiography. *Circulation* **51**, 535–540.

Demakis, J. G., Proskey, A., Rahimtoola, S. H., Jamil, M., Sutton, G. C., Rosen, K. M., Gunnar, R. M., and Tobin, J. R., Jr. (1974). The natural course of alcoholic cardiomyopathy. *Ann. Intern. Med.* **80**, 293–297.

Editorial. (1977). Alcohol and hypertension. *Lancet* **ii**, 122.

Eliaser, M., and Giansiracusa, F. J. (1956). The heart and alcohol. *Calif. Med.* **84**, 234–236.

Ettinger, P. O., Wu, C. F., De la Cruz, C., Jr., Weisse, A. B., and Regan, T. J. (1976). Tachyarrhythmias associated with the preclinical cardiomyopathy of alcoholism. *Am. J. Cardiol.* **37,** 134.

Evans, W. (1959). The electrocardiogram of alcoholic cardiomyopathy. *Br. Heart J.* **21,** 445–456.

Evans, W. (1961). Alcoholic cardiomyopathy. *Am. Heart J.* **61,** 556–567.

Ferrans, V. J. (1966). Alcoholic cardiomyopathy. *Am. J. Med. Sci.* **252,** 123/89–138/104.

Friedman, G. D., Klatsky, A. L., and Siegelaub, A. B. (1974). Kaiser–Permanente epidemiologic study of myocardial infarction. Study design and results for standard risk factors. *Am. J. Epidemiol.* **99,** 101–116.

Friedman, G. D., Klatsky, A. L., and Siegelaub, A. B. (1975). Predictors of sudden cardiac death. *Circulation* **51/52,** Suppl. 3, 164–169.

Friedreich, N. (1861). "Handbuch der speziellen Pathologie und Therapie," 5th Sect. Krankheiten des Herzens, Ferdinand Enke, Erlangen.

Ganz, V. (1963). The acute effect of alcohol on the circulation and on the oxygen metabolism of the heart. *Am. Heart J.* **66,** 494–497.

Gimeno, A. L., Gimeno, M. F., and Webb, J. L. (1962). Effects of ethanol on cellular membrane potentials and contractility of isolated rat atrium. *Am. J. Physiol.* **203,** 194–196.

Goodwin, J. F. (1972). Clarification of the cardiomyopathies. *Mod. Concepts Cardiovasc. Dis.* **41**(9), 41–46.

Gould, L., Zahir, M., Mahmood, S., and Di Lieto, M. (1969). Cardiac hemodynamics in alcoholic heart disease. *Ann. Intern. Med.* **71,** 543–553.

Gould, L., Zahir, M., DeMartino, A., and Gomprecht, R. F. (1971). Cardiac effects of a cocktail. *J. Am. Med. Assoc.* **218,** 1799–1802.

Gould, L., Zahir, M., DeMartino, A., Gomprecht, R. F., and Jaynal, F. (1972). Hemodynamic effects of ethanol in patients with cardiac disease. *Q. J. Stud. Alcohol* **33,** 714–721.

Gowers, W. R. (1901). *In* Royal Medical and Chirurgical Society. Epidemic of arsenical poisoning in beer-drinkers in the north of England during the year 1900. *Lancet* **i,** 98–100.

Grollman, A. (1930). The action of alcohol, caffeine, and tobacco on the cardiac output (and its related functions) of normal man. *J. Pharmacol. Exp. Ther.* **39,** 313–327.

Hamby, R. I. (1970). Primary myocardial disease. *Medicine (Baltimore)* **49,** 55–78.

Heberden, W. (1786). Some account of a disorder of the breast. *Med. Trans. R. Coll. Physicians (London)* **2,** 59–67.

Horwitz, L. D., and Atkins, J. M. (1974). Acute effects of ethanol on left ventricular performance. *Circulation* **49,** 124–128.

Hultgen, J. F. (1910). Alcohol and nephritis: Clinical study of 460 cases of chronic alcoholism. *J. Am. Med. Assoc.* **55,** 279–281.

Isselbacher, K. J. (1977). Metabolic and hepatic effects of alcohol. *N. Engl. J. Med.* **296,** 612–619.

Jones, R. H. (1959). Beriberi heart disease. *Circulation* **19,** 275–281.

Juchems, R., and Klobe, R. (1969). Hemodynamic effects of ethyl alcohol in man. *Am. Heart J.* **78,** 133–135.

Kannell, W. B., and Sorlie, P. (1974). Hypertension in Framingham. *In* "Epidemiology and Control of Hypertension" (O. Paul, ed.), pp. 553–592. Stratton Intercont. Med. Book Corp., New York.

Keefer, C. S. (1930). The beri-beri heart. *Arch. Intern. Med.* **45,** 1–22.

Kesteloot, H., Roelandt, J., Willems, J., Claes, J. H., and Joossens, J. V. (1968). An enquiry into the role of cobalt in the heart disease of chronic beer drinkers. *Circulation* **37,** 854–864.

Kikuchi, T., and Kako, K. J. (1970). Metabolic effects of ethanol on the rabbit heart. *Circ. Res.* **26,** 625–634.

Klatsky, A. L., Friedman, G. D., and Siegelaub, A. B. (1974). Alcohol consumption before myocardial infarction: Results from the Kaiser-Permanente epidemiologic study of myocardial infarction. *Ann. Intern. Med.* **81**, 294-301.

Klatsky, A. L., Friedman, G. D., and Siegelaub, A. B. (1976). Medical history questions predictive of myocardial infarction: Results from the Kaiser-Permanente epidemiologic study of myocardial infarction. *J. Chron. Dis.* **29**, 683-696.

Klatsky, A. L., Friedman, G. D., Siegelaub, A. B., and Gérard, M. J. (1977a). Alcohol consumption among white, black or oriental men and women: Kaiser-Permanente multiphasic health examination data. *Am. J. Epidemiol.* **105**, 311-323.

Klatsky, A. L., Friedman, G. D., Siegelaub, A. B., and Gérard, M. J. (1977b). Alcohol consumption and blood pressure. *N. Engl. J. Med.* **296**, 1194-1200.

Kozam, R. L., Esguerra, O. E., and Smith, J. J. (1972). Cardiovascular beriberi. *Am. J. Cardiol.* **30**, 418-422.

Leary, T. (1931). Therapeutic value of alcohol, with special consideration of relations of alcohol to cholesterol, and thus to diabetes, to arteriosclerosis, and to gallstones. *N. Engl. J. Med.* **205**, 231-242.

Levine, S. A. (1951). "Clinical Heart Disease," 4th ed., p. 98. Saunders, Philadelphia, Pennsylvania.

Liedtke, A. J., and DeMuth, W. E. (1975). Effects of alcohol on cardiovascular performance after experimental nonpenetrating chest trauma. *Am. J. Cardiol.* **35**, 243-250.

Limas, C. J., Guiha, N. H., Lekagul, O., and Cohn, J. N. (1974). Impaired left ventricular function in alcoholic cirrhosis with ascites: Ineffectiveness of ouabain. *Circulation* **49**, 755-760.

Lochner, A., Cowley, R., and Brink, A. J. (1969). Effect of ethanol on metabolism and function of perfused rat heart. *Am. Heart J.* **78**, 770-780.

McDermott, P. H., Delaney, R. L., Egan, J. D., and Sullivan, J. F. (1966). Myocardosis and cardiac failure in men. *J. Am. Med. Assoc.* **198**, 163-166.

MacKenzie, J. (1902). "The Study of the Pulse," p. 237. Y. J. Pentland, Edinburgh and London.

Maines, J. E., and Aldinger, E. E. (1967). Myocardial depression accompanying chronic consumption of alcohol. *Am. Heart J.* **73**, 55-63.

Mendoza, L. C., Hellberg, K., Rickart, A., Tillich, G., and Bing, R. J. (1971). The effect of intravenous ethyl alcohol on the coronary circulation and myocardial contractility of the human and canine heart. *J. Clin. Pharmacol.* **11**, 165-176.

Mitchell, J. H., and Cohen, L. S. (1970). Alcohol and the heart. *Mod. Concepts Cardiovasc. Dis.* **39**(7), 109-113.

Morin, Y., and Daniel, P. (1967). Quebec beer-drinkers' cardiomyopathy: Etiologic considerations. *Can. Med. Assoc. J.* **97**, 926-928.

Myrhed, M. (1974). Blood pressure. *In* Alcohol consumption in relation to factors associated with ischemic heart disease. *Acta Med. Scand., Suppl.* No. 567, pp. 40-46.

Newman, W. H., and Valicenti, J. F., Jr. (1971). Ventricular function following acute alcohol administration: Strain-gauge analysis of depressed ventricular dynamics. *Am. Heart J.* **81**, 61-68.

Orlando, J., Aronow, W. S., Cassidy, J., and Prakash, R. (1976). Effect of ethanol on angina pectoris. *Ann. Intern. Med.* **84**, 652-655.

Osler, W. (1899). "The Principles and Practice of Medicine," 3rd ed. Appleton, New York.

Pachinger, O. M., Tillmans, H., Mao, J. C., Fauvel, J.-M., and Bing, R. J. (1973). The effect of prolonged administration of ethanol on cardiac metabolism and performance in the dog. *J. Clin. Invest.* **52**, 2690-2696.

Parker, B. M. (1974). The effects of ethyl alcohol on the heart. *J. Am. Med. Assoc.* **228**, 741-742.

Parrish, H. M., and Eberly, A. L., Jr. (1961). Negative association of coronary atherosclerosis with liver cirrhosis and chronic alcoholism—a statistical fallacy. *J. Indiana State Med. Assoc.* **54**, 341-347.

Paul, O., Lepper, M. H., Phelan, W. H., Dupertuis, G. W., MacMillan, A., McKean, H., and Park, H. (1963). A longitudinal study of coronary heart disease. *Circulation* **28**, 20–36.

Perkoff, G. T., Dioso, M. M., Bleisch, V., and Klinkerfuss, G. (1967). A spectrum of myopathy associated with alcoholism. I. Clinical and laboratory features. *Ann. Intern. Med.* **67**, 481–492.

Pintar, K., Wolanskyj, B. M., and Gubbay, E. R. (1965). Alcoholic cardiomyopathy. *Can. Med. Assoc. J.* **93**, 103–107.

Ramsay, L. E. (1977). Liver dysfunction in hypertension. *Lancet* **ii**, 111–114.

Regan, T. J., Koroxenidis, G., Moschos, C. B., Oldewurtel, H. A., Lehan, P. H., and Hellems, H. K. (1966). The acute metabolic and hemodynamic responses of the left ventricle to ethanol. *J. Clin. Invest.* **45**, 270–280.

Regan, T. J., Levinson, G. E., Oldewurtel, H. A., Frank, M. J., Weisse, A. B., and Moschos, C. B. (1969). Ventricular function in noncardiacs with alcoholic fatty liver: Role of ethanol in the production of cardiomyopathy. *J. Clin. Invest.* **48**, 397–407.

Regan, T. J., Khan, M. I., Ettinger, P. O., Haider, B., Lyons, M. M., and Oldewurtel, H. A. (1974). Myocardial function and lipid metabolism in the chronic alcoholic animal. *J. Clin. Invest.* **54**, 740–752.

Reynolds, E. S. (1901). An account of the epidemic outbreak of arsenical poisoning occurring in beer-drinkers in the North of England and the Midland Counties in 1900. *Lancet* **i**, 166–170.

Rhoads, G. G., Kagan, A., and Yano, K. (1976). Associations between dietary factors and plasma lipoproteins. *Circulation* **54**, Suppl. 2, II–53.

Riff, D. P., Jain, A. C., and Doyle, J. T. (1969). Acute hemodynamic effects of ethanol on normal human volunteers. *Am. Heart J.* **78**, 592–597.

Royal Commission Appointed to Inquire into Arsenical Poisoning from the Consumption of Beer and other Articles of Food or Drink. (1903). "Final Report," Part I. Wyman and Sons, London.

Royal Commission on Arsenical Poisoning. (1901a). First Report. *Lancet* **ii**, 218.

Royal Commission on Arsenical Poisoning. (1901b). First Report. *Lancet* **i**, 672–673.

Ruebner, B. H., Miyai, K., and Abbey, H. (1961). The low incidence of myocardial infarction in hepatic cirrhosis—a statistical artefact? *Lancet* **ii**, 1435–1436.

Russek, H. I., Naegele, C. F., and Regan, F. D. (1950). Alcohol in the treatment of angina pectoris. *J. Am. Med. Assoc.* **143**, 355–357.

Sanders, V. (1963). Idiopathic disease of myocardium. *Arch. Intern. Med.* **112**, 661–676.

Schenk, E. A., and Cohen, J. (1970). The heart in chronic alcoholism. *Pathol. Microbiol.* **35**, 96–104.

Schmitthenner, J. E., Hafkenschiel, J. H., Forte, I., Williams, A. J., and Riegel, C. (1958). Does alcohol increase coronary blood flow and cardiac work? (abstract). *Circulation* **18**, 778.

Schwartz, L., Sample, K. A., and Wigle, E. D. (1975). Severe alcoholic cardiomyopathy reversed with abstention from alcohol. *Am. J. Cardiol.* **36**, 963–966.

Segel, L. D., Rendig, S. V., Choquet, Y., Chacko, K., Amsterdam, E. A., and Mason, D. T. (1975). Effects of chronic graded ethanol consumption on the metabolism, ultrastructure, and mechanical function of the rat heart. *Cardiovasc. Res.* **9**, 649–663.

Shugoll, G. I., Bowen, P. J., Moore, J. P., and Lenkin, M. L. (1972). Follow-up observations and prognosis in primary myocardial disease. *Arch. Intern. Med.* **129**, 67–72.

Smith, J. J., and Furth, J. (1943). Fibrosis of the endocardium and the myocardium with mural thrombosis. *Arch. Intern. Med.* **71**, 602–619.

Spann, J. F., Jr., Mason, D. T., Beiser, G. D., and Gold, H. K. (1968). Actions of ethanol on the contractile state of the normal and failing cat papillary muscle (abstract). *Clin. Res.* **16**, 249.

Spodick, D. H., Pigott, V. M., and Chirife, R. (1972). Preclinical cardiac malfunction in chronic alcoholism: Comparison with matched normal controls and with alcoholic cardiomyopathy. *N. Engl. J. Med.* **287**, 677–680.

Stason, W. B., Neff, R. K., Miettinen, O. S., and Jick, H. (1976). Alcohol consumption and nonfatal myocardial infarction. *Am. J. Epidemiol.* **104,** 603–608.

Steell, G. (1893). Heart failure as a result of chronic alcoholism. *Med. Chron. Manchester* **18,** 1–22.

Steell, G. (1906). "Textbook on Diseases of the Heart," p. 79. Blakiston, Philadelphia, Pennsylvania.

Strümpel, A. (1890). "A Textbook of Medicine," p. 294. Appleton, New York.

Sullivan, J. F., George, R., Bluvas, R., and Egan, J. D. (1969). Myocardiopathy of beer drinkers: Subsequent course. *Ann. Intern. Med.* **70,** 277–282.

Tunnicliffe, F. W., and Rosenheim, O. (1901). Selenium compounds as factors in the recent beer-poisoning epidemic. *Lancet* **i,** 318.

U.S. National Heart Institute. (1966). "The Framingham Heart Study: Habits and Coronary Heart Disease," U.S. Public Health Serv. Publ. No. 1515, p. 11. U.S. Gov. Print. Off., Washington, D.C.

Vaquez, H. (1921). "Maladies du Coeur," p. 308. Baillière et Fils, Paris.

Various authors. (1967). Quebec beer-drinkers' cardiomyopathy *Can. Med. Assoc. J.* **97**(15), 881–931.

Walsche, W. H. (1873). "Disease of the Heart and Great Vessels," 4th ed. Smith, Elder, London.

Webb, W. R., and Degerli, I. U. (1965). Ethyl alcohol and the cardiovascular system. *J. Am. Med. Assoc.* **191,** 77/1055–80/1058.

Webb, W. R., Degerli, I. U., Cook, W. A., and Unal, M. O. (1966). Alcohol, digitalis and cortisol, and myocardial capacity in dogs. *Ann. Surg.* **163,** 811–817.

Weiss, S., and Wilkins, R. W. (1937). The nature of the cardiovascular disturbances in nutritional deficiency states (beriberi). *Ann. Intern. Med.* **11,** 104–148.

Wendt, V. E., Stock, T. B., Hayden, R. O., Bruce, T. A., Gudbjarnason, S., and Bing, R. J. (1962). The hemodynamics and cardiac metabolism in cardiomyopathies. *Med. Clin. North Am.* **46,** 1445–1469.

Wendt, V. E., Ajluni, R., Bruce, T. A., Prasad, A. S., and Bing, R. J. (1966). Acute effects of alcohol on the human myocardium. *Am. J. Cardiol.* **17,** 804–812.

Whereat, A. F., and Perloff, J. K. (1973). Ethyl alcohol and myocardial metabolism. *Circulation* **47,** 915–917.

White, P. D. (1931). "Heart Disease," p. 436. Macmillan, New York.

Wilens, S. L. (1947). The relationship of chronic alcoholism to atherosclerosis. *J. Am. Med. Assoc.* **135,** 1136–1139.

Wilhelmsen, L., Wedel, H., and Tibblin, G. (1973). Multivariate analysis of risk factors for coronary heart disease. *Circulation* **48,** 950–958.

Wong, M. (1973). Depression of cardiac performance by ethanol unmasked during autonomic blockade. *Am. Heart J.,* **86,** 508–515.

Wu, C. F., Sudhakar, M., Jaferi, G., Ahmed, S. S., and Regan, T. J. (1976). Preclinical cardiomyopathy in chronic alcoholics. A sex difference. *Am. Heart J.* **91,** 281–286.

Yano, K., Rhoads, G. G., and Kagan, A. (1977). *N. Engl. J. Med.* **297,** 405–409.

22

EFFECTS OF ALCOHOL ON THE NERVOUS SYSTEM

Pierre M. Dreyfus

I. General Considerations	342
II. Disorders Associated with High Blood Alcohol Levels	343
A. Inebriation Leading to Coma	343
B. Blackouts	343
C. Combativeness	343
III. Disorders Associated with Zero or Diminishing Blood Alcohol Levels	344
A. Tremulousness	344
B. Rum Fits	344
C. Delirium Tremens	345
D. Hallucinosis	345
IV. Neurological Disorders of Nutritional Cause Associated with Chronic Alcoholism	347
A. Nutritional Neuropathy	348
B. Wernicke-Korsakoff Syndrome	349
C. Amblyopia	350
V. Neurological Disorders of Undetermined Cause Associated with Chronic Alcoholism	351
A. Cerebellar Degeneration	352
B. Central Pontine Myelinolysis	353
C. Marchiafava-Bignami Disease	354
D. Alcoholic Dementia	354
References	356

I. GENERAL CONSIDERATIONS

It has been claimed that the ingestion of a single alcoholic beverage destroys at least 10,000 nerve cells and that devastation of the brain follows excessive and prolonged intake of alcohol. To date, reports of critical experiments that either support or refute this contention cannot be found in the scientific literature. However, it has been established that ethanol does affect the function of the nervous system in the same manner as it does that of other organ systems. In general, its effect on membranes of neurons and their synapses is most noticeable. As an anesthetic agent, alcohol depresses neural activity. Small doses, by suppressing inhibitory neurons, stimulate neuronal activity, while higher doses result in overall depression. Alcohol is reported to influence the permeability of excitable membranes, decreasing so-called ionic conductance and thus affecting ionic fluxes and the release of neurohumeral substances at synapses (Eidelberg, 1970). There is increasing evidence that the chronic and abusive intake of alcohol significantly alters vitamin, mineral, carbohydrate, protein, lipid, and biogenic amine metabolism within the nervous system (Leevy and Baker, 1968). In addition, the metabolic breakdown products of alcohol, such as acetaldehyde, may be harmful to the nervous system (Majchrowicz and Mendelson, 1970; Truitt and Walsh, 1971). How these physiological and biochemical changes lead to neurological and psychiatric disorders remains to be demonstrated.

This chapter presents an overview of the clinical aspects of the most common neurological and psychiatric disorders engendered by or associated with the excessive consumption of alcoholic beverages. It aims at setting the syndromes caused by elevated blood levels of alcohol apart from those engendered by withdrawal from alcohol and from those caused by malnutrition and other possible metabolic insults. No attempt will be made to review in detail the neuropharmacology of alcohol and the investigations carried out to elucidate the effects of alcohol on the nervous system at the cellular level.

The most frequently encountered neuropsychiatric complications of excessive alcohol consumption are those associated with elevated blood alcohol levels. These are followed in order of frequency by disorders caused by alcohol withdrawal and by a group of metabolic disturbances related to the nutritional depletion frequently associated with protracted and steady drinking (in contrast to spree or periodic drinking). Finally, one can identify a relatively rare group of neurological and psychiatric diseases of as yet undetermined cause that are associated with alcoholism but are not encountered exclusively in the chronic alcoholic population.

Whereas the disorders associated with inebriation and withdrawal are generally reversible, the nutritionally determined diseases and those of undetermined etiology may lead to serious, frequently irreversible, damage to parts of the central and the peripheral nervous systems.

II. DISORDERS ASSOCIATED WITH HIGH BLOOD ALCOHOL LEVELS

A. Inebriation Leading to Coma

Inebriation is a clinical state that requires very little introduction, having been experienced by a large segment of the population at one time or another. Although the behavioral and neurological status of the inebriated individual vary considerably, they tend to correlate well with the levels of alcohol in the blood. Thus an individual whose blood alcohol level has reached 20 mg% may feel pleasantly relaxed while talking freely and loosely. As the level reaches 30 mg%, the individual tends to become increasingly euphoric and at 50 mg%, he may be "sitting on top of the world." As the blood alcohol level reaches 100 mg% (a level usually associated with being "officially" drunk) the individual may have slurred speech, stagger about, sing loudly, and be involved in antisocial acts for which he will have only partial recollection when sober. When the blood alcohol level reaches 300 mg%, most people become stuporous and may "pass out." If the level reaches 400 mg% or more, coma and death may ensue. However, the chronic ingestion of alcohol may build up a tolerance in the habituated individual, leading to an ability to sustain higher blood alcohol levels before exhibiting significant behavioral or neurological signs (Victor and Adams, 1953).

The basic biochemical mechanisms involved in states of inebriation and tolerance remain to be demonstrated. Alcohol, or one of its breakdown products, appears to affect certain parts of the nervous system selectively, causing incoordination, slurred speech, and nystagmus (Duritz and Truitt, 1966).

B. Blackouts

Chronic drinkers may at times suffer from episodic, total loss of memory for events that have occurred during a spree of heavy drinking. The individual may perform highly complex acts but have absolutely no recall of the events, although remaining conscious throughout the episode. He may have spotty memory loss, forgetting discrete parts of the events. On rare occasions amnesia recovers spontaneously or upon prompting. The so-called lost weekend is typical of this alcohol-induced amnesic state. This as yet poorly understood clinical state has been labeled alcoholic blackout and it may represent an alcohol-related failure of short-term memory retrieval or retention. Both pharmacological and psychodynamic mechanisms have been invoked as the underlying cause of the disorder (Goodwin, 1970).

C. Combativeness

Occasionally an individual, while under the influence of alcohol, may display a sudden onset of combative, impulsive, irrational, and destructive behavior.

This condition is commonly referred to as alcoholic combativeness or "patho-logic intoxication." It can be triggered by very small doses of alcohol. The few clinical studies that have been carried out on patients who have reported this reaction have suggested that psychological factors in combination with al-cohol trigger the behavioral aberration.

III. DISORDERS ASSOCIATED WITH ZERO OR DIMINISHING BLOOD ALCOHOL LEVELS

Chronic and spree drinking result in the adaptation of the nervous system to elevated blood levels of alcohol. The sudden cessation of drinking and the conse-quent drop in blood alcohol levels result in well-recognized and distinct neuro-logical and psychiatric states that can be placed under the more general term of "withdrawal" or "abstinence syndrome."

A. Tremulousness

The most common neurological manifestation of alcohol withdrawal is a state of tremulousness popularly referred to as "the shakes." Its onset usually follows a drinking spree of several days and a period of abstinence of 24 hours or more. Symptoms of nausea, vomiting, restlessness, nervousness, and insomnia accom-pany a coarse, generalized tremulousness of eyelids, lips, and fingers, aggra-vated by motion and ameliorated by rest. The tremulousness may be almost imperceptible, the patient complaining of being "shaky inside," or it may be sufficiently violent to incapacitate speech, coordination, stance, and locomotion. Mild forms of tremulousness may occur the morning after a "night on the town" (a few hours of sleep constituting a period of abstinence) in individuals not considered to be chronic alcoholics. All forms of tremulousness tend to be worse in the morning and they are invariably ameliorated by the ingestion of alcohol (Victor, 1973).

B. Rum Fits

Withdrawal from alcohol (and other drugs) may lead to generalized convul-sions and loss of consciousness. The terms withdrawal seizures and "rum fits" are commonly used to describe this very characteristic clinical state. In the majority of cases seizures begin 10–48 hours after the cessation of drinking, most commonly between 13 and 24 hours. The seizures are usually generalized in type. In most instances only a single seizure occurs but occasionally there may be a flurry of two, six, or more seizures in quick succession; on rare occasions the patient may develop status epilepticus. The majority of patients with rum fits show a marked sensitivity to photic stimulation, which generates generalized myoclonic activity (so-called photomyoclonus, or muscle jerking). This distinguishes rum fits from other seizure states, such as idiopathic epilepsy. In most cases the electroencephalogram of patients who have suffered one or several rum fits is normal between seizures. Because of the transient nature of the reduced seizure threshold that occurs during withdrawal from alcohol, anticonvulsant drugs are ineffective because a therapeutic drug level cannot be attained soon enough. Whereas the majority of patients recover rapidly and completely from the attendant postictal confusional state, one-third of the patients go on to develop full-blown delirium tremens within a matter of hours following recovery from seizures. Occasionally this occurs after a lucid period of a day or two (Victor, 1973).

The etiology of rum fits seems to be related to respiratory alkalosis and hypomagnesemia, which occur rapidly following the withdrawal from alcohol (Wolfe and Victor, 1969). Seizures are directly related to the serum magnesium level: the lower the level, the lower the seizure threshold.

C. Delirium Tremens

The most serious and life-threatening of all the clinical states caused by alcohol withdrawal is delirium tremens, a disorder characterized by severe confusion, disorientation, delusions, agitation, hallucinosis, sleeplessness, tremulousness, increased startle, and evidence of hyperactivity of the autonomic nervous system: profuse sweating, fever, tachycardia, and dilated pupils. In most cases, delirium tremens begins 2 or 3 days after cessation of drinking, always following a bout of rum fits; tremulousness may also precede the advent of delirium tremens. The delirious state usually lasts 2–3 days, ending suddenly. Although delirium tremens usually is a relatively benign condition when appropriately treated, 5–15% of patients succumb to the illness, the cause of death being overwhelming sepsis, hyperthermia, electrolyte imbalance, circulatory collapse, and cardiac decompensation (Victor, 1973).

D. Hallucinosis

Withdrawal from alcohol can engender a variety of disturbances of perception, ranging from the mildest form of misinterpretation of sounds and objects to the most vivid visual and auditory hallucinations. As part of the acute phase of abstinence, accompanying either tremulousness or fully developed delirium tremens, the patient may complain of vivid, animate hallucinations that can take the form of animals, insects, or other human beings. These hallucinations tend to be episodic and of relatively brief duration, but they may occur.

As a consequence of prolonged inebriation and subsequent withdrawal, the alcoholic patients can develop a paranoid schizophrenia-like state, characterized by auditory hallucinations in the presence of a clear sensorium and in the absence of memory disturbance, confusion, or disorientation. At first the patient may have no appreciation for the unreal nature of the hallucinations. As recovery takes place, he may begin to express doubt concerning the form and content of the hallucinations; eventually the imaginary quality of the episode is fully realized and the patient becomes reluctant to discuss it further. The auditory hallucinations can consist of unstructured or musical sounds; sometimes they take the form of human voices—frequently those of members of the family, of friends, or employer, or of neighbors. The voices may be aimed directly at the patient or may be discussing the patient. They are frequently reproachful, threatening, disturbing, maligning, or derogatory. Usually they are sufficiently vivid to evoke an appropriate response, such as seeking help from the police or blocking the doors and windows to prevent assault. While auditory hallucinations associated with chronic alcoholism resemble those seen in typical schizophrenia, review of the natural course of the illness reveals that prior to the advent of hallucinosis, patients could not be considered schizophrenic and that the average age of onset of the disease was generally later than that of typical schizophrenia (Victor and Hope, 1958).

A number of significant biochemical observations have been used to explain the phenomenon of alcohol dependence and the subsequent symptoms and signs of abstinence in the habitual drinker. It is generally believed that the neurological manifestations of withdrawal are a reflection of neuronal excitability, which returns to normal as the symptoms vanish. It has been demonstrated that the chronic ingestion of alcohol causes changes in hypothalamic, pituitary, and adrenal function (Noble, 1970). Thus, serum cortisol levels rise with increasing ethanol consumption and, in certain individuals, a sudden drop occurs with cessation of drinking (Mendelson *et al.*, 1970). Aldosterone excretion increases rapidly upon the ingestion of alcohol, returning to normal in some individuals while remaining elevated in others (Stokes, 1973). It has been shown that alcohol inhibits the normal oxidative metabolism of the intermediate aldehyde products that result from the deamination of serotonin and norepinephrine. This is probably due to competitive inhibition between these products and acetaldehyde produced from alcohol. The result is abnormal biogenic amine metabolism and the production of aberrant and potentially addictive metabolites that are the result of the condensation of the parent amine with intermediary aldehydes. These substances are not present under normal circumstances. The administration of alcohol has resulted in the accumulation of acetaldehyde and methanol in tissues and blood (Majchrowicz and Mendelson, 1970; Davis *et al.*, 1967; Davis and Walsh, 1970). Whether these and other potentially noxious substances ever reach physiologically significant levels remains to be demonstrated. The literature is replete with biochemical observations purported to form the biological basis for the abstinence syndrome. Despite all of these observations, there continue to be large gaps in our understanding of this unique phenomenon.

IV. NEUROLOGICAL DISORDERS OF NUTRITIONAL CAUSE ASSOCIATED WITH CHRONIC ALCOHOLISM

During periods of heavy drinking, the chronic alcoholic patient may sharply decrease his intake of food because of decreased appetite. A large intake of alcohol decreases the absorption, intestinal transport, tissue storage, and utilization of vitamins and other essential nutrients. In addition to causing abnormal vitamin metabolism, alcohol has been shown to affect mineral, carbohydrate, protein, and lipid metabolism (Leevy and Baker, 1968).

A number of the neurological disorders of nutritional origin that have been described in the chronic alcoholic patient have also been encountered in individuals who are malnourished and debilitated for reasons other than chronic alcoholism. These diseases all share the common neuropathological features of metabolic and nutritional disorders of the nervous system. As such they appear to have a predilection for specific areas of the nervous system, and the lesions have a symmetrical and bilateral distribution. In contrast with neurological disorders engendered by too much alcohol or withdrawal therefrom, nutritional disorders associated with alcoholism are relatively rare when one considers the overall size of the chronic alcoholic population and the magnitude of the sociological, psychological, and general medical problems caused by chronic alcoholism. It is estimated that nutritional problems constitute no more than 3% of the alcohol-related neurological problems requiring hospitalization. The various alcoholic nutritional syndromes that will be described in this section present either separately, in relatively pure form, or in varying combinations. Why, under seemingly identical circumstances, one nutritionally depleted chronic alcoholic patient develops one or several neurological syndromes while another seems to remain essentially untouched remains to be explained (Victor and Adams, 1961; Dreyfus, 1974).

The chronic alcoholic patient quite often presents with symptoms and signs of abstinence that tend to mask those which can be attributed to malnutrition. The relatively frequent coexistence of these two alcoholic complications should be borne in mind when therapeutic intervention is being considered. In a patient who is on the verge of nutritional depletion, a full-blown nutritional syndrome can be precipitated by an overload of the calorie-rich parenteral fluids used in the management of withdrawal symptoms. It is therefore essential that adequate doses of the B vitamins be administered despite the fact that withdrawal symptoms are not caused by vitamin deficiency.

A. Nutritional Neuropathy

The two most common nutritional disorders of the nervous system associated with chronic alcoholism are polyneuropathy and the Wernicke–Korsakoff syndrome. The polyneuropathy that affects the alcoholic patient is usually referred to as nutritional polyneuropathy. Because this disorder is quite common in the chronic alcoholic population the term alcoholic neuropathy is used, although the disorder of peripheral nerve function sometimes referred to as "dry beriberi" that occurs endemically in underdeveloped parts of the world and a disease that results from nutritional depletion associated with the excessive intake of alcohol are probably identical conditions sharing a common etiology.

Polyneuropathy is characterized by progressive weakness and muscle wasting, involving, in a symmetrical fashion, the legs to a greater extent than the arms, and the distal muscles more than the proximal ones. In addition, complaints of abnormal sensations occurring in a similar distribution are very common. These consist of aching, coldness, hotness, deadness, numbness, prickliness, and tenderness, most commonly localized to the toes and fingers. In the most advanced cases of polyneuropathy muscle wasting is very pronounced. Deep tendon reflexes are usually greatly diminished or totally absent. The sensory loss is usually symmetrical and is most evident in the distal parts of the limbs, diminishing gradually toward the more proximal parts, where it fades into normal sensations. Although all of the sensory modalities are involved (pain, temperature, vibratory, and position sense), not all are affected to the same degree. In rare instances, nutritional polyneuropathy may be accompanied by vertigo, deafness, aphonia, and blindness (amblyopia). Evidence of disturbed function of the autonomic peripheral nervous system (sexual impotence, bladder atony, bouts of nocturnal diarrhea, and abnormalities of pupillary function) has been described in cases of polyneuropathy (Victor, 1975; Mayer and Garcia-Mullin, 1972).

The mode of development, the evolution, and the severity of polyneuropathy vary considerably from case to case. Whereas the onset is usually insidious and the progression slow, in some cases the onset can be abrupt and the course crippling to the patient in a matter of weeks. Recovery is often incomplete and in most cases it is very slow, occurring over a period of months.

Polyneuropathy associated with chronic alcoholism may be caused by the deficiency of a single or of several B vitamins in combination. The vitamins that are most commonly lacking are thiamine, pyridoxine, pantothenic acid, and vitamin B_{12}.

The pathological changes that are noted in polyneuropathy consist of parenchymal damage of the nerves, segmental destruction of myelin, predominantly in the peripheral part of the nerves, and destruction of axis cylinders in the most severe and advanced cases (Victor, 1975).

B. Wernicke-Korsakoff Syndrome

Perhaps one of the most dramatic syndromes encountered in the malnourished alcoholic patient is the Wernicke-Korsakoff's syndrome, which is generally regarded as the most extreme form of thiamine deficiency in man. Whereas Wernicke's syndrome can occur alone, Korsakoff's psychosis represents a continuum of Wernicke's disease.

Frequently Wernicke's disease comes on rapidly. At first the patient presents with disorientation and confusion while suffering at the same time from double vision and unsteadiness of gait (ataxia). When examined, he often shows either bilateral weakness or paralysis of eye movements; on occasion there may also be drooping of the eyelids. If the eyes are not totally paralyzed, nystagmus (symmetrical, to and fro or up and down jerking of the eyes when the patient fixes on an object) can be detected. In many cases, however, nystagmus cannot be evoked until the paralysis of the eyes has improved. Ataxia may be so profound that the patient may be incapable of either walking or standing without help. It is not unusual to find that individuals afflicted with Wernicke's disease, when first seen, exhibit symptoms of alcohol withdrawal. This confuses the clinical picture. More often, however, patients are apathetic, indifferent, and mentally dull, making it difficult to evaluate completely their neurological status. Improved nutrition and specific vitamin replenishment lead to increasing alertness and attentiveness, resulting in improved testability. Some of the patients may then show signs characteristic of Korsakoff's psychosis, a remarkable syndrome characterized by a severe deficit in retentive memory and in the ability to learn newly presented material. This renders the patient virtually incapable of performing any but the simplest of tasks dependent upon memory. He may be unable to retain such things as the examiner's name, a sequence of simple test words, or a list of objects. These patients can also suffer from abnormalities of perceptual function and concept formation. Finally they may confabulate—fabricate fictitious or improperly sequenced stories. This symptom frequently disappears in the more chronic stages of the illness.

Wernicke-Korsakoff's syndrome occurs more frequently in steady drinkers than in spree drinkers. The dietary history, when obtainable, points to severe nutritional depletion. Thus, the majority of patients show clinical evidence of chronic malnutrition.

In all untreated cases of Wernicke-Korsakoff's syndrome, blood transketolase activity, a sensitive biochemical test of thiamine deficiency, is significantly abnormal. This test, although specific for thiamine deficiency, is not considered definitive for the syndrome (Embree and Dreyfus, 1963).

When left untreated, most patients afflicted with Wernicke's disease go on to develop full-blown Korsakoff's psychosis. It is therefore extremely important to recognize the disease early and to treat it with improved nutrition and thiamine supplementation. Ocular symptoms tend to respond dramatically (3–6 hours) to the administration of as little as 50 mg of thiamine administered by the parenteral route. In three-fourths of the patients, unsteadiness of gait improves remarkably within a matter of a few days after nutritional replenishment. Sometimes complete recovery occurs in a matter of months (Victor *et al.,* 1971).

Korsakoff's psychosis may improve within a matter of days after specific vitamin or nutritional therapy has begun. In one-fourth of the patients, the memory disturbance is completely reversible, improvement occurring over a time span ranging from a few days to several months. In 50% of cases, improvement ranges from slight to significant; however, it is never complete and, therefore, the memory loss is totally incapacitating. In the other one-fourth of patients suffering from Korsakoff's psychosis, the memory disturbance is completely irreversible. The salient pathological changes in cases of Wernicke–Korsakoff's syndrome tend to be remarkably constant. The lesions are always bilaterally symmetrical, being found in the mammillary bodies, the terminal fornices, the periaqueductal region of the midbrain, the floor of the fourth ventricle, and certain nuclei of the thalamus (anteromedial, medial dorsal, and pulvinar). The anterior superior parts of the middle of the cerebellum are almost always affected, which probably correlates with the disturbance of stance and gait. The brain stem lesions most likely account for the abnormalities of ocular motility. The lesions in the thalamus, the hypothalamus, the mammillary bodies, and fornices, may explain some of the psychological abnormalities, particularly those affecting learning and retentive memory. The most advanced and pronounced lesions seen in the central nervous system consist of severe parenchymal necrosis involving both nerve cells and fibers. The hemorrhages that are so frequently encountered in the lesions were for many years regarded as the principal hallmarks of the disease. They most likely represent a nonspecific pathological change seen in the end stages of the disease.

The lesions observed in the brains of patients afflicted with Wernicke–Korsakoff's are very similar to those found in the brains of experimental animals that have been deprived of thiamine. While the distribution of the lesions in the nervous system of thiamine-deficient animals varies from species to species, bilateral and symmetrical areas of necrosis remain a striking and common feature (Victor *et al.,* 1971).

C. Amblyopia

It is generally recognized that chronic malnutrition caused by the abusive intake of alcohol over long periods of time can cause a special form of blindness (nutritional amblyopia). The visual disturbance is remarkably uniform from case to case. It develops gradually over a period of weeks or months. The patient usually complains of blurring or dimness of vision, difficulty in reading, discomfort behind the eyes. Examination shows diminution of visual acuity and bilaterally symmetrical areas of visual loss (scotomata). These areas of blindness are more striking for red and green than for white test objects. The vision in the periphery is usually intact. In general, no discernible abnormality is noted on visualization of the eye grounds (Victor *et al.,* 1960).

This type of visual disturbance has been observed in undernourished populations throughout the world, especially in times of famine, and among civilian and military prisoners during wars. The disease continues to be endemic in certain parts of Africa, Asia, and South America, among the malnourished, impoverished populations. In the more affluent parts of the world, the disease is seen most frequently in individuals who are chronically addicted to alcohol and occasionally to tobacco and who have neglected their nutrition. Because the illness is sometimes associated with excessive drinking and smoking, it is commonly known as "tobacco–alcohol amblyopia," implying that either or both of these toxic agents can be directly implicated in its etiology. The term nutritional amblyopia seems more appropriate, however, since there is overwhelming clinical and scientific evidence that favors the notion that the visual disorder has a nutritional rather than a toxic etiology. On occasion a similar syndrome has been encountered in patients suffering from vitamin B_{12} deficiency and diabetes mellitus. In the chronic alcoholic patient, the syndrome of blindness may occur in conjunction with Wernicke–Korsakoff's syndrome, peripheral neuropathy, and other alcohol-related neurological syndromes. The essential pathological changes that are seen in amblyopia are bilateral, symmetrical loss of myelinated fibers in the central parts of the optic nerves, the chiasm, and the optic tracts (papillomacular bundle). The changes are very similar to the white matter lesions noted in other nutritional disorders of the nervous system (Dreyfus, 1976).

Treatment with oral or parenteral B vitamins and improved nutrition bring about improvement of vision. Improvement is, of course, related to the severity of the disease and its duration before therapy is instituted.

V. NEUROLOGICAL DISORDERS OF UNDETERMINED CAUSE ASSOCIATED WITH CHRONIC ALCOHOLISM

A number of neurological disorders of as yet ill-defined etiology have been observed among chronic alcoholic patients. It seems most probable that the underlying cause of some of these illnesses is a combination of nutritional and metabolic factors, while in some of the others a mixture of metabolic and toxic causes may be invoked. Although the clinical features and the pathological attributes of the diseases that will be described in this section are well known, clinical or biochemical studies aimed at elucidating their specific causes are lacking.

A. Cerebellar Degeneration

Degeneration of the cerebellar cortex is a syndrome most commonly encountered in men under the age of 50 who have been drinking excessively for many years and who show evidence of nutritional neglect. The syndrome has often been referred to as alcoholic cerebellar degeneration. It is of interest that the disease and its pathological changes have occasionally been described in nutritionally depleted individuals who were not alcoholic. Progressive unsteadiness of stance and gait, occasionally severe enough to confine the patient to bed, represents the most frequent complaint. When he does walk, it is in an unsteady manner, on a broadened base, and he frequently requires assistance or support; rapid postural adjustments, such as sudden changes in direction, are made with difficulty. In addition to unsteadiness, mild tremulousness of the hands and incoordination of finger movements may be present. Speech may be slowed and slightly slurred. The disease evolves in a number of ways. In most instances it progresses rapidly, the maximum severity being reached in a matter of days or weeks, followed by years of relative stability. However, in some cases the disease may be mild at its onset, only to worsen suddenly during a period of metabolic stress (increased alcohol consumption, fever, or debilitating illness). On occasion, with abstention from alcohol and nutritional improvement, symptoms abate (Victor *et al.,* 1959).

The pathological changes seen in cerebellar degeneration associated with chronic alcoholism are remarkably constant and restricted. The degeneration involves predominantly the anterior and superior portions of the central part of the cerebellar cortex (vermis), where there is a marked loss of Purkinje cells. In the more severe and advanced cases, nerve cells of other layers of the cerebellum cortex (molecular and granular cell layers) may also be affected. Because of these pathological changes, the cerebellar hemisphere may look atrophic (Victor *et al.,* 1959).

Although it seems most probable that cerebellar degeneration in the alcoholic patient is due to nutritional depletion, a toxic–metabolic etiology has not been entirely excluded. In favor of a nutritional etiology is the fact that many patients afflicted with the disease give a history of progressive weight loss prior to the onset of their symptoms. In addition, signs of malnutrition and cirrhosis of the liver are not uncommon (Allsop and Turner, 1966; Mancall and McEntee, 1965). It is of interest, however, to note that cerebellar degeneration could be due in part to the chronic exposure of the nervous system to toxic levels of acetaldehyde, the first intermediate metabolite of alcohol. It has been shown that the administration of alcohol to rats results in the accumulation of acetaldehyde in most organs of the body, including the brain. Furthermore, the highest concentrations of this metabolite have been detected in the cerebellum, where mitochondria seem to be unusually sensitive to the *in vitro* addition to acetaldehyde. Finally, the administration of acetaldehyde to animals and man appears to produce both unsteadiness of gait and nystagmus (Truitt and Walsh, 1971).

B. Central Pontine Myelinolysis

Central pontine myelinolysis is an unusual and relatively rare neurological disease that has been encountered principally in chronic alcoholics suffering from a combination of malnutrition and disturbed electrolyte metabolism. The disease has also been described in nonalcoholic patients suffering from a variety of debilitating illnesses complicated by malnutrition and electrolyte imbalance. Because of the protean nature of its presentation, the diagnosis of central pontine myelinolysis is frequently missed in the living patient and is detected only at postmortem examination. The disease affects mainly adults in the middle decades of life, but it has also been described in children. Its principal clinical feature appears to be progressive weakness of the bulbar muscles (muscles of the tongue and pharynx), causing difficulties with speech and swallowing, sometimes totally abolishing these functions within a matter of days. Partial or complete paralysis of eye movements, fixed or dilated pupils, and absent facial sensations have also been described. Weakness of all four extremities (quadriparesis) on the onset of the illness is usually followed rapidly by total paralysis (quadriplegia) and pathologically active reflexes. Urinary incontinence, abnormal postures (decerebration), respiratory paralysis, and a total lack of response to pain throughout the body have been reported. As the illness evolves the patient becomes drowsy, then stuporous, lethargic, and, eventually, comatose. As best as can be established, the duration of the illness varies from a few days to several weeks. The patient usually succumbs to medical complications, such as pneumonia, overwhelming infection, or uremia (Adams *et al.*, 1959). In rare instances, a gradual, slow, and sometimes complete recovery can occur.

As the name implies, the primary pathological changes noted in this illness are restricted to the pons (part of the brain stem). A characteristic focus of demyelination, variable in size and extent, is usually limited to the central parts of the base of the mid to upper pons. The overall size and severity of the lesion as well as its anatomic confines generally correlate well with the clinical syndrome and signs during life. Recently it has been shown that in some cases demyelination may involve other areas of the brain, i.e., the midbrain, basal ganglia, and cerebral white matter (Victor, 1977).

The etiology of this curious disorder remains completely obscure. It has been suggested that it may be caused by a toxic substance to which the patient is particularly sensitive. When first described, the causes of the condition seemed to be restricted to malnutrition and alcoholism. While malnutrition and alcoholism have afflicted mankind for centuries, central pontine myelinolysis is a disease which has either been overlooked or was discovered very recently. It has been suggested (Poser, 1973) that it represents a "disease of medical progress," that it may reflect the effects of the treatment of dehydration, electrolyte imbalance, and malnutrition associated with chronic alcoholism rather than the results of malnutrition or alcoholism per se. It is of interest to note that, in most instances, patients have died after a period of intensive treatment (Klavins, 1963; Cadman and Rorke, 1969; Rosman *et al.*, 1966; Goebel and Herman-Ben Zur, 1976).

C. Marchiafava–Bignami Disease

Marchiafava–Bignami's disease is an exceedingly rare ailment characterized by neurological symptoms and signs that suggest bilateral involvement of the frontal lobes. The symptoms include a disturbance of language function (dysphasia and echolalia), abnormal gait and motor skills, generalized seizures, incontinence of urine, grasping, sucking, perseveration, tremulousness of hands and tongue, and dysarthria. In addition the patients display psychological symptoms, such as agitation, confusion, hallucinations (visual, auditory, and gustatory), disturbed memory, negativism, impaired judgment, and disorientation. The course of the illness is quite variable, developing over a period of days or months. Sudden and complete recovery, although rare, has been reported (Leventhal *et al.*, 1965; Brion, 1976; Ironside *et al.*, 1961).

The principal lesions in this disease consist of symmetrical zones of demyelination located in the central parts of the corpus callosum and the anterior commissure, both white matter bundles that connect one side of the brain with the other. In the most advanced cases, white matter of the frontal lobes adjacent to the corpus callosum may also be involved (Brion, 1976).

The cause of this unusual complication of chronic alcoholism is totally unknown. Originally it was believed that the disease occurred exclusively in men of Italian descent who consumed excessive amounts of crude red wine. However, the illness has been described in patients of varied ancestry who consumed all sorts of alcoholic beverages (Brion, 1976).

D. Alcoholic Dementia

Progressive dementia (alcoholic dementia) resulting from chronic alcoholism has been described as a specific and separate disease entity with distinct clinical and pathological features. Since very few careful clinical studies have been published, the clinical hallmarks of the disease, if indeed it is a distinct illness, remain ill defined. It is generally believed that the long-time drinker may show signs of dementia after 16 to 20 years of fairly steady alcohol consumption. The majority of patients show an impairment of abstract thinking, amnesia for recent events, disorientation to time and space, impairment of judgment, emotional lability, intellectual enfeeblement, and sociopathic behavior. In addition, patients may exhibit a variety of disturbances of language function (Horvath, 1973; Dreyfus, 1976).

In some cases of alcoholic dementia, evidence of cerebral degeneration can be obtained by using computerized axial tomography that shows enlargement of the lateral ventricles and widening of the sulci of the cerebral hemispheres.

Careful and systematic pathological studies of brains of patients afflicted with alcoholic dementia have been difficult for several reasons. For the most part this is due to inadequate and highly inaccurate historical data and to a paucity of clinical information about the patient's alcoholic intake and general state of ill health, which has usually spanned several decades. Morel (1939) claims that alcoholic dementia is a distinct pathological entity in which the third layer of the cerebral cortex reveals spongy, laminar necrosis characterized by nerve cell loss and appropriate glial response. Similar observations have been made in the brains of chronic alcoholic patients afflicted with Marchiafava–Bignami's disease, and therefore it has been speculated that the restricted laminar necrosis observed in the cortex of some demented chronic alcoholic patients is in some way related to the demyelination of the commisural structures affected in Marchiafava–Bignami's disease (Brion, 1976). Despite extensive pathological studies of the nervous system of chronic alcoholics of long standing, laminar necrosis of the cerebral cortex appears to be an exceedingly rare pathological alteration, considering the large number of chronic alcoholic patients (for example, skid row alcoholics) who, during life, have shown unequivocal evidence of intellectual impairment. Whether the alcoholic dementia and cerebral degeneration associated with chronic alcoholism is a separate clinical and pathological entity or the result of the accumulation of a host of medical complications incurred over the years remains totally unknown.

This chapter has reviewed briefly the salient clinical and pathological features of the various neurological disorders engendered by the consumption of excessive amounts of alcohol. It has attempted to highlight differences between the syndrome caused by excessive blood alcohol levels and the syndromes caused by abstinence after excessive ingestion, the malnutrition that so frequently complicates chronic alcoholism, and other as yet unknown factors. It is hoped that a better understanding of what alcohol does to the nervous system will result in a more effective approach to the treatment of the neurological complications of alcoholism.

Much is known about the effects of alcohol on a number of physiological functions of several organ systems, including the nervous system, and numerous detailed clinical and pathological studies exist. However, there continue to exist serious gaps in our understanding of the specific mechanisms by which the nervous system reacts to the drug—its acute ingestion, its withdrawal, and long-term exposure to it. Although it is generally assumed that small doses of an alcoholic beverage, regardless of its type or alcohol content, are relatively harmless, this contention has never been tested by scientific means. This is one of the many important questions still to be answered.

REFERENCES

Adams, R. D., Victor, M., and Mancall, E. L. (1959). Central pontine myelinolysis: A hitherto undescribed disease occurring in alcoholic and malnourished patients. *Arch. Neurol. Psychiatry* **81**, 154–172.

Allsop, J., and Turner, B. (1966). Cerebellar degeneration associated with chronic alcoholism. *J. Neurol. Sci.* **3**, 238–258.

Brion, S. (1976). Marchiafava–Bignami syndrome. *In* "Handbook of Clinical Neurology" (P. Vinken and G. Bruyn, eds.), p. 317. Elsevier/North-Holland Press, Amsterdam.

Cadman, T. E., and Rorke, L. B. 1969. Central pontine myelinolysis in childhood and adolescence. *Arch. Dis. Child.* **44**, 342–350.

Davis, V. E., and Walsh, M. J. (1970). Alcohol, amines, and alkaloids: A possible biochemical basis for alcohol addiction. *Science* **167**, 1005–1007.

David, V. E., Brown, H., Huff, J. A., and Cashaw, J. L. (1967). The alteration of serotonin metabolism to 5-hydroxytryptophol by ethanol ingestion in man. *J. Lab. Clin. Med.* **69**, 132–140.

Dreyfus, P. M. (1974). Diseases of the nervous system in chronic alcoholics. *In* "The Biology of Alcoholism" (B. Kissin and H. Begleiter, eds.), p. 265. Plenum, New York.

Dreyfus, P. M. (1976). Amblyopia and other neurological disorders associated with chronic alcoholism. *In* "Handbook of Clinical Neurology" (P. Vinken and G. Bruyn, eds.), p. 331. Elsevier/North-Holland Press, Amsterdam.

Duritz, G., and Truitt, E. B. (1966). Importance of acetaldehyde in the action of ethanol on brain norephinephrine and 5-hydroxy-tryptamine. *Biochem. Pharmacol.* **15**, 711–721.

Eidelberg, E. (1970). Effects of ethanol upon central nervous system neurons. *In* "Recent Advances in Studies of Alcoholism" (N. Mello and J. Mendelson, eds.), p. 274. U.S. Gov. Print. Off., Washington, D.C.

Embree, L. J., and Dreyfus, P. M. (1963). Blood transketolase determinations in nutritional disorders of the nervous system. *Trans. Am. Neurol. Assoc.,* pp. 36–42.

Goebel, H. H., and Herman-Ben Zur, P. (1976). Central pontine myelinolysis. *In* "Handbook of Clinical Neurology" (P. Vinken and G. Bruyn, eds.), p. 285. Elsevier/North-Holland Press, Amsterdam.

Goodwin, D. W. (1970). Blackouts and alcohol induced memory dysfunction. *In* "Recent Advances in Studies of Alcoholism" (N. Mello and J. Mendelson, eds.), p. 508. U.S. Gov. Print. Off., Washington, D.C.

Horvath, T. (1973). Clinical spectrum and epidemiological features of alcoholic dementia. *Proc. Int. Symp. Eff. Chron. Use Alcohol Other Psychoactive Drugs Cereb. Funct., Toronto.*

Ironside, R., Bosanquet, F. D., and McMenemey, W. H. (1961). Central demyelination of the corpus collosum (Marchiafava–Bignami Disease). With a report of a second case in Great Britain. *Brain* **84**, 212–230.

Klavins, J. V. (1963). Central pontine myelinolysis. *J. Neuropathol. Exp. Neurol.* **22**, 302–317.

Leevy, C. M., and Baker, H. (1968). Vitamins and alcohol. *J. Clin. Nutr.* **21**, 1325–1328.

Leventhal, C. M., Baringer, J. R., Arnason, B. G., and Fisher, C. M. (1965). A case of Marchiafava–Bignami disease with clinical recovery. *Trans. Am. Neurol. Assoc.* **90**, 87–91.

Majchrowicz, E., and Mendelson, J. (1970). Blood levels of acetaldehyde and methanol during chronic ethanol ingestion and withdrawal. *In* "Recent Advances in Studies of Alcoholism" (N. Mello and J. Mendelson, eds.), p. 200. U.S. Gov. Print. Off., Washington, D.C.

Mancall, E. L., and McEntee, W. J. (1965). Alterations of the cerebellar cortex in nutritional encephalopathy. *Neurology* **15**, 303–313.

Mayer, R. F. and Garcia-Mullin, R. (1972). Peripheral nerve and muscle disorders associated with alcoholism. *In* "The Biology of Alcoholism" (B. Kissin and H:. Begleiter, eds.), p. 29. Plenum, New York.

Mendelson, J. H., Ogata, M., and Mello, N. K. (1970). Adrenal function and alcoholism: I. Serum Cortisol. *In* "Recent Advances in Studies of Alcoholism" (N. Mello and J. Mendelson, eds.), p. 72. U.S. Gov. Print. Off., Washington, D.C.

Morel, F. (1939). Une forme anatomo-clinique particuliere de l'alcoolisme chronique: Sclerose cortical laminaire alcoolique. *Rev. Neurol.* **71**, 280–288.

Noble, E. P. (1970). Ethanol and adrenocortical stimulation in inbred mouse strains. *In* "Recent Advances in Studies of Alcoholism" (N. Mello and J. Mendelson, eds.), U.S. Gov. Print. Off., Washington, D.C.

Poser, C. M. (1973). Demyelination in the central nervous system in chronic alcoholism: Central pontine myelinolysis and Marchiafava–Bignami's disease. *Ann. N.Y. Acad. Sci.* **215**, 373–381.

Rosman, N. P., Kakulas, B. A., and Richardson, E. P. (1966). Central pontine myelinolysis in a child with leukemia. *Arch. Neurol.* **14**, 273-280.

Stokes, P. E. (1973). Adrenocortical activation in alcoholics during chronic drinking in alcoholism and the central nervous system. *Ann. N.Y. Acad. Sci.,* **215**, 77.

Truitt, E. B., Jr., and Walsh, M. J. (1971). The role of acetaldehyde in the actions of ethanol. *In* "The Biology of Alcoholism" (B. Kissin and H. Begleiter, eds.), pp. 161–195. Plenum, New York.

Victor, M. (1973). The alcohol withdrawal syndrome. *Ann. N.Y. Acad. Sci.* **215**, 210.

Victor, M. (1975). Polyneuropathy due to nutritional deficiency and alcoholism. *In* "Peripheral Neuropathy" (P. J. Dyck, P. K. Thomas, and E. H. Lambert, eds.), pp. 1030–1066. Saunders, Philadelphia, Pennsylvania.

Victor, M. (1977). Personal communication.

Victor, M., and Adams, R. D. (1953). The effect of alcohol on the nervous system in metabolic and toxic diseases of the nervous system. *Res. Publ. Assoc. Res. Nerv. Ment. Dis.* **32**, 526.

Victor, M., and Adams, R. D. (1961). On the etiology of the alcoholic neurologic diseases with special reference to the role of nutrition. *Am. J. Clin. Nutr.* **9**, 379–397.

Victor, M., and Hope, J. M. (1958). The phenomenon of auditory hallucinations in chronic alcoholism. A critical evaluation of the status of alcoholic hallucinosis. *J. Nerv. Ment. Dis.* **126**, 451–481.

Victor, M., Adams, R. D., and Mancall, E. L. (1959). A restricted form of cerebellar cortical degeneration occurring in alcoholic patients. *Arch. Neurol.* **1**, 597–688.

Victor, M., Mancall, E. L., and Dreyfus, P. M. (1960). Deficiency amblyopia in the alcoholic patient. *Arch. Ophthalmol.* **64**, 1–33.

Victor, M., Adams, R. D., and Collins, G. H. (1971). "The Wernicke-Kosakoff Syndrome." Davis, Philadelphia, Pennsylvania.

Wolfe, S. M., and Victor, M. (1969). The relationship of hypomagnesemia and alkalosis to alcohol withdrawal symptoms. *Ann. N.Y. Acad. Sci.* **162**, 973–984.

23

ALCOHOLISM: HOW DO YOU GET IT?

R. M. Morse

I. Common Concepts in Etiology of Alcoholism	359
II. Is Alcoholism Associated with Underlying Psychopathology?	361
III. Role of the Pleasurable Experience from Alcohol	364
IV. Habituation, Dependence, Problems	365
V. Tolerance and Need	366
VI. Is Everyone Vulnerable?	367
References	367
Editorial Comment	369

I. COMMON CONCEPTS IN ETIOLOGY OF ALCOHOLISM

Of the many controversial aspects of alcoholism, none is more likely to create argument than its etiology. Two recent presentations serve to highlight the opposite extremes of this issue. A well-known psychiatrist–author spoke to a group of his colleagues, including a number of resident physicians. The topic had to do with psychologic distress within practicing psychotherapists, which he thought may account for higher than average suicide rates among psychiatrists. The speaker believed the physicians' conflicts were related to what he termed the "sad soul of the psychiatrist," or something akin to losing one's sense of purpose. Because of the consequent anguish, the soul may attempt to self-destruct in various ways, including the rather prevalent abuse of or dependence upon alcohol and other drugs. Clearly, this psychiatrist believed alcoholism to be the direct result of rather severe emotional stress and its effects on the vulnerable personality.

About 2 weeks later a paper was presented at a national meeting (Vaillant, 1977) by a psychiatrist known for his continuing follow-up of 202 college men first studied extensively in 1940 after being selected for their mental health.

Attempting to shed some light on the many suggested causes of alcoholism, including the "oral personality," this researcher found 26 (14%) of the 185 surviving men to be alcoholic. After having the childhood of each subject rated by persons unaware of later experiences, the researcher compared 34 men from the worst childhood environments with 41 men from the best. Surprisingly, the rates of alcoholism were 18% in the former and 17% in the latter. So this group of men now 55 years old, originally chosen for their mental health, suffered at least as much alcoholism as the general population but with no difference in incidence between those from the best and the worst childhood environments. There was also no correlation between oral personality traits and alcoholism. The author concluded:

> The view that alcoholism is a mere symptom of underlying distress may rest upon erroneous retrospective reconstructions. The middle class alcoholic does not drink because his childhood was unhappy; he is now unhappy because he cannot control how much he drinks.

This psychiatrist apparently does not see the roots of alcoholism in severe psychopathology.

As a rule, specificity and effectiveness of treatment are inversely proportional to the complexity of the concept of the disease confronted. Certainly this is true of alcoholism, which continues to be debated as to its "real" quality—i.e., is it a disease, an addiction, a symptom, a learned response, continued self-indulgence, or just a bad habit? Without attempting to review the multiple theories of its causation, I would like to examine briefly arguments for and against considering alcoholism a symptom of an underlying psychiatric disorder. By choice, this discussion will omit any attention to the valid points made by geneticists, sociologists, biochemists, nutritionists, or pathologists. And although I subscribe to the theory of multicausation of alcoholism, the present chapter will concentrate only upon psychologic aspects.

Any student of alcoholism is aware that, despite the general lack of theoretic consensus, we do know a great deal about alcohol as a sedative drug, its psychobiologic effects, the psychology of the alcoholic, and the sociology of alcoholism. We have not been able, however, to fit it all together to make a sensible explanation of the whole. I would like to contend that the process of becoming alcoholic (how do you get it?) is reasonably well known and generally agreed upon—and that this process should be a focus of our attention to preventive measures, rather than the many-faceted research efforts which now attempt to elucidate small segments of the multicausal chain.

Perhaps epitomizing the traditional psychiatric position are the following statements from a standard psychiatric textbook, its chapter on alcoholism written by a knowledgeable psychiatrist (Chafetz, 1967).

> Alcoholism results from a disturbance and deprivation in early life experiences and the associated related alterations in basic physicochemical responsiveness, from the iden-

tification . . . with significant figures who deal with life problems through an unhealthy preoccupation with alcohol, and from a sociocultural milieu that causes ambivalence, conflict, and
guilt in the use of alcohol. . . . Earliest deprivation can only be gratified by oral satisfaction
through incorporation and destruction of the love object. . . . For this reason, people with
alcohol problems look for love in whatever, wherever, or whoever will provide it.

Most representative of psychoanalytic thinking, the above statements* clearly
describe the alcoholic as predisposed by significant emotional disorder caused by
childhood deprivation.

Even more pointed is a recent psychiatric report of the consultation and evaluation of 1000 alcoholic patients, all of whom were eventually given a psychiatric
diagnosis: 58% were diagnosed as neurotic, 6% as psychotic, and 36% as having
personality disorders (Tyndel, 1974). The author lamented the practice of some
treatment centers that apply the label "alcoholism" without elucidating associated psychopathology, and he concluded:

The paradoxical and puzzling fact that the alcoholic patient continues to drink despite his
awareness of the past, present and future disastrous consequences of his drinking, is almost
invariably associated with the psychopathology and the individual defenses which were in
existence during earlier periods of the patient's development.

Certainly this position is attractive and consistent with some popular and professional opinion.

II. IS ALCOHOLISM ASSOCIATED WITH UNDERLYING PSYCHOPATHOLOGY?

What of the opposite viewpoint—that alcoholism need not be associated with
underlying psychopathology? Although all clinicians are aware of the sometimes
dramatic onset of alcoholic drinking in response to a personal loss, grief, or
traumatic event, the psychiatrist assessing a large number of alcoholics is faced
with another reality. There seem to be a substantial number of alcoholic people
who do not show signs of psychopathology apart from the effects of alcoholism.
In many it becomes increasingly difficult to differentiate cause from effect, or the
preexisting psychic state from the emotional, behavioral, and cognitive changes
that alcoholism has produced. It is not uncommon, for example, to see the
suicidally depressed alcoholic return to affective normality with no other treatment than 3 days' abstinence from alcohol and supportive counseling—or to see
the sleepless, anxious, and guilt-ridden woman improve in all respects over a
period of 3 to 4 weeks with abstinence and treatment focused solely upon her
alcohol dependence. A case history may be of interest.

*From Chafetz (1967), reprinted by permission.

The 44-year-old owner and president of a successful business was referred for psychiatric evaluation under pressure by the management of an associated business with whom he was subcontracting. Referral was occasioned because of suspected drinking during the day, which seemed to be affecting his judgment and reducing his effectiveness. Excessive weekend drinking was also known, and reports of marital discord secondary to drinking had been heard. The patient, although initially defensive, acknowledged excessive and problem drinking for the past 3 years. Since becoming president of his company 2 years before, he had been associating increasingly with friends who met daily to discuss their mutual interests and problems over alcoholic beverages; and this had become a habitual pattern. He acknowledged marital difficulty connected with his drinking, chiefly because of his irritability and mood change while drinking. During the initial interview he was defensive, rationalizing and minimizing the significance of his drinking and intellectualizing many aspects of the problem.

Further history from the patient and his wife indicated an alcoholic drinking pattern with increased tolerance, occasional memory blackouts, attempts to abstain or cut down, beginnings of loss of control over drinking, secret drinking, remorse after drinking. There was no history of withdrawal symptoms or severe hangovers. There was no family history of alcoholism. The patient's developmental history was unremarkable with relatively normal milestones and sexual and personality development. School adjustment had been satisfactory, with good grades and popularity among his peers. College years were similarly described as having been pleasant and productive. The 19-year marriage and three children were unremarkable, aside from a developmental handicap of one child. There were no current symptoms of psychiatric disorder or a history of such. The patient had, in general, been a productive businessman with community interests.

There was no clinical evidence of other overt psychopathology. The Minnesota Multiphasic Personality Inventory (MMPI) was completely within normal limits and showed no suggestion of psychiatric disorder. The patient was seen individually and conjointly with his wife for a total of six outpatient interviews, during which time the nature of alcoholism was discussed as well as an marital communication problem related to his drinking. After a couple of brief attempts to drink "socially," the patient accepted the fact that he must abstain completely. Four years later, he continued to abstain from alcohol.

This alcoholic man apparently had developed the disorder without underlying psychiatric disturbance and without "symptomatic" use of alcohol to "treat" an emotional condition.

Apart from clinical experience and anecdotal reflection, there are other interesting bits of knowledge that support the contention that psychopathology is not a necessary condition for the development of alcoholism. Syme (1957), after reviewing the available psychologic and psychiatric literature from the years

1936 to 1956, concluded that there was no reason to believe persons of one type of personality are more likely to become alcoholics than persons of another type. And in accord with this opinion, one sees in practice alcoholic persons who are neurotic, psychotic, suffering from psychophysiologic problems or personality disorders—and some without any obvious psychopathology.

The rather good overall results of Alcoholics Anonymous in helping alcoholics recover may also indicate the absence of serious psychopathology in many alcoholics, or at least that treatment of other psychopathology is not necessary to recover. This spiritually oriented, generally supportive organization does not attempt to elicit psychopathology but conversely to suppress many overt conflicts in the interest of "sobriety."

Another indication that alcoholism (as dependence on the drug ethyl alcohol) may not require underlying psychologic factors is derived from the results of recent animal experiments. Much as with other addicting drugs, it has now been shown that addiction, defined as increase in tolerance to alcohol and frank withdrawal reactions when it is discontinued, can be induced in the rhesus monkey, beagle dog, and mouse (Ellis and Pick, 1973; Freund, 1973). Whether alcohol dependence may be produced similarly in man, simply by chronic and excessive ingestion over a long enough period, has yet to be demonstrated and for obvious reasons may never be.

If we compare the development of a more mundane drug dependence— tobacco smoking—to alcoholism, again we are forced to acknowledge that psychopathology is not necessary to become drug dependent. As Bejerot (1972) indicated, the underlying reason a teenager begins smoking (i.e., to look grown-up) cannot explain his craving for and inability to abstain from cigarettes 30 years later, after he has become nicotine dependent. He may show no evidence of psychopathology but yet be unable to relinquish his dependence.

Finally, another recent psychiatric report tends to support the earlier-mentioned findings of Vaillant (1977). Nussbaum *et al.* (1976), investigating the prevalence of hidden alcoholism among applicants for Social Security disability payments, discovered a negative relationship between severity of alcohol involvement and severity of psychiatric manifestations. Psychiatric impairment decreased, in other words, as degree of alcoholism increased. Hard pressed to explain these unexpected findings, the authors entertained at least two possibilities. Drinking may in effect be "self-medication" and hence mask existing psychiatric symptoms. However, they wondered as well whether their findings demonstrated truly that no relationship exists between psychiatric disorder and degree of alcoholism and that perhaps alcoholism "may not be primarily a psychiatric disease."

As Reinert (1968) suggested in his interesting paper describing alcoholism as a "bad habit," addictions seem too widespread to justify looking for esoteric psychologic or physiologic explanations. The pertinent factors must be closer to a universal human condition. Perhaps a more rewarding strategy would be to ask why alcohol is not abused more or why everyone is not alcoholic.

III. ROLE OF THE PLEASURABLE EXPERIENCE FROM ALCOHOL

Clearly, alcohol can serve many purposes through its pharmacologic and symbolic characteristics. It gives pleasure, reduces pain, eliminates fears, raises self-esteem, solves conflict, and so on. But basically the pleasurable experience from alcohol underlies all alcohol problems and perhaps all alcohol use. The alcoholic drinks to relieve tension, to celebrate, to become brave, to be sociable, to handle boredom, to unwind, to get drunk, to feel good, to drown his sorrows, to get high—in other words, for the same reasons everyone else drinks. In its nature and quality, however, his drinking has changed from that of the nonalcoholic. In most cases this process of becoming alcoholic (how do you get it?) can be traced without complete understanding of etiologic factors. Perhaps we should make this knowledge more available. Alcoholism may be a disorder to which many are vulnerable if they show but one simple characteristic, that of obtaining pleasure (a good feeling) from alcohol, for that fact indicates that both psychologic and physiologic systems are "go" for a potential problem. Some people, often from certain ethnic or racial groups (e.g., Oriental), do not experience pleasure from alcohol. In fact, some have quite unpleasant and aversive results and therefore are quite unlikely to proceed along the pathways to alcoholism.

One of the most interesting, detailed, and, I believe, accurate descriptions of the route to alcoholism was given by Bell (1970) in an appropriately titled chapter, The Psychological Labyrinth, in his book "Escape From Addiction." Drawing somewhat from his comments as well as from our clinical experience, I shall attempt to describe the path to alcoholism followed by the great majority of alcoholic persons, allowing for their individual differences. In general, there seem to be three rather clear phases in this process. First, of course, there must be drinking, for the process to be initiated. The second phase is termed "habituation or psychologic dependence," and is followed in most cases by physical dependence or what Bell terms "defensive dependence."

As already indicated, the reasons for initial drinking are as numerous as are individuals. Alcoholics, from their drinking histories, seem to have begun their drinking for exactly the same reasons given by normal drinkers. Wellman (1955), who conducted an interesting study of alcoholic men, stated that "not one of them began to drink or get drunk because of personal problems." He emphasized that their reasons for drinking usually had more to do with peer pressure, social custom, or the ubiquitous rebellion of adolescence. Even in the early stages, however, the eventual alcoholic often will describe a different sort of euphoric or pleasurable experience than the social drinker. It seems that his euphoria is more intense, persistent, and consistent. The eventual alcoholic may also initially suffer fewer of the "hangover" effects from excessive drinking than his counterpart. For example, hangover headache is frequently a problem for the periodically excessive social drinker but is often not suffered by the eventual alcoholic. In other words, almost from the beginning, some drinkers seem to reap a more regular and qualitatively intense pleasurable experience from alcohol without suffering its adverse psychophysiologic consequences. In early non-pathologic drinking, patterns are determined by many factors, including the cultural, social, ethnic, emotional, and physical characteristics of the drinker. This pursuit of pleasure is determined as well, however, by the desire or need to "feel different" and as such is to be expected more frequently in those who have experienced some sort of regular or periodic dysphoria (bad feeling). One would, therefore, expect a greater desire to drink (feel different) in emotionally troubled persons who initially experience euphoria from alcohol, and this probably is so. Certainly a sizable segment of eventual alcoholics—but again by no means all—begin drinking "symptomatically" after discovering the ameliorating effects of doses of ethanol upon their anxiety, feelings of insecurity, etc.

IV. HABITUATION, DEPENDENCE, PROBLEMS

Like any pleasurable experience repeated over a considerable time, the initial use of alcohol may become a habit. Habituation or psychologic dependence, in one sense, is simply the feeling that one must use alcohol or have it available to be comfortable and to cope with everyday life and its problems. Habituation to alcohol, of course, is not always dangerous. Many regular drinkers, by our definition, can be considered habitual drinkers; and the majority of them have no significant problem related to their use of alcohol. If alcohol use is interwoven into the total life experience with balanced activities of work, love, and recreation, its habitual use may be no more damaging than the habitual use of any other substance. The danger in using alcohol habitually is that it is an effective drug. Most habitual drinkers are aware of the sedating and tranquilizing effects of alcohol upon their tensions and anxieties, although they may prefer not to recognize this as a "drug effect." It is not difficult for the habitual drinker, whether suffering from significant psychopathology or not, to begin to use alcohol as a medication with gradually increasing doses to handle greater or lesser life problems. Usually unaware of his changing use of alcohol and its increasing importance in his life, the drinker becomes more dependent upon it, with some increase in tolerance to its effects, and begins to manipulate persons and events to maintain a steady supply. As the habitual dependence upon alcohol increases, it becomes an increasingly important part of all aspects of life. The drinker is unwittingly developing an underlying sense of urgency to make certain alcohol is always available.

As he begins to cross the gray area from chiefly psychologic to physical or "defensive" dependence (addictive alcoholic drinking), the drinker is becoming aware that all is not well in his world. Problems begin to surface, at first usually interpersonal problems between himself and those most close to him such as his wife, family, friends, and fellow workers. Although he is not aware of it, most of these problems are results of his increasingly excessive drinking, which is beginning to affect his personality, reliability, and judgment. The drinker now shows many of the signs and symptoms of alcoholism, including preoccupation with drinking or the next opportunity to drink, an increasing tolerance for alcohol, solitary drinking, secret drinking, unpremeditated drinking. He has suffered a few alcoholic blackouts or amnesic spells during periods of excessive use. He has begun to gulp drinks rapidly prior to any drinking occasion. He may even try a period of abstinence or "going on the wagon," in part to reassure himself and others that alcohol is really not a problem. Problems that launch this phase—the entrance into true alcoholic drinking—can be physical, psychological, or social. Or habituation may blend into addictive drinking with no obvious precipitating cause. We have seen habitual drinkers begin their addictive drinking after gastrectomy (which may change alcohol tolerance in that alcohol is much more rapidly absorbed following this procedure), after traumatic life experiences (a businessman suffering severe flood damage to his business), or a change in social environment (an executive's wife who moved from Europe to an American metropolis and immediately became involved with heavy-drinking acquaintances).

V. TOLERANCE AND NEED

Concurrently with increase of tolerance to alcohol, there is an increase of nervous system overactivity, a homeostatic attempt to adapt to the repeated depressant effects of alcohol. As a result of this overactivity and the secondary symptoms of tension, anxiety, and insomnia, the alcoholic now finds he needs regular doses of alcohol for sedation. Thus has begun the vicious addictive cycle that often results in morning drinking. Now forced by his psychophysiologic dependence to drink despite some knowledge that he is in trouble—though not necessarily awareness that he is in trouble because of his drinking—the alcoholic begins to insist that his drinking is no problem. In attempts to explain his behavior to himself and others, he erects a rather elaborate defense system consisting chiefly of denial, rationalization (alibis), and projection (blaming others).

The fluctuating levels of sedation and overactivity of his nervous system produce various states of anxiety and depression. His defense system labels him as unreliable and dishonest in the eyes of those about him. Alcohol-related transient memory loss as well as difficulty with thinking, reasoning, and concentration begin to affect everyday adjustment. By this stage of alcoholism, most drinkers begin to resemble each other as their dependence has shifted from harmless habituation to harmful drug dependence and the results of the addictive process have become difficult to discern from its antecedents. What began as a rather deliberate and pleasurable experience, or even "abuse" for whatever motivation, has been transformed over an indefinite period to a condition quite different in quality from initial use of alcohol. The reason that the drinker began to drink now has in principle nothing to do with the reason he continues to do so. Conversely, treatment of the initiating condition will not have much effect on the course of this well-developed addiction. The desire or craving for alcohol has become a condition of its own (a "disease"?) with a dynamic force of its own (Bejerot, 1972).

VI. IS EVERYONE VULNERABLE?

Whether anyone exposed long enough to alcohol will develop alcoholism may never be known. However, we have just reviewed some of the evidence that many if not most of our population may be potentially alcohol dependent. Clinical experience, psychiatric studies, success of Alcoholics Anonymous, animal experiments, other drug dependencies, and recent psychiatric research have all supported the concept that vulnerability to alcohol addiction is widespread. We do know about the development of the alcoholic process, even though specific etiologic factors remain somewhat obscure. Perhaps efforts in prevention and in treatment should be focused more on the process of alcoholism, which is treatable and perhaps preventable. Maybe one of our chief slogans for prevention should be "Everyone is vulnerable."

REFERENCES

Bejerot, N. (1972). "Addiction: An Artificially Induced Drive." Thomas, Springfield, Illinois.
Bell, R. G. (1970). "Escape From Addiction." McGraw-Hill, New York.
Chafetz, M. E. (1967). Alcoholism. *In* "Comprehensive Textbook of Psychiatry" (A. M. Freedman and H. I. Kaplan, eds.), pp. 1011–1026. Williams & Wilkins, Baltimore, Maryland.
Ellis, F. W., and Pick, J. R. (1973). Animal models of ethanol dependency. *Ann. N.Y. Acad. Sci.* **215**, 215–217.
Freund, G. (1973). Alcohol, barbiturate, and bromide withdrawal syndromes in mice. *Ann. N.Y. Acad. Sci.* **215**, 224–234.
Nussbaum, K., Schneidmuhl, A., Kacsur, A. R., and Shaffer, J. W. (1976). "Hidden" alcoholism among disability insurance applicants: Prevalence and degree of impairment. *Mil. Med.* **141**, 596–599.

Reinert, R. E. (1968). The concept of alcoholism as a bad habit. *Bull. Menninger Clin.* **32,** 35–46.

Syme, L. (1957). Personality characteristics and the alcoholic: A critique of current studies. *Q. J. Stud. Alcohol* **18,** 288–302.

Tyndel, M. (1974). Psychiatric study of one thousand alcoholic patients. *Can. Psychiatr. Assoc. J.* **19,** 21–24.

Vaillant, G. E. (1977). Antecedents of alcoholism and orality (abstract). *Sci. Proc. Summ. Form,, Am. Psychiatr. Assoc. Annu. Meet., 130th, Toronto,* pp. 110–111.

Wellman, M. (1955). Towards an etiology of alcoholism: Why young men drink too much. *Can. Med. Assoc. J.* **73,** 717–725.

EDITORIAL COMMENT

There is remarkably little objective evidence for or against genetic factors in susceptibility to alcoholism, although renewed interest in the subject has been manifested recently. Studies of racial or ethnic differences in the metabolism of alcohol have shown conflicting results [Bennion, L. J., and Li, T.-K. (1976). Alcohol metabolism in American Indians and whites. *N. Engl. J. Med.* **294,** 9–13; Farris, J. J., and Jones, B. M. (1978). Ethanol metabolism in male American Indians and whites. *Alcoholism, Clin. Exp. Res.* **2,** 77–81; Reed, T. E. (1978). Racial comparisons of alcohol metabolism: Background, problems, and results. *Alcoholism, Clin. Exp. Res.* **2,** 83–87]. Most studies of siblings or close relatives of alcoholics, however, show a higher incidence of alcoholism, regardless of environment, than do controls [Goodwin, D. W. (1978). Hereditary factors in alcoholism. *Hosp. Pract.* **13,** 121–130; Rutstein, D. D., and Veech, R. L. (1978). Genetics and addiction to alcohol. *N. Engl. J. Med.* **298,** 1140–1141]. Laboratory strains of mice have been bred that are eight to ten times more susceptible to alcohol as judged by sleep time than controls [McClearan, G. E. (1976). Experimental behavioural genetics. *Postgrad. Med. J. Suppl.* **2,** 31–43].

NATURAL HISTORY OF ALCOHOL DEPENDENCE

Jack H. Mendelson and Nancy K. Mello

I. Introduction 371
II. Procedures 373
    A. Subjects 373
    B. Research Ward Facilities 374
    C. Sequence of Procedures 374
    D. Alcohol Administration Procedures 374
III. Results 376
    Volume of Alcohol and Withdrawal Onset 378
IV. Discussion 386
    A. Temporal Development of Alcohol Dependence 386
    B. Alcohol Dosage and Alcohol Withdrawal 387
    C. Drinking Patterns and Alcohol Withdrawal 389
V. Research Needs 390
VI. Summary 391
    Appendix I: Social and Drinking History 392
    Appendix II: Neurological Rating Form 393
    References 394

I. INTRODUCTION

Alcohol withdrawal phenomena were first described in the medical literature after 1780 (Lettsom, 1787; Pearson, 1813; Sutton, 1813). Over the ensuing 170 years, the casual role of alcohol in producing abstinence phenomena was questioned, and it was generally believed that an interaction between malnutrition, intercurrent illness, and toxic effects of alcohol abuse accounted for the abstinence syndrome (Victor and Adams, 1953). In 1953, Victor and Adams demonstrated that cessation of drinking per se resulted in the alcohol withdrawal syndrome. These findings were confirmed in clinical experimental studies in man in 1955 (Isbell *et al.*, 1955) and in 1964 (Mendelson, 1964). Today there is little question that termination of drinking by an alcohol addict may result in a characteristic pattern of signs and symptoms known as the alcohol abstinence syndrome.

The common abstinence syndrome is usually characterized by tremor, sweating, gastrointestinal symptoms, hyperreflexia, anxiety, sleep disturbances, and occasional disorientation and hallucinosis (Isbell *et al.*, 1955; Mello and Mendelson, 1978; Mendelson, 1964; Victor and Adams, 1953). It usually begins within 8 to 12 hours after the last drink, reaches peak severity at about 24 hours, and spontaneously remits within 48 to 72 hours (Mello and Mendelson, 1978; Victor and Adams, 1953). The common abstinence syndrome should not be confused with delirium tremens, a separate, infrequent, and potentially lethal consequence of alcohol withdrawal (Mello and Mendelson, 1978; Victor and Adams, 1953). Delirium tremens may occur independently of or following apparent recovery from the common abstinence syndrome and is characterized by profound confusion and disorientation, delusions, and vivid hallucinations, as well as autonomic hyperreactivity.

Although the clinical phenomena associated with alcohol withdrawal are well known, our understanding of the basic mechanisms involved is decidedly limited (Gross *et al.*, 1974; Mello and Mendelson, 1978; Mendelson, 1971). Little is known about the critical contributing factors or the time course of development of the alcohol withdrawal syndrome in man. It is usually assumed that physical dependence on alcohol develops after some years of heavy drinking, and it is impossible to confirm or disaffirm this impression by direct experimental observation. The only method available to examine these questions is the analysis of self-report data. We have frequently commented on the unreliability of self-report data obtained from sober alcohol addicts (Mello, 1972), and all of the self-evident reservations apply to the present report. However, we felt it might be useful to examine the case records of 129 alcohol addicts studied over a 4-year period on the clinical research ward of the Laboratory of Alcohol Research, National Institute on Alcohol Abuse and Alcohol, ADAMHA, at the St. Elizabeths Hospital in Washington, D.C.

Our primary goal was to determine if there were discernible relationships between the reported time of onset of withdrawal signs and symptoms as a function of (1) years of problem drinking; (2) reported amount of alcohol consumed each day during drinking sprees; or (3) the type of alcohol preferred, i.e., whiskey, wine, or beer.

A second objective was to observe the actual pattern of withdrawal signs and symptoms that occurred after a period of experimentally induced intoxication. If some concordance between the empirically observed and reported withdrawal signs was found, this would lend some greater credence to the self-report data.

Finally, we attempted to determine if the experimentally imposed pattern of drinking produced differences in type and number of withdrawal signs and symptoms. We previously reported that spontaneous drinking was followed by more severe, varied, and prolonged withdrawal signs and symptoms than programmed (Q^4H) alcohol administration in eight subjects (Mello and Mendelson, 1970a). Those data suggested that the pattern of drinking was more important than duration of drinking in accounting for the expression of the alcohol abstinence syndrome. In this chapter the effects on alcohol withdrawal signs and symptoms of four drinking conditions—programmed drinking, spontaneous drinking, and work contingent drinking with immediate or delayed reinforcement—are compared.

II. PROCEDURES

A. Subjects

One hundred and twenty-nine male volunteers with a history of alcoholism of 1–40 years were subjects for a series of clinical studies of the behavioral and biological factors of alcoholism. Subjects were recruited from the Occoquan Alcohol Rehabilitation Center, Lorton, Virginia, and each was fully informed of the nature and duration of the clinical study. No subject was under any legal constraint during the course of the study.

Subjects ranged in age from 21 to 51. Some had completed high school and all had a recent history of sporadic employment in semiskilled and nonskilled jobs. Most subjects were spree drinkers, accustomed to consuming about one quart of whiskey each day or an equivalent amount of wine or beer. Each subject had abstained from alcohol for at least 7 days prior to admission to the clinical research ward.

Each subject was given a complete physical examination and mental status assessment prior to his participation in the study. All volunteers selected were in good health and none had a history of neurological disease, seizure disorders, hepatic, renal, pulmonary, cardiac or gastrointestinal disease, nutritional or metabolic disorder. No subject showed evidence of alcohol-induced liver disease, as measured by hepatic enzymes and a bromsulphalein clearance test (BSP) administered prior to, during, and following the experimental drinking period. No subject had a history of opiate or barbiturate addiction, or was using any form of medication at the time of the study. A more complete description of the medical screening procedures is available in a previous report (Mendelson, 1964).

Throughout the study, the medical status of each subject was monitored daily by a resident physician. Caloric intake was recorded and subjects were encouraged to eat all of their meals and were given multiple vitamin capsules daily.

After completion of the study, each subject was given an opportunity to return to the alcohol rehabilitation center. The social service department and the rehabilitation unit then arranged for job placement or vocational training.

B. Research Ward Facilities

Subjects were studied in groups of three to six. Each subject had a private bedroom and free access to a spacious dayroom which contained color television, books, cards, etc. No other patients were present on the research ward. The subjects were restricted to the research ward for the course of the study (30–60 days).

C. Sequence of Procedures

Each subject served as his own control for the behavioral and biochemical assessments. A predrinking baseline period of 5–10 days preceded an experimental drinking period of 7–60 days, which was followed by a postintoxication withdrawal period of 5–10 days.

D. Alcohol Administration Procedures

Data obtained under four different alcohol administration procedures are summarized in this chapter.

1. Programmed Drinking ($N = 11$)

Beverage alcohol (bourbon) was given in divided doses every 4 hours around the clock. Alcohol dosage was gradually increased and during the first 3 days the six daily doses were as follows: 0.5, 1.0, and 1.5 ml/kg of 43% alcohol. This dosage resulted in the following approximate daily volume totals: 8, 16, and 24 oz. Subjects were maintained on the maximum daily dose of 2.0 ml/kg unless contraindicated by gastritis or other intercurrent illness. The maximum dose resulted in daily total volumes which ranged between 23 and 32 oz.

2. Spontaneous Drinking ($N = 81$)

Each subject was given 32 or 33 tokens each day that he could use to buy alcohol at any time during the experimental drinking period. Each token purchased 1 oz. of alcohol that was directly dispensed from an apparatus located on the wall of the dayroom. There were no constraints as to how much alcohol a subject could buy at any one time.

3. Work Contingent Drinking ($N = 33$); Immediate Reinforcement

Twenty subjects earned alcohol by performance on a titrated delayed visual matching-to-sample task designed to assess short-term memory function. Completion of four correct matching-to-sample trials before four incorrect trials yielded a single token which could be used immediately to purchase alcohol from an automatic dispenser. Details of this procedure have been described previously (Mello, 1973).

The 13 remaining subjects earned alcohol by producing a preprogrammed sequence of eight numbers on a numbered keyboard. The subjects were required to work in pairs and one subject had to replicate the number choices of a second subject within 2 seconds. Once the subjects arrived at the correct sequence of eight numbers, a token was automatically dispensed into a receptacle in the dayroom. Details of this procedure have been described previously (Steinglass, 1975).

4. Work Contingent Drinking (*N* = 4); Delayed Reinforcement

Each subject earned all of his alcohol during the experimental drinking period by working at a simple operant task. One thousand responses on a portable button box earned a single token which could be spent to obtain a single cigarette or a single ounce of bourbon. Subjects were able to earn one token in about 5 minutes of maximally rapid performance or about 12 oz. of alcohol per hour. Points earned during a 24-hour period were exchanged for alcohol and cigarette tokens once each day, at 8 A.M. (Mello and Mendelson, 1972).

5. Experimental Assessments

The following assessments were completed during each study.

a. Self-Report Assessments. Upon admission to the clinical research ward a number of psychological and screening tests were performed on each subject. Drinking and withdrawal history was evaluated in a standard interview conducted by Dr. Mello (Appendix I). The answers to this questionnaire were compared for consistency with responses given during a second interview by a resident physician. Data are presented for 123 subjects from a total sample of 133 subjects. All interviews were conducted during the predrinking baseline period.

b. Neurological Examinations. A neurological evaluation of the occurrence of tremor, nystagmus, and disorientation was completed three times daily by the attendant staff during the postdrinking withdrawal period. The neurological rating scale employed in these studies is attached as Appendix II.

c. Blood Alcohol Levels. Throughout each study, blood alcohol levels were determined three times a day, at 8 A.M., 4 P.M., and 12 midnight, with an instrument designed to measure the concentration of ethanol in the breath (Breathalyzer,* Stevenson Corporation, Redbank, New Jersey). Blood alcohol levels also were measured during the baseline period to ensure that subjects were not

---

*Now available only through the Law Enforcement Maintenance Co., Framingham, Massachusetts.

obtaining alcohol from any outside source. To assess the accuracy of these blood alcohol measurements, breath alcohol values were compared with serum alcohol determinations carried out by an enzymatic method. The breath sample values showed no greater variance than ± 10 mg/100 ml from the blood alcohol levels as determined by enzymatic techniques.

d. Withdrawal Medication. In studies where observations of behavioral or biological correlates of withdrawal phenomena were not the primary experimental focus, subjects were occasionally given paraldehyde (1 to 2 oz., Q⁴H) or chlordiazepoxide (25 to 50 mg i.m., Q⁴H). Sometimes subjects received an anti-insomnia medication, Noludar (300–600 mg) (Mello and Mendelson, 1970b). Medications were not administered routinely, but only in response to the development of severe withdrawal signs and symptoms and reports of discomfort by the patient.

III. RESULTS

Subjects were asked when they first experienced withdrawal signs and symptoms after they began problem drinking. This question had been preceded by a question about the age and onset of heavy drinking and a question about the occurrence of withdrawal symptoms and their nature. The reported time of onset of withdrawal signs and symptoms after the onset of problem drinking for 106 whiskey drinkers is shown in Fig. 1.

The latency of development of withdrawal signs and symptoms, after the subject identified drinking as a problem (0–34 years) is shown on the abscissa of Fig. 1. The withdrawal onset latency is plotted against the number of years of problem drinking reported at the time of the clinical interview. The ordinate shows the duration of each subject's drinking history (1–40 years) at the time of the interview. Consequently, all subjects who reported 10 years of problem drinking at the time of the interview must have had a withdrawal syndrome onset latency of 10 years or less. The age of each subject at the time of the interview is shown in the lower right-hand corner of Fig. 1.

Fifty percent of the subjects studied reported that withdrawal signs and symptoms began after at least 5 years and within 15 years of heavy drinking (cf. columns 2 and 3, Fig. 1). Twenty-three subjects or 22% of the sample reported that withdrawal signs and symptoms began after 15 or more years of heavy drinking (cf. columns 4 to 7). Early development of withdrawal signs and symptoms during the first five years of problem drinking was reported by 30 subjects, or 28% of this sample. Twelve of these subjects (11% of the total sample) reported that withdrawal signs and symptoms began within the first year of heavy drinking. These data indicate that a minimum of 2 years of heavy drinking usually preceded the development of withdrawal signs and symptoms and, for the

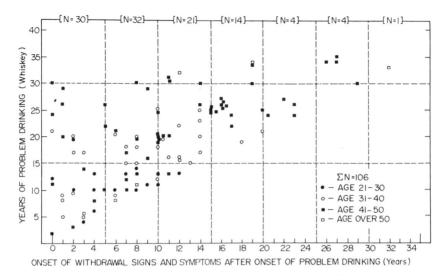

Fig. 1. Self-report data from 106 alcohol addicts describing the latency between the initiation of problem drinking (whiskey) and the development of withdrawal signs and symptoms. The circles and squares indicate the age of the subject at the time of the interview (lower right corner). The total number of subjects appearing in each 5-year column is indicated at the top of the figure. (From Mello and Mendelson, 1976, reproduced by permission.)

vast majority (72%), at least 5 or more years of heavy drinking were required for physical dependence upon alcohol to evolve.

It might be anticipated that the consistent consumption of a low alcohol concentration beverage such as wine or beer, in contrast to whiskey, could result in a postponement of the development of withdrawal signs and symptoms. In Fig. 2, withdrawal latency data are presented for 17 subjects who drank wine or beer rather than whiskey. The onset of withdrawal signs and symptoms is displayed as a function of years after the onset of problem drinking. The age and beverage preference of each subject are shown at the lower right.

The meaningfulness of percentage comparisons between whiskey and wine or beer drinkers is limited by the small sample size. However, the number of wine or beer drinkers who reported more than 15 years of heavy drinking before the onset of withdrawal signs and symptoms was 35%, a 13% increase over whiskey drinkers. Fifty-three percent reported the onset of withdrawal signs and symptoms after at least 5 years and within 15 years of heavy drinking. This percentage figure is equivalent to that reported by the whiskey drinkers. However, within the first 5 years of problem beer or wine drinking, only two subjects reported withdrawal signs and symptoms. Therefore, the frequency of early onset of withdrawal signs and symptoms is 17% less in the wine and beer drinkers. These data provide some tentative support for the notion that abuse of wine or beer may

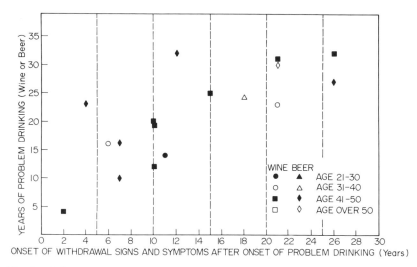

Fig. 2.   Self-report data for 17 alcohol addicts describing the latency between the initiation of problem drinking of wine or beer and the development of withdrawal signs and symptoms. Circles and squares indicate the age of the subject at the time of the interview (lower right). (From Mello and Mendelson, 1976, reproduced by permission.)

result in some delay in the development of physical dependence upon alcohol as compared with the abuse of whiskey. However, it is apparent that alcohol addiction does result if sufficient alcohol is consumed in any beverage form regardless of the alcohol concentration.

Volume of Alcohol and Withdrawal Onset

It is obvious that data relating the latency of withdrawal syndrome onset to years of heavy drinking can be meaningfully interpreted only if the volume of alcohol consumed is also considered. Figure 3 presents data for 94 of the 106 whiskey-drinking subjects described in Fig. 1. Only subjects who gave the most consistent estimates of usual daily consumption were included. It is important to note that most subjects were spree drinkers and their daily intake estimates were based upon whiskey consumption during a usual drinking episode.

The development of physical dependence is probably a function of both volume consumed and years of drinking. It might be expected that those whiskey drinkers who consumed only 1 pint of alcohol per day would develop withdrawal signs and symptoms somewhat later than those subjects who consumed 4/5 quart or more each day. However, the data presented in Fig. 3 indicate that the majority of subjects developed withdrawal signs and symptoms during the first

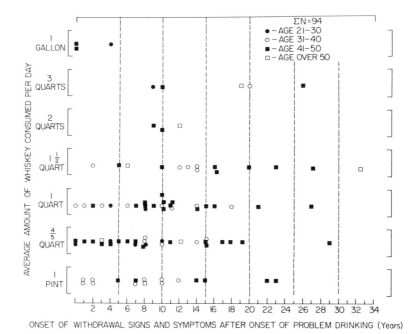

Fig. 3. Self-report data for 94 alcohol addicts describing the relationship between the average amount of whiskey consumed each day during a drinking spree and the time between onset of problem drinking and the development of withdrawal signs and symptoms. Circles and squares indicate the age of the subject at the time of the interview (upper right). (From Mello and Mendelson, 1976, reproduced by permission.)

10 years of heavy drinking, independently of the absolute amount of alcohol that they drank during sprees.

Fifty-three of the 94 subjects drank from 4/5 to 1 quart of whiskey per day. Thirty-seven percent of these 53 heavy drinkers did not develop withdrawal signs until after at least 10 years of heavy drinking. Sixty-two percent of those heavy drinkers developed withdrawal signs within 10 years of heavy drinking. Twenty-six percent of these heavy-drinking subjects developed withdrawal signs and symptoms during the first 5 years.

Approximately the same temporal distribution of withdrawal syndrome onset was observed for the 16 individuals who drank only 1 pint of whiskey per day. No clear temporal distribution pattern emerged for those subjects who reported drinking 1½–3 quarts of whiskey per day. The three subjects who claimed to drink 1 gallon of whiskey daily all developed physical dependence quite rapidly, within less than 1–4 years of problem drinking.

Since the absolute volume of alcohol consumed over time is also a function of the frequency of drinking sprees, and since drinking spree frequency could not be assessed with any degree of reliability from a retrospective report, these data must be interpreted with some caution. However, these findings testify to the enormous interindividual variability in resistance to the induction of physical dependence on alcohol.

Data presented in Fig. 4 indicate that the absolute volume of wine and beer

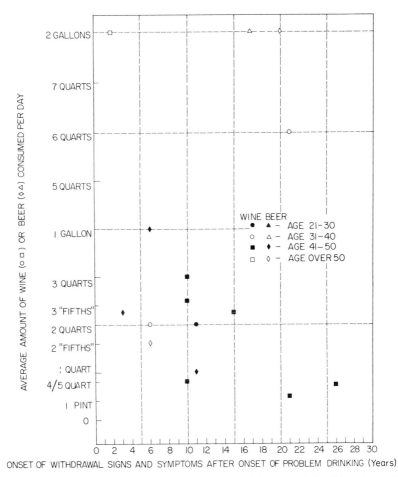

Fig. 4. Self-report data for 16 alcohol addicts describing the relationship between the average amount of beer or wine consumed each day during a drinking spree and the time between onset of problem drinking and the development of withdrawal signs and symptoms. Circles and squares indicate the age of the subject at the time of the interview (middle right). (From Mello and Mendelson, 1976, reproduced by permission.)

consumed per day during a drinking spree also did not predict the time of onset of withdrawal signs and symptoms with any degree of reliability. Some subjects who reported consuming about 1 pint per day and some subjects who reported consuming as much as 2 gallons per day of wine or beer were both able to drink for 15–20 years before the onset of signs of physical dependence. Once again, the limitations of retrospective reports of drinking behavior must be emphasized. However, insofar as these data may be valid reflections of actual drinking patterns of alcohol addicts, it appears that the onset of physical dependence on alcohol is a complex interaction of both the years of drinking and the average volume of ingestion, which varies greatly from individual to individual. Those factors that are most important in accounting for a differential susceptibility to or resistance to alcohol addiction remain to be determined.

The self-report data were supplemented where possible by examining the frequency of occurrence of various kinds of withdrawal signs and symptoms as a function of the actual average blood alcohol levels maintained throughout a period of experimental drinking on the clinical research ward. Data presented for 52 subjects in Fig. 5 indicate the frequency of occurrence of tremor, nystagmus, profuse sweating, gastrointestinal symptoms, disorientation, hallucinosis, and insomnia as a function of the average blood alcohol levels maintained throughout the drinking period. Blood alcohol levels are divided into successive increments of 50 mg/100 ml in a range from 50 to 300 mg/100 ml. The range of values for each average blood alcohol level is shown at the top of each histogram. Only subjects who did not receive medication for withdrawal signs and symptoms are included. Composite data are based on one occurrence of any sign or symptom for each subject. Consequently, the severity or persistence of these signs and symptoms cannot be determined from these data.

It was somewhat surprising to find that tremor, nystagmus, and gastrointestinal symptoms were observed in 20% or more of the sample, even when very low average blood alcohol levels (below 50 mg/100 ml) were maintained. Neither insomnia nor profuse sweating was ever observed at blood alcohol levels below 50 mg/100 ml. Five subjects showed withdrawal symptoms usually associated with the most severe expression of alcohol abstinence, i.e., disorientation and hallucinosis, at low blood alcohol levels.

Most subjects showed tremor, the most common alcohol withdrawal sign, when average blood alcohol levels exceeded 500 mg/100 ml. Nystagmus occurred most frequently in subjects who maintained blood alcohol levels above 200 mg/100 ml. Profuse sweating occurred in 60% or more of the subjects who maintained blood alcohol levels above 50 mg/100 ml. An increasing number of subjects complained of gastrointestinal symptoms and hallucinosis as a direct function of increases in the average blood alcohol level maintained during the experimental drinking period. The occurrence of disorientation was extremely

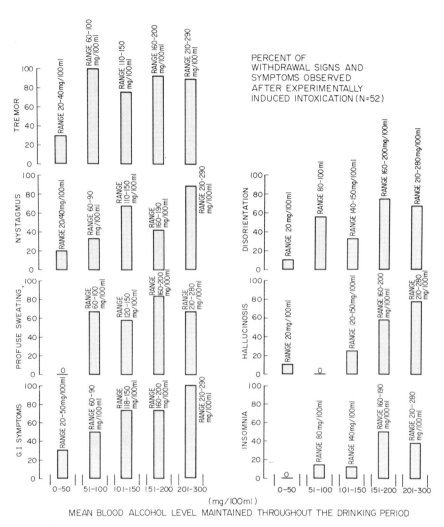

Fig. 5. Frequency of occurrence of seven withdrawal signs and symptoms in 52 subjects as a function of the average blood alcohol levels maintained throughout the drinking period. Blood alcohol levels are shown in 50 mg/100 ml units across the abscissa. The range of blood alcohol levels is shown above each histogram. Histograms are percent of withdrawal signs and symptoms in the samples studied at a particular average blood alcohol level. (From Mello and Mendelson, 1976, reproduced by permission.)

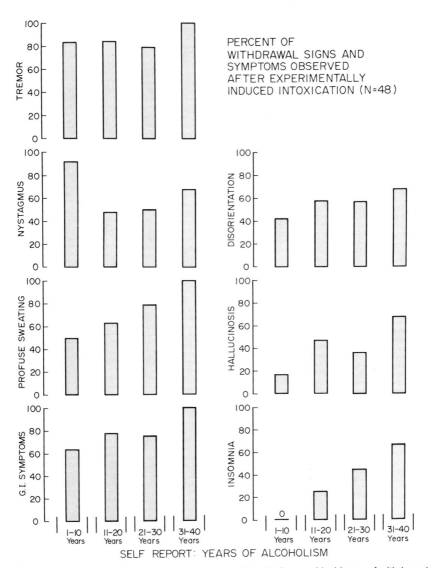

PERCENT OF
WITHDRAWAL SIGNS AND
SYMPTOMS OBSERVED
AFTER EXPERIMENTALLY
INDUCED INTOXICATION (N=48)

Fig. 6. The relationship of self-report of years of alcoholism and incidence of withdrawal symptoms. (From Mello and Mendelson, 1976, reproduced by permission.)

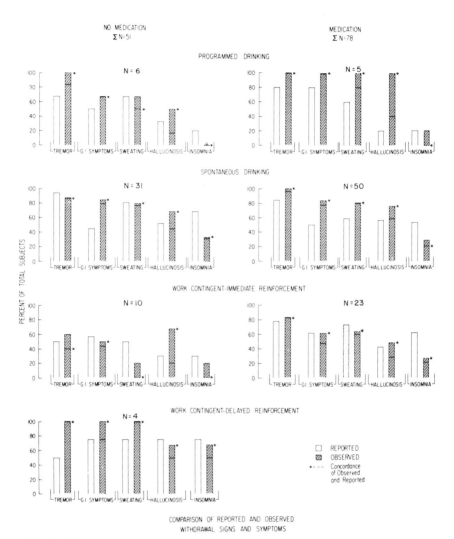

Fig. 7. The percent incidence of each of five common withdrawal signs and symptoms is shown as hatched histograms. Percent of withdrawal signs and symptoms in each category is shown as a function of the type of drinking paradigm used. Row one describes withdrawal signs and symptoms after a *programmed* drinking regimen for subjects who were not medicated (left) and subjects who were medicated (right). The second row describes withdrawal signs and symptoms for subjects in a *spontaneous* drinking paradigm and further compares the frequency distribution as a function of medication or absence of medication. The third row presents the same data for subjects run in a *work-contingent immediate reinforcement* drinking paradigm. The last row shows comparable data for subjects run in a *work-contingent delayed reinforcement* drinking paradigm who did not receive medication at the conclusion of drinking.

The hatched histogram indicating the observed percent occurrence of withdrawal signs variable and did not appear to be linearly related to blood alcohol maintenance levels. Insomnia was an infrequently reported symptom in this subject group, irrespective of blood alcohol level.

Data from 48 subjects were also analyzed to determine the relationship between subject's self-report of years of alcoholism and the frequency of occurrence of common withdrawal signs and symptoms—tremor, nystagmus, profuse sweating, gastrointestinal symptoms, disorientation, hallucinosis, and insomnia (Fig. 6). Most subjects showed tremor upon cessation of drinking. A greater number of individuals reporting 1–10 years of alcoholism showed nystagmus than did those reporting 11–40 years of alcoholism. The frequency of profuse sweating and reports of insomnia and hallucinosis increased as a function of years of alcoholism. The proportion of individuals reporting gastrointestinal symptoms increased by 20% after 30 years of alcoholism. Over half the subjects showed some disorientation after 11–30 years of alcoholism, and the frequency of occurrence of this sign increased by 10% in the 30–40 year group.

A comparison of the frequency of occurrence of five withdrawal signs and symptoms as a function of the type of experimental drinking paradigm is shown in Fig. 7. The percentage of the subjects who showed tremor, and who reported gastrointestinal symptoms, profuse sweating, hallucinosis, and insomnia under each experimental drinking condition at least once, is indicated by the striped histograms. Data are reported for a total of 129 subjects, 51 of whom received no medication following cessation of drinking and 78 of whom received either paraldehyde or chlordiazepoxide in response to development of severe withdrawal signs and symptoms.

The percentage of each subject group who reported that they usually experienced a particular withdrawal sign or symptom is shown by the open histograms. The percent concordance of the observed pattern of withdrawal signs and symptoms with the subject's retrospective self-report of withdrawal signs and symptoms in each category is shown as an asterisk (*) and a dotted line across each

---

Fig. 7 (*Continued*)
and symptoms is compared with the percent of this sample who previously reported experiencing these five withdrawal signs and symptoms. Open histograms denote the percent of the subjects in each subcategory who previously reported each withdrawal sign and symptom. The precent concordance between those that reported a particular symptom, e.g., tremor, and those that actually showed that symptom following cessation of drinking is indicated as a dotted line and an asterisk on the crosshatched histograms. For example, if 100% of those subjects who reported usually experiencing tremor in fact showed tremor following the experimental drinking period, then the coefficient of concordance would be 100%. This is, in fact, the case for tremor in row 1 under both medicated and unmedicated conditions. However, in row 1, column 1, only 85% of the entire sample actually showed tremor. Consequently, the coefficient of concordance is higher than the actually observed value. (From Mello and Mendelson, 1976, reproduced by permission.)

hatched histogram. These data were statistically analyzed by computing a coefficient of concordance for each variable for each subject subgroup.

In some instances, the percent concordance between the reported and observed signs and symptoms exceeded the actual frequency of occurrence in the subject group. For example, 70% of the nonmedicated programmed drinking group previously reported tremor on cessation of drinking and all of those subjects actually showed tremor in the current study, yielding a coefficient of concordance of 100%. However, only 80% of the entire group actually had tremor. In the case of gastrointestinal symptoms, 50% of the nonmedicated programmed drinking group previously reported gastrointestinal distress and only 70% actually reported gastrointestinal symptoms. The coefficient of concordance between reported and observed happened to coincide with the actual percent of the entire group reporting gastrointestinal symptoms.

Statistical evaluation of these data with the nonparametric Mann–Whitney U test revealed no significant differences in the percent occurrence of any withdrawal sign or symptom as a function of the experimental drinking condition. For example, tremor in the unmedicated, programmed drinking group did not differ significantly from tremor in the medicated group or from the incidence of tremor in any of the other five experimental groups. The lack of statistical significance probably reflects the small number of subjects in four of the seven categories.

On the basis of visual inspection, it is apparent that the percent frequency of occurrence of all five signs and symptoms was higher in the spontaneous drinking groups than in either the programmed or the work-contingent groups. This observation is consistent with our previous findings on only eight subjects (Mello and Mendelson. 1970a). The delayed reinforcement condition appeared to produce a higher frequency of withdrawal signs and symptoms than the immediate reinforcement condition. This is consistent with our previous observations (Mello and Mendelson, 1972) and speculation that work contingent alcohol acquisition under delayed reinforcement conditions may be concordant with conditions in the real world.

In most instances, the frequency of occurrence of withdrawal signs and symptoms was greater in the medicated than in the unmedicated group in any experimental drinking condition. This reflects the fact that medication was usually administered in response to the emergence of a relatively severe pattern of withdrawal signs and symptoms.

IV. DISCUSSION

A. Temporal Development of Alcohol Dependence

Despite the several limitations of self-report data, some consistent impressions can be derived concerning the contributions of the duration and volume of alcohol consumption to the development of physical dependence upon alcohol. Consumption of low alcohol content beverages, such as beer or wine, did not delay the onset of physical dependence in comparison with the reported time course of development for whiskey drinkers. The volume of alcohol consumed during drinking sprees did not reliably predict the time course of development of alcohol dependence for either whiskey or wine and beer drinkers. This could be in part a function of the fact that the length of the usual drinking spree and the interspree interval could not be determined from these retrospective data. These factors undoubtedly contributed to the wide variability in the development of physical dependence, a range of 1–32 years. Although almost one-third of the sample studied reported the onset of alcohol dependence within the first 5 years of heavy drinking, over half of this sample developed withdrawal signs and symptoms within the fifth to fifteenth year of heavy drinking. This prolonged time course confirms the usual clinical impression that problem drinking gradually merges into alcoholism as defined by physiological dependence.

It is difficult to evaluate the apparent differences in the time course of development of alcohol dependence in man and in experimental animals, because of the considerable differences in life span. However, it is important to note that alcohol dependence can be produced in as little as 3 days in mice (Freund, 1969; Goldstein, 1974), 5 days in rats (Majchrowicz, 1975), and 10–14 days in primates (Ellis and Pick, 1970). Other centrally active drugs, such as opiates, have produced acute tolerance and physical dependence in experimental animals. Similarly, in the human alcohol addict, a brief exposure to alcohol followed by a rapid fall in blood alcohol levels may result in the appearance of "partial" withdrawal signs and symptoms (Mello and Mendelson, 1972; Victor and Adams, 1953). These partial withdrawal signs, usually characterized by tremor and some gastrointestinal distress, may occur even when blood alcohol levels remain about 100 mg/100 ml, and appear to depend upon the rate of blood alcohol level fall rather than the absolute blood alcohol levels (Mello and Mendelson, 1972). The extent to which comparable phenomena may occur in normal or heavy drinkers has not been systematically studied. It is not known whether the experience of a complete alcohol abstinence syndrome is a necessary prerequisite for subsequent partial withdrawal signs and symptoms.

B. Alcohol Dosage and Alcohol Withdrawal

It is almost universally accepted that the ingestion of enough alcohol over a long enough period of time will produce alcohol dependence. There has been minimal attention to how little alcohol may produce the same result. The strategy followed by most investigators attempting to produce animal models of alcoholism has been to provide or induce the animal to self-administer consistently high doses of alcohol, e.g., above 6–8 gm/kg in primates (Ellis and Pick, 1970). Pursuit of high blood alcohol levels has diverted attention from examining the effects of low blood alcohol levels. In the literature on animal studies, French and Morris (1972) have been virtually alone in reporting that nonintoxicating doses of alcohol given over 2–3 weeks can result in measurable signs of alcohol dependence. These data on mice are supported by the clinical observations reported in this study. Goldstein (1972) has also noted mild withdrawal signs in mice after a single injection of alcohol. However, the intensity of withdrawal convulsions did increase as a function of both blood alcohol levels and duration of alcohol exposure (Goldstein, 1972).

In the present study, both the retrospective data (Figs. 3 and 4) and the empirical data (Fig. 5) converge to suggest that maintenance of a chronic *high* dose of alcohol may not be essential for the expression of alcohol dependence. As little as a pint of whiskey per day resulted in the development of withdrawal signs and symptoms (Fig. 3). Maintenance of a blood alcohol level of 50 mg/100 ml also resulted in the expression of withdrawal signs and symptoms following an experimental drinking period (Fig. 5).

It is difficult to explain the occurrence of withdrawal phenomena such as disorientation and hallucination following extremely low doses of alcohol (Fig. 5). Over 50% incidence of disorientation was observed in subjects whose average blood alcohol levels during drinking were in the range of 51–100 mg/100 ml (Fig. 5). It is possible that some withdrawal phenomena, such as disorientation and hallucinosis, may also reflect learning and conditioning factors which are independent of alcohol dose. In experimental animal studies, it has been shown that severity of withdrawal to centrally acting substances, such as opiates, may be affected by environmental conditions (Wikler, 1965; Wikler and Pescor, 1967). The extent to which conditioning factors may affect the abstinence syndrome in man has not been systematically studied.

It is possible to argue that the increases in six of the seven withdrawal signs and symptoms studied as a function of years of alcoholism may also reflect experiential or learning factors (Fig. 6). However, it would be virtually impossible to dissect out the contribution of experience from naturally occurring changes as a function of aging. For example, changes in sleep patterns, including decreases in REM activity and stage IV and increases in insomnia, are common concomitants of aging (Feinberg, 1969; Feinberg and Carlson, 1968). Profuse sweating increased during alcohol withdrawal as a function of both duration of alcoholism and age. Profuse sweating reflects heightened autonomic reactivity following alcohol withdrawal, which in turn may increase vulnerability to intercurrent illness. For example, marked dehydration can occur as a consequence of profuse sweating and may cause disorders of fluid and electrolyte balance. Clearly, the dimensions of expression of an alcohol abstinence syndrome are a function of the overall status of the alcoholic at the time of observation.

The percent incidence of occurrence of the most severe withdrawal signs and symptoms, i.e., disorientation and hallucinations, did tend to increase as a function of the mean blood alcohol levels maintained by subjects during the drinking period (Fig. 5). Gastrointestinal symptoms also occurred with increasing frequency as a function of progressive increments in mean blood alcohol levels. This finding might be anticipated, since gastrointestinal symptoms during withdrawal are due not only to CNS factors regulating gastrointestinal function but also to the cumulative toxic effects of alcohol in the gastrointestinal tract. Gastrointestinal disease is common among alcohol addicts and these individuals may respond to pain and discomfort associated with an alcohol-induced gastrointestinal inflammatory reaction by consuming more alcohol to produce central nervous depression and pain reduction.

Tremor is generally acknowledged to be the most commonly observed alcohol withdrawal sign. Most subjects who reported the previous occurrence of tremor during alcohol withdrawal also showed tremor following cessation of drinking in this study. However, it is of interest that not all subjects developed tremor during alcohol withdrawal even though they had maintained high average blood alcohol levels in the range of 210–290 mg/100 ml (Fig. 7).

A series of complicated mechanisms appears to underlie the development of tremor during alcohol withdrawal. It has been shown that both tremor and convulsions following cessation of drinking may be related to intracellular-extracellular levels of divalent cations such as magnesium (Mendelson *et al.*, 1969). Changes in electrolyte balance may depend upon alterations in acid–base balance which, in turn, are related in part to changes in the sensitivity of the CNS respiratory center during both acute alcohol intoxication and alcohol withdrawal (Wolfe and Victor, 1971; Wolfe *et al.*, 1969). Finally, changes in acid–base balance that may be induced by hyperventilation (respiratory alkalosis) may also be affected by internal and environmental contingencies that modify the patient's level of anxiety. Thus, the occurrence of the most commonly observed withdrawal phenomenon, tremor, is dependent on a number of physiological mechanisms which are difficult to correlate with dose and duration of alcohol ingestion.

C. Drinking Patterns and Alcohol Withdrawal

Data obtained in this study tend to support our previous observations (Mello and Mendelson, 1970a) that spontaneous drinking patterns result in a greater incidence of withdrawal signs and symptoms than programmed drinking paradigms. These data could be interpreted to suggest that alcohol dosage (Fig. 5) may be less important than the duration and pattern of alcohol exposure (Fig. 7) and experiential factors (Fig. 6) in determining the expression of alcohol withdrawal signs and symptoms. However, this impression may be quite misleading. The data presented in Figs. 5–7 indicate only the frequency of occurrence of the five withdrawal signs and symptoms considered, not the relative severity or persistence of these signs and symptoms. For example, a single occurrence of tremor in one subject contributed one point to the data pool, but subsequent occurrences of tremor in the same subject or exceptionally severe tremor, as judged by the neurological ratings, did not contribute any additional points. The attempts to quantify the occurrence of these signs and symptoms have, of necessity, eliminated qualitative information. Given this limitation, it still appears that a work-related delayed reinforcement paradigm is most comparable to a spontaneous drinking paradigm, an impression that has been confirmed in previous reports (Mello and Mendelson, 1972). A work-contingent immediate reinforcement paradigm, in which some sequence of behavior must be completed before acquisition of a single reinforcement, appears most similar to a programmed drinking regimen.

V. RESEARCH NEEDS

There have been relatively few systematic studies of the alcohol withdrawal syndrome following an experimental drinking paradigm, and consequently, it is difficult to compare data presented in this report with other investigations. Gross and co-workers (1971a,b) have attempted to quantify the alcohol withdrawal syndrome by evaluating 100 consecutive male alcoholic admissions to a general psychiatric hospital with a rating scale, then factor-analyzing the scale patterns. Three factors were identified: (1) nausea, tinnitus, visual disturbances, pruritus, parasthesias, muscle pain, hallucinations (tactile, visual, and auditory) and agitation, which accounted for 27% of the total variance; (2) tremor, sweating, depression, and anxiety, which accounted for 19% of the total variance; and (3) level of consciousness, quality of contact, disturbance of gait and nystagmus, which accounted for 20% of the total variance. Gross (1975) subsequently studied the relationship between blood alcohol levels and withdrawal symptom clusters in 13 subjects given programmed doses of alcohol for 5 or 7 days, for 10 hours each day. A minimum daily blood alcohol level of approximately 50 mg/100 ml was necessary for the development of withdrawal signs and symptoms, a finding comparable to our observations (Fig. 5). Withdrawal factors (1) and (2) were significantly correlated with cumulative blood alcohol concentrations during the 3-day withdrawal period, but factor (3) was not (Gross, 1975). Consequently, blood alcohol levels could account for 46% of the variance in this sample using these procedures and methods of data analysis. The variables that accounted for the remaining 54% of the variance could not be identified.

Clarification of the determinants of the alcohol withdrawal syndrome in its several modes of expression remains to be determined. Blood alcohol levels, duration of drinking, pattern of drinking, medical status, conditioning and learning factors, and expectancy are perhaps only part of the total matrix that affects the expression of alcohol abstinence. These data testify to the fact that alcohol dependence is not an all-or-none phenomenon.

VI. SUMMARY

Aspects of the contribution of years of drinking and volume of alcohol consumed to the development of withdrawal signs and symptoms were examined in 129 alcohol addicts on the basis of self-report data. These findings were compared with the observed pattern of withdrawal signs and symptoms following a period of experimentally induced intoxication. The contribution of average blood alcohol levels and the type of experimental drinking paradigm to actual expression of the withdrawal syndrome was studied. Frequency of occurrence of major withdrawal signs and symptoms as a function of years of alcoholism was also noted.

Volume of alcohol consumed was not a consistent predictor of either the time course of development of alcohol dependence or acute expression of most withdrawal signs and symptoms. Alcohol dependence developed after at least 5 years of problem drinking in the majority of subjects and the time course of development was independent of the alcohol concentration of the preferred beverage, e.g., whiskey vs. wine or beer. Exposure to a spontaneous experimental drinking paradigm resulted in more withdrawal signs and symptoms than the other paradigms. These data suggest that the duration of alcoholism and the pattern of drinking may be more important in determining the expression of withdrawal signs and symptoms than the volume of alcohol consumed. This is inconsistent with most data on animal models of alcoholism.

APPENDIX I

SOCIAL AND DRINKING HISTORY   Name: _____
                                                Date: _____
                                                Interviewer: _____

Age _____ Weight _____   Current Job _____
Race _____   Previous Job _____
Birthplace _____   Marital Stauts _____ No. of Marriages _____
Parent's Birthplace: Children _____ Religion _____
Father _____   Wife's Drinking Pattern: Social, Heavy, Alcoholic, Abstinent
Mother _____   Mother's Drinking Pattern: Social, Heavy, Alcoholic, Abstinent
Residence at time of   Father's Drinking Pattern: Social, Heavy, Alcoholic, Abstinent
admission: _____   Sibs. Drinking Pattern: Social, Heavy, Alcoholic, Abstinent
Education _____

Age of first drink _____   Beverage first drunk _____
Onset of heavy drinking _____ Duration of drinking (Yrs.) ___ Preferred beverage _____
Frequency of drinking: Daily, Weekend, Spree _____ Starts drinking _____
Amount of alcohol drunk per day_____ Finishes drinking _____
Pattern of drinking within a day _____
Usually drinks: Alone, with friends, with strangers _____
Usually drinks: At home, at bar or club, in street or park _____
Any change in drinking pattern? _____ What? _____ When? _____
Memory problems? _____
Intercurrent illness during drinking _____ Every use: Toxic beverage_____ Drugs _____

Withdrawal symptoms? _____ Usually last for _____ days: First began _____ years ago
During withdrawal usually: At home, hospital EW, Jail equivalent _____
Withdrawal symptoms: tremulous, sweating, insomnia, nausea, vomiting, hallucinosis:
  (Visual, auditory _____), seizures, delirium _____
Usual severity of withdrawal symptoms: Mild _____ Moderate _____ Severe _____
Withdrawal usually after what kind of drinking? _____
Attitude about withdrawal symptoms: _____
_____

Attitude about own drinking: _____
Usual effect of alcohol on behavior: _____
Usual effect of alcohol on feelings: _____
Usual effect of alcohol on sleep pattern: _____
Usual effect of alcohol on food intake: _____
While drinking: Mood   + - 0 Socialization + - 0 Self-esteem + - 0 Sleep + - 0
                 Sexuality + - 0 Guilt   + - 0 Anxiety   + - 0
                 Aggression + - 0 Memory   + - 0 Emotionality + - 0
Effects constant throughout drinking spree? _____ Describe _____

Reports more alcohol required to produce these effects than previously? _____
Expectancy about effects of alcohol during this study _____

Attributes drinking problem to specific initiating event?_____
Usually begins drinking episode because _____
Usually terminates drinking episode because _____
Longest period of voluntary abstinence _____ When? _____
Why began? _____ Why ended? _____

APPENDIX II

NEUROLOGICAL RATING FORM: WITHDRAWAL SIGNS AND SYMPTOMS

ORIENTATION:

TIME: Normal   Impaired

  eg: _____

PLACE: Normal   Impaired

  eg: _____

MEMORY:

Recent: Normal   Impaired

  eg: _____

Remote: Normal   Impaired

  eg: _____

HALLUCINATIONS:

Visual: 0 1 2 3

Auditory: 0 1 2 3

Tactile: 0 1 2 3

0 = absence of hallucinations
1 = reported as experienced since
    last exam, but not reported
2 = reported as present during interview
3 = subject distracted by hallucina-
    tions, eg: _____

NYSTAGMUS:

Lateral: 0 1 2 3 sustained

Vertical: 0 1 2 3 sustained

0 = absence; 1 = mild; 2 = major

REFLEXES	LEFT	RIGHT
Knee Jerk	0 1 2 3 4	0 1 2 3 4
Suprapatellar	0 1 2 3 4	0 1 2 3 4
Ankle Jerk	0 1 2 3 4	0 1 2 3 4
Ankle Clonus	0 1 2 3 4	0 1 2 3 4
Biceps	0 1 2 3 4	0 1 2 3 4
Plantar	Normal ABN	Normal ABN

0-1 = hypoactive; 2 = normal;
3-4 = hyperactive

Name: _____
Date: _____
Hour: _____ Exp. Day: _____
Physician: _____
Rater: _____

TREMOR:

Tongue: 0 1 2 3

Hand and Arm: 0 1 2 3

Trunk: 0 1 2 3

0 = absent; 1 = mild; 2 = moderate;
3 = major

FINGER-NOSE (L) Normal   Impaired
            (R) Normal   Impaired

SWEATING: 0 1 2

SYSTOLIC BLOOD PRESSURE: _____

RESPIRATORY RATE: _____

TEMPERATURE: _____

GAIT: Normal   Impaired

TANDEM WALK: Normal   Impaired

RAPID ALTERNATION: (L) Normal
                  (R) Impaired

GASTRITIS: _____

Signature: _____

ANXIETY: 0 1 2 3 self-report
        0 1 2 3 examiner's report

0 = calm
1 = slightly nervous or frightened
2 = very nervous or frightened

INTENSITY OF WITHDRAWAL SYNDROME

Subject Rating: (0 - 10) _____
Rank Order:
#/# of subjects: _____

ACKNOWLEDGMENTS

This research for this chapter was conducted in the Intramural Laboratory of Alcohol Research supported by the National Institute of Mental Health and later by the National Institute on Alcohol Abuse and Alcoholism, ADAMHA. Data analysis and preparation of this chapter were supported by the National Institute on Drug Abuse under Grant DA4RG010. Portions of this chapter have been published in the *McLean Hospital Journal,* 1, 64–88, 1976. We are grateful to the publishers of the Journal for permission to reproduce this material.

REFERENCES

Ellis, F. W., and Pick, J. R. (1970). Experimentally induced ethanol dependence in rhesus monkeys. *J. Pharmacol. Exp. Ther.* **175,** 88–93.

Feinberg, I. (1969). Effects of age on human sleep patterns. *In* "Sleep: Physiology and Pathology, A Symposium" (A. Kales, ed.), pp. 39–52. Lippincott, Philadelphia, Pennsylvania.

Feinberg, I., and Carlson, V. R. (1968). Sleep variables as a function of age in man. *Arch. Gen. Psychiatry* **18,** 239–259.

French, S. W., and Morris, J. R. (1972). Ethanol dependence in the rat induced by non-intoxicating levels of ethanol. *Res. Commun. Pathol. Pharmacol.* **4,** 221–233.

Freund, G. (1969). Alcohol withdrawal syndrome in mice. *Arch. Neurol.* **21,** 315–320.

Goldstein, D. B. (1972). Relationship of alcohol dose to intensity of withdrawal signs in mice. *J. Pharmacol. Exp. Ther.* **180,** 203–215.

Goldstein, D. B. (1974). Rates of onset and decay of alcohol physical dependence in mice. *J. Pharmacol. Exp. Ther.* **190,** 377–383.

Gross, M. M. (1975). Physical dependence and alcohol withdrawal syndrome in man. *Proc. Int. Congr. Pharmacol. 6th, Helsinki.*

Gross, M. M., Rosenblatt, S. M., Chartoff, S., Herman, A., Schachter, E., Sheinkin, D., and Broman, M. (1971a). Evaluation of the acute alcoholic psychoses and related states. The daily clinical course rating scale. *Q. J. Stud. Alcohol* **32,** 611–619.

Gross, M. M., Rosenblatt, S. M., Malenowski, B., Broman, M., and Lewis, E. (1971b). A factor analytic study of the clinical phenomena in the acute alcohol withdrawal syndromes. *Alkohologia* **2,** 1–7.

Gross, M. M., Lewis, E., and Hastey, J. (1974). Acute alcohol withdrawal syndrome. *In* "The Biology of Alcoholism, Vol. 3, Clinical Pathology" (B. Kissin and H. Begleiter, eds.), pp. 191–263. Plenum, New York.

Isbell, H., Fraser, H., Wikler, A., Belleville, R., and Eisenman, A. (1955). An experimental study of the etiology of rum fits and delirium tremens. *Q. J. Stud. Alcohol* **12,** 1–33.

Lettsom, J. C. (1787). Some remarks on the effects of Lignum Quassil Amare. *Mem. Med. Soc. London* **1,** 151–165.

Majchrowicz, E. (1975). Induction of physical dependence on alcohol and the associated metabolic and behavioral changes in the rat. *Psychopharmacologia* **43,** 245–254.

Mello, N. K. (1972). Behavioral studies of alcoholism. *In* "The Biology of Alcoholism, Vol. 2, Physiology and Behavior" (B. Kissin and H. Begleiter, eds.), pp. 219–291. Plenum, New York.

Mello, N. K. (1973). Short-term memory function in alcohol addicts during intoxication. *In* "Alcohol Intoxication and Withdrawal: Experimental Studies" (M. M. Gross, ed.), pp. 333–344. Plenum, New York.

Mello, N. K., and Mendelson, J. H. (1970a). Experimentally induced intoxication in alcoholics: A comparison between programmed and spontaneous drinking. *J. Pharmacol. Exp. Ther.* **173,** 101–116.

Mello, N. K., and Mendelson, J. H. (1970b). Behavioral studies of sleep patterns in alcoholics during intoxication and withdrawal. *J. Pharmacol. Exp. Ther.* **175**, 94–112.

Mello, N. K., and Mendelson, J. H. (1972). Drinking patterns during work-contingent and non-contingent alcohol acquisition. *Psychosom. Med.* **34**, 139–164.

Mello, N. K., and Mendelson, J. H. (1976). The development of alcohol dependence: A clinical study. *McLean Hosp. J.* **1**(2), 64–88.

Mello, N. K., and Mendelson, J. H. (1978). Clinical aspects of alcohol dependence. *In* "Handbook of Experimental Pharmacology: Drug Addiction" (W. R. Martin, ed.). Springer-Verlag, Berlin and New York.

Mendelson, J. H. (ed.). (1964). Experimentally induced chronic intoxication and withdrawal in alcoholics." *Q. J. Stud. Alcohol* Suppl. No. 2.

Mendelson, J. H. (1971). Biochemical mechanisms of alcohol addiction. *In* "The Biology of Alcoholism, Vol. 1, Biochemistry" (B. Kissin and H. Begleiter, eds.), pp. 513–544. Plenum, New York.

Mendelson, J. H., Ogata, M., and Mello, N. K. (1969). Effects of alcohol ingestion and withdrawal on magnesium states of alcoholics: Clinical and experimental findings. *Ann. N.Y. Acad. Sci.* **162**, 918–933.

Pearson, S. B. (1813). Observations on brain fever; delirium tremens. *Edinburgh Med. Surg. J.* **9**, 326–332.

Steinglass, P. (1975). The simulated drinking gang: An experimental model for the study of a systems approach to alcoholism. A description of the model. *J. Nerv. Ment. Dis.* **161**, 100–109.

Sutton, T. (1813). "Tracts on Delirium Tremens on Periodonitis and Other Inflammatory Afflictions." Thomas Underwood, London.

Victor, M., and Adams, R. D. (1953). The effect of alcohol on the nervous system. *Res. Publ. Assoc. Nerv. Ment. Dis.* **32**, 526–573.

Wikler, A. (1965). Conditioning factors in opiate addiction and relapse. *In* "Narcotics" (D. M. Wilner and G. G. Kassebaum, eds.), pp. 85–100. McGraw-Hill, New York.

Wikler, A., and Pescor, F. T. (1967). Classical conditioning of a morphine abstinence phenomenon, reinforcement of opioid-drinking behavior and "relapse" in morphine-addicted rats. *Psychopharmacologia* **10**, 255–284.

Wolfe, S. M., and Victor, M. (1971). The physiological basis of the alcohol withdrawal syndrome. *In* "Recent Advances in Studies of Alcoholism" (N. K. Mello and J. H. Mendelson, eds.), pp. 188–199. U.S. Gov. Print. Off., Washington, D.C.

Wolfe, S. M., Mendelson, J., Ogata, M., Victor, M., Marshall, W., and Mello, N. (1969). Respiratory alkalosis and alcohol withdrawal. *Trans. Assoc. Am. Physicians* **82**, 344–352.

NUTRITIONAL STATUS OF ALCOHOLICS BEFORE AND AFTER ADMISSION TO AN ALCOHOLISM TREATMENT UNIT

Richard D. Hurt, Ralph A. Nelson, E. Rolland Dickson, John A. Higgins, and Robert M. Morse

I. Introduction: Skid Row Alcoholic versus the More Common Alcoholic   397
II. The Present Study 399
III. Methods for Nutritional Data 400
IV. Results 401
    A. Alcohol Intake 401
    B. Blood Alcohol Concentration 401
    C. Ideal Weight 402
    D. Nutrient Intake on Admission to ADDU Observed Nutrient Intake
       While in ADDU 402
    E. Recommended Dietary Allowances 405
V. Conclusions 407
    References 407
    Editorial Comment 408

I. INTRODUCTION: SKID ROW ALCOHOLIC VERSUS THE MORE COMMON ALCOHOLIC

The general topic of alcoholism has become somewhat more in vogue in recent years as funds and efforts at the national, state, and local levels have increased. The fact that alcohol abuse accounted for 10% of the total health care expenditures in the United States in 1971 would seem to justify practically any research effort in this field (1). Of particular interest to us has been the nutritional status of the alcoholic patient. Until recently it was felt, and is still felt by many, that the average alcoholic is malnourished. Most investigators and practitioners take this even one step further and state that many alcohol-associated illnesses are, in part, a result of nutritional deficiencies. Reports in the literature allude to the poor nutritional status of alcoholics and cite the incidence of deficiency-associated illnesses as the basis for the assumption that the alcoholic population is under-nourished. Unfortunately, almost all of these types of studies were done on patients with evidence of alcohol-associated illnesses and, therefore, are a group selected from the much larger total alcoholic patient population. The skid row alcoholic comprises only 5% of the total alcoholic population in the United States, nevertheless this has been the most thoroughly studied subgroup of alcoholics.

Patek *et al.* in 1941 reported on 54 patients with alcoholic cirrhosis that were noted to be a derelict patient population (2). They concluded that (i) There was a significant relationship between the occurrence of nutritional deficiency and cirrhosis, and (ii) the clinical course of these cirrhotic patients was more favorable to the group treated with diet high in protein and B vitamins. Almost 25 years later Patek *et al.* reported on 304 hospitalized, alcoholic patients and concluded that the diets of these patients were poor when judged by traditional standards (3). They found that protein constituted roughly 6% of the total calories, which is approximately one-third to one-half the recommended intake. The mean total caloric intake was between 3200 and 3500 with alcohol contributing 51–58%. Again, the majority of these patients was noted to be from a "lower-middle socioeconomic class." This group was selected from patients admitted to a general hospital and had other medical problems that prompted their admission. Almost two-thirds of the patients in this study were found to be cirrhotic, indicating that the patients were highly selected. Lelbach has shown that only 12% of alcoholic patients develop cirrhosis (4).

In reviewing the literature, there have been few studies performed on the average alcoholic patient, who comprises 90–95% of the 9 million alcoholics in the United States. There have been even fewer studies done on the nutritional status of this group of patients. Nelville *et al.* studied a group of alcoholics that were admitted to the hospital for treatment of their alcoholism (5). This group was felt to be more representative of the usual alcoholic because their entrance into the hospital was voluntary for alcoholism treatment and not because of severe medical problems. Out of this group, 34 patients were intensely studied with regard to their historical nutritional status. The mean caloric intake was 2710 ± 933 for men and 2578 ± 1459 for females, with alcohol representing 36.4 and 22% of calories, respectively. Males consumed a mean of 68 ± 32 gm of protein per day and females 72 ± 47 gm of protein per day, which accounted for 10 and 11.2% of total calories, respectively. The authors concluded "results do not support the view that the nutritional status of alcoholics is markedly inferior to that of nonalcoholics, particularly those of similar economic and health histories.''

The reliability of the alcoholic patient's dietary history was investigated by Eagles and Longman (6). It is generally felt that because the alcoholic cannot be relied on to give an accurate assessment of his alcohol intake, this would apply to food intake as well. Eagles and Longman found this not to be the case in a group of 28 alcoholic patients that had careful dietary histories taken by a nutritionist or research dietition. A dietary history was obtained by identical means from the spouse or other family member involved in the patient's food preparation. When the dietary histories of these two groups were compared, there was no evidence from which to conclude that the groups were different. One would conclude from this that the alcoholic patient was not defensive about his food intake and was able to give a fairly accurate assessment of his own dietary intake.

At the same time, one cannot make a similar statement regarding the alcohol intake of the alcoholic patient. The alcoholic patient's history of his alcohol intake is almost certainly an underestimate of the actual intake. Unfortunately, because of the nature of the illness, there is no one else that can give a better assessment of the alcoholic's alcohol intake than the patient himself. Therefore, we must utilize the available information concerning the daily alcohol intake, keeping in mind its questionable accuracy. In our study, we tried to minimize this reporting error by the method that the information was acquired. However, we still view these data as being very subject to reporting error.

The direct, toxic effect of alcohol on various body tissues, regardless of the presence of adequate nutrition, has been well documented by the work of Lieber and his associates over the past 10 years. In review, Lieber alludes to the direct toxicity of alcohol on human tissues, but the nagging question concerning the contribution of the alcoholic's nutritional status continues to arise (7). It was our impression that the nutritional status of the usual alcoholic patient had not been thoroughly investigated and, therefore, we undertook this study.

II. THE PRESENT STUDY

The patients in our study were admitted for treatment of their alcoholism. They were admitted to the Alcoholism and Drug Dependence Unit (ADDU) of the Rochester Methodist Hospital, which is the major chemical dependency treatment facility for Mayo Clinic patients. I might begin this chapter with what we feel is one of the key points of this study and of major importance. That is, that this group of consecutively admitted alcoholics were not ''skid row'' alcoholics, but rather represented a broad cross section of middle class patients. Table I is a breakdown by occupation of our group of alcoholic patients. Only three of our patients were unemployed, though several were in jeopardy of losing their jobs dependent on the outcome of their treatment for alcoholism.

Table I. Occupational Classification of 62 Alcoholic
Patients Admitted to ADDU

Blue collar ($N = 23$)
White collar ($N = 15$)
Housewife ($N = 10$)
Professional ($N = 7$)
Retired ($N = 3$)
Unemployed ($N = 3$)
Student ($N = 1$)

III. METHODS FOR NUTRITIONAL DATA

Sixty-two consenting alcoholic patients admitted consecutively to the ADDU were included in the study. The only factors that excluded a patient from entry to the study were the following: (i) patient was on a special diet, i.e., diabetic, low salt, etc.; and (ii) the patient was transferred from another hospital after hospitalization of greater than 24 hours.

After entry to the study, the patient had a complete medical history and examination. One patient left the Unit against medical advice before the examination could be completed. A dietary history was obtained on 58 of the 62 patients. The dietary histories were obtained on each patient by the same clinical dietition. The patients had their dietary history taken within 48–72 hours after admission. From the dietary history, an average daily intake was calculated for the various nutrients utilizing USDA Handbook No. 8 (8). While the patients were in the Unit, their dietary intake was recorded for each meal and snack. Based on the food portion weight and USDA Handbook No. 8, a detailed breakdown of the nutrients was made, and from this an average daily in Unit intake was calculated.

The history of the patient's alcohol intake was obtained from the patient by several people including the principal investigator, the psychiatrist, the psychiatric fellow, the counselor, and the dietition. From this pooled information, an estimate was made for each patient regarding the daily intake of alcohol and the duration of the intake. A further breakdown was made estimating the patient's intake 0–3 months before admission, 4–6 months, and 7–12 months prior to admission. We did this not only to see if there was a trend toward heavier drinking prior to admission, but also to try to determine if the patient's overall assessment was accurate. As noted in the Rand Report, the reliability of reported frequency of drinking is probably good in the alcoholic patient (9). These authors, also, felt that the reliability of the quantity of alcohol intake for the alcoholic population was adequate, especially on entry to treatment. However, they point out their reservations regarding these data. We share these reservations, particularly when the authors note that "perhaps 10–15 percent of alcoholics who have been drinking underreport to such an extent that they might be incorrectly classified as non-alcoholic." It would be our conclusion that the alcoholic's reported alcohol intake is usually underreported by the patient and that the only way to arrive at an accurate figure for this patient group would be constant, direct observation over a substantial period of time. Since this is an impossibility, we must take the available information, but keep in mind that it may be underestimated by as much as 50%.

IV. RESULTS

Of the 62 patients admitted to the study, 17 (27%) were female and 45 (73%) were male with a mean age of 45 years. From a clinical standpoint, there was no evidence of nutritional deficiency in any of the patients. However, one patient had a severe organic brain syndrome that was felt to be due to diffuse cerebral atrophy. Sixty-three percent of the patients had evidence of alcoholic liver disease on laboratory testing, and nine patients had heptomegaly. Twenty-six percent had an elevated mean corpuscular volume (MCV) without a concomitant anemia. Twenty-two percent of the patients had abnormally elevated serum triglyceride levels with only 3% having abnormally elevated cholesterol.

A. Alcohol Intake

Beverage selection showed a wide range of alcoholic beverages used. The beverage of choice in over three-fourths of the patients was either beer, vodka, or whiskey. Table II shows a breakdown of this by percentages.

Of the 62 patients, 57 were felt to give a reasonable and consistent history of their alcohol intake over the 12 months prior to admission. Table III is a breakdown of the alcohol intake history for the year leading up to the admission to the ADDU.

Table II. Beverage Selection of 62 ADDU Patients

Selection	Percentage
Beer	27
Vodka	27
Whiskey	24
Scotch	3
Brandy	2
Gin	2
Combinations (Two or more beverages per day)	10
Totally unreliable or no information	5

Table III. Alcohol Intake History for 57 Patients for the Year Preceding ADDU Admission[a]

	0–3 Months	4–6 Months	7–12 Months
Mean	199	149	134
Median	178	134	107

[a] Data in grams per day.

B. Blood Alcohol Concentration

On entry to the Unit, blood alcohol concentration was obtained on 59 patients, and in 32 (54%), the blood alcohol concentration was zero. Of the 27 patients with detectable blood alcohol concentration on admission, the mean value was 193.6 mg/100 ml with the highest recorded value being 413 mg/100 ml. Since less than 50% of those patients with blood alcohol concentrations available had detectable levels, no conclusion can be drawn relating the blood alcohol concentration to the reliability of alcohol intake history, as was noted in the Rand Report (9).

C. Ideal Weight

When we compared our group of patients with the ideal weight ranges based on age, sex, and frame from the Metropolitan Life Insurance Company, we found that 57% (35) were over their ideal weight, 31% (19) were within accepted ranges, and 12% (7) were under their ideal weight (10). This result is presented in Table IV.

Table IV. Comparison of Weight of 61 ADDU Patients to "Ideal Weight" from Metropolitan Life Insurance Company[a]

	Below weight	Normal weight	Above weight	Total
Females N	3	3	10	16
Row (%)	19	19	62	
Males N	4	16	25	45
Row (%)	9	36	55	
Total N	7	19	35	61
Row (%)	12	31	57	

[a] One patient's weight is missing.

Table V. Protein, Fat, and Carbohydrate Intake in Grams

Total intake[a]	Means			t	$P\ (\mu_\Delta=0)$
	Hx	ADDU	$\Delta=$ADDU-Hx		
Protein (gm)	86.25	99.50	13.25	2.9	0.01
Fat (gm)	104.89	123.44	18.55	3.4	0.001
Carbohydrate (gm)	177.75	299.75	122.00	9.7	<0.001

[a] $N = 58$.

D. Nutrient Intake on Admission to ADDU/Observed Nutrient Intake While in ADDU

Table V shows the evaluation of our dietary data on 58 patients. Four patients either left the Unit against medical advice prior to the accumulation of the necessary information or were unable to give a dietary history because of an organic brain syndrome.

Another way of looking at the same data is to observe the calories provided by each of the nutrient categories as depicted in Table VI.

When the mean of the nonalcohol total calories from the dietary history is compared to the actual observed mean total caloric intake while the patients were in the Unit, there is a significant difference. In fact, all categories of food show a significantly higher intake in the ADDU. The largest increase takes place in the CHO foods and this accounts for greater than two-thirds of the caloric gain. Bebb and Howser have noted similar trends even in nonalcoholic drinkers (11). That is, ingested alcohol seemed to have little effect on the amount of protein and fat ingested but did have a fairly profound effect on carbohydrate intake.

When one adds alcohol to the total calories, we get a somewhat different picture. The conversion factor 7.1 kcal/gm of alcohol is used and added to the calories from other foods to obtain the data in Table VII.

From the 53 patients on whom we had reasonably reliable information on their calories from alcohol, as well as dietary histories, we calculated the contribution that alcohol made to their total caloric intake. In our group, alcohol had accounted for 34.9% of the patient's daily caloric intake. This is similar to the results noted by Neville *et al.*, but significantly less than that noted by Patek. Again, we must note that the patient population included in Patek's study more closely aligned the "skid row" alcoholic, whereas Nelville *et al.* and our group of patients is more representative of the general alcoholic population. Table VIII summarizes the percentage of daily calories due to alcohol in our patients.

Table VI.   Total (Nonalcohol) Calories, Protein Calories, Fat Calories, and Carbohydrate Calories

Calorie intake[a]	Means			t	P ($\mu_\Delta = 0$)
	Hx	ADDU	Δ = ADDU-Hx		
Total (nonalcohol)	2029	2661	632	6.5	<0.001
Protein	345	398	53	2.9	0.01
Fat	944	1111	167	3.4	0.001
Carbohydrate	711	1199	488	9.7	<0.001

[a] $N = 58$.

Table VII.   Total (Alcohol + Food) Calories[a]

	Means			t	P ($\mu_\Delta = 0$)
	Hx	ADDU	Δ = ADDU-Hx		
Means for total calories: ($N = 53$)	3111	2633	−478	−3.9	<0.001

[a] Fifty-three patients had information on their past calories from alcohol and also had dietary histories.

Table VIII.   Percentage of Average Daily Caloric Intake from Alcohol in 53 Patients on Admission to ADDU[a]

Percentage of daily calories due to alcohol[b]	Frequency	Percentage	Cumulative frequency	Cumulative (%)
0–9	1	1.9	1	1.9
10–19	7	13.2	8	15.1
20–29	13	24.5	21	39.6
30–39	13	24.5	34	64.1
40–49	9	17.0	43	81.1
50–59	8	15.1	51	96.2
60–69	0	0.0	51	96.2
70–79	2	3.8	53	100.0

[a] $N = 53$; $\bar{x} = 34.9\%$; SD = 15.0; range, 3–73.
[b] Percentage of daily calories due to alcohol = alcohol calories/(alcohol + food calories).

One must remember that the grams per day of alcohol intake is an estimate given by the alcoholic patient and most likely represents an underestimate. However, even with this high probability of underreporting, our patients show they ingest an adequate number of calories per day. We looked selectively at protein intake in our patients and expressed this in terms of grams per body weight in kilograms. When we analyzed the data in this manner, there were six patients who, according to their dietary history, were eating less than 0.6 gm of protein per kilogram of body weight per day prior to admission. There were three other patients whose dietary history indicates they were eating between 0.6 and 0.8 gm/kg body weight per day. However, 40 patients noted that they ate greater than 0.8 gm of protein per kilogram of body weight per day. While in the ADDU, all patients ate quantities of protein exceeding 0.6 gm/kg of body weight per day. Therefore, looking selectively at the average daily intake from dietary history does not tell the entire story. Because of this we elected to compare the nutrient intake of our patients to the Recommended Dietary Allowances (RDA) in order that we might get a better idea of the nutritional status of the individual patients.

E. Recommended Dietary Allowances

"The Recommended Dietary Allowances are the levels of intake of essential nutrients considered, in the judgement of the Food and Nutrition Board on the basis of available scientific knowledge, to be adequate to meet the known nutritional needs of practically all healthy persons" (12). We compared the nutrient intake of our patients with the Recommended Dietary Allowances as prepared by the Food and Nutrition Board of the National Research Council (12). When we compared the mean values of our patients to the RDA for the separate nutrient categories, the means were in acceptable ranges. However, as is noted by the authors of the RDA, this is an inappropriate use of the RDA since there is a wide range of individual requirements. We, therefore, have taken the RDA for the nutrient categories and divided them into the following percentages. Table IX points out several interesting aspects of our patients. There was only a small percentage of patients that were below 50% RDA for protein and calories on their dietary history. Overall, there was a trend of the patients toward the higher percentage of RDA side of the table. The wide variability of the individual nutrients is not totally unexpected, since there is a wide variation of nutrient intake among individual patients. It was surprising to find such a large number of patients below 50% of RDA for vitamin A, ascorbic acid, thiamine, and niacin. We have no explanation for this, but further evaluation of the data is in progress. There was a shift to the higher percentage of RDA when we made the same analysis of the in Unit intake.

Table IX. Distribution of Nutrient Intake Expressed as a Percentage of RDA Historically and While in the ADDU

Nutrient	N	Nutrient intake (percent of RDA)					
		<25	25-49	50-74	75-99	100-149	≥150
Total calories							
Hx[a]	53	0	0	8	20	42	30[b]
ADDU	61	0	0	8	28	62	2
Protein							
Hx	58	2	2	3	3	33	57
ADDU	61	0	0	0	0	20	80
Calcium							
Hx	58	5	5	21	10	28	31
ADDU	61	0	2	11	8	27	52
Phosphorous							
Hx	58	2	2	2	18	24	52
ADDU	61	0	0	0	2	21	77
Iron							
Hx	58	0	2	24	15	33	26
ADDU	61	0	2	13	6	38	41
Vitamin A							
Hx	58	2	19	17	21	25	16
ADDU	61	0	0	6	2	18	74
Thiamine							
Hx	58	2	7	25	19	33	14
ADDU	61	0	0	5	23	65	7
Riboflavin							
Hx	58	0	3	7	12	37	41
ADDU	61	0	0	0	10	23	67
Niacin							
Hx	58	0	9	20	37	27	7
ADDU	61	0	0	5	46	47	2
Ascorbic acid							
Hx	58	2	17	9	10	21	41
ADDU	61	0	0	0	0	3	97

[a] Calories from alcohol are included.
[b] Percentage of 53 patients with total caloric intake ≥150% of RDA.

V. CONCLUSIONS

1. The 62 alcoholic patients in our study are representative of the general alcoholic patient population.

2. Over 80% of our patients met or exceeded their ideal weight.

3. The mean values for calories, fat, carbohydrate, protein, and other nutrients calculated from the dietary histories were in acceptable ranges.

4. By comparing the individual dietary intake to the RDA, we found there are a significant number of our patients who fall below acceptable levels for various nutrients.

5. The nutrient intake of the majority of our alcoholic patients met or exceeded the nutrient requirements as defined by RDA.

ACKNOWLEDGMENT

We are most grateful to Cheryl Bodding Stensvad, who as our research assistant made the project possible. We also are most appreciative of the statistical work performed by Kenneth Offord and Erik Bergstralh.

REFERENCES

1. Berry, R. E., Jr. Estimating the economic costs of alcohol abuse. *N. Engl. J. Med.* **295,** 620–621 (1976).
2. Patek, A. J., Jr., and Post, J. Treatment of cirrhosis of the liver by a nutritious diet and supplements in vitamin B complex. *J. Clin. Invest.* **20,** 481–505 (1941).
3. Patek, A. J., Jr., Toth, I. G., Saunders, J. G., Castro, G. A. M., and Engel, J. J. Alcohol and dietary factors in cirrhosis. *Arch. Intern. Med.* **135,** 1053–1057 (1975).
4. Lelbach, W. K. Cirrhosis in the alcoholic and its relation to the volume of alcohol abuse. *Ann. N.Y. Acad. Sci.* **252,** 85–105 (1975).
5. Nelville, J. N., Eagles, J. A., Samson, G., and Olson, R. E. Nutritional status of alcoholics. *Am. J. Clin. Nutr.* **21,** 1329–1340 (1968).
6. Eagles, J. A., and Longman, D. Reliability of alcoholics' reports of food intake. *J. Am. Diet. Assoc.* **42,** 136–139 (1963).
7. Lieber, C. S. The metabolic basis of alcohol's toxicity. *Hosp. Pract.* pp. 73–80 (1977).
8. USDA. Composition of foods. *U.S. Dep. Agric., Agric. Handb.* No. 8, 1964.
9. Armor, D. J., Polich, J. M., and Stambul, H. B. "Alcoholism and Treatment," pp. 165–166. Rand Corp. Santa Monica, 1976.
10. *Stat. Bull. Metrop. Life Insur. Co.* **40,** Nov.–Dec. (1959).
11. Bebb, H. T., and Howser, H. B. Caloric and nutrient contribution of alcoholic beverages to the usual diets of 155 adults. *Am. J. Clin. Nutr.* **24,** 1042 (1971).
12. Food and Nutrition Board, National Research Council. "Recommended Dietary Allowances." Natl. Acad. Sci., Washington, D.C., 1974.

EDITORIAL COMMENT

The contrast in nutriture of the groups of alcoholics studied by Hurt *et al.* and by Wood and Breen in the following chapter, "Thiamine Status of Australian Alcoholics," clearly underscores the essentiality of assessing the nutriture of individual patients and those within various settings, in order appropriately to manage their rehabilitation. Alcoholics exhibit a wide range of nutritures, but the indigent alcoholic or the chronic alcoholic with clinically evident complications very frequently exhibits clear evidence of one or more nutrient deficiencies. Such nutrient deficiencies may reflect the deficit of food intake or deficit of the intake of a specific group of foodstuffs resulting from the generally poor quality of the individual's diet and its further deterioration through displacement of an increased percentage of the calories by nutrient-poor alcoholic beverages. Classic examples of this mechanism include alcoholic beriberi, alcoholic pellagra, many cases of Wernicke's syndrome, and some instances of nutritional anemia among alcoholics. Excessive alcohol intake and the consequences of it may also markedly alter absorption, excretion, tissue storage, or transport of a variety of nutrients and thereby contribute to clinical and/or biochemical evidences of deficiencies of inorganic as well as organic nutrients. These nutrients include magnesium, zinc, calcium, iron, folic acid, retinol, and glucose. These phenomena are well reviewed by Halsted (1), by Somogyi and Kopp (2,3), and, in relations to trace elements, in the monograph edited by Ananda S. Prasad (4).

Because of these phenomena, it is essential critically to define the relative roles of possible specific nutrient deficiencies and of the toxic effects of alcohol per se relating to phenomena manifest by heavy drinkers such, for example, as in assessing the role of alcohol vs. malnutrition in the so-called fetal alcohol syndrome.

REFERENCES

1. Halsted, C. H. Nutritional implications of alcohol. *In* "Present Knowledge in Nutrition," Nutrition Reviews, 4th ed., pp. 467–477. Nutr. Found., New York, 1976.
2. Somogyi, J. C., and Kopp, P. M. Alkoholismus und Ernährungsstatus. *Bibl. Nutr. Dieta* **24,** 17–31 (1976).
3. Somogyi, J. C., and Kopp, P. M. Alcohol and nutritional status. *Proc. Int. Congr. Nutr., 9th, Mexico, D.F.* **1,** 212–224 (1972).
4. Prasad, A. S., ed. "Trace Elements in Human Health and Disease," Vols. 1 and 2. Academic Press, New York, 1976.
5. Turner, T. B., Mezey, E., and Kimball, A. W. Measurement of alcohol-related effects in man: Chronic effects in relation to levels of alcohol consumption. Part A. *Johns Hopkins Med. J.* **141,** 235–248 (1977).

26

THIAMINE STATUS OF AUSTRALIAN ALCOHOLICS

Beverley Wood and Kerry J. Breen

I.	Introduction	409
II.	Clinical Studies	411
III.	Biochemical Studies	412
IV.	Etiology of Thiamine Deficiency in Alcoholism	419
	A. Low Thiamine Intake	419
	B. Impaired Thiamine Absorption	419
	C. Impaired Thiamine Utilization	419
V.	Thiamine Status of the Australian Population	421
VI.	Summary	422
	References	423
	Editorial Comment	426

I. INTRODUCTION

The original Australian, the aborigine, was a hunter–gatherer of food and did not know alcohol, which was introduced in 1788 when the First Fleet landed in Sydney Cove, New South Wales. The first settlers (800 convicts and 160 seamen and officers) sought refuge in alcohol presumably to forget the loneliness, the heat, and other deprivations of a way of life very different to what they had known in Britain. Grapes were one of the few crops to flourish in the harsh conditions and were first picked from the Governor's garden in 1791. Rum was used as a currency in the early days of the colony and by 1823 four breweries were operating in Sydney when the population was only 31,000. Now, nearly two hundred years later, chronic alcoholism is a major public health problem. It has been ranked fourth in importance after heart disease, mental disorders, and

Table I. Apparent per Capita Alcohol Consumption in Australia[a]

	1958–1959	1971–1972	1974–1975
Beer (liters)	103.2	127.5	142.7
Wine (liters)	5.0	9.0	12.5
Spirits (liters of alcohol)	0.8	1.1	1.2

[a] Australian Bureau of Statistics (1977).

cancer (Dax, 1968). A recent survey indicated that alcoholism affects 5% of adult males and 1% of adult females in the population (Krupinski and Stoller, 1971). As the per capita consumption of alcohol by Australians has been increasing over the last 25 years (Table I), it is almost certain that the problems relating to alcoholism are also increasing (Lederman, 1954; Rankin, 1974).

The alcohol problem is reflected in the admission rates of alcoholics to large city general hospitals, estimated in 1965 to be 11.3% at this institution in Melbourne (Green, 1965) and the same in Brisbane (Smithurst, 1965). In 1976, a 1-day survey of the inpatient population at this institution revealed that of 469 inpatients, 5% admitted to drinking 80–120 gm ethanol per day, and 12.4% admitted drinking more than 120 gm ethanol per day (P. Rennie, unpublished work). These levels of alcohol intake represent excessive drinking.

A detailed profile of 1000 alcoholics attending the Alcoholism Clinic at this hospital has been obtained (Wilkinson *et al.,* 1969). This profile indicates that in Australia male alcoholics start to drink regularly in their late teens, excessively in the mid- or late twenties, and present in their early forties consuming an average of 220 gm ethanol per day. By comparison, the female alcoholics start to drink later and initially moderately, but then drink excessively for a shorter period of time before presenting for treatment, at which time they were consuming an average of 155 gm ethanol daily. Most subjects were continuous drinkers of beer (containing 3.8 gm ethanol per 100 ml). All social classes were represented and one in seven were women.

A later profile obtained by Wilkinson *et al.* (1971) also provided data on physical complications of alcohol abuse. These data suggested that the problems of "gastritis," peptic ulcer, and automobile accidents were associated with relatively short histories of heavy alcohol intake, while alcoholic liver disease and more especially dementia, were seen much later in the course of the alcoholism. Nutritional status was not studied in Wilkinson's survey, but it is clear that the excessive drinker is at risk of developing primary and/or secondary nutrient deficiencies (Wood, 1975; Halsted, 1976). Indeed, nutrient deficiency is probably inevitable in severe chronic alcoholism where normal social functioning has been destroyed.

In individual subjects a precise diagnosis of nutritional status is very difficult to make because of the multiple nature of the deficiencies. Biochemical investigation is rarely undertaken and treatment is rapidly started, particularly if there is evidence of Wernicke's encephalopathy, which is potentially the most catastrophic of the nutrient deficiencies seen in chronic alcoholics in Australia.

This cerebral disorder, which occurs in association with severe thiamine deficiency (Dreyfus, 1962; Embree and Dreyfus, 1963), is characterized by altered mental state (particularly confusion and apathy), ophthalmoplegia, and ataxia (Jolliffe *et al.*, 1941). It is closely associated with and frequently heralds the onset of Korsakoff's psychosis and the two are now usually referred to as the Wernicke–Korsakoff syndrome (Victor *et al.*, 1971).

Our studies have focused particularly on the nutritional status of a similar alcoholic population to that described above from the same institution. Several detailed studies of clinical and biochemical assessment of thiamine status have been undertaken and are summarized below. The majority of the subjects studied were alcoholics admitted to hospital for medical reasons and most had drunk more than 200 gm ethanol daily in the form of beer for more than 20 years (Wood *et al.*, 1977).

II. CLINICAL STUDIES

The clinical diagnosis of thiamine deficiency is difficult since the early symptoms are nonspecific (Williams, 1961; Sauberlich, 1967) and it is often accompanied by other B vitamin deficiencies in people living in industrialized countries (Leevy *et al.*, 1965). The World Health Organization has recommended the use of the physical signs of loss of ankle and knee jerks, sensory loss and motor weakness, calf muscle tenderness, cardiovascular dysfunction, and edema in nutrition surveys (World Health Organization, 1963), but these are of little value in chronic alcoholics where all these abnormalities can have other explanations. In this population one has to resort to grosser evidence of thiamine deficiency such as Wernicke's encephalopathy, beriberi heart disease, or peripheral neuropathy (the last interpreted cautiously in view of its multiple etiologies).

Using these markers, clinical thiamine deficiency was diagnosed in 82 (38.1%) of 215 alcoholic patients (Wood *et al.*, 1977). Of these, 24 patients (11.1%) had Wernicke's encephalopathy,* 15 (7.0%) had beriberi heart disease† (presumed to be beriberi but alcoholic cardiomyopathy could not be completely excluded), and 43 (20.0%) had peripheral neuropathy alone. Since peripheral neuropathy may have been due to other causes, these figures viewed alone may represent an overestimate of the incidence of thiamine deficiency. However,

---

*Thirteen patients also had peripheral neuropathy.
†One patient also had peripheral neuropathy.

biochemical assessment (see below) strongly supports the existence of thiamine deficiency of this order of magnitude.

In a clinical study of dementia in alcoholics, also conducted at this hospital, 100 of 1100 alcoholics admitted had dementia. Of the 100, 26 showed evidence of Wernicke's encephalopathy and 20 had Korsakoff's psychosis preceded by Wernicke's encephalopathy (Horvath *et al.,* 1969). The majority of the remainder were shown to have cerebral atrophy. These demented patients had drunk more heavily and had a higher incidence of malnutrition when compared to the nondemented alcoholics in the same group. However, biochemical assessment of vitamin status was not undertaken.

Data regarding the incidence of the Wernicke-Korsakoff syndrome in the Australian population cannot be obtained because the national classification of morbidity and mortality follows that of the World Health Organization (1967), which categorizes Wernicke's encephalopathy under vitamin B deficiency—unspecified, 263.9. There is an urgent need to reclassify the Wernicke-Korsakoff syndrome as a separate item under thiamine deficiency, enabling collection of data on the incidence of clinical thiamine deficiency in Australian alcoholics and comparison of these statistics with those of other countries.

III. BIOCHEMICAL STUDIES

Assessment of thiamine status by biochemical means is more objectively made than by clinical means (Sauberlich, 1967). The most widely accepted method of assessment is by the erythrocyte transketolase assay first developed by Brin *et al.* (1960). In using this assay we chose a slight modification (Wood and Penington, 1973) of the sedoheptulose method published by Schouten and colleagues (1964). The assay procedure measures erythrocyte transketolase activity (ETKA), erythrocyte transketolase activity following the addition of thiamine pyrophosphate (TPP) *in vitro* (ETKA + TPP), and the percentage enhancement of erythrocyte transketolase activity by TPP *in vitro* (the TPP effect).

The data obtained in 215 alcoholic patients* are shown in Tables II and III and Figs. 1–3. Of the 24 patients with Wernicke's encephalopathy, a high TPP effect was seen in only 13 (Table III, Fig. 3). The striking difference between this clinical diagnosis and the laboratory assessment by the TPP effect stimulated closer analysis of the data. Thus, it was found that in 21 of the 24 patients with Wernicke's encephalopathy, ETKA levels were low. There was no statistical difference between the clinical evidence of Wernicke's encephalopathy and the incidence of thiamine deficiency as indicated by ETKA levels. Because of this

---

*The results in a further 50 alcoholics are included in Table II but are not discussed as they were found to have had recent vitamin therapy.

Table II. Results of Erythrocyte Transketolase Assay and Urinary Thiamine Excretion in Alcoholic Patients[a]

Values[b] in apparently healthy hospital staff	All patients studied (265)	Patients with known recent vitamin therapy (50)	Patients without recent vitamin therapy[c] (215)
ETKA (mean ± SD)	80.7[d] ± 31.4	99.0 ± 27.4	76.7 ± 30.9
Low (<59 IU)	72 (27.2%)	2 (4.0%)	70 (32.6%)
Normal (59-125.4 IU)	171 (64.6%)	39 (78.0%)	132 (61.5%)
High (>125.4 IU)	22 (8.3%)	9 (18.0%)	13 (6.0%)
ETKA + TPP (mean ± SD)	90.2[d] ± 30.8	103.9 ± 27.2	87.3 ± 30.8
Low (<61.5 IU)	49 (18.5%)	2 (4.0%)	47 (21.9%)
Normal (61.5-140.3 IU)	201 (75.8%)	43 (86.0%)	158 (73.5%)
High (>140.3 IU)	15 (5.7%)	5 (10.0%)	10 (4.7%)
Percent TPP effect median	8.9	4.7	14.3
Range	−8.5-739	−2.1-42.7	−8.5-739
Values <24.0%	210 (79.2%)	48 (96.0%)	162 (75.3%)
Values >24.0%	55 (20.8%)	2 (4.0%)	53 (24.7%)
Thiamine excretion per gram creatinine[e]			
Median (μg)	80	400	72
Range (μg)	3-11364	51-11364	3-8621
Low (<65)	32 (40.5%)	2 (16.7%)	30 (44.8%)
Normal (65-129)	13 (16.5%)	1 (8.3%)	12 (17.9%)
High (>130)	34 (43.0%)	9 (75.0%)	25 (37.3%)

[a] Reproduced from Wood et al. (1977), with permission.
[b] See text for definition and source of values.
[c] No known vitamin ingestion in preceding 90 days.
[d] In International Units.
[e] Urinary thiamine excretion was determined in only 79 persons, 12 of whom had documented recent vitamin therapy.

Table III. Biochemical Assessment of Thiamine Status in 24 Patients with Wernicke's Encephalopathy Including Responses to Thiamine Therapy[a]

Group[b]	Subject	AST[c]	AP[d]	Bil.[e]	ETKA/ETKA + TPP (IU)	TPP effect (%)	After thiamine in vivo[f] — ETKA/ETKA + TPP (IU)	After thiamine in vivo[f] — TPP effect (%)	Clinical response[j] to thiamine in vivo[f]
I	J.R.	244	5	0.9	177/180	1.8	160/157	-2.2	—[k]
	A.P.	32	7	—	136/132	-3.2	—	—	>7 days
II	W.S.	54	11	0.5	97/102	5.7	129/134	5.1	—[k]
IIIA	M.P.[g]	120	7	0.4	33/66	100.0	95/94	-1.1	—
	S.B.[g]	20	10	0.5	53/100	90.1	—	—	—
	W.B.	73	10	0.9	56/67	20.8	—	—	36 hours
IIIB	R.S.[h]	78	8	1.8	24/49	105.2	71/73	2.7	<7 days
	C.M.[i]	70	8	0.5	27/49	81.4	—	—	>7 days
	A.T.[i]	—	—	—	28/40	44.9	—	—	24 hours
	P.P.[i]	63	7	6.9	43/55	27.8	—	—	<7 days
	B.L.	137	10	1.5	32/34	6.4	—	—	72 hours
	S.J.	120	12	0.4	35/56	61.1	—	—	<7 days
	J.L.	115	11	3.1	32/35	8.3	65/66	2.7	<7 days
	T.L.	73	10	1.4	48/52	7.2	94/102	8.4	24 hours

L.P.	56	13	0.5	28/48	74.4	79/80	1.3	None
G.L.	48	7	0.8	33/42	25.0	—	—	>7 days
H.M.	48	15	0.6	56/56	0	91/91	0	>7 days
R.H.	42	6	0.7	36/57	55.8	66/76	15.7	48 hours
M.F.	40	9	0.6	38/38	0.7	70/71	0.9	<7 days
G.C.	39	5	0.9	21/36	68.1	58/59	1.9	48 hours
P.H.	37	12.5	0.8	29/27	-8.3	56/66	19.3	>7 days
R.W.	25	7	0.9	27/45	64.1	106/100	-5.8	72 hours
W.R.	24	5	0.8	29/58	101.3	79/82	3.2	Death[l]
E.D.	20	2	0.6	45/45	1.0	—	—	<7 days

[a] Reproduced from Wood et al. (1977), with permission. Italics indicate abnormal results.
[b] Grouped according to ETKA (see Figs. 1–3).
[c] Serum asparate-aminotransferase in international units per milliliter (normal range 5–20).
[d] Serum alkaline phosphate in King-Armstrong units per milliliter (normal range 3–13).
[e] Total serum bilirubin in mg/100 ml (normal, <1.0).
[f] Data for patients after treatment with parenteral thiamine hydrochloride for 1 to 9 days (median, 2 days).
[g] Alcoholic hepatitis on liver biopsy.
[h] Alcoholic cirrhosis on liver biopsy.
[i] Clinical evidence of cirrhosis (no liver biopsy).
[j] Resolution of lateral rectus palsy.
[k] Lost to immediate follow-up.
[l] Examination of the brain at autopsy showed changes consistent with past Wernicke's encephalopathy.

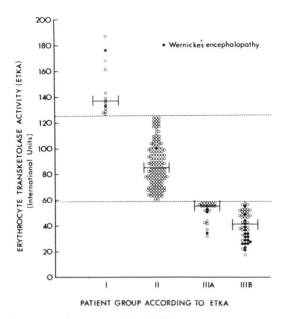

Fig. 1. Erythrocyte transketolase activity (ETKA) in 215 alcoholic patients without evidence of recent vitamin therapy. The broken lines represent the upper and lower limit of values for ETKA in apparently healthy hospital staff (see Table II). The patients have been grouped according to whether ETKA was high (group I, *n* = 13), normal (group II, *n* = 132), or low (group III, *n* = 70). See Fig. 2 for explanation of subgroups IIIA and IIIB. The horizontal solid bars represent the median value for each group. (Reproduced from Wood *et al.*, 1977, with permission.)

better correlation, we chose to use low ETKA as the best index of thiamine status. In the entire group of 215 subjects, 70 (32.6%) had a low ETKA and were considered to have biochemical thiamine deficiency (Table III, Fig. 1).

Disparity between biochemical and clinical assessment of thiamine status in people with disease, in normal individuals, and in population groups has been previously recognized (Plough and Bridgforth, 1960; Leevy *et al.*, 1965). In experimental thiamine deficiency in human volunteers, a clear relationship has been demonstrated between thiamine intake and urinary thiamine excretion and between ETKA and the derived TPP effect (Brin, 1962; Bamji, 1970). However, these studies were necessarily short term and clinical thiamine deficiency was not permitted to develop. Population studies also support the above relationships between these biochemical parameters of thiamine status (Plough and Bridgforth, 1960; Reuter *et al.*, 1967; Brubacher *et al.*, 1972; Sauberlich *et al.*, 1973). However, with the possible exception of our own data regarding ETKA in 24 patients with Wernicke's encephalopathy (Wood *et al.*, 1977), no single

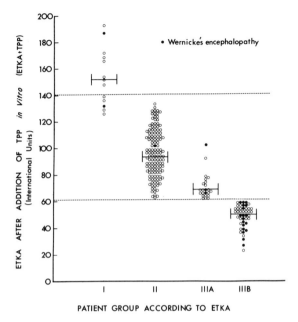

Fig. 2. Erythrocyte transketolase activity after the addition of TPP *in vitro* (ETKA + TPP) in 215 alcoholic patients. The broken lines represent the upper and lower limit of values for ETKA + TPP in apparently healthy hospital staff (see Table II). The patients have been grouped according to ETKA (see Fig. 1). Twenty-three subjects in group III had normal ETKA + TPP values and formed group IIIA. Forty-seven subjects in group III had low ETKA + TPP values and formed group IIIB. The horizontal solid bars represent the median value for each group. One patient in group II had a slightly elevated ETKA + TPP value and is not included in the figure. (Reproduced from Wood *et al.*, 1977, with permission.)

biochemical measure of thiamine status has been found to correlate completely with the clinical assessment of thiamine deficiency. The reasons for this lack of correlation are not clear but may include metabolic adaptation to gradual thiamine depletion, secondary thiamine deficiency (for example, impaired storage or activation of thiamine), or imprecision in clinical diagnosis.

Two of the major difficulties in the interpretation of thiamine status data provided by the erythrocyte transketolase assay is the selection of the parameter of the assay to be used in assessment and the normal values for this parameter. The incidence (32.6%) of biochemical thiamine deficiency in our study may be influenced by this. Our "normal" range for the assay was taken from a group of apparently healthy staff members (Table II) (Wood and Penington, 1974a; Wood *et al.*, 1977). This range differs from that described in normal populations from other countries, but as our normal population was not thiamine saturated, we

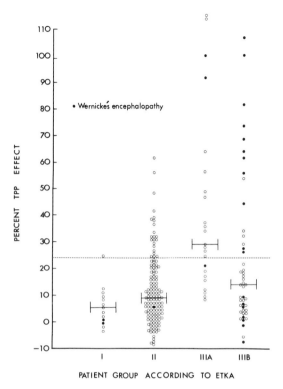

Fig. 3. The percent TPP effect in 215 alcoholic patients. The broken line represents the upper limit of values for the TPP effect in apparently healthy hospital staff (see Table II). The patients have been grouped according to ETKA (see Fig. 1). The horizontal solid bars represent the median value for each group. One patient in group IIIB had a TPP effect of 739% and is not represented in the figure. (Reproduced from Wood *et al.,* 1977, with permission.)

would expect our results to underestimate rather than to overestimate the incidence of biochemical thiamine deficiency. The arbitrary choices of normal and abnormal relate to whether one believes that thiamine saturation is or is not essential for "optimal" health. While this difference of opinion exists, it is unlikely that a consensus will be reached regarding normal values for the assay.

Urinary thiamine excretion is another means of assessment of thiamine status. However, in our study of 215 patients, urinary thiamine excretion did not correlate with clinical thiamine deficiency or with any of the methods of expression of the ETKA assay (Table II). It is our opinion that this is not a useful index of thiamine status in alcoholism, although it may be of value as a screening test for recent thiamine supplementation.

IV. ETIOLOGY OF THIAMINE DEFICIENCY IN ALCOHOLISM

There are several possible explanations for the high incidence of biochemical and clinical thiamine deficiency in Australian alcoholics, including low dietary intake of thiamine, impaired thiamine absorption, and impaired thiamine utilization.

A. Low Thiamine Intake

Thiamine intake may be reduced because of decreased food intake from anorexia and/or vomiting related to gastritis and peptic ulcer. Excessive alcohol consumption also leads to displacement of thiamine from the diet. In a detailed dietary assessment made of a small group of alcoholics it was observed that daily thiamine intake fell below the physiological requirement of 0.33 mg of thiamine per 1000 kcal per day when alcohol contributed 30% or more of the total energy supply, i.e., approximately 100 gm of alcohol per day (Fig. 4) (Wood, 1972). Interestingly, thiamine appeared to be displaced before protein from the diet of these drinkers. Australian alcoholics appear to have characteristic food habits (Joske and Turner, 1952; Wood, 1972). Another earlier study in this city indicated that 50% of 96 chronic alcoholics had a thiamine intake below 0.22 mg of thiamine per 1000 kcal per day (Australian National Health and Medical Research Council, 1959). This problem of thiamine displacement could, of course, be exaggerated should the national dietary supply of thiamine be suboptimal (see below).

B. Impaired Thiamine Absorption

Thiamine is absorbed from the small intestine by active transport in the rat and probably by a saturable mechanism in man (Hoyumpa *et al.*, 1975b; Desmond *et al.*, 1976). Two studies have indicated that thiamine absorption is impaired in chronic malnourished alcoholics (Tomasulo *et al.*, 1968; Thomson *et al.*, 1970). These studies utilized relatively crude techniques and should be repeated using a method such as jejunal perfusion. The cause of thiamine malabsorption does not appear to be a simple effect of alcohol on intestinal mucosal function, as in normal volunteers studied by jejunal perfusion, alcohol added to the perfusate did not reduce thiamine absorption (Desmond *et al.*, 1976). These data are strikingly different from findings in the rat, wherein the active transport system for thiamine is significantly impaired by acute exposure to alcohol (Hoyumpa *et al.*, 1975a).

C. Impaired Thiamine Utilization

On the basis of studies in rats with carbon tetrachloride liver injury and in humans with liver disease, Leevy and co-workers have suggested that severe

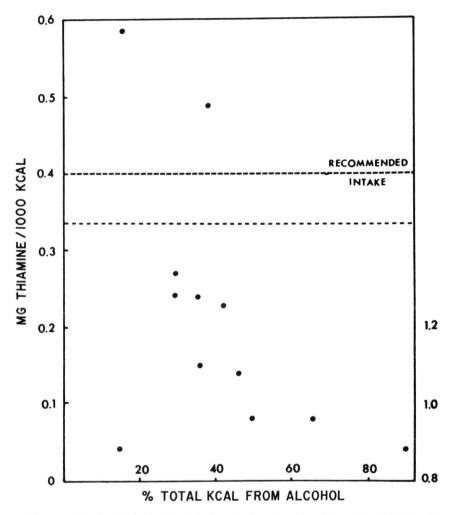

Fig. 4. Thiamine intake in 12 alcoholic patients. (Reproduced from Wood, 1972, with permission.)

liver disease affects the synthesis of apotransketolase (as determined by the failure of thiamine added *in vitro* and given *in vivo* to alter erythrocyte trans-ketolase activity). They also feel that this invalidates the use of the erythrocyte transketolase assay in the assessment of thiamine status in patients with liver injury (Fennelly *et al.*, 1963, 1964, 1967; Baker, 1967). Impaired phosphoryla-tion *in vivo* of thiamine to its biologically active form (TPP) has been demon-strated in patients with liver disease (Ocho and Peters, 1938; Williams and

Bissell, 1944; Baker, 1967; Fennelly *et al.*, 1967). Decreased hepatic storage of thiamine in the presence of severe fatty liver but not in cirrhosis has also been reported (Baker *et al.*, 1964).

We have been unable to confirm that severe liver disease interferes with thiamine status as assessed by the erythrocyte transketolase assay (Wood *et al.*, 1977), nor does it appear to interfere with the reversal of the clinical evidence of Wernicke's encephalopathy or with biochemical return to normality. Admittedly, not all the patients studied by us were subjected to liver biopsy and further work in this area is indicated. Truswell *et al.* (1972) also found that florid liver disease did not appear to affect erythrocyte transketolase levels in alcoholic patients.

The apparent impaired synthesis of apotransketolase ascribed to liver injury by Leevy and colleagues could have an alternate explanation. There appears to be a discrepancy between the clear restoration by TPP *in vitro* of progressively declining ETKA in the early stages of thiamine deficiency and less complete restoration in the later stages of thiamine deficiency; the latter is reflected by a decrease in TPP effect apparent in the data of several workers (Brin, 1962; Dreyfus, 1962; Fennelly *et al.*, 1964, Bamji, 1970). Dreyfus (1962) suggested that thiamine deficiency per se may be responsible for apparent deficiency of apotransketolase. Our data in alcoholic patients would be compatible with this hypothesis. However, the possibility that TPP fails to combine with apotransketolase for some other reason cannot be excluded. Indeed, very recently, altered affinity of apoenzyme for TPP in patients with the Wernicke–Korsakoff syndrome has been proposed as an important predisposing factor to thiamine deficiency (Blass and Gibson, 1977).

V. THIAMINE STATUS OF THE AUSTRALIAN POPULATION

Whether flour and bread should be enriched with thiamine (as occurs in the United Kingdom and the United States of America) was carefully considered by an advisory committee of the Australian National Health and Medical Research Council (1959) and is again under consideration by that body. In 1959, the advisory committee reported that it "was unable to locate any evidence that any significant section of the population apart from the alcoholics is suffering from an insufficient intake of thiamine" and considered that "it would be wrong in principle to add a substance to bread in order to treat a minority of the population which is, in fact, suffering from a disease—alcoholism." Thus in 1959 Australia chose not to enrich flour and bread with thiamine.

In the intervening 19 years, bread intake has declined (Wood, 1973; Butcher and Venn Brown, 1975). The extraction rate of white flour from wheat is presently 75% (Venn Brown, 1975). The Australian food pattern is rapidly changing (Wood, 1977) and there are indications that the thiamine supply in the national diet may not be sufficient to meet the needs of those groups in the population taking a poor diet. From data from 1959 to 1974 of apparent thiamine supply per capita, dietary surveys of thiamine intake, and the use of the erythrocyte transketolase assay in a number of population groups (Wood and Penington, 1974b; Nobile, 1974; Woodhill *et al.,* 1974; Kamien *et al.,* 1974, 1975; Araya *et al.,* 1975), it appears that in addition to alcoholics there may be considerable numbers of Australian adults with depleted thiamine reserves. The precise importance of depleted thiamine reserves or "subclinical" thiamine deficiency remains unknown. It is nevertheless clear that the absence of overt thiamine deficiency in the population cannot be regarded as sufficient evidence for the adequacy of the national diet with respect to thiamine supply. It is difficult to judge whether the national diet has a satisfactory safety margin. If it does not then the heavy drinker and alcoholic are at even greater risk of developing thiamine deficiency.

The arguments for and against measures aimed at increasing thiamine supply in the national diet (including increasing the extraction rate of white flour from bread, promotion of bread as a good food, and thiamine enrichment of flour) are complex. Whether an increase in thiamine intake by alcoholics resulting from such measures would prevent thiamine deficiency is unknown, although there is indirect evidence that beriberi in American alcoholics has largely disappeared since thiamine enrichment of flour was enforced by law in most states (Bradley, 1962). Should a decision ever be made to supplement flour with thiamine in this country, the opportunity should also be taken to document as carefully as possible the effects of this step on the clinical and biochemical incidence of thiamine deficiency in alcoholics and other groups in the population. The introduction of fortified foods, unless for specific reasons and carefully controlled, may lead to abuses such as unwarranted claims for the health advantages of such foods (FAO/WHO, 1970) and may result in a slackening of efforts directed at the basic problem, mainly, excessive alcohol consumption.

VI. SUMMARY

A study of a large group of alcoholic subjects in Melbourne has shown a high incidence of both biochemical and clinical thiamine deficiency. This problem may be related to low thiamine intake, impaired thiamine absorption, and/or impaired thiamine utilization in alcoholism. The thiamine status of the Australian population at large may also be suboptimal, thereby contributing to the frequency of thiamine deficiency in alcoholics.

Many areas remain unexplored. Further data are required on the incidence of clinical thiamine deficiency (particularly Wernicke's encephalopathy) in Australian alcoholics. More work needs to be done to define the normal range for the parameters of the erythrocyte transketolase assay in the Australian population.

Studies are already in progress in this laboratory to determine, by partial dietary deprivation of thiamine, the significance of any elevated TPP effects seen in apparently healthy persons and to determine if there are specific early symptoms of thiamine deficiency. The question of the effect of liver disease on thiamine status needs to be further examined and the role of malabsorption as a major cause of thiamine deficiency in alcoholism should be more carefully evaluated. It must not be forgotten that the basic problem is excessive consumption of alcohol, which needs to be tackled seriously by the community, the health professions and the government. Beyond this general recommendation, any other specific action regarding thiamine deficiency in Australian alcoholics should be preceded by a clearer resolution of the many facets of the problem.

ACKNOWLEDGMENTS

We gratefully acknowledge the valuable contribution that Professor D. G. Penington has made to this work and thank him for his support, encouragement, and continued assistance. This work has been supported by Hoffmann-La Roche (Australia) and the Australian National Health and Medical Research Council.

REFERENCES

Araya, M., Silink, S. J., Nobile, S., and Walker-Smith, J. A. (1975). Blood vitamin levels in children with gastroenteritis. *Aust. N.Z. J. Med.* **5,** 239.

Australian Bureau of Statistics. (1977). ''Apparent Consumption of Foodstuffs and Nutrients: Australia, 1974–75.'' Australian Bureau of Statistics, Canberra, A.C.T.

Australian National Health and Medical Research Council. (1959). ''Report of a Committee Appointed to Review the Nutritive Significance of Australian Bread,'' Spec. Rep. Ser. No. 9. Aust. Natl. Health Med. Res. Counc., Canberra.

Baker, H. (1967). Discussion. Following Sauberlich, H. E. (1967). Biochemical alterations in thiamine deficiency—their interpretation. *Am. J. Clin. Nutr.* **20,** 528.

Baker, H., Frank, O., Ziffer, H., Goldfarb, S., Leevy, C. M., and Sobotka, H. (1964). Effect of hepatic disease on liver-B complex vitamin titres. *Am. J. Clin. Nutr.* **14,** 1.

Bamji, M. S. (1970). Transketolase activity and urinary excretion of thiamine in the assessment of thiamine-nutrition status in Indians. *Am. J. Clin. Nutr.* **23,** 52.

Blass, J. P., and Gibson, G. E. (1977). Abnormality of a thiamine requiring enzyme in patients with Wernicke–Korsakoff Syndrome. *N. Engl. J. Med.* **297,** 136.

Bradley, W. B. (1962). Thiamine enrichment in the United States. *Ann. N.Y. Acad. Sci.* **98,** 602.

Brin, M. (1962). Erythrocyte transketolase in early thiamine deficiency. *Ann. N.Y. Acad. Sci.* **98,** 528.

Brin, M., Tai, M., Ostashever, A. S., and Kalinsky, H. (1960). The effect of thiamine deficiency on the activity of erythrocyte haemolysate transketolase. *J. Nutr.* **71,** 273.

Brubacher, G., Haenel, A., and Ritzel, G. (1972). Transketolaseaktivität, Thiaminausscheidung und Blutthiamingehalt beim Menschen zur Beurteilung der Vitamin-B₁-Versorgung. *Int. Z. Vitaminforsch.* **42,** 190.

Butcher, J., and Venn Brown, U. (1975). Bread consumption patterns in Australia. *Food Nutr. Notes Rev.* **32,** 41.

Dax, E. C. (1968). ''Responsibility and Alcoholism,'' Inaugural Leonard Ball Oration. Alcohol. Found. Victoria, Victoria, Australia.

Desmond, P. V., Lourensz, C. R., and Breen, K. J. (1976). Thiamine hydrochloride absorption in man: Normal kinetics and absence of acute effect of ethanol. *Aust. N.Z. J. Med.* **6,** 264. (Abstr.)

Dreyfus, P. M. (1962). Clinical application of blood transketolase determinations. *N. Engl. J. Med.* **267,** 596.

Embree, L. J., and Dreyfus, P. M. (1963). Blood transketolase determinations in nutritional disorders of the nervous system. *Trans. Am. Neurol. Assoc.* **88,** 36.

FAO/WHO. (1970). Food fortification. *Jt. FAO/WHO Exp. Comm. Nutr., 8th Rep.*

Fennelly, J. J., Baker, H., Frank, O., and Leevy, C. M. (1963). Deficiency of thiamine pyrophosphate apoenzyme in liver disease. *Clin. Res.* **11,** 182. (Abstr.)

Fennelly, J. J., Frank, O., Baker, H., and Leevy, C. M. (1964). Transketolase activity in experimental thiamine deficiency and hepatic necrosis. *Proc. Soc. Exp. Biol. Med.* **116,** 875.

Fennelly, J. J., Frank, O., Baker, H., and Leevy, C. M. (1967). Red blood cell transketolase activity in malnourished transketolase activity in malnourished alcoholics with cirrhosis. *Am. J. Clin. Nutr.* **20,** 946.

Green, J. R. (1965). The incidence of alcoholism in patients admitted to medical wards of a public hospital. *Med. J. Aust.* **1,** 465.

Halsted, C. H. (1976). ''Present Knowledge in Nutrition,'' 4th ed., p. 467. Nutr. Found., New York.

Horvath, T. B., Wilkinson, P., Santamaria, J. N., and Rankin, J. G. R. (1969). Dementia in alcoholics. *Aust. Ann. Med.* **18,** 165.

Hoyumpa, A. M., Breen, K. J., Schenker, S., and Wilson, F. A. (1975a). Thiamine transport across the rat intestine. II. Effect of ethanol. *J. Lab. Clin. Med.* **86,** 803.

Hoyumpa, A. M., Middleton, H. M., III, Wilson, F. A., and Schenker, S. (1975b). Thiamine transport across the rat intestine. I. Normal characteristics. *Gastroenterology* **68,** 1218.

Jolliffee, N., Wortis, H., and Fein, H. D. (1941). The Wernicke Syndrome. *Arch. Neurol. Psychiatry* **46,** 569.

Joske, R. A., and Turner, C. N. (1952). Studies in chronic alcoholism. 1. The clinical findings in 78 cases of chronic alcoholism. *Med. J. Aust.* **1,** 729.

Kamien, M. Nobile, S., Cameron, P., and Rosevear, P. (1974). Vitamin and nutritional status of a part aboriginal community. *Aust. N.Z. J. Med.* **4,** 127.

Kamien, M., Woodhill, J. M., Nobile, S., Cameron, P., and Rosevear, P. (1975). Nutrition in the Australian aborigine. Effects of the fortification of white flour. *Aust. N.Z. J. Med.* **5,** 123.

Krupinski, J., and Stoller, A. (1971). ''The Health of a Metropolis.'' Heinemann, Educational Australia Pty. Ltd., Victoria.

Lederman, S. (1954). ''Alcohol, Alcoholism, Alcoholisation,'' Inst. Natl. Etud. Demogr. Trav. Doc. Cah. No. 29. Presses Univ. Fr., Paris.

Leevy, C. M., Cardi, L., Frank, O., Gellene, R., and Baker, H. (1965). Incidence and significance of hypovitaminaemia in a randomly selected municipal hospital population. *Am. J. Clin. Nutr.* **17,** 259.

Nobile, S. (1974). Blood vitamin levels in aboriginal children and their mothers in western New South Wales. *Med. J. Aust.* **1,** 601.

Ocho, S., and Peters, R. A. (1938). Vitamin B-1 and cocarboxylase in animal tissues. *Biochem. J.* **32,** 1501.

Plough, I. C., and Bridgforth, E. B. (1960). Relations of clinical and dietary findings in nutrition surveys. *Public Health Rep.* **75,** 699.

Rankin, J. G. (1974). Alcohol—a specific toxin or nutrient displacer. *Miles Symp., Ontario, Can.*

Reuter, H., Gassmann, B., and Erhardt, V. (1967). Beitrag zur Frage des menschlechen Thiaminbedarfs. *Int. Z. Vitamin forsch.* **37,** 315.

Sauberlich, H. E. (1967). Biochemical alterations in thiamine deficiency—their interpretation. *Am. J. Clin. Nutr.* **20,** 528.

Sauberlich, H. E., Dowdy, R. P., and Skala, J. H. (1973). Laboratory tests for the assessment of nutritional status. *Crit. Rev. Clin. Lab. Sci.* **4,** 236.

Schouten, H., van Eps Statius, L. W., and Struyker Boudier, A. M. (1964). Transketolase in blood. *Clin. Chim. Acta* **10,** 474.

Smithurst, B. A. (1965). The incidence of alcoholism in patients committed to medical wards of a public hospital. *Med. J. Aust.* **1,** 738.

Thomson, A. D., Baker, H., and Leevy, C. M. (1970). Patterns of ^{35}S-thiamine hydrochloride absorption in the malnourished alcoholic patient. *J. Lab. Clin. Med.* **76,** 34.

Tomasulo, P. A., Kater, R. M. H., and Iber, F. L. (1968). Impairment of thiamine absorption in alcoholism. *Am. J. Clin. Nutr.* **21,** 1340.

Truswell, A. S., Konno, T., and Hansen, J. D. L. (1972). Thiamine deficiency in adult hospital patients. *S. Afr. Med. J.* **46,** 2097.

Venn Brown, U. (1975). "Go . . . , eat thy bread with joy." *Med. J. Aust.* **1,** 720.

Victor, M., Adams, M. D., and Collins, G. D. (1971). "The Wernicke–Korsakoff Syndrome." Davis, Philadelphia, Pennsylvania.

Wilkinson, P. M., Santamaria, J. N., Rankin, J. G., and Martin, D. (1969). Epidemiology of alcoholism: Social data and drinking patterns of a sample of Australian alcoholics. *Med. J. Aust.* **1,** 1020.

Wilkinson, P. M., Kornacewski, A., Rankin, J. G., and Santamaria, J. N. (1971). Physical disease in alcoholism. *Med. J. Aust.* **1,** 1217.

Williams, R. H., and Bissell, G. W. (1944). Thiamine metabolism with particular reference to the role of liver and kidneys. *Arch. Intern. Med.* **73,** 203.

Williams, R. R. (1961). "Toward the Conquest of Beri Beri." Harvard Univ. Press, Cambridge, Massachusetts.

Wood, B. (1972). A dietary study of alcoholism. *Food Nutr. Notes Rev.* **29,** 33.

Wood, B. (1973). The contribution of bread to the thiamine intake of Australians. A review. *J. Diet. Assoc. Victoria* **24,** 9.

Wood, B. (1975). Nutritional factors in alcoholism. *Food Nutr. Notes Rev.* **32,** 127.

Wood, B. (ed.). (1977). "Tucker in Australia." Hill of Content Pub. Co. Pty. Ltd. Melbourne.

Wood, B., and Penington, D. G. (1973). Biochemical assessment of thiamine status in adult Australians. *Int. Z. Vitaminforsch.* **43,** 12.

Wood, B., and Penington, D. G. (1974a). Objective measurement of thiamine status by biochemical assay in adult Australians. *Med. J. Aust.* **1,** 95.

Wood, B., and Penington, D. G. (1974b). The thiamine status of Australians. *Food Technol. Aust.* **26,** 278.

Wood, B., Breen, K. J., and Penington, D. G. (1977). Thiamine status in alcoholism. *Aust. N.Z. J. Med.* **7,** 475.

Woodhill, J. M., Nobile, S., Silink, S., and Winston, J. M. (1974). Case studies on the nutritional status of socially deprived children in Sydney. *Aust. Paediatr. J.* **10,** 199.

World Health Organization (1963). Expert committee on medical assessment of nutritional status. *W.H.O. Tech. Rep. Ser.* No. 258.

World Health Organization. (1967). "International Classification of Diseases" (1965 Revision). W.H.O., Geneva.

EDITORIAL COMMENT

Why only rare individuals with low thiamine intakes and a heavy intake of alcohol get the Wernicke or Wernicke–Korsakoff syndrome has not been adequately explained. Blass and Gibson (1977) have now reported the existence of abnormalities of a thiamine-dependent enzyme in four patients with clinical abnormalities characteristic of this syndrome. These abnormalities include profound and chronic memory disorder—particularly affecting recent memory, nystagmus, and ataxia—as well as biochemical evidences of thiamine deficiency. Their subjects all responded to therapy with thiamine and all had, in cells cultured from skin biopsies, an abnormality of transketolase, an enzyme necessary for the conversion of pentoses. In these four cases, the apparent K_m for binding of thiamine pyrophosphate was between 10 and 20 times higher in the cell extracts than in six controls. The authors suggest that the decreased activity of this enzyme was involved in the pathogenesis of this form or manifestation of thiamine deficiency. Further studies of this phenomenon will be awaited with interest.

REFERENCE

Blass, J. P., and Gibson, G. E. (1977). Abnormality of a thiamine-requiring enzyme in patients with Wernicke–Korsakoff Syndrome. *N. Engl. J. Med.* **297**, 1367–1370.

27

CANCER AND ALCOHOLIC BEVERAGES

A. J. Tuyns

I. Introduction   427
II. Alcohol and Carcinogenesis   428
   A. Ethanol and Other Alcohols   428
   B. Alcohol as a Solvent for Carcinogens   428
III. Human Cancer in Relation to Alcohol Consumption: The
   Epidemiological Evidence   428
   A. Correlation Studies   428
   B. Cohort Studies of Alcoholics   429
   C. The Retrospective Case-Control Studies   429
   D. The Case of Primary Liver Cancer   431
   E. Alcohol and Tobacco   431
IV. Summary   434
   References   435
   Editorial Comment   437

I. INTRODUCTION

The use and abuse of alcoholic beverages have, for a long time, been known to be associated with cancer of the buccal cavity, larynx, esophagus and liver. It is common clinical experience that patients with one or more of these cancers are often heavy drinkers and also often heavy smokers when the cancer is located in the respiratory and upper digestive tract.

There is substantial epidemiological evidence to confirm these associations, and this will be reviewed in Section III of this chapter. However, the mechanisms by which alcohol* produces cancer are still controversial.

*For practical purposes the term ''alcohol,'' as used throughout this chapter, must be interpreted as the equivalent of ''alcoholic beverages,'' while ''ethanol'' means ''ethyl alcohol'' in the chemical sense.

II. ALCOHOL AND CARCINOGENESIS

A. Ethanol and Other Alcohols

The main chemical constituent of alcoholic beverages is ethanol and this has been extensively used in experiments on various species of animals. Provided one can persuade the animal to drink ethanol—which is not so easy—various kinds of lesions can be obtained, but no cancer.

Alcoholic beverages, however, contain much more than pure ethanol; methanol, propanol, butanol are often present at various concentrations. The toxic effects of these substances are known and Gibel *et al.* (1968, 1970) described the carcinogenicity of these "fusel oils" in rats. The characteristic flavor of a given drink is due, to a certain extent, to the presence of large numbers of other congeners, the action of which is often unknown.

B. Alcohol as a Solvent for Carcinogens

Alcohol could also be a *solvent* or a *vector* for other carcinogens. Walker and Castegnaro (IARC, 1976) showed the existence of various nitrosamines in samples of cider distillates from Brittany and Normandy, a province of France where esophageal cancer is very frequent (Tuyns, 1970; Tuyns and L. M. F. Massé, 1973; Tuyns and G. Massé, 1975). Although these chemicals are potent animal carcinogens, there is as yet no evidence of human carcinogenicity; these were found, moreover, at concentrations of a few parts per billion, and one may wonder about the real significance of the finding.

Experiments bearing more resemblance to what might happen in man were carried out by Japanese investigators studying polycyclic aromatic hydrocarbons known to be carcinogens widespread in the human environment. When applied to rats in ethanolic solution, the substances were found to be present in the esophagus (Kuratsune *et al.*, 1965) and even to cause cancer there (Horie *et al.*, 1965), an effect that was not obtained by aqueous solutions of the same carcinogens.

It should be noted, however, that such animal experiments normally oversimplify the problem. The carcinogenic cocktails man prepares for himself are much more varied and tasty than the ones he offers to rats.

III. HUMAN CANCER IN RELATION TO ALCOHOL CONSUMPTION: THE EPIDEMIOLOGICAL EVIDENCE

A. Correlation Studies

Several studies have demonstrated the correlation existing in human groups between alcohol sales and various cancers. One of the most recent and most intriguing studies (Breslow and Enstrom, 1974) refers to cancer of the rectum in relation to beer drinking. When Jensen (1977), however, studied the fate of Danish brewery workers who have the privilege of a free ration of some 4 pints of beer per day, he did not observe any increased risk of rectal cancer. The matter is not settled, and more studies of the same kind are presently underway.

In another type of correlation study, geographical distribution of cancer mortality has been compared to that of "alcoholism" or cirrhosis of the liver, known to be associated with an excessive consumption of alcohol (Tuyns, 1970). In France, for example, the distribution of esophageal cancer mortality resembles very much that of the two other groups, with a region in the West showing the highest rates (Fig. 1). Even though this is no real proof of an association, it is, nevertheless, a suggestion that it may exist.

B. Cohort Studies of Alcoholics

More convincing are the prospective studies conducted on groups of individuals who had some medical or social problem in relation to their drinking. Such groups have a rather well-defined mortality pattern. This is characterized by an excess mortality from alcoholism, cirrhosis of the liver, suicide, accidents, and indeed cancer. Studies in various parts of the world, including Canada (Schmidt and De Lint, 1972), Norway (Sundby, 1967), Finland (Hakulinen *et al.*, 1974), and Japan (Hirayama, 1975) all showed an excess mortality for cancer of the respiratory and upper digestive tract and for lung. Equally instructive is the observation that Mormons (Enstrom, 1975; Lyon *et al.*, 1976) and Seventh Day Adventists (Lemon *et al.*, 1964), who abstain from alcohol drinking, experience a lesser risk for the same cancer sites.

The advantage of these cohort studies is to give a good overall picture of the ill effects of alcohol consumption. Their disadvantage is that individual alcohol intake is usually imprecise as to quantity and kind of drink: it is only known to be very high or very low in the groups considered. Another difficulty arises from the fact that other factors are ignored in that kind of analysis. We know, however, that the risk of smoking, for example, is far from being negligible in the cancer sites concerned.

C. The Retrospective Case-Control Studies

To distinguish between the respective weight of drinking, smoking, and possibly other factors in these cancers, more detailed information needs to be collected and compared in patients and in suitable controls, preferably taken from the general population (Tuyns *et al.*, 1977a). There are a great number of such case-control studies available today. In this review we have deliberately ignored those showing controversial or ambiguous results and have focused on the most significant findings.

Fig. 1. Geographic distribution of mortality in France (1960–1963) for males aged 45–64, from alcoholism (left), esophageal cancer (center), and cirrhosis of liver (right).

In one of the first systematic reviews of the role of alcoholic beverages in human cancer, Schwartz *et al.* (1962) determined the average daily consumption in milliliters of ethanol among 3938 cancerous patients and 1807 controls in a predominantly wine-consuming French population. They were also examined for the presence of clinical symptoms of alcoholic intoxication, such as liver enlargement and trembling. The various cancer sites were then ranked by increasing average consumption and by increasing frequency of symptoms (Table I). In both series, tongue, hypopharynx, larynx, esophagus, buccal cavity, and oropharynx came first. An apparent association was also found with lung cancer, but this was an indirect one, smoking being the main factor for this site: for the others, the effect of alcohol drinking was still observed after correction for smoking.

D. The Case of Primary Liver Cancer

One may wonder why primary liver cancer does not often appear in overall studies—either retrospective or prospective—in relation to alcohol drinking in spite of the widely recognized association of primary liver cancer with cirrhosis of the liver and of the latter with alcohol consumption—at least in the so-called Western civilizations. The explanation lies in the relative rarity of this form of cancer in the Western world. Increased risks, whenever observed, are usually below the level of statistical significance and thus not commented on by the authors. It has been observed, however, that when the autopsy rate is high enough in a population with a sizable alcohol consumption, as is the case in Geneva (Switzerland), the incidence rate for primary liver cancer may be fairly elevated (Tuyns and Obradovic, 1975). One may suspect that more extensive practice of postmortem examinations in areas of heavy drinking might result in more realistic estimates concerning hepatomas.

E. Alcohol and Tobacco

For cancers of the respiratory and upper digestive tract, the action of alcohol seems to be more direct than in liver cancer, as it clearly affects the mucosa exposed to it. This is also true for smoking. All anatomical sites influenced by alcohol are also influenced by tobacco. Whether one or the other plays the greater role in causing cancer at one site, as opposed to another, is still uncertain. Recent studies have contributed to clarifying the interaction of alcohol and tobacco.

In a study on cancer of the mouth and of the pharynx, Rothman and Keller (1972) demonstrated the individual effect of drinking at every level of exposure to smoking and vice versa as can be seen in Table II, which suggests that the combined effect equals the sum of two strong individual effects.

We have described an almost identical situation for esophageal cancer in the

Table I. Alcohol Consumption and Percentage of Persons with Alcoholism among Cancer Patients and Noncancer Control Group (Workers in the Paris Area)[a,b]

Site of cancer	Number of cases	Quantity consumed				Percentage with alcoholism			
		Ounces of absolute alcohol consumed daily	Gross significance	Significance level after adjustment for		%	Gross significance	Significance level after adjustment for	
				Tobacco	Tobacco and age			Tobacco	Tobacco and age
Tongue	43	5.2	***	*		74	***	***	***
Buccal cavity (other locations)	23	5.4	**	*	*	83	***	***	**
Oropharynx	34	4.9	**		*	56			
Hypopharynx	63	5.5	***	*	*	62	**	*	*
Larynx	100	5.4	***	**	*	61	**	*	**
Esophagus		5.3	***	**	**	58	**	*	*
Control group	366	4.0				43			

a From Schwartz et al. (1962).
b *p < 0.05; **p < 0.01; ***p < 0.001.

Table II. Relative Risk[a] or Oral Cancer According to Level of Exposure to Smoking and Alcohol[b]

Alcohol (oz./day)	Smoking (cigarette equivalents per day)			
	0	<20	20-39	40+
0	1.00	1.52	1.43	2.43
<0.4	1.40	1.67	3.18	3.25
0.4-1.5	1.60	4.36	4.46	8.21
1.6+	2.33	4.13	9.59	15.5

[a] Risks are expressed relative to a risk of 1.00 for persons who neither smoked nor drank.

[b] From Rothman and Keller (1972), reprinted by permission.

French department of Ille et Vilaine, in Brittany. Here again, the risk of developing the disease increases with the amount of alcohol consumed at each level of smoking and vice versa. For each of these factors, the logarithm of the relative risk of contracting an esophageal cancer is a linear function of the amount consumed. The risks observed for the various combinations of exposures to both factors are consistent with the hypothesis that the risks are multiplied (Tuyns *et al.*, 1977b) (Fig. 2).

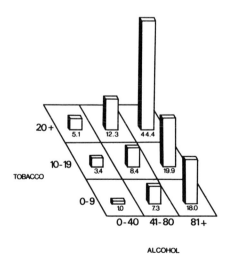

Fig. 2. Cancer of the esophagus. Relative risks in relation to daily consumption of alcohol and tobacco (in grams per day).

For laryngeal cancer the situation is not so clear. Alcohol consumption enhances the effect of tobacco—the principal risk factor for this cancer—but in the
absence of tobacco, alcohol does not seem to increase the risk (Wynder *et al.*,
1976).

IV. SUMMARY

In human beings, the consumption of alcoholic beverages is thus clearly associated with cancer of the buccal cavity, pharynx, larynx, and esophagus. In
animal experiments, however, ethanol does not appear to be carcinogenic. This
apparent contradiction certainly calls for further work on the mechanisms by
which these drinks induce cancer in man, the species in which we are most
particularly interested.

In the meantime, even though it is not clear whether one kind of alcoholic
beverage may be more carcinogenic than another, our present knowledge provides arguments for active preventive measures. From Fig. 3, one can easily

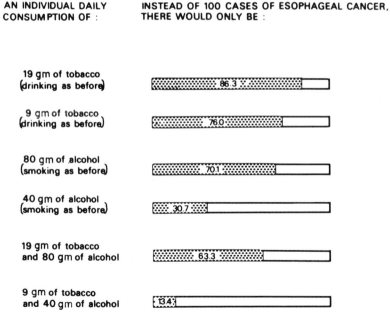

Fig. 3. Potential reduction in numbers of cases of esophageal cancer if given consumption limits were not exceeded.

derive what could be gained if a given population did not consume more alcohol or tobacco beyond a certain threshold. In the case of Ille et Vilaine, if people kept within the reasonable limits of 10 cigarettes and half a liter of wine (i.e., 40 gm of ethanol) per day, the incidence rate of esophageal cancer would be cut by 6/7 of what it is now. A reduction in alcohol consumption would definitely mean a decreased risk for cancer of the esophagus.

REFERENCES

Breslow, N. E., and Enstrom, J. E. (1974). Geographical correlations between cancer mortality rates and alcohol-tobacco consumption in the United States. *J. Natl. Cancer Inst.* **53**(3), 631–639.

Enstrom, J. E. (1975). Cancer mortality among Mormons. *Cancer (Philadelphia)* **36**, 825–841.

Gibel, W., Wildner, G. P., and Lohs, K. (1968). Untersuchungen zur Frage einer kanzerogenen und hepatotoxischen Wirkung von Füselal. *Arch. Geschwulstrforsch.* **32**, 115–125.

Gibel, W., Lohs, K., Schremmer, K., and Wildner, G. P. (1970). Experimentelle Untersuchungen über toxische Wirkungen von Alkoholbeistoffen. *Dtsch. Gesundheitswes.* **25**, 573–579.

Hakulinen, T., Lehtimaki, L., Lehtonen, M., and Teppo, L. (1974). Cancer morbidity among two male cohorts with increased alcohol consumption in Finland. *J. Natl. Cancer Inst.* **52**, 1711–1714.

Hirayama, T. (1975). Prospective studies on cancer epidemiology based on census population in Japan. *Proc. Int. Cancer Cong., 11th, Florence, 1974* **3**, 26–35.

Horie, A., Kohchi, S., and Kuratsune, M. (1965). Carcinogenesis in the esophagus. II. Experimental production of esophageal cancer by administration of ethanolic solution of carcinogens. *Gann* **56**, 429–441.

International Agency for Research on Cancer (IARC). (1976). "Annual Report," pp. 52–54. IARC, Lyon.

Jensen, O. M. (1977). Personal communication.

Kuratsune, M., Kohchi, S., and Horie, A. (1965). Carcinogenesis in the esophagus. I. Penetration of benzo(a)pyrene and other hydrocarbons into the esophageal mucosa. *Gann* **56**, 177–187.

Lemon, F. R., Walden, R. T., and Woods, R. W. (1964). Cancer of the lung and mouth in Seventh Day Adventists. A preliminary report on a population study. *Cancer (Philadelphia)* **17**(4), 486–497.

Lyon, J. L., Klauber, M. R., Gardner, J. W., and Smart, C. R. (1976). Cancer incidence in Mormons and non-Mormons in Utah, 1966–1970. *N. Engl. J. Med.* **294**(3), 129–133.

Rothman, K., and Keller, A. (1972). The effect of joint exposure to alcohol and tobacco on risk of cancer of the mouth and pharynx. *J. Chron. Dis.* **25**(12), 711–716.

Schmidt, W., and De Lint, J. (1972). Causes of death of alcoholics. *Q. J. Stud. Alcohol* **33**(1), 171–185.

Schwartz, D., Lasserre, O., Flamant, R., and Denoix, P. F. (1962). Alcool et Cancer. Résultats d'une enquête rétrospective. *Rev. Fr. Etud. Clin. Biol.* **7**, 590–604.

Sundby, P. (1967). "Alcoholism and Mortality," Natl. Inst. Alcohol Res., Publ. No. 6, pp. 82–114. Oslo Univ. Press, Oslo.

Tuyns, A. J. (1970). Cancer of the oesophagus: Further evidence of the relation to drinking habits in France. *Int. J. Cancer* **5**, 152–156.

Tuyns, A. J., and Massé, L. M. F. (1973). Mortality from cancer of the oesophagus in Brittany. *Int. J. Epidemiol.* **2**(3), 242–245.

Tuyns, A. J., and Massé, G. (1975). Cancer of the oesophagus in Brittany. An incidence study in Ille et Vilaine. *Int. J. Epidemiol.* **4**(1), 55–59.

Tuyns, A. J., and Obradovic, M. (1975). Brief communication: Unexpected high incidence of primary liver cancer in Geneva, Switzerland. *J. Natl. Cancer Inst.* **54**(1), 61–64.

Tuyns, A. J., Jensen, O. M., and Péquignot, G. (1977a). Le choix difficle d'un bon groupe de témoins dans une enquête rétrospective. *Rev. Epidemiol. Med. Soc. Sante Publique* **25**, 67–84.

Tuyns, A. J., Péquignot, G., and Jensen, O. M. (1977b). Le cancer de l'oesophage en Ille et Vilaine en fonction des niveaux de consommation d'alcool et de tabac. Des risques qui se multiplient. *Bull. Cancer* **64**(1), 45–60.

Wynder, E. L., Covey, L. S., Mabuchu, K., and Mushinski, M. (1976). Environmental factors in cancer of the larynx. *Cancer (Philadelphia)* **38,** 1591–1601.

EDITORIAL COMMENT

The occurrence of upper respiratory and esophageal cancer in Western Europe and North America shows a positive association with consumption of alcoholic beverages and smoking. Tuyns presents data (Fig. 2) which suggest a strong synergistic effect between the two, although each alone also shows a weaker positive association. The relationship in each instance appears to be dose related. Several important and unanswered questions remain, however.

One question is whether the positive association is with a particular type of alcoholic beverage. Only limited evidence is available on this point. Wynder *et al.* (1) in a study of cancer of the larynx state, "there was no increased risk for those who consumed beer or wine," but it "rises among those who consume seven or more units of whiskey either alone or predominantly." Tuyns (2) in correlating the occurrence of esophageal cancer with drinking habits in the various departments of France found suggestive evidence that the positive association was more with spirit consumption than with wine or beer. Esophageal cancer shows a particularly high incidence in Brittany and Normandy where the consumption of a type of brandy, Calvados, is high [Tuyns and Massé (3)].

A second unanswered question relates to whether it is the ethanol per se that enhances the risk of cancer of the upper respiratory and digestive tract or some other ingredient in the alcoholic beverage. The highest recorded incidences of cancer of the esophagus in the world are in certain areas of Iran and two provinces of China, all areas in which the consumption of alcohol is reported to be very low. Morton (4), on the basis of correlation studies, has suggested that ingestion of certain types of naturally occurring tannins, in whatever vehicle contained, is associated with a high incidence of esophageal cancer.

REFERENCES

1. Wynder, E. L., Bross, I. J., and Day, E. A study of environmental factors in cancer of the larynx. *Cancer (Philadelphia)* **9**, 86–110 (1956).
2. Tuyns, A. J. Cancer of the esophagus: Further evidence of the relation to drinking habits in France. *Int. J. Cancer* **5**, 152–156 (1970).
3. Tuyns, A. J., and Massé, G. Cancer of the esophagus in Brittany—An incidence study in Ille et Villaine. *Int. J. Epidemiol.* **4**, 55–59 (1975).
4. Morton, J. F. Tentative correlations of plant usage and esophageal cancer zones. *Econ. Bot.* **24**, 217–226 (1970).

28

ALCOHOL IN PREGNANCY AND ITS EFFECTS ON OFFSPRING

Eileen M. Ouellette

I.	Introduction	439
II.	Historical Review	440
	A. Clinical	440
	B. Laboratory Investigations	441
III.	Modern Studies	441
	A. Clinical Investigations	441
	B. Current Animal Investigations	443
IV.	Discussion	445
	A. Alcohol and Its Metabolites	445
	B. Hypoglycemia	446
	C. Vitamin Deficiencies	446
	D. Amino Acid Deficiencies	447
	E. Trace Metal Deficiencies	447
	F. Malnutrition	447
	G. Drug Abuse	448
	H. Smoking	448
	I. Pathology	448
V.	Research Needs	448
	References	449
	Editorial Comment	454

I. INTRODUCTION

Alcohol is our most widely used drug. Excessive alcohol ingestion has long been known to damage the mature nervous system, but its effects on the developing nervous system are still unknown. It may be a significant cause of fetal malformations and developmental delay in offspring of alcoholic women.

FERMENTED FOOD BEVERAGES IN NUTRITION

439

If alcohol is teratogenic to the fetus, it is vital that this be proved since the proportion of women in the United States who drink and drink heavily has been rising rapidly in the past few years (3,12). There are approximately 4–5 million women alcoholics in the United States and the group from 21 to 29 years appears to have the highest proportion of heavy drinkers according to the most recent report by the National Institute on Alcohol Abuse and Alcoholism to the U.S. Congress (44).

The role of maternal alcohol abuse during pregnancy in the production of congenital anomalies, growth abnormalities, and developmental delay in off-spring has recently become a subject of great interest (22,28–34,48–51). This constellation of abnormalities in offspring of heavy drinking women has been termed "the Fetal Alcohol Syndrome" (29,32). Numerous clinical and labora-tory studies are underway to determine the frequency and scope of abnormalities seen (8,9,16,54,55,57,60–62,74–77). It is important to recognize that although this topic is currently generating much interest, there are centuries of anecdotal and historical information pertaining to it (6,17,63,70–72,80,84,89).

II. HISTORICAL REVIEW

A. Clinical

1. Anecdotal

The first descriptive information concerning this syndrome comes from Greek mythology where the god Hephaestus, son of Hera and Zeus, was said to be deformed because of the intoxication of his parents at the time of his conception (89). Indeed, in the ancient world, both Sparta and Greece had rules forbidding the ingestion of alcoholic beverages by newly married couples at their wedding reception in order to prevent abnormalities in their offspring.

Throughout the Middle Ages, a folklore developed concerning the deleterious effects of alcohol ingestion at the time of conception. During this time, alcohol ingestion was primarily in the form of beer and wine. With the invention of gin in Holland in the late seventeenth century and the introduction of distilled beverages into England a marked rise in consumption took place and the so-called gin epidemic occurred.

2. Historical

Parliament became interested in the problem of alcoholism because of its rapid increase in England, and as early as 1726 reports began to appear concerning the occurrence of epilepsy, mental deficiency, and other abnormalities in offspring of alcoholic parents (6). Reports of this type continued to appear in the European and American literature (67,80,84). In the 1950s and 1960s, several reports were published in France citing increased abnormalities in offspring of alcoholic parents (10,18). Lemoine *et al.* in 1968 reported for the first time an identifiable syndrome in 127 offspring of alcoholics, mainly women in Marseilles (37). They stated that the resemblance of the children to each other was so striking that maternal alcoholism could be diagnosed from viewing the child.

B. Laboratory Investigations

Meanwhile, embryologists and teratologists began scientific studies into the production of monstrosities in the early twentieth century, and ethanol was one of the agents used. Early studies focused on its effects in decreasing litter size, number of litters, and increasing sterility in a number of laboratory animals fed ethanol before and during pregnancy (39). Stockard and Papanicolaou first reported a number of cerebral and limb abnormalities in such animals but because of poor controls their work was criticized and disbelieved (70–72).

III. MODERN STUDIES

A. Clinical Investigations

In 1972, Ulleland reviewed newborn records in Seattle, Washington looking for causes of small-for-gestational-age babies (86). She discovered that a very high proportion of infants born to chronic alcoholic women were small for gestational age. Jones *et al.* in 1973 then examined these babies and discovered that most of them showed a recognizable pattern of congenital anomalies as well as prenatal and postnatal growth retardation (28). They termed this "the fetal alcohol syndrome" (29). Since that time, numerous investigators mainly in retrospective studies have reported on the occurrence of a number of similar abnormalities in offspring of heavy drinking and chronic alcoholic women (Table I).

Jones *et al.* also reviewed the data from the Collaborative Perinatal Project of the National Institute of Neurologic Diseases and Stroke study of 55,000 infants followed up to 7 years of age to assess the frequency of this abnormality (31). They discovered that questions concerning maternal alcohol use had not been asked in this study. From data given, however, they were able to identify a small number of women who clearly showed advanced stages of alcoholism at the time of delivery and noted that offspring of the 23 women thus identified had a perinatal mortality rate of 17% versus 2% among controls carefully matched for age, race, education, parity, and socioeconomic level.

These data were all retrospective, and their publication proved to be controversial. It is felt by many that the identification of only 23 women out of 55,000

Table I.　Abnormal Features Compared in the Affected Children

Cases	Numbers affected of those evaluated		
	Jones et al. (28)	Jones and Smith (29)	Palmer et al. (51)
Growth			
Prenatal growth deficiency	8/8	3/3	3/3
Postnatal growth deficiency	8/8	2/2	3/3
Neurological			
Developmental delay	8/8	2/2	3/3
Microcephaly	7/8	3/3	3/3
Fine motor disorders	5/6	2/2	3/3
Gross motor disorders	7/7	2/2	3/3
Facial			
Short palpebral fissures	8/8	3/3	3/3
Epicanthal folds	4/8	0/3	0/3
Asymmetric ptosis	1/8	0/3	1/3
Strabismus	2/8	—	1/3
Myopia	1/8	—	1/3
Maxillary hypoplasia	7/8	0/3	?2/3
Micrognathia and/or cleft palate	0/8	2/3	0/3
Deficient superior helix of ear	3/8	1/3	0/3
Limbs			
Hip dislocation	2/8	2/3	?1/3
Elbow limitation	2/8	1/3	0/3
Phalangeal anomalies	4/8	1/3	1/3
Altered palmar creases	6/8	1/3	3/3
Other			
Cardiac anomalies	5/8	3/3	?1/3
Anomalous external genitalia	2/5	2/2	3/3
Capillary hemangioma	3/8	1/3	1/3
Accessory nipple	1/8	0/3	1/3
Pectus excavatum	2/8	0/3	0/3

pregnancies represented a sample so deviant that little could be said about the actual risk to offspring. Three prospective studies are currently under way to examine this problem in different socioeconomic groups, and data from these studies are not yet completely analyzed.

In the study we carried out at the Boston City Hospital from May 1974 to June 1976 we found that 10% of women living in an inner-city ghetto were classified as heavy drinking at the first prenatal clinic visit, using the criteria of Cahalan et al. (7,49,50,57). They had a consistent daily average of at least 45 ml of absolute alcohol and consumed five or more drinks on occasion. All the women were poor with two-thirds of them having a monthly income of less than $400. Nearly half of them were not living with the father of the baby at the time of the first prenatal clinic interview, and therefore data on fathers could not be obtained.

A detailed questionnaire was given to each woman at the time of the first prenatal clinic visit in which their drinking and nutritional habits were ascertained (56). Additional sociometric data and history concerning drug abuse and smoking were obtained. Counseling was provided for women who were determined to be heavy drinkers. At delivery, the infants were examined by one pediatric neurologist who had no prior knowledge of either the mother's drinking or pregnancy history. Following recording of data, additional historical information was sought.

The results of this study showed that infants born to heavy drinking women had twice the risk of abnormality as those born to abstinent, rare, and moderate drinking women. Seventy-one percent of the infants born to heavy drinking women were found to be abnormal as compared to 35% in the other two groups. The Boston City Hospital is a high risk population and approximately 35% of all infants are admitted to the special care unit following delivery. This high rate of abnormality in the control groups correlated well with the overall percentage of abnormal infants found generally in the nursery.

The abnormalities found in these offspring consisted of intrauterine growth retardation and an increase risk of prematurity, occurring in 20% of the infants at risk compared to 5% of controls, but there was no increase in fetal or neonatal deaths. Infants tended to be small for gestational age and five of 42 babies born to heavy drinking women were microcephalic as compared to one of 274 infants born to abstinent or moderate drinkers (14,38,78).

Congenital anomalies were present in 32% of the infants born to heavy drinking women as compared to 9% of other babies (40). Both single and multiple, major and minor anomalies were increased. A reproducible syndrome could not be identified (Table II). There were no differences in offspring which could be accounted for by race, maternal age, parity, or nutrition. The most disturbing feature found was the high incidence of microcephaly, which was defined as head circumference at birth which fell beyond 2 standard deviations below the mean for gestational age.

B. Current Animal Investigations

In 1954, Papara-Nicholson and Telford and later Sandor *et al.* reported on the appearance of abnormalities in chickens and guinea pigs following exposure to alcohol in fetal life (2,60–62). Subsequently, more extensive work by Randall

Table II. Congenital Anomalies

Group	Craniofacial		Limb		Cardiac		Other	
I[a] (N = 151)	Microcephaly	(1)	Abnormal palmar crease	(2)	Patent ductus arteriosus	(1)	Ectodermal dysplasia	(1)
	Abnormal ear	(7)	Polydactyly	(3)			Ventral hernia	(1)
	Micrognathia	(1)	Clinodactyly	(1)			Vaginal skin tag	(1)
			Syndactyly	(2)				
			Phocomelia	(1)				
II[b] (N = 128)	Microcephaly	(0)	Abnormal palmar crease	(3)	Patent ductus arteriosus	(3)	Bilateral hip dislocation	(3)
	Abnormal ear	(9)	Polydactyly	(0)	Ventricular septal defect	(2)	Hypoplastic penis	(1)
	High-arched palate	(1)	Clinodactyly	(1)	Pulmonary artery coarctation	(1)	Accessory nipple	(1)
			Club foot	(2)			Renal anomaly	(1)
III[c] (N = 41)	Microcephaly	(5)	Abnormal palmar crease	(1)	Patent ductus arteriosus	(3)	Renal anomaly	(2)
	Abnormal ear	(3)	Polydactyly	(2)	Ventricular septal defect	(1)	Absent rib	(1)
	Micrognathia	(1)	Rocker bottom feet	(1)	Pulmonary artery coarctation	(1)		
	Asymmetric facies	(1)						
	Cortical hypoplasia	(1)						
	Short neck	(1)						
	Redundant neck skin	(1)						
	Beak nose	(1)						

[a] Major anomalies (5), minor anomalies (8), multiple anomalies (5).
[b] Major anomalies (3), minor anomalies (15), multiple anomalies (7).
[c] Major anomalies (7), minor anomalies (6), multiple anomalies (8).

has demonstrated the occurrence of teratogenic effects of alcohol fed to pregnant rats, and the types of abnormalies noted have been remarkable for their similarities to those seen in humans (2,54,55).

A mouse model of the fetal alcohol syndrome has been developed by Chernoff, and he has also found that the number of abnormalities discovered varies proportionally to the percentage of alcohol consumed in the diet (8,9). Neural malformations occur with maternal blood alcohol levels well below those diagnostic of alcoholism. In both of these studies, the nutritional input has been well controlled, and it seems quite clear that alcohol or one of its metabolites is acting as the offending agent. Although alcohol passes directly via the placenta into the fetus, there is no evidence that acetaldehyde enters any mammalian fetus, and it is believed that the placenta itself may act as a barrier by metabolizing acetaldehyde (13,25-27,35,36,53,65).

Chernoff in enzyme assays on maternal mouse liver found an inverse relationship between alcohol dehydrogenase activity and maternal blood alcohol levels (9). A positive relationship of microsomal ethanol oxidizing system (MEOS) and maternal blood alcohol levels was discovered, raising the possibility that metabolic products of MEOS induction may act as secondary teratogens.

Behavioral changes in progeny born to mice and rats given ethanol when pregnant have been noted in several recent studies (1,4,5,42). Difficulties in learning mazes, decreased emotionality, and the occurrence of seizures in previously resistant strains have been seen. A great many other laboratory studies are underway but many results are still unpublished (81,85).

IV. DISCUSSION

It appears quite clear from both clinical and laboratory studies that there is a definite increased risk of growth and morphological and neurological abnormality in offspring of animals and humans who consume excessive amounts of alcohol during pregnancy. The existence of a definitive, reproducable syndrome has not yet been conclusively proved. There are a number of factors which might produce these effects.

A. Alcohol and Its Metabolites

Alcohol may be acting as a direct teratogen. It passes the placenta and readily enters the fetus (13,25-27). Studies done on premature and term infants show that clearance rates are prolonged and sometimes doubled in humans (27,88). No data are available for the fetus. The central nervous system depressant effects of alcohol may play some role in altering the activity of enzyme systems and thus produce abnormalities.

Alternatively or additively, any of the metabolic breakdown products of alcohol may also be teratogenic. Acetaldehyde is a known toxin and a sympathomimetic agent, but does not cross the placenta (35,65). Chernoff has found that levels of the microsomal ethanol oxidizing system (MEOS) in the mouse fetus rise with the maternal blood alcohol level and suggests this might be acting as a secondary teratogen (9).

B. Hypoglycemia

Alcohol ingestion produces hypoglycemia. Infants of diabetic mothers have an increased risk of congenital anomalies. Up to 15% of infants of diabetic mothers have shown major congenital anomalies in some series. It may be that a bolus of alcohol consumed by a mother early in pregnancy results in hypoglycemia which in turn damages the fetal brain and other organs. If this were so, it would appear likely that a large bolus of alcohol consumed on infrequent occasions might prove to be more deleterious to the baby than a steady consistent dose of alcohol consumed daily where the peak blood alcohol levels remained lower. At the present time, there is no information on this subject.

C. Vitamin Deficiencies

Deficiency of thiamine and folic acid may be potential contributors to the development of fetal abnormalities (45,46,73). These vitamins are often deficient in adult alcoholics (15,21,87). Folic acid deficiency has resulted in spontaneous and habitual abortions, prematurity, and fetal abnormalities in the rat (21,23,24,73,82). Aminopterin, a folic acid antagonist, is also embryotoxic (19,82).

Blood levels of vitamins A, C, and folic acid have been measured in several mothers of infants with the fetal alcohol syndrome and found to be normal (22,29). Possible effects of ethanol on intestinal absorption of folic acid have been reported (21). Hematologic response to folic acid therapy was repeatedly prevented by the concomitant administration of whiskey, wine, or ethanol. Eichner and Hillman found that when a folate-poor diet was given to alcoholics along with ethanol, megaloblastic changes developed much more rapidly than when the same diet was given to these subjects without ethanol (15). These findings suggest that alcohol administration causes megaloblastic changes only when body vitamin stores are decreased and dietary intake is poor. Under these circumstances, ethanol may act as a weak folate antagonist.

Little appears to be known about the fetal effects of maternal thiamine deficiency.

D. Amino Acid Deficiencies

Amino acid stores decrease in women during pregnancy. It is possible that women who have chronically consumed large amounts of alcohol prior to becoming pregnant may enter pregnancy with diminished stores of these vital nutrients and that this may in turn contribute to the poor outcome in their offspring. Zamenhoff *et al.* have shown that separate omission of some single amino acids in pregnant rats produces decreased body and brain weights in their offspring and decreased levels of cerebral DNA (93).

E. Trace Metal Deficiencies

Trace metals deficiencies, particularly zinc and magnesium, have been shown to produce by themselves abnormalities which are strikingly similar to those seen in offspring of alcoholic women (20,47). Many alcoholics have deficiencies in these particular trace metals, and they may be playing a role in the production of this syndrome (79,83).

F. Malnutrition

A voluminous literature, both clinical and experimental, has evolved from the study of malnutrition (11,43,66,68,69,90–92). Smith studied the effects of acute severe malnutrition on pregnancy in several hundred pregnant women in Holland during the final days of World War II (66). This work demonstrated an increase in small and premature infants, but malformed infants accounted for only 0.5% of the deliveries. More recent work in Latin America demonstrated diminished stature, head size, and intellectual accomplishments in offspring of chronically malnourished mothers, but an increased occurrence of malformations was not noted (11). Stoch and Smythe in South Africa also found that malnutrition in early life impaired body growth, especially body weight, to a greater extent than head growth and psychomotor development (68,69). If such children are fed adequately postnatally, catch-up growth occurs.

By contrast, in infants previously described with the fetal alcohol syndrome, the head circumference and body length are affected more than body weight, and this discrepancy persists postnatally, even when the diet is known to be adequate. In addition, many of the mothers of the patients with the fetal alcohol syndrome have been well followed throughout their pregnancies, and malnutrition has not been noted to be present (22,29,50).

Such clinical data are always suspect because historical data concerning nutritional and alcohol intake may not be accurate. Recent studies, well-controlled for nutrition, have produced in animals many of the features seen in human offspring of heavy drinking and alcoholic women (54,55,85). These data are more convincing in ruling out malnutrition alone as the causative factor.

G. Drug Abuse

Infants of drug-addicted mothers often show withdrawal symptoms in the neonatal period, but growth abnormalities and an increase in congenital anomalies have not been noted (58).

H. Smoking

Smoking during pregnancy is associated with intrauterine growth retardation, as well as an increase in prematurity and perinatal mortality (59,64). Heavy drinkers tend to be heavy smokers (49,50,57). Clinical studies to date have not been able to separate out the effects of smoking from alcohol on the production of smaller sized babies. There is no evidence, however, that smoking is a cause of fetal malformations.

I. Pathology

To date, there is only one known autopsied case of the fetal alcohol syndrome (29). The brain in this case showed lissencephaly, or smooth brain. This is a rare abnormality resembling a fetal brain of 4 months' gestation and may be inherited as an autosomal recessive (41). The production of this abnormality by alcohol abuse remains problematical.

V. RESEARCH NEEDS

Studies of the embryopathic effects of alcohol are in their infancy. At the present time, the risk of abnormality to offspring and the scope of these effects are unknown. Extensive well-controlled clinical epidemiologic studies are currently underway in several parts of the country to assess these issues among women in differing ethnic and socioeconomic groups.

Follow-up studies of offspring at risk should be carried out since many abnormalities are not identified at birth, and other findings such as hyperactivity and learning disorders may not appear until several years later.

Many laboratory investigations are currently underway with regard to changes in growth and biochemical composition of the brain in animals exposed to ethanol *in utero*. More detailed analyses of the effects of different patterns of ethanol exposure are needed and far more extensive biochemical measures of maternal and fetal levels of vitamins, trace metals, and metabolic breakdown products of alcohol should be done.

If these deficiencies are found to exert an influence on the production of fetal abnormalities, there remains the potential for intervention and ultimately the hope of prevention of these effects.

Smith believes that maternal alcohol abuse may be the third most common cause of mental retardation. As such, it is totally preventable. Educational programs not only during pregnancy, but for young women of junior high school age, will be vital if these risks to the fetus are to be minimized and ultimately eliminated.

REFERENCES

1. Arlitt, A. The effect of alcohol on the intelligent behavior of the white rat. *Psychol. Monogr.* **26,** 1–50 (1919).
2. Barrow, M. V., and Taylor, W. J. A rapid method for detecting malformations in rat fetuses. *J. Morphol.* **127,** 291–306 (1969).
3. Belfer, M. L., Shader, R. K., Carroll, M., and Harmatz, J. S. Alcoholism in women. *Arch. Gen. Psychiatry* **25,** 540–545 (1971).
4. Bond, N. W., and Digusto, E. L. Effects of prenatal alcohol consumption on open-field behavior and alcohol preference in rats. *Psychopharmacology* **46,** 163–168 (1976).
5. Branchey, L., and Friedhoff, A. J. Biochemical and behavioral changes in rats exposed to ethanol in utero. *Ann. N.Y. Acad. Sci.* **273,** 328–330 (1976).
6. Burton, R. "The Anatomy of Melancholy," Vol. 1. William Tegg, London, 1806.
7. Cahalan, D., Cisin, I. H., and Crossley, H. M. "American Drinking Practices: A National Study of Drinking Behavior and Attitudes," Monogr. No. 6. Rutgers Cent. Alcohol Stud., New Brunswick, New Jersey.
8. Chernoff, G. F. A mouse model of the fetal alcohol syndrome. *Teratology* **11,** 14a (1975).
9. Chernoff, G. F. The fetal alcohol syndrome in mice: An animal model. *Teratology* **15,** 223–229 (1977).
10. Christiaens, L., Miron, L. P., and Demarle, G. Sur la descendance des alcooliques. *Ann. Pediatr. (Paris),* 36, 37–42 (1960).
11. Cravioto, J., DeLicardie, E. R., and Birch, H. C. Nutrition, growth and neurointegrative development: An experimental and ecologic study. *Pediatrics* **38,** Suppl. 2, 319–372 (1966).
12. Criteria Committee, National Council on Alcoholism. Criteria for the diagnosis of alcoholism. *Ann. Intern. Med.* **77,** 249–258 (1972).
13. Dilts, P. V. Placental transfer of ethanol. *Am. J. Obstet. Gynecol.* **107,** 1195–1198 (1970).
14. Dubowitz, L. M. S., Dubowitz, V., and Goldberg, C. Clinical assessment of gestational age in the newborn infant. *J. Pediatr.* **77,** 1–10 (1970).
15. Eichner, E. R., and Hillman, R. S. The evolution of anemia in alcoholic patients. *Am. J. Med.* **50,** 218–232 (1971).
16. Ellis, F. W., and Pick, J. R. Beagle model of the fetal alcohol syndrome (abstract). *Pharmacologist* **18,** 190 (1976).
17. Fielding, H. "An Inquiry Into the Causes of the Late Increase of Robbers, etc., With Some Proposals for Remedying This Growing Evil," pp. 19–20. A. Millar, London, 1751.
18. Giroud, A., and Tuchmann-Duplessis, H. Malformations congenitales, roles des facteurs exogenes. *Pathol. Biol.* **10,** 141–145 (1962).

19. Goestch, C. An evaluation of aminopterin as an abortifacient. *Am. J. Obstet. Gynecol.* **83,** 1474–1477 (1962).
20. Gunther, T., Dern. F., and Merker, H. J. Embryo-toxic effects produced by magnesium deficiency in rats. *Chem. Klin. Biochem.* **11,** 87–92 (1973).
21. Halsted, C. H., Griggs, R. C., and Harris, J. W. The effect of alcoholism on the absorption of folic acid (H³-PGA) evaluated by plasma levels and urine excretion. *J. Lab. Clin. Med.* **69,** 116–131 (1967).
22. Hanson, J. W., Jones, K. L., and Smith, D. W. Fetal alcohol syndrome: Experience with 41 patients. *Am. Med. Assoc.,* **235,** 1458–1460 (1976).
23. Hibbard, B. M. The role of folic acid in pregnancy with particular reference to anemia, abruption and abortion. *J. Obstet. Gynecol. Br. Commonw.* **71,** 529–542 (1964).
24. Hibbard, E. D., and Smithells, R. W. Folic acid metabolism and human embryopathy. *Lancet* **i,** 1254 (1965).
25. Ho, B. T., Fritchie, G. E., Idanpaan-Heikkila, J. E., and McIsaac, W. M. Placental transfer and tissue distribution of ethanol-1-¹⁴C. *Q. J. Stud. Alcohol* **33,** 485–493 (1972).
26. Idanpaan-Heikkila, J. E., Ho, B. T., and McIsaac, W. M. Placental transfer of C-14 ethanol. *Am. J. Obstet. Gynecol.* **110,** 426–428 (1971).
27. Idanpaan-Heikkila, J. E., Jouppila, P., Aberblum, H. K., Isoaho, R., Kauppila, E., and Koiuisto, M. Elimination and metabolic effects of ethanol in mother and fetus, and newborn infant. *Am. J. Obstet. Gynecol.* **112,** 387–393 (1972).
28. Jones, K. L. Smith, D. W. Ulleland, C., and Streissguth, A. Pattern of malformation in offspring of chronic alcoholic mothers. *Lancet* **i,** 1267–1271 (1973).
29. Jones, K. L., and Smith, D. W. Recognition of the fetal alcohol syndrome in early infancy. *Lancet* **ii,** 999–1001 (1973).
30. Jones, K. L., Smith, D. W., Streissguth, A. P., and Myrianthopoulos, N. C. Incidence of the fetal alcohol syndrome in offspring of chronically alcoholic women. *Pediatr. Res.* **8,** 440–446 (1974).
31. Jones, K. L., Smith, D. W., Streissguth, A. P., and Myrianthopoulos, N. C. Outcome of offspring of chronic alcoholic women. *Lancet* **i,** 1076–1078 (1974).
32. Jones, K. L. The fetal alcohol syndrome. *Addict. Dis.* **2**(1), 79–88 (1975).
33. Jones, K. L., and Smith, D. W. Fetal alcohol syndrome. *Teratology* **12,** 1–10 (1975).
34. Jones, K. L., Smith, D. W., and Hanson, J. W. Fetal alcohol syndrome: Clinical delinations. *Ann. N.Y. Acad. Sci.* **273,** 130–137 (1976).
35. Kesaniemi, Y. A. Ethanol and acetaldehyde in the milk and peripheral blood of lactating women after ethanol administration. *J. Obstet. Gynecol. Br. Commonw.* **81,** 84–86 (1974).
36. Kronick, J. B. Teratogenic effects of ethyl alcohol administered to pregnant mice. *Am. J. Obstet. Gynecol.* **124,** 676–680 (1976).
37. Lemoine, P., Harousseau, H., Borteyru, J. P., and Menur, J. C. Les enfants de parents alcooliques. Anomalies observees. *Ouest Med.* **25,** 476–482 (1968).
38. Lubchenco, L., Hansman, C., and Boyd, E. Intrauterine growth in length and head circumference as estimated from live births at gestational ages from 26 to 42 weeks. *Pediatrics* **37,** 403–408 (1966).
39. MacDowell, E. C. The influence of alcohol in the fertility of white rats. *Genetics* **7,** 117–141 (1922).
40. Marden, P. M., Smith, D. W., and McDonald, M. J. Congenital anomalies in the newborn infant, including minor variations. *J. Pediatr.* **64,** 357–371 (1964).
41. Miller, J. Lissencephaly in two siblings. *Neurology* **13,** 841–850 (1963).
42. Morra, M. Ethanol and maternal stress on rat offspring behaviors. *J. Gen. Psychol.* **114,** 77–83 (1969).
43. Naeye, R. L., Blanc, W., and Paul, C. Effects of maternal nutrition on the human fetus. *Pediatrics* **52,** 494–503 (1973).

44. National Institute on Alcohol Abuse and Alcoholism. "Second Report to U.S. Congress." U.S. Dep. Health, Educ. Welfare, Washington, D.C., 1974.

45. Nelson, M. M., and Evans, H. M. Reproduction in rats on purified diets containing succinylfulfathizole. *Proc. Soc. Exp. Biol. Med.* **66,** 289–291 (1947).

46. Nelson, M. M., Asling, C. W., and Evans, H. M. Production of multiple congenital abnormalities in young by pteroyl glutamic acid deficiency during gestation. *J. Nutr.* **48,** 61–79 (1952).

47. Oberleas, D., Caldwell, D. F., and Prasad, A. S. Trace elements and behavior. *In* "Neurobiology of the Trace Metals Zinc and Copper" (C. C. Pfeiffer, ed.), pp. 83–103. Academic Press, New York, 1972.

48. Ouellette, E. M. The fetal alcohol syndrome, additional familial cases. *Proc. Natl. Meet. Child. Neurol. Soc., 3rd, Madison, Wis.,* p. 5 (1974).

49. Ouellette, E. M., and Rosett, H. L. A pilot prospective study of the fetal alcohol syndrome at the Boston City Hospital. Part II: The infants. *Ann. N.Y. Acad. Sci.* **273,** 123–129 (1976).

50. Ouellette, E. M., Rosett, H. L., Rosman, N. P., and Weiner, L. The adverse effects of maternal alcohol abuse during pregnancy in offspring. *N. Engl. J. Med.* **297,** 528–530 (1977).

51. Palmer, R. H., Ouellette, E. M., Warner, L., and Leichtman, S. Congenital malformations in offspring of a chronic alcoholic mother. *Pediatrics* **53,** 490–494 (1974).

52. Papara-Nicholson, D., and Telford, I. R. Effects of alcohol on reproduction and fetal development in the guinea pig. *Anat. Rec.* **127,** 438–439 (1957).

53. Raiha, N. C. R., Koskinen, M., and Pikkarainen, P. Developmental changes in alcoholdehydrogenase activity in rat and guinea pig liver. *Biochem. J.* **103,** 623–626 (1967).

54. Randall, C. M., Taylor, J. W., and Walker, D. W. Teratogenic effects of prenatal ethanol exposure. *Proc. Natl. Counc. Alcohol. Meet., Washington, D.C., 1976* (1977).

55. Randall, C. M., Taylor, W. J., and Walker, D. W. Teratogenic effects of prenatal ethanol exposure. *In* "Alcohol and Opiates" (K. Blum, ed.), pp. 92–107. Academic Press, New York, 1977.

56. "Recommended Dietary Allowances." Food Nutr. Board, Natl. Acad. Sci., Natl. Res. Counc., Washington, D.C., 1973.

57. Rosett, H. L., Ouellette, E. M., and Weiner, L. A prospective study of the fetal alcohol syndrome at Boston City Hospital. Part I: Maternal drinking. *Ann N.Y. Acad. Sci.* **273,** 118–122 (1976).

58. Rothstein, P., and Gould, J. B. Born with a habit: Infants of drug addicted mothers. *Pediatr. Clin. North Am.* **21,** 307–321 (1974).

59. Rush, D., and Kass, E. H. Maternal smoking: A reassessment of the association with perinatal mortality. *Am. J. Epidemiol.* **96,** 183–196 (1972).

60. Sandor, S., and Elias, S. The influence of aethyl-alcohol on the development of the chick embryo. *Rev. Roum. Embryol. Cytol., Ser. Embryol.* **5,** 51–76 (1968).

61. Sandor, S. The influence of aethyl-alcohol on the developing chick embryo. *Rev. Roum. Embryol. Cytol., Ser. Embryol.* **5,** 167–171 (1968).

62. Sandor, S., and Amels, D. The action of aethanol on the prenatal development of albino rats. *Rev. Roum. Embryol. Cytol., Ser. Embryol.* **8,** 105–118 (1971).

63. Sedgewick, J. "A New Treatise on Liquors, Wherein the Use and Abuse of Wine, Malt Drinks, Water, etc., are Particularily Considered in Many Diseases, Constitutions and Ages, with the Proper Manner of Using Them, Hot or Cold, either Physick, Diet or Both." Charles Rivington, London, 1725.

64. Simpson, W. J. A preliminary report on cigarette smoking and the incidence of prematurity. *Am. J. Obstet. Gynecol.* **73,** 808–815 (1957).

65. Sippel, H. W., and Kesaniemi, Y. A. Placental and foetal metabolism of acetaldehyde in rat. II. Studies on metabolism of acetaldehyde in the isolated placenta and foetus. *Acta Pharmacol. Toxicol.* **37,** 49–55 (1975).

66. Smith, C. Effects of maternal undernutrition upon the newborn infant in Holland (1944-1945). *J. Pediatr.* **30,** 229-243 (1947).
67. Stevens, J.'P. Some of the effects of alcohol upon the physical constitution of man. *South. Med. Surg. J.* **13,** 451-462 (1857).
68. Stoch, M. B., and Smythe, P. M. Does undernutrition during infancy inhibit brain growth and subsequent intellectual development? *Arch. Dis. Child.* **38,** 546-552 (1963).
69. Stoch, M. B., and Smythe, P. M. The effect of undernutrition during infancy on subsequent brain growth and intellectual development. *S. Afr. Med. J.* **41,** 1027-1030 (1967).
70. Stockard, C. R. An experimental study of racial degeneration in mammals treated with alcohol. *Arch. Intern. Med.* **10,** 369-398 (1912).
71. Stockard, C. R., and Papanicolaou, G. A further analysis of the hereditary transmission of degeneracy and deformities by the descendants of alcoholized mammals. *Am. Nat.* **50,** 65-88 (1916).
72. Stockard, C. R., and Papanicolaou, G. U. Further studies in modification of the germ cells in mammals: The effect of alcohol on treated guinea pigs and their descendants. *J. Exp. Zool.* **26,** 119-226 (1918).
73. Stone, M. L. Effects on the fetus of folic acid deficiency in pregnancy. *Clin. Obstet. Gynecol.* **11,** 1143-1153 (1968).
74. Streissguth, A. P. Psychologic handicaps in children with fetal alcohol syndrome. *Ann. N.Y. Acad. Sci.* **273,** 140-145 (1976).
75. Streissguth, A. P., Martin, D. C., and Buffington, V. E. Test-retest reliability of three scales derived from a quantity-frequency-variability assessment of self-reported alcohol consumption. *Ann. N.Y. Acad. Sci.* **273,** 458-466 (1976).
76. Streissguth, A. P. Maternal alcoholism and the outcome of pregnancy. A review of the fetal alcohol syndrome. *In* "Alcoholism Problems in Women and Children" (M. Greenblatt, and M. D. Schuchit, eds.), pp. 251-274. Grune & Stratton, New York, 1976.
77. Streissguth, A. P., Martin, D. C., and Buffington, V. E. Identifying heavy drinkers: A comparison of eight alcohol scores obtained on the same sample. *Ann. N.Y. Acad. Sci.* **273,** 458-466 (1976).
78. Stuart, H. C. Dept. Maternal Child Health, Harvard Sch. Public Health, Children's Hosp. Med. Cent., Boston, Massachusetts.
79. Sullivan, J. F., and Lankford, H. G. Zinc metabolism in chronic alcoholism. *Am. J. Clin. Nutr.* **17,** 57-63 (1965).
80. Sullivan, W. C. A note on the influence of maternal inebriety in the offspring. *J. Ment. Sci.* **45,** 489-503 (1899).
81. Sze, P., Yanai, J., and Ginsberg, B. E. Effects of early ethanol input on the activities of ethanol-metabolizing enzymes in mice. *Biochem. Pharmacol.* **25,** 215-217 (1976).
82. Thiersch, J. B. Therapeutic abortions with folic acid antagonists, 4 amino-pteroylglutamic acid (4-amino PGA) administered by oral route. *Am. J. Obstet. Gynecol.* **63,** 1298-1304 (1952).
83. Traviesa, D. C. Magnesium deficiency: A possible cause of thiamine refractoriness in Wernicke-Kosakoff encephalopathy. *J. Neurol., Neurosurg. Psychiatry* **35,** 959-962 (1974).
84. Triboulet, H., Matthieu, F., and Mignot, R. "Traite de l'Alcoolisme." Masson, Paris, 1905.
85. Tze, W. J., and Lee, M. Adverse effects of maternal alcohol consumption on pregnancy and foetal growth in rats. *Nature (London)* **257,** 479-480 (1975).
86. Ulleland, C. The offspring of alcoholic mothers. *Ann. N.Y. Acad. Sci.* **197,** 167-169 (1972).
87. Victor, M., Adams, R. D., and Collins, C. H., eds. Experimentally induced thiamine deficiency. *In* "The Wernicke-Korsakoff Syndrome," pp. 147-154. Davis, Philadelphia, Pennsylvania, 1971.
88. Wagner, L., Wagner, G., and Guerrero, J. Effect of alcohol on premature newborn infants. *Am. J. Obstet. Gynecol.* **108,** 308-315 (1970).

89. Warner, R. H., and Rosett, H. L. The effects of drinking in offspring: An historical survey of the American and British Literature. *Q. J. Stud. Alcohol* **36**(11), 1395–1420 (1975).
90. Winick, M. Malnutrition and brain development. *J. Pediatr.* **74,** 667–679 (1969).
91. Winick, M., and Rosso, P. Head circumference and cellular growth of the brain in normal and marasmic children. *J. Pediatr.* **74,** 774–778 (1969).
92. Winick, M., and Rosso, P. The effect of severe early malnutrition on cellular growth of the human brain. *Pediatr. Res.* **3,** 181–184 (1969).
93. Zamenhoff, S., Hall, S. M., Grauel, L., Von Marthens, E., and Donahue, M. J. Deprivation of amino acids and prenatal brain development in rats. *J. Nutr.* **104,** 1002–1007 (1974).

EDITORIAL COMMENT

For the most part the fetal alcohol syndrome has been recognized with confidence only in children born of mothers who were consuming substantial amounts of alcohol daily or intermittently during the pregnancy. One estimate has been that the equivalent of six drinks daily will constitute a major risk to the fetus but that one cannot say that there is an absolutely safe level of ethanol consumption during pregnancy. One can speculate that small amounts of alcohol might cause some blunting of intelligence, perhaps the most prominent characteristic of the fetal alcohol syndrome. Convincing evidence of harm to the offspring is seen only with substantial consumption of alcohol, however [Clarren, S. K., and Smith, D. W. The fetal alcohol syndrome. *N. Engl. J. Med.* **298,** 1063–1067 (1978)].

VI

An Experimental Model

29

MINIATURE SWINE AS A MODEL FOR THE STUDY OF HUMAN ALCOHOLISM: THE WITHDRAWAL SYNDROME

M. E. Tumbleson, D. P. Hutcheson, J. D. Dexter, and C. C. Middleton

I. Introduction 457
II. Experiments on Physical Dependence 460
   A. Experiment I 460
   B. Experiment II 463
   C. Experiment III 463
III. Discussion 471
IV. Summary and Conclusions 472
   References 473

I. INTRODUCTION

Definition of an animal model for the study of human alcoholism has been successful only to a limited extent. As models, miniature swine satisfy many of the criteria for the diagnosis of alcoholism (Criteria Committee, 1972), i.e., (1) voluntary consumption equivalent to one-fifth of a gallon whiskey, for more than 1 day, by an 82 kg individual (Dexter *et al.*, 1976); (2) blood ethanol level at any time of more than 300 mg/dl or level of more than 100 mg/dl in routine examination (Dexter *et al.*, 1976); (3) hypoglycemia (Burke *et al.*, 1976, 1978); (4) elevation of serum lactic acid concentration (Dienhart *et al.*, 1975; Burke *et al.*, 1978); (5) a blood ethanol level of more than 150 mg/dl without gross evidence of intoxication (Dexter *et al.*, 1977); and (6) physiologic dependence as manifested by evidence of a withdrawal syndrome (Dexter *et al.*, 1977; Tumbleson *et al.*, 1978).

FERMENTED FOOD BEVERAGES IN NUTRITION

457

Mello (1973) reviewed the methods used to develop an animal model, but concluded "Uniform criteria for evaluating the adequacy and potential applicability of the various techniques for inducing physical dependence upon ethanol have not been developed." However, minature swine meet many of the suggested (Lester and Freed, 1973) criteria for an animal model of alcoholism: (1) oral ingestion of ethanol without food deprivation (Dexter *et al.*, 1977); (2) substantial ingestion of ethanol with competing fluids available (Dexter *et al.*, 1977); (3) ingestion directed to the central intoxicating character of ethanol, substantiated by determination of circulating blood ethanol levels (Dexter *et al.*, 1976); (4) intoxication sustained over a long period (Dexter *et al.*, 1976); (5) production of a withdrawal syndrome and physical dependence (Dexter *et al.*, 1977; Tumbleson *et al.*, 1978); and (6) after abstinence, reacquisition of drinking to intoxication and reproducibility of the alcoholic process (Dexter *et al.*, 1976).

Many of the pressing questions pertaining to the biology of alcoholism as briefly mentioned below might be approached by use of the miniature swine model. The role of ethanol as an hepatotoxic agent has been reviewed (Lieber and Rubin, 1968; Scheig, 1970; Takeuchi and Takada, 1975) and discussed by numerous investigators. The view that ethanol is an hepatotoxic drug and that chronic ethanol ingestion will result in liver injury has much scientific support but awaits definitive elucidation.

Ethanol oxidation occurs primarily in hepatic tissue (Lieber, 1977; Lundquist, 1971; Lundsgaard, 1938; Winkler *et al.*, 1969) with only small amounts eliminated via the lungs and kidneys. Conversion of ethanol to acetaldehyde, generally considered to be the rate-limiting step in ethanol metabolism, occurs mainly in hepatic cell cytoplasm. The role of vitamins in ethanol catabolism may be direct; however, the indirect relationship, via the respiratory chain intermediates, is well known. Therefore, availability of cofactors at specific reaction sites is critical and intestinal absorption of vitamins becomes biologically important. Ethanol elimination rate, consequently nutrient utilization, is crucial to decreasing the deleterious effects of sustained high levels of ethanol in the liver (Lieber and DeCarli, 1977) and brain (Noble and Tewari, 1977).

Sharma and Moskowitz (1977) reported that elimination of blood ethanol was more rapid when male human subjects ingested ethanol with full stomachs as compared to those with empty stomachs. As the normal human diet contains considerable sucrose, the "fructose effect" (Stuhlfauth and Neumaier, 1951; Tygstrup *et al.*, 1965) may play an important role. The increased rate of blood ethanol removal has been documented both in human beings (Patel *et al.*, 1969; Pawan, 1968; Soterakis and Iber, 1975) and in animals (Burke *et al.*, 1976). Also, the presence of foodstuffs in the digestive tract may influence blood ethanol levels by physical, physiochemical, or biochemical means which are yet to be identified.

Chronic alcoholism has been reported to result in vitamin deficiencies due to impaired intestinal absorption (Baker *et al.*, 1975; Halsted *et al.*, 1971; Lindenbaum and Lieber, 1975; Roggin *et al.*, 1969; Tomasulo *et al.*, 1968), decreased hepatic storage (Cherrick *et al.*, 1965; Sorrell *et al.*, 1974), and increased degradation in the liver (Barak *et al.*, 1973). In rats, French (1966) reported that ethanol ingestion enhanced pyridoxine deficiency, and Barak *et al.* (1971) found that ethanol administration increased the choline requirement. Also, Hermos *et al.* (1972) reported that ethanol ingestion produced histologic lesions of the duodenojejunal mucosa in human beings. Derr *et al.* (1970) found that a combination of ethanol and cobalt feeding to rats caused a synergistic effect which was more than expected from the additive detrimental consequences.

Administration of ethanol results in increased gastric acid secretion (Chey *et al.*, 1972), decreased gastric mucosal potential (Shanbour *et al.*, 1973), and decreased gastric emptying (Barboriak and Meade, 1969, 1970). When correlating the various findings, it is suggestive that the increased time nutrients remain in the stomach, decreased gastric pH, synergistic deleterious effects, and increased hepatic degradation of nutrients may result in injurious effects on numerous mammalian tissues.

The interrelationship between alcoholism and malnutrition is acknowledged and has been discussed by numerous authors. A few of the nutritional aspects reviewed include nutritional status and dietary factors (Heller, 1955; Neville *et al.*, 1968; Porta *et al.*, 1967), vitamins (Frank *et al.*, 1976; Leevy *et al.*, 1970; Levander *et al.*, 1973), lipotropes (Hartroft *et al.*, 1969), and liver injury as a result of alcohol intake concomitant with nutrient deficiencies (Lieber, 1975; Lelbach, 1974; Patek *et al.*, 1975).

The request for a single animal model to provide all possible similarities to human beings is unrealistic and should be discarded, whereas particular animal models should be selected to answer specific, relevant questions as they pertain to human alcoholism. One of the critical factors in selecting an animal model is that considerable basic information must be known, i.e., effects of age and sex on longitudinally, clinically obtainable parameters such as serum biochemical and hematological values (Burks *et al.*, 1977; Tumbleson *et al.*, 1976a,b). Also, to assess the interactions among alcoholism and nutrient deficiencies, data must be collected which result from the nutrient deficiencies without the presence of ethanol.

For example, studies on the effect of protein–calorie malnutrition in miniature swine have been initiated to ascertain changes in growth and development (Badger *et al.*, 1972; Tumbleson *et al.*, 1970, 1972c), serum biochemical and hematological values (Badger and Tumbleson, 1974b; Burks *et al.*, 1974; Tumbleson, 1972; Tumbleson *et al.*, 1972a; Tumbleson *et al.*, 1972b; Tumbleson and Hutcheson, 1972), and brain constituents (Badger and Tumbleson, 1974a; Fishman *et al.*, 1972; Sun and Tumbleson, 1972; Tumbleson and Badger, 1974). Hepatic alcohol dehydrogenase activity was depressed when rats were fed a low protein diet (Goebell and Bode, 1971). Also, blood ethanol elimination was inhibited when the rats were fed the low protein diet for 3 weeks. Preston *et al.* (1972) reported that miniature swine fed an 8% protein diet *ad libitum* consumed less ethanol than did control pigs fed a 16% protein diet *ad libitum*. Siegel *et al.* (1964) reported a decrease in plasma methionine, leucine, valine, and isoleucine concentrations with a concomitant increase in plasma glutamic acid and taurine concentrations when adult human males were administered ethanol. Serum albumin concentration was decreased and serum-free amino acids were increased when rats were fed ethanol for 2 months (Ramakrishnan, 1972); this was suggestive of an impairment of protein metabolism. Hypoalbuminemia has been noted (Kyosola and Salorinne, 1975) as a consistent finding in "skid row" human alcoholics.

Burke *et al.* (1975) reported a decrease in hepatic mitochondrial protein synthesis as a result of feeding ethanol to miniature swine for 6 weeks. Subsequent to that experiment, we fed ethanol to miniature swine for a prolonged period. After 6 weeks of ingesting in excess of 4 gm ethanol per kg body weight per day, there was a suppression in hepatic mitochondrial protein synthesis; however, after 9 months on test, there was no difference between control and ethanol-fed boars. Perhaps there is sufficient evidence to support the concept that ethanol ingestion has a deleterious effect on protein metabolism; therefore, enzyme systems may be affected.

II. EXPERIMENTS ON PHYSICAL DEPENDENCE

A. Experiment I

To illustrate some of the ways in which miniature swine have been used experimentally, three of our studies will be described. The purpose of this experiment was to ascertain the maximum amount of daily ethanol consumption that did not lead to development of physical dependence in miniature swine. We utilized 36 three-month-old Sinclair (S-1) miniature gilts. Twelve pigs were allotted randomly to each of three groups. Pigs in Groups I, II, or III were fed 1, 2, or 3 gm of ethanol per kg body weight per day, respectively. A switchback design was used to feed ethanol for 2, 4, 6, or 8 weeks, with 4-week intervals, to each of the gilts; the duration of the experiment was 32 weeks. Also, there was a 1-week withdrawal observation period during week 33. Each pig was fed 65% of the feed and ethanol ration at 0800 and 35% at 1600 hours daily. During the three 4-week interval periods, an equivalent caloric quantity of sucrose was substituted for ethanol.

Diets were formulated using corn, soybean meal, meat and bone meal, cellulose, alfalfa meal, wheat shorts, calcium carbonate, dicalcium phosphate, trace mineral salt, and vitamin premix. Quantities of feed and ethanol to be fed were calculated biweekly to adjust for body weight gains. Water was provided *ad libitum*. Body weights were recorded weekly. Each pig was fed a ration to equalize the daily intakes of calories, protein, minerals, and vitamins. For the 32-week period, the average maintenance energy requirement was 1170 kcal per pig per day. The pigs were fed an average of 1722 kcal per pig per day; therefore, 552 kcal per pig per day was available for growth. Also, the diet was formulated to provide 100 gm of protein per 1000 kcal energy. Based on previous studies with young miniature swine, we anticipated a body weight gain of 80 gm per pig per day.

Depicted in Table I are body weights and energy consumptions for the 32-week study. As they were selected randomly prior to the initiation of the experiment, there were differences in mean initial body weights. The recommended dietary intakes (National Research Council, 1968) were for rapidly growing market swine; therefore, we used a lesser amount of dietary protein as would be anticipated for growing miniature swine. The slightly lower rate of gain exhibited

Table I. Mean (±SEM) Body Weights, Ethanol Consumption, and Energy Consumption for Young Sinclair (S-1) Minature Swine on Test for 32 Weeks

Parameter	Group		
	I	II	III
Beginning body weight (kg)	8.2 ± 0.7	7.7 ± 0.6	6.9 ± 0.7
Ending body weight (kg)	26.3 ± 1.5	25.8 ± 1.6	24.2 ± 1.2
Body weight gain (gm/pig/day)	81	81	77
Ethanol consumed (gm/kg body wt/day)	1	2	3
Feed consumed (gm/pig/day)	580	580	580
Total energy (kcal/pig/day)	1722	1722	1722
Energy from ethanol (kcal/pig/day)	119	238	357
Energy from feed (kcal/pig/day)	1653	1484	1365
Calories from ethanol (%)	7	14	21
Caloric density of feed (kcal/kg)	2850	2550	2350
Protein/calorie ratio (gm/1000 kcal)	110	110	110

by those pigs receiving 3 gm of ethanol per kg body weight per day may have been a result of the lower initial mean body weight. Listed in Table II are the average daily quantities of protein, as well as selected minerals and vitamins, consumed and recommended (National Research Council, 1968).

Plasma ethanol concentrations were determined weekly from venous blood samples collected at 1000 hours on Wednesdays. For the 20 weeks which the gilts were fed ethanol, mean plasma ethanol concentrations were 5, 26, and 57 mg/dl for those pigs receiving 1, 2, and 3 gm of ethanol per kg body weight per day, respectively.

During some of the withdrawal periods, some of those gilts receiving 3 gm of ethanol per kg body weight per day exhibited minimal clinical signs of withdrawal. Using an activity box from 4 days prior to withdrawal through 4 days of withdrawal, it was possible to detect behavioral differences in those gilts receiving 3 gm of ethanol per kg body weight per day as compared with gilts receiving either 1 or 2 gm of ethanol per kg body weight per day. Each $1.2 \times 1.2 \times 1.2$ meter activity box was supported on rubber and affixed with a three-direction accelerometer transducer in the center of the link-wire lid.

Compared to prewithdrawal activity, gilts receiving 1 gm of ethanol per kg

Table II. Average Daily Quantities of Protein, Minerals, and Vitamins Consumed and Recommended (National Research Council, 1968) for Young Sinclair (S-1) Miniature Swine on Test for 32 Weeks

Nutrient	Consumed	Recommended	Consumed/recommended (%)
Protein (gm)	191	225	85
Calcium (gm)	20.5	8.1	253
Phosphorus (gm)	11.7	6.3	185
Salt (gm)	2.9	1.4	207
Iron (mg)	90	64	140
Zinc (mg)	81	40	203
Copper (mg)	9.1	4.8	190
Manganese (mg)	31	16	195
Selenium (μg)	88	80	110
Vitamin A (IU)	3830	2200	174
Vitamin D (IU)	1020	250	409
Vitamin E (mg)	19	14	136
Thiamine (mg)	2.6	1.4	188
Riboflavin (mg)	10.2	3.8	269
Nicotinic acid (mg)	30	22	135
Pantothenic acid (mg)	20	14	148
Pyridoxine (mg)	2.4	1.9	128
Vitamin B_{12} (μg)	88	19	469
Choline (mg)	1310	1125	117

body weight per day exhibited -3.5, $+5.2$, -0.31, or $+6.2\%$ change in activity on days 1, 2, 3, or 4 of withdrawal. Those gilts receiving 2 gm of ethanol per kg body weight per day exhibited -3.2, -2.8, $+4.9$, or $+3.8\%$ change in activity during days 1, 2, 3, or 4 of withdrawal. Gilts receiving 3 gm of ethanol per kg body weight per day exhibited $+12.6$, $+40.3$, $+51.2$, or $+37.8\%$ change in activity during withdrawal days 1, 2, 3, or 4.

After 8 weeks of ethanol ingestion, those gilts receiving 3 gm of ethanol per kg body weight per day had increased serum glutamic–pyruvic transaminase activities; serum glutamic–oxalacetic transaminase, lactic acid dehydrogenase, and γ-glutamyltranspeptidase activities were not altered as a result of ethanol ingestion. Administration of 1 or 2 gm of ethanol per kg body weight per day for 8 weeks did not produce any change in serum glutamic–pyruvic transaminase activities.

B. Experiment II

The purpose of this experiment was to determine plasma ethanol clearance, subsequent to chronic intragastric ethanol infusion. Each adult Sinclair (S-1) miniature boar was implanted surgically with a cranial vena caval catheter (Wingfield *et al.*, 1974) and an intragastric catheter (Tumbleson *et al.*, 1978) 2 weeks prior to the experimental period. Feed and water were supplied *ad libitum* prior to and during the study. Using an infusion schedule similar to that reported previously (Tumbleson *et al.*, 1978), two boars were infused intragastrically, at 6-hour intervals, with 1, 2, 3, or 4 gm of ethanol per kg body weight per day on days 1, 2, 3, or 4, respectively. Immediately prior to infusion of 1 gm/kg at 0800 on day 5, a 4 ml venous blood sample was collected from each boar. Similarly, at ½, 1, 2, 4, 6, 8, 12, 16, 20, 24, and 28 hours, postinfusion samples were collected to follow clearance of ethanol from the blood.

Plasma ethanol clearance curves are depicted in Fig. 1. Plasma ethanol concentrations, at 6 hours postinfusion, from the 0200 infusion on day 4 and the 0800 infusion on day 5, were 230 and 222 mg/dl and 129 and 142 mg/dl for boars 1682 and 1644, respectively. From 2–16 hours postinfusion, boar 1682 exhibited a plasma ethanol disappearance rate of 19.5 mg/dl/hr; whereas, boar 1644 had a plasma ethanol disappearance rate of 16.9 mg/dl/hr from 2–12 hours postinfusion.

C. Experiment III

This investigation was designed to (1) determine quantities of ethanol, infused at 8 and 16 hour intervals, that could be tolerated by adult miniature swine and (2) document withdrawal signs during a 7-day withdrawal period. The animals used were four (two boars and two gilts) 21-month-old Sinclair (S-1) miniature

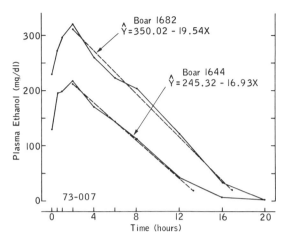

Fig. 1. Plasma ethanol concentrations and clearance rates for two adult Sinclair (S-1) miniature boars infused intragastrically with ethanol with 1 gm/kg per body weight, following 4 days of ethanol administration.

pigs. A 16% protein, corn–soybean meal diet was fed *ad libitum* prior to the study. During the experimental period, a 32% protein, corn–soybean meal diet was fed *ad libitum*. Fresh drinking water was available *ad libitum*. Each pig was confined to a 1.2 × 1.2 meter pen with 18-gauge galvanized sheet metal on the sides and expanded metal floors. Seven days prior to initiation of the study, each pig was implanted surgically with an intragastric catheter (Tumbleson *et al.*, 1978).

Ethanol was administered intragastrically twice daily at 0800 and 1600. The schedule for infusions and venous blood sample collections is depicted in Table III. The administered aqueous ethanol solution contained 500 mg of ethanol/ml. Feed and water consumption were quantitated daily at 0730. Venous blood samples were collected from the forelimb (Tumbleson *et al.*, 1968). For 2 days prior to termination of ethanol infusions, boar 1786 and gilt 1803 were administered 3.0 gm of ethanol per kg body weight per day, and gilt 1813 and boar 2138 received 4.0 gm of ethanol per kg body weight per day.

Two hour postinfusion blood samples were collected at 1000 on days 2, 8, 14, and 20; 7 hour postinfusion samples were collected at 1500 on days 4, 10, 16, and 22; and 15 hour postinfusion samples were collected at 0700 on days 6, 12, 18, and 24 (Table III). Packed cell volumes were determined by the microhematocrit method, and plasma ethanol concentrations were assayed enzymatically (Bonnichsen and Theorell, 1951).

The infusion schedule (Table III) was followed only until days 12, 11, 10, and 18 for pigs 1786, 1807, 1813, and 2138, respectively (Figs. 2, 3, 4, and 5). Less

Table III. Infusion and Sampling Schedule [a]

Hour	0	1	2	3	4	5	6	7	8	9	10	11	12	13	14	15	16	17	18	19	20	21	22	23	24
0700	X [b]						X						X						X						X
0800		A	B	B	C	C	C	D	D	D	E	E	E	E	E	E	F	F	F	F	F	F	?	?	?
1000			X						X						X						X				
1500					X						X						X						X		
1600		A	B	B	C	C	C	D	D	D	E	E	E	E	E	E	F	F	F	F	F	F	?	?	?

(Day)

[a] A–F, ethanol (gm/kg body weight per infusion). A, 0.5; B, 1.0; C, 1.5; D, 2.0; E, 2.5; F, 3.0.
[b] X, Collection of 4 ml blood samples for plasma ethanol and packed cell volume determinations.

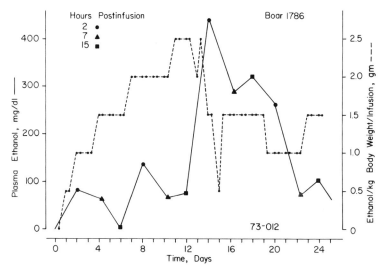

Fig. 2. Levels (----) of ethanol (gm/kg body weight) infused intragastrically and plasma ethanol concentrations (⎯⎯) for boar 1786.

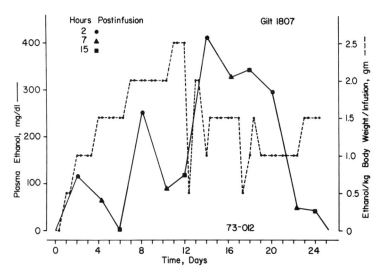

Fig. 3. Levels (----) of ethanol (gm/kg body weight) infused intragastrically and plasma ethanol concentrations (⎯⎯) for gilt 1807.

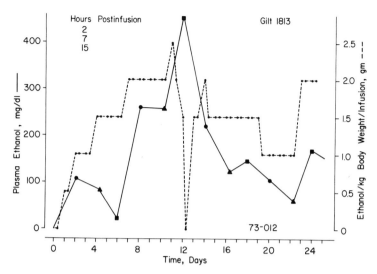

Fig. 4. Levels (----) of ethanol (gm/kg body weight) infused intragastrically and plasma ethanol concentrations (——) for gilt 1813.

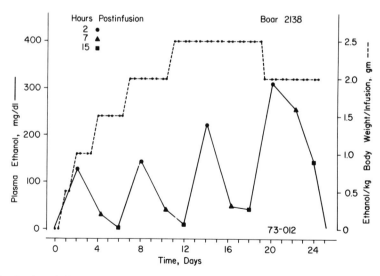

Fig. 5. Levels (----) of ethanol (gm/kg body weight) infused intragastrically and plasma ethanol concentrations (——) for boar 2138.

quantities of ethanol were infused any time a pig appeared unable to tolerate the administered ethanol. Intolerance was characterized by the absence of voluntary feed and water intake, inability to stand, indifference to external stimuli, and Cheyne–Stokes respiration.

At 2 hours postinfusion (1000) on day 14, boar 1786 had a plasma ethanol concentration (Fig. 2) of 440 mg/dl even though the quantity of ethanol infused at 0800 was only 1.5 gm/kg body weight. Likewise, at 0700 on day 18, gilt 1807 exhibited a plasma ethanol level (Fig. 3) of 343 mg/dl even though the quantity of ethanol infused 15 hours previously was only 0.5 gm/kg body weight. At 0700 (15 hours postinfusion) on day 12, gilt 1813 had a plasma ethanol concentration (Fig. 4) of 453 mg/dl; therefore, no ethanol was infused at 1600 on day 12. Boar 2138 was able to tolerate the quantities of ethanol infused until day 19 (Fig. 5). However, on day 19, it was necessary to decrease the quantity of ethanol infused.

Mean packed cell volumes for pigs 1786, 1807, 1813, and 2138 were 38.3 ± 1.1, 37.3 ± 1.0, 38.1 ± 1.1, and 37.8 ± 1.1%, respectively, during the experimental period. There were no biologically important changes in packed cell volumes as a result of ethanol infusions.

Body weights, voluntary water consumptions, and voluntary feed consumptions are depicted in Figs. 6, 7, 8, and 9 for pigs 1786, 1807, 1813, and 2138,

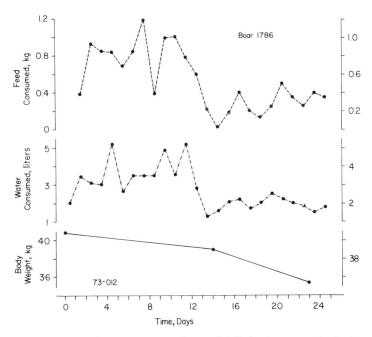

Fig. 6.   Body weight, daily water consumption, and daily feed consumption for boar 1786.

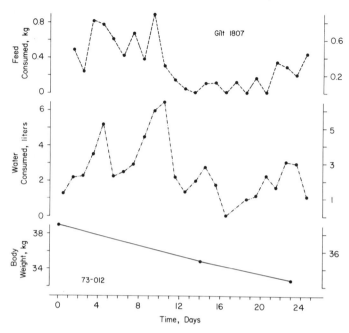

Fig. 7. Body weight, daily water consumption, and daily feed consumption for gilt 1807.

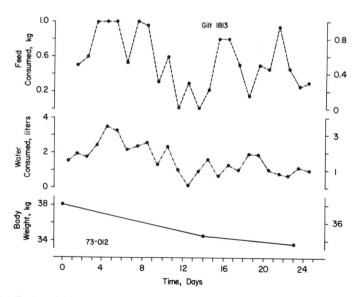

Fig. 8. Body weight, daily water consumption, and daily feed consumption for gilt 1813.

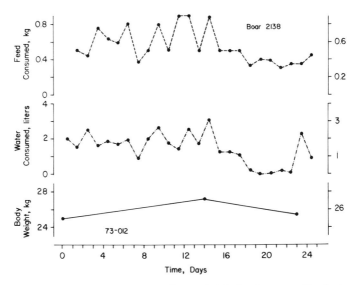

Fig. 9. Body weight, daily water consumption, and daily feed consumption for boar 2138.

respectively. During the first 2 weeks on test, body weights decreased 4, 10, and 9% for pigs 1786, 1807, and 1813, respectively. Marked decreases in feed and water consumption were recorded concomitant with periods when the pigs first exhibited inabilities to clear ethanol from the plasma.

Subsequent to the time ethanol was cleared from the plasma on day 25, the pigs were irritable, hyperactive, and aggressive. Between 24 and 72 hours postinfusion, they exhibited periods of hypoactivity interspersed with periods of aggressiveness. Piloerection, increased palpebral fissures, dilated pupils, ataxic steppage gait, defensive posturing, muscle fasciculations, bruxism, yawning, multifocal myoclonus, reacting to stimuli not observed by the investigators, and static and volitional tremors were observed from 72 to 96 hours after the last ethanol infusions. Five and 6 days postinfusion, nearly constant movement of the ears and tail, tremors, muscle fasciculations in many areas of the body and severe myoclonic jerks were observed.

Despite the allowance of 16 hours between infusions, the pigs used in this experiment were intolerant to intragastrically infused ethanol in lower amounts than those consumed voluntarily (Dexter *et al.*, 1976). One animal (boar 2138) had 2, 7, and 15 hours postinfusion plasma ethanol concentrations which were suggestive of an animal capable of metabolizing ethanol sufficiently to prevent clinical signs of intolerance. However, all four pigs exhibited signs of intolerance at various times during the experimental period. The severe muscle tremors, muscle fasciculations, and myoclonic jerks, recorded 5 and 6 days after the last ethanol infusions, were indicative of a complete withdrawal syndrome.

III. DISCUSSION

Feeding 1 or 2 gm to ethanol per kg body weight per day did not result in overt withdrawal signs, whereas feeding 3 gm of ethanol per kg body weight per day produced minimal clinical signs of withdrawal. We successfully quantitated changes in activity during withdrawal using an activity box. There were no biologically important nor statistically significant changes in serum glutamic-pyruvic transaminase, glutamic–oxalacetic transaminase, lactic acid dehydrogenase, or γ-glutamyltranspeptidase activities as a result of feeding 1 or 2 gm of ethanol per kg body weight per day. However, those gilts receiving 3 gm of ethanol per kg body weight per day did exhibit increased serum glutamic–pyruvic transaminase activities.

Plasma ethanol clearance rates of 17 and 19 mg/dl hour were in the range which investigators should use for a biomedical research subject selected for the study of human alcoholism. The rates of clearance were greater than reported (Burke *et al.*, 1976) for naive miniature pigs infused intragastrically with either 1 gm of ethanol per kg body weight (9.1 mg/dl/hour) or 2 gm of ethanol per kg body weight (10.1 mg/dl/hour).

Intragastric infusion of ethanol (Experiment III) with 8 and 16 hour intervals resulted in intolerance after 10–18 days on test. Perhaps the most important finding of this study was the variability among pigs with respect to tolerance to quantity of ethanol infused. That plasma ethanol levels were 340–450 mg/dl at 15 hours postinfusion may have been indicative of a marked inability of the mammalian system to maintain metabolic normalcy during times of ethanol excess.

Withdrawal signs exhibited by all four pigs in this study were similar to those reported previously (Dexter *et al.*, 1976) for adult Sinclair (S–1) miniature gilts consuming ethanol voluntarily. Many of the withdrawal signs reported (Criteria Committee, 1972) for human beings were observed. The severe muscle tremors, muscle fasciculations, and myoclonic jerks, recorded 5 and 6 days after the last ethanol infusions, were indicative of a complete withdrawal syndrome.

Experimental design is of paramount importance when attempting to differentiate those effects due to ethanol ingestion from those resulting from concomitant malnutrition. Research studies often are designed meticulously with respect to quantity, duration, and time of ethanol administration, time interval between the last ethanol administered and tissue collection, tissue preparation, laboratory analyses, and mathematical manipulation and statistical evaluation of the resultant data. However, meaningful conclusions may be elusive as a result of alteration of nutrient intake, subclinical unrelated infections, and management of the experimental subjects.

Care must be taken to provide similar rations, as well as similar ratios of dietary ingredients. Nutrient ingestion, absorption, utilization, turnover, and excretion should be considered. Determinants must be selected to circumvent those which could be altered as a result of sample collection procedures.

Based on our findings reported herein, as well as from previously reported studies (Preston *et al.*, 1972; Dienhart *et al.*, 1975; Burke *et al.*, 1975, 1976, 1978; Dexter *et al.*, 1976, Dexter *et al.*, 1977; Tumbleson *et al.*, 1978), we suggest that miniature swine will be important animal models for the study of human alcoholism. Nutrient requirements of pigs and human beings are similar. Adult miniature pigs weigh 70 to 100 kg. The gestation period of pigs is 16 weeks, which allows sufficient time for evaluation of fetal development. Litter size ranges from 4 to 10 offspring, weighing 500 to 800 gm at birth. Miniature swine, which are tractable, trainable, and manageable, require less space and diet than do conventional swine. Also, perinatal pigs can be weaned at 3–5 days of age, enabling investigators to manipulate diets and stress factors throughout life. Females and males are mature sexually at 3 and 5 months of age, respectively, therefore allowing for subsequent generation studies during a short time period. Miniature swine live approximately 13–15 years, which provides a life span of sufficient length to conduct longitudinal studies. An important consideration is that miniature swine will repeat the alcoholic process subsequent to withdrawal. We are able to maintain blood ethanol levels of 150 mg/dl. Plasma ethanol clearance rates are similar to those for human beings. Withdrawal signs are analogous to those reported for human alcoholics. Miniature swine will consume voluntarily in excess of 4 gm of ethanol per kg body weight per day. Also, miniature swine exhibit intoxication at blood ethanol levels similar to those observed in human alcoholics.

IV. SUMMARY AND CONCLUSIONS

1. Suggested criteria for the development of an animal model for the study of alcoholism are reviewed in relation to certain characteristics and attributes of miniature swine.

2. Three experiments related to physical dependence on alcohol are cited to illustrate the use of miniature swine in alcohol research. Special attention was given to the maintenance of a balanced diet throughout the experiments. (a) In the first experiment three groups of pigs consumed 1, 2, or 3 gm of ethanol per kg body weight per day for 2, 4, 6, or 8 weeks, with alcohol-free periods of 4 weeks, for 32 weeks. Minimal withdrawal signs using an activity box were noted at

the 3 gm level, but not at lower amounts. After 8 weeks pigs given alcohol at the highest level showed increased serum GPT values; SGOT, LDH, and γGT values were normal. (b) In the second experiment, increasing amounts of ethanol were infused daily in 2 pigs for 4 days and blood clearance rates followed for 20 hours following the last infusion. Clearance rates were 16.9 and 19.5 mg/dl/hour, respectively, between 2 and 12–16 hours. (c) The third experiment was designed to determine the limits of tolerance to twice daily intragastric infused ethanol for 24 days, in progressive amounts from 1.0 to 6.0 gm/kg body weight per day. Intolerance as manifested by inability to eat and drink developed to 10–18 days but varied considerably from animal to animal as did blood alcohol levels reached; intolerance was roughly correlated with sustained high blood levels. No biologically significant changes were noted in packed red cell volumes. All pigs developed marked withdrawal signs when infusions were terminated.

3. The nutritional requirements of miniature pigs are similar to those of human beings; pigs will consume ethanol voluntarily up to 4 gm/kg body weight per day; and withdrawal signs similar to many of those observed in man occur in pigs.

4. It is concluded that miniature swine provide an important animal model for the study of human alcoholism.

ACKNOWLEDGMENT

Supported in part by a grant from the United States Brewers Association, Inc.

REFERENCES

Badger, T. M., and Tumbleson, M. E. (1974a). Protein–calorie malnutrition in young miniature swine: Brain free amino acids. *J. Nutr.* **104,** 1329–1338.

Badger, T. M., and Tumbleson, M. E. (1974b). Protein–calorie malnutrition in young miniature swine: Serum free amino acids. *J. Nutr.* **104,** 1339–1347.

Badger, T. M., Tumbleson, M. E., and Hutcheson, D. P. (1972). Protein–calorie malnutrition in young Sinclair(S-1) miniature swine. *Growth* **36,** 235–245.

Baker, H., Frank, O., Zetterman, R. K., Rajan, K. S., ten Hove, W., and Leevy, C. M. (1975). Inability of chronic alcoholics with liver disease to use food as a source of folates, thiamin and vitamin B$_6$. *Am. J. Clin. Nutr.* **28,** 1377–1380.

Barak, A. J., Tuma, D. J., and Beckenhauer, H. C. (1971). Ethanol feeding and choline deficiency as influences on hepatic choline uptake. *J. Nutr.* **101,** 533–538.

Barak, A. J., Tuma, D. J., and Sorrell, M. F. (1973). Relationship of ethanol to choline metabolism in the liver: A review. *Am. J. Clin. Nutr.* **26,** 1234–1241.

Barboriak, J. J., and Meade, R. C. (1969). Impairment of gastrointestinal processing of fat and protein by ethanol in rats. *J. Nutr.* **98,** 373–378.

Barboriak, J. J., and Meade, R. C. (1970). Effect of alcohol on gastric emptying in man. *Am. J. Clin. Nutr.* **23,** 1151–1153.

Bonnichsen, R. K., and Theorell, H. (1951). An enzymatic method for the determination of ethanol. *Scand. J. Clin. Lab. Invest.* **3,** 58–62.

Burke, J. P., Tumbleson, M. E., Hicklin, K. W., and Wilson, R. B. (1975). Effect of chronic ethanol ingestion on mitochondrial protein synthesis in Sinclair(S-1) miniature swine. *Proc. Soc. Exp. Biol. Med.* **148**, 1051-1056.

Burke, J. P., Tumbleson, M. E., Burks, M. F., and Hicklin, K. W. (1976). Plasma glucose in miniature swine infused with ethanol and fructose. *Q. J. Stud. Alcohol* **37**, 1-10.

Burke, J. P., Tumbleson, M. E., and Zatzman, M. L. (1978). Effect of intravenous infusion of ethanol on plasma glucose, lactic acid and pyruvic acid concentrations. *Q. J. Stud. Alcohol* **39**, 1195-1205.

Burks, M. F., Evans, P. S., and Tumbleson, M. E. (1974). Hematologic values of malnourished Sinclair(S-1) miniature swine. *Lab Anim. Sci.* **24**, 84-89.

Burks, M. F., Tumbleson, M. E., Hicklin, K. W., Hutcheson, D. P., and Middleton, C. C. (1977). Age and sex related changes of hematologic parameters in Sinclair(S-1) miniature swine. *Growth* **41**, 51-62.

Cherrick, G. R., Baker, H., Frank, O., and Leevy, C. M. (1965). Observations on hepatic acidity for folate in Laennec's cirrhosis. *J. Lab. Clin. Med.* **66**, 446-451.

Chey, W. Y., Kosay, S., and Lorber, S. H. (1972). Effects of chronic administration of ethanol on gastric secretion of acid in dogs. *Am. J. Dig. Dis.* **17**, 153-159.

Criteria Committee, National Council on Alcoholism. (1972). Criteria for the diagnosis of alcoholism. *Ann. Intern. Med.* **77**, 249-258.

Derr, R. F., Aaker, H., Alexander, C. S., and Nagasawa, H. T. (1970). Synergism between cobalt and ethanol on rat growth rate. *J. Nutr.* **100**, 521-524.

Dexter, J. D., Tumbleson, M. E., Hutcheson, D. P., and Middleton, C. C. (1976). Sinclair(S-1) miniature swine as a model for the study of human alcoholism. *Ann. N.Y. Acad. Sci.* **273**, 188-193.

Dexter, J. D., Tumbleson, M. E., Hutcheson, D. P., and Middleton, C. C. (1977). The relationship between social dominance and ethanol consumption in Sinclair(S-1) miniature swine. *In* "Currents in Alcoholism" (F. A. Seixas, ed.), pp. 255-264. Grune & Stratton, New York.

Dienhart, G. B., Tumbleson, M. E., Hicklin, K. W., and Hutcheson, D. P. (1975). Plasma lactic acid and pyruvic acid concentrations following intragastric infusion of ethanol in adult miniature swine (*Sus scrofa*). *Int. J. Biochem.* **6**, 211-218.

Fishman, M. A., Prensky, A. L., Tumbleson, M. E., and Daftari, B. (1972). Relative resistance of the later phase of myelination to severe undernutrition in miniature swine. *Am. J. Clin. Nutr.* **25**, 7-10.

Frank, O., Luisada-Opper, A., Sorrell, M. F., Zetterman, R., and Baker, H. (1976). Effects of a single intoxicating dose of ethanol on the vitamin profile of organelles in rat liver and brain. *J. Nutr.* **106**, 606-614.

French, S. W. (1966). Effect of chronic ethanol ingestion on liver enzyme changes induced by thiamine, riboflavin, pyridoxine, or choline deficiency. *J. Nutr.* **88**, 291-302.

Goebell, H., and Bode, C. (1971). Influence of ethanol and protein deficiency on the activity of alcohol dehydrogenase in the rat liver. *In* "Metabolic Changes Induced by Alcohol" (G. A. Martini and C. Bode, eds.), pp. 23-30. Springer-Verlag, Berlin and New York.

Halsted, C. H., Robles, E. A., and Mezey, E. (1971). Decreased jejunal uptake of labeled folic acid (H^3-PGA) in alcoholic patients. Roles of alcohol and nutrition. *N. Engl. J. Med.* **285**, 701-706.

Hartroft, W. S., Porta, E. A., and Sugioka, G. (1969). Protein and lipotropic requirements for protection of the liver at various intakes of alcohol by experimental animals. *In* "Biochemical and Clinical Aspects of Alcohol Metabolism" (V. M. Sardesai, ed.), pp. 161-175. Thomas, Springfield, Illinois.

Heller, H. (1955). The nutritional aspect of chronic alcoholism. *Br. J. Addict.* **52,** 45–52.

Hermos, J. A., Adams, W. H., Liu, Y. K., Sullivan, L. S., and Trier, J. S. (1972). Mucosa of the small intestine in folate-deficient alcoholics. *Ann. Intern. Med.* **76,** 957–965.

Kyosola, K., and Salorinne, Y. (1975). Liver biopsy and liver function tests in 28 consecutive long-term alcoholics. *Ann. Clin. Res.* **7,** 80–84.

Leevy, C. M., Thompson, A., and Baker, H. (1970). Vitamins and liver injury. *Am. J. Clin. Nutr.* **23,** 493–499.

Lelbach, W. K. (1974). Organic pathology related to volume and pattern of alcohol use. *In* "Research Advances in Alcohol and Drug Problems" (R. J. Gibbins, Y. Israel, H. Kalant, R. E. Popham, W. Schmidt, and R. G. Smart, eds.), pp. 93–198. Wiley, New York.

Lester, D., and Freed, E. X. (1973). Criteria for an animal model of alcoholism. *Pharmacol. Biochem. Behav.* **1,** 103–107.

Levander, O. W., Morris, V. C., Higgs, D. J., and Varma, R. N. (1973). Nutritional interrelationships among vitamin E, selenium, antioxidants and ethyl alcohol in the rat. *J. Nutr.* **103,** 536–542.

Lieber, C. S. (1975). Alcohol and malnutrition in the pathogenesis of liver disease. *J. Am. Med. Assoc.* **233,** 1077–1082.

Lieber, C. S. (1977). Metabolism of ethanol. *In* "Metabolic Aspects of Alcoholism" (C. S. Lieber, ed.), pp. 1–29. Univ. Park Press, Baltimore, Maryland.

Lieber, C. S., and DeCarli, L. M. (1977). Metabolic effects of alcohol on the liver. *In* "Metabolic Aspects of Alcoholism" (C. S. Lieber, ed.), pp. 31–79. Univ. Park Press, Baltimore, Maryland.

Lieber, C. S., and Rubin, E. (1968). Ethanol—a hepatotoxic drug. *Gastroenterology* **54,** 642–646.

Lindenbaum, J., and Lieber, C. S. (1975). Effects of chronic ethanol administration on intestinal absorption in man in the absence of nutritional deficiency. *Ann. N.Y. Acad. Sci.* **252,** 228–234.

Lundquist, F. (1971). The metabolism of alcohol. *In* "Biological Basis of Alcoholism" (Y. Israel and J. Mardones, eds.), pp. 1–52. Wiley (Interscience), New York.

Lundsgaard, E. (1938). Alcohol oxidation as a function of the liver. *Ser. Chim.* **22,** 333–337.

Mello, N. K. (1973). A review of methods to induce alcohol addiction in animals. *Pharmacol. Biochem. Behav.* **1,** 89–101.

National Research Council. (1968). "Nutrient Requirements of Swine," pp. 1–69. Natl. Acad. Sci., Washington, D.C.

Neville, J. N., Eagles, J. A., Samson, G., and Olson, R. E. (1968). Nutritional status of alcoholics. *Am. J. Clin. Nutr.* **21,** 1329–1340.

Noble, E. P., and Tewari, S. (1977). Metabolic aspects of alcoholism in the brain. *In* "Metabolic Aspects of Alcoholism" (C. S. Lieber, ed.), pp. 149–185. Univ. Park Press, Baltimore.

Patek, A. J., Jr., Toth, I. G., Saunders, M. G., Castro, G. A. M., and Engel, J. J. (1975). Alcohol and dietary factors in cirrhosis. *Arch. Intern. Med.* **135,** 1053–1057.

Patel, A. R., Paton, A. M., Rowan, T., Lawson, D. H., and Linton, A. L. (1969). Clinical studies on the effect of laevulose on the rate of metabolism of ethyl alcohol. *Scott. Med. J.* **14,** 268–271.

Pawan, G. L. S. (1968). Vitamins, sugars and ethanol metabolism in man. *Nature (London)* **220,** 374–376.

Porta, E. A., Hartroft, W. S., Gomez-Dumm, C. L. A., and Koch, O. R. (1967). Dietary factors in the progression and regression of hepatic alterations associated with experimental chronic alcoholism. *Fed. Proc., Fed. Am. Soc. Exp. Biol.* **26,** 1449–1457.

Preston, A. M., Tumbleson, M. E., Hutcheson, D. P., and Middleton, C. C. (1972). Alcohol consumption and vehicle preference in young Sinclair(S-1) miniature swine fed two levels of dietary protein. *Proc. Soc. Exp. Biol. Med.* **141,** 585–589.

Ramakrishnan, S. (1972). Effect of ethyl alcohol intake on blood cholesterol, serum albumin and free amino acids in adult rats. *Indian J. Biochem. Biophys.* **9**, 203-205.

Roggin, G. M., Iber, F. L., Kater, R. M. H., and Tabon, F. (1969). Malabsorption in the chronic alcoholic. *Johns Hopkins Med. J.* **125**, 321-330.

Scheig, R. (1970). Effects of ethanol on the liver. *Am. J. Clin. Nutr.* **23**, 467-473.

Shanbour, L. L., Miller, J., and Chowdhury, T. K. (1973). Effects of alcohol on active transport in the rat stomach. *Am. J. Dig. Dis.* **18**, 311-316.

Sharma, S., and Moskowitz, H. (1977). Food effects on blood alcohol levels in humans. *Alcohol. Clin. Exp. Res.* **1**, 162.

Siegel, F. L., Roach, M. K., and Pomeroy, L. R. (1964). Plasma amino acid patterns in alcoholism: The effects of ethanol loading. *Proc. Natl. Acad. Sci. U.S.A.* **51**, 605-611.

Sorrell, M. F., Baker, H., Barak, A. J., and Frank, O. (1974). Release by ethanol of vitamins into rat liver perfusates. *Am. J. Clin. Nutr.* **27**, 743-745.

Soterakis, J., and Iber, F. L. (1975). Increased rate of alcohol removal from blood with oral fructose and sucrose. *Am. J. Clin. Nutr.* **28**, 254-257.

Stuhlfauth, K., and Neumaier, H. (1951). Die Wirkung der Laevulose auf Alkohol-Intoxikationen. *Med. Klin. (Munich)* **46**, 591-593.

Sun, G. Y., and Tumbleson, M. E. (1972). Levels of brain lipids in white matter from undernourished Sinclair(S-1) miniature swine. *J. Neurochem.* **19**, 909-912.

Takeuchi, J., and Takada, A. (1975). Alcohol and its hepatotoxic effect. *In* "Alcoholic Liver Pathology" (J. M. Khanna, Y. Israel, and H. Kalant, eds.), pp. 199-224. Addict. Res. Found., Toronto.

Tomasulo, P. A., Kater, R. M. H., and Iber, F. L. (1968). Impairment of thiamine absorption in alcoholism. *Am. J. Clin. Nutr.* **21**, 1341-1344.

Tumbleson, M. E. (1972). Protein-calorie undernutrition in young Sinclair(S-1) miniature swine: Serum biochemic and hematologic values. *Adv. Autom. Anal.* **7**, 51-71.

Tumbleson, M. E., and Badger, T. M. (1974). DNA, RNA and protein concentrations in forebrain, cerebellum and brain stem of young malnourished Sinclair(S-1) miniature swine (*Sus scrofa*). *Int. J. Biochem.* **5**, 57-62.

Tumbleson, M. E., and Hutcheson, D. P. (1972). Effect of maternal dietary protein deprivation on serum biochemic and hematologic parameters of miniature piglets. *Nutr. Rep. Int.* **6**, 321-329.

Tumbleson, M. E., Dommert, A. R., and Middleton, C. C. (1968). Techniques for handling miniature swine for laboratory procedures. *Lab. Anim. Care* **18**, 584-587.

Tumbleson, M. E., Tinsley, O. W., Mulder, J. B., and Flatt, R. E. (1970). Undernutrition in young Sinclair(S-1) miniature swine: Body and organ weights. *Growth* **34**, 401-418.

Tumbleson, M. E., Burks, M. F., Evans, P. S., and Hutcheson, D. P. (1972a). Serum electrolytes in undernourished Sinclair(S-1) miniature swine. *Am. J. Clin. Nutr.* **25**, 476-479.

Tumbleson, M. E., Burks, M. F., and Hook, R. R., Jr. (1972b). Serum proteins in undernourished Sinclair(S-1) miniature swine. *Clin. Biochem.* **5**, 51-61.

Tumbleson, M. E., Tinsley, O. W., Hicklin, K. W., and Mulder, J. B. (1972c). Fetal and neonatal development of Sinclair(S-1) miniature piglets effected by maternal dietary protein deprivation. *Growth* **36**, 373-387.

Tumbleson, M. E., Hicklin, K. W., and Burks, M. F. (1976a). Serum cholesterol, triglyceride, glucose and total bilirubin concentrations, as functions of age and sex, in Sinclair(S-1) miniature swine. *Growth* **40**, 293-300.

Tumbleson, M. E., Hutcheson, D. P., and Middleton, C. C. (1976b). Serum protein concentrations and enzyme activities, as functions of age and sex, in Sinclair(S-1) miniature swine. *Growth* **40**, 53-68.

Tumbleson, M. E., Dexter, J. D., Hutcheson, D. P., and Middleton, C. C. (1978). Intolerance of miniature swine to intragastrically infused ethanol. *Q. J. Stud. Alcohol* **39** (in press).

Tygstrup, N., Winkler, K., and Lundquist, F. (1965). The mechanism of the fructose effect on the ethanol metabolism of the human liver. *J. Clin. Invest.* **44,** 817–830.

Wingfield, W. E., Tumbleson, M. E., Hicklin, K. W., and Mather, E. C. (1974). An exteriorized cranial vena caval catheter for serial blood sample collection from miniature swine. *Lab. Anim. Sci.* **24,** 359–361.

Winkler, K., Lundquist, F., and Tygstrup, N. 1969. The hepatic metabolism of ethanol in patients with cirrhosis of the liver. *Scand. J. Lab. Invest.* **23,** 59–69.

VII

*Socioeconomic
Considerations*

30

THE MEDICAL COSTS OF EXCESSIVE USE OF ALCOHOL

Milton Terris

I. Cirrhosis of the Liver 481
II. Accidents 482
III. Suicide, Homicide, and Assault 483
IV. Cancer 484
V. Mental Disorders 485
VI. Other Disorders 485
VII. Conclusions... 485
    References 486
    Editorial Comment 488

The medical costs of excessive use of alcohol, like those incurred by the excessive use of tobacco, result from a fairly wide range of diseases and injuries. We shall start with some of the more important ones.

I. CIRRHOSIS OF THE LIVER

In a major study of urban mortality by the Pan American Health Organization, data were collected in ten Latin American cities, San Francisco, California, and Bristol, England, for deaths occurring from 1962 to 1964.

Every attempt was made to obtain accurate information. For each death, a home visit was made by a public health nurse or social worker who obtained further information. This served as the starting point for the investigating physicians who assembled all the available clinical evidence. Hospital records were traced and surgical and autopsy findings were recorded, together with the results of ancillary studies such as laboratory investigations, radiology, and electrocardiography. The physicians having knowledge of the patient were interviewed. In each city, the principal collaborating physician summarized the findings and expressed his opinion as to the cause of death. All reports were then sent to the central office of the Pan American Health Organization, where reviewing physicians and expert medical referees made the final determination.

In the 12 cities, 1649 of the 2051 deaths from cirrhosis of the liver, or 80%, were found to be associated with alcoholism. In San Francisco, 93% of the cirrhosis deaths were found to be related to alcoholism (1).

It is not surprising, therefore, that there exists an uncanny parallel between the cirrhosis death rate and the per capita consumption of alcohol. Data from Canada, the United Kingdom, and the United States indicate that as the consumption rate goes up, so does the cirrhosis death rate, and as it decreases the cirrhosis rate falls (2–4).

In the United States, the cirrhosis death rate varied between 13 and 15 per 100,000 population in the period from 1900 to 1914. Effective wartime prohibition lowered the rate to 7 in 1920, and the partially effective peace time prohibition maintained the rate at 7 until 1933 when the Volstead Act was repealed. Since then both consumption and the cirrhosis death rate have risen steadily, so that the rate in 1975 stood at 15 deaths per 100,000 population (3,5).

In 1965, cirrhosis of the liver did not appear among the ten leading causes of death in the United States. Only a decade later, in 1975, it had become the seventh leading cause of death. There were 31,623 deaths from cirrhosis of the liver that year. If, instead of a cirrhosis death rate of 15, the United States had achieved the remarkably low rate of 3 per 100,000 existing in the United Kingdom, 25,298 lives would have been saved (3). The high death rate was not necessarily restricted to the elderly; 73% of the cirrhosis deaths occurred before the age of 65 (5).

The medical costs of cirrhosis include not only the deaths, but the costs of illness as well. These are difficult to calculate, but we may be certain that the number of cases far outnumbers the deaths, and that they receive a great amount of hospital as well as ambulatory care.

II. ACCIDENTS

Accidents are the fourth leading cause of death in the United States and the leading cause under age 35. In 1975 the death rate for accidents was 48 per 100,000, and for motor vehicle accidents 22 per 100,000 (5).

A series of brilliantly controlled studies of fatal automobile accidents in New York City demonstrated significant blood alcohol concentrations in 54% of fatally injured drivers and 47% of fatally injured pedestrians (6,7). Equivalent or higher proportions with high blood alcohol levels have been reported from other areas of the country (8,9).

We may estimate, therefore, that of the 45,853 deaths from motor vehicle accidents in 1975, there were about 23,000 in which alcohol played an important role. In addition to these deaths there were numerous injuries of varying severity. Their magnitude is hard to quantify, and so are the resultant medical costs, but these are substantial and need to be included in the total accounting.

Other injuries are also related to the use of alcohol. A study in the emergency service of the Massachusetts General Hospital found alcohol involved in 22% of the accident patients compared with 9% of the nonaccident patients. The percentages varied for different types of accidents: home, 22%; transportation, 29%; occupation, 16%; and other, 24% (10).

Similarly a study of accidental deaths in Sacramento, California, showed that blood alcohol levels of 100 mg% or more occurred in 82% of deaths by ingestion, 64% of deaths by fire, 60% of deaths by falls, 24% of deaths by submersion, and 12% of deaths by other injury. This contrasts with about 9% for patients dying of cardiovascular or respiratory disease (11).

Other investigations have demonstrated positive blood alcohol concentrations in 42% of a consecutive series of patients with mild head injuries with concussion (12) and 47% of drownings (13). A study of self-poisonings found that 72% of the men and 40% of the women had been drinking heavily before taking the overdose of drug; the percentages with significant blood alcohol concentrations were even higher (14).

In 1975 there were 57,177 deaths from accidents excluding those involving motor vehicles. There were, of course, many more such accidents without fatal outcomes. If we make a conservative estimate that alcohol played a significant contributing role in 20% of these, the medical costs of alcohol use would amount to 11,435 deaths, a far greater number of injuries, and the substantial treatment costs incurred in these non-motor vehicle accidents.

III. SUICIDE, HOMICIDE, AND ASSAULT

The suicide rate in chronic alcoholics is many times greater than that in the general population (15). Furthermore, studies of consecutive suicides in defined communities have shown that a high proportion of suicides occurs in chronic alcoholics. In St. Louis City County, for example, 23% of suicides occurred among chronic alcoholics (16), while in Seattle—King County—the proportion was 31% (17).

In 1975 suicide was the ninth leading cause of death, responsible for 27,063 deaths (5). Accepting a conservative estimate that a fifth of these suicides occurred in chronic alcoholics, we would calculate that approximately 5400 deaths were related to alcoholism.

Homicide is also strongly associated with alcohol use. In a study at the Office of the Chief Medical Examiner of Maryland, where blood alcohol concentrations are determined for all persons who die within 6 hours of the time of injury, it was found that of 14 homicide deaths in robbery victims or other "uninvolved" individuals, only two were alcohol positive. There were, on the other hand, 42 individuals who were "involved" in the homicide by attempting robbery or engaging in arguments or altercations; 33 of these, or 79%, were alcohol positive (18).

In 1975, there were 21,310 deaths from homicide in the United States. Making the conservative assumptions that half of these deaths occurred in "involved" persons, and that alcohol is involved in the death of half this group, we may estimate that somewhat more than 5000 of the homicide deaths were alcohol-related.

It must be recognized that suicide and homicide deaths represent endpoints of continua of injury. For every successful suicide there are many more unsuccessful ones, and these have also been found to be related to alcoholism (19,20).

Assault injuries show an even greater association with alcohol. In the previously quoted Massachusetts General Hospital study, alcohol was found to be involved in 22% of the accident patients. However, the corresponding figure for patients with injuries from fights or assaults was 56%, more than double the proportion for accidents (10).

IV. CANCER

Recent epidemiologic studies have established a definite association of alcohol consumption with cancer of the mouth, pharynx, larynx, and esophagus. These are primary associations which do not disappear when the case-control series are matched or adjusted for tobacco use (21–26).

It is unclear at this point whether alcohol acts as a carcinogen or cocarcinogen. In any case, the action appears to be a direct one affecting the tissues of contact. The hypothesis that the action may be indirect, through damage to the liver's capacity to metabolize carcinogens, can be questioned by the finding that there is no increase in liver cirrhosis in these cases (27).

Primary cancer of the liver, on the other hand, is related not only to excessive alcohol consumption but also to cirrhosis of the liver (27).

In 1975, in the alcohol-related sites, there occurred in the United States somewhat less than 40,000 new cases and somewhat more than 20,000 deaths (28). How many of these cases and deaths may be attributed to alcohol is difficult to determine. A cautious estimate would judge alcohol to be a significant factor in perhaps a quarter of the total, namely, 10,000 cases and 5000 deaths.

V. MENTAL DISORDERS

In 1946, the rate of first admissions to state and county mental hospitals for alcohol disorders was 6.1 per 100,000 population; by 1972 it had risen to 17.8 per 100,000 (29).

In 1971, there were 353,000 patient care episodes for alcohol disorders in the United States. Of these, 125,000 occurred in state and county mental hospitals, 8800 in private mental hospitals, 32,500 in VA hospitals, 45,500 in other general hospitals' inpatient psychiatric units, 15,800 in inpatient services of community mental health centers, 39,500 in such centers' outpatient services, and 86,000 in other outpatient psychiatric services.

In 1971, the total number of episodes for alcohol disorders represented 9% of all patient care episodes for mental disorders in the United States (29).

VI. OTHER DISORDERS

A study of sudden deaths in Baltimore disclosed that in the 20- to 39-year-old age group the leading cause of sudden, unexpected death was alcoholism and fatty liver. This accounted for 28% of the sudden deaths in this age group, while 22% were due to arteriosclerotic heart disease (30).

A fetal alcohol syndrome has been described recently among the offspring of alcoholic mothers. The syndrome comprises craniofacial, limb, and cardiovascular defects, prenatal and postnatal growth deficiency, and small head size with mental subnormality (31–33).

Among the protean manifestations of alcohol toxicity are neurologic disorders, myopathy and cardiomyopathy, gastritis and pancreatitis (34). Whether alcohol depresses resistance to infection, and if so how, remains unclear (35).

VII. CONCLUSIONS

The available data on the medical costs of excessive use of alcohol are still far from adequate, and estimates must often be based on single or limited studies. It is evident, nevertheless, that these costs are high. A cautious estimate is that in 1975, alcohol-related deaths included 90% of those from cirrhosis of the liver, or 28,400 deaths; 23,000 from motor vehicle and 11,400 from other accidents; 5400 from suicide and 5000 from homicide; and 5000 from cancer of specific sites. The total number of alcohol-related deaths was somewhat more than 78,000.

The emphasis in this chapter has been skewed toward mortality, partly because of its importance but partly also because the data are more accessible. It should be obvious, however, that mortality represents only a small part of the medical costs of alcohol, most of which result from a wide spectrum of illness and disability. The recent statement by the head of the Veterans Administration, Max Cleland, that one-fourth of his agency's hospitalized patients "are there because they drink too much" (36) indicates a state of affairs which is by no means limited to VA hospitals, although they tend to have a greater problem because their patients are primarily male. Alcoholism is a major factor among patients of city and county hospitals, in part because voluntary hospitals often exclude such patients. Nevertheless, since alcoholism affects all social classes, the voluntary hospitals also carry a good part of the burden of alcohol-induced medical care and costs.

It is interesting that Cleland's response to his discovery is to make treatment of alcohol abuse a priority in his budget planning. This is praiseworthy, but in all likelihood will have little impact on the problem. The only approach that can be effective is a program for prevention—not a token program, but one which is of sufficient scale to match the huge and growing burden of death, disability, and illness caused by alcohol (37).

REFERENCES

1. Puffer, R. R., and Griffith, G. W. "Patterns of Urban Mortality." Pan Am. Health Organ., Washington, D.C., (1967).
2. Seeley, J. R. Death by liver cirrhosis and the price of beverage alcohol. *Can. Med. Assoc. J.* **83,** 1361–1366 (1960).
3. Terris, M. Epidemiology of cirrhosis of the liver: National mortality data. *Am. J. Public Health* **57,** 2076–2088 (1967).
4. Alcohol: A Growing Danger. *W.H.O. WHO Chron.* **29,** 102–105 (1975).
5. Advance Report, Final Mortality Statistics, 1975. *Mon. Vital Stat. Rep.* **25**(11), Suppl. (1977).
6. Haddon, W., Jr., *et al.* A controlled investigation of the characteristics of adult pedestrians fatally injured by motor vehicles in Manhattan. *J. Chron. Dis.* **14,** 655–678 (1961).
7. McCarroll, J. R., and Haddon, W., Jr. A controlled study of fatal automobile accidents in New York City. *J. Chron. Dis.* **15,** 811–826 (1962).
8. Waller, J. A. Use and misuse of alcoholic beverages as factor in motor vehicle accidents. *Public Health Rep.* **81,** 591–597 (1966).
9. Laessig, R. H., and Waterworth, B. S. Involvement of alcohol in fatalities of Wisconsin drivers. *Public Health Rep.* **85,** 535–549 (1970).
10. Wechsler, H., *et al.* Alcohol level and home accidents. *Public Health Rep.* **84,** 1043–1050 (1969).
11. Waller, J. A. Nonhighway injury fatalities. I. The role of alcohol and problem drinking, drugs and medical impairment. *J. Chron. Dis.* **25,** 33–45 (1972).
12. Rutherford, W. H. Diagnosis of alcohol ingestion in mild head injuries. *Lancet* **i,** 1021–1023 (1977).
13. Dietz, P. E., and Baker, S. P. Drowning: Epidemiology and prevention. *Am. J. Public Health* **64,** 303–312 (1974).

14. Patel, A. R., *et al.* Self-poisoning and alcohol. *Lancet* **ii,** 1099–1103 (1972).
15. "Alcohol Control Policies in Public Health Perspective." Finn. Found. Alcohol Stud., Helsinki, 1975.
16. Robins, E., Murphy, G. E., Wilkinson, R. H., Jr., Gassner, S., and Kayes, J. Some clinical considerations in the prevention of suicide based on a study of 134 successful suicides. *Am. J. Public Health* **49,** 888–899 (1959).
17. Dorpat, T. L., and Ripley, H. S. A study of suicide in the Seattle area. *Comp. Psychiatry* **1,** 349–359 (1960).
18. Baker, S. P., Robertson, L. S., and Spitz, W. U. Tattoos, alcohol, and violent death. *J. Forensic Sci.* **16,** 219–225 (1971).
19. Batchelor, I. R. C. Alcoholism and attempted suicide. *J. Ment. Sci.* **100,** 451–461 (1954).
20. Kessel, N. Self-poisoning. Part 1. *Br. Med. J.* **ii,** 1265–1270 (1965).
21. Wynder, E. L., Bross, I. J., and Day, E. Epidemiological approach to the etiology of cancer of the larynx. *J. Am. Med. Assoc.* **160,** 1384–1391 (1956).
22. Wynder, E. L., Bross, I. J., and Feldman, R. M. A study of the etiological factors in cancer of the mouth. *Cancer (Philadelphia)* **10,** 1300–1323 (1957).
23. Wynder, E. L., and Bross, I. J. A study of etiological factors in cancer of the esophagus. *Cancer (Philadelphia)* **14,** 389–413 (1961).
24. Schwartz, D., Lellouch, J., Flamant, R., and Denoix, P. F. Alcool et cancer: Resultats d'une enquete retrospective. *Rev. Fr. Etud. Clin. Biol.* **7,** 590–604 (1962).
25. Keller, A. Z., and Terris, M. The association of alcohol and tobacco with cancer of the mouth and pharynx. *Am. J. Public Health* **55,** 1578–1585 (1965).
26. Martinez, I. Factors associated with cancer of the esophagus, mouth, and pharynx in Puerto Rico. *J. Natl. Cancer Inst.* **42,** 1069–1094 (1969).
27. Terris, M. Unpublished data.
28. Silverberg, E., and Holleb, A. I. Cancer statistics, 1975—25-year cancer survey. *Ca* **25,** 2–21 (1975).
29. Kramer, M. "Historical Tables on Changes in Patterns of Use of Psychiatric Facilities, 1946–1971." Biom. Branch, Natl. Inst. Ment. Health, Washington, D.C., 1973.
30. Kuller, L., Lilienfeld, A., and Fisher, R. Sudden and unexpected deaths in young adults: An epidemiological study. *J. Am. Med. Assoc.* **198,** 158–162 (1966).
31. Jones, K. L., Smith, D. W., Ulleland, C. N., and Streissguth, A. P. Pattern of malformation in offspring of chronic alcoholic mothers. *Lancet* **i,** 1267–1271 (1973).
32. Hanson, J. W., Jones, K. L., and Smith, D. W. Fetal alcohol syndrome: Experience with 41 patients. *J. Am. Med. Assoc.* **235,** 1458–1460 (1976).
33. Warner, R. H., and Rosett, H. L. The effects of drinking on offspring: An historical survey of the American and British literature. *Q. J. Stud. Alcohol* **36,** 1395–1420 (1975).
34. "Alcohol and Health." U.S. Dep. Health, Educ. Welfare, Washington, D.C., 1971.
35. Smith, F. E., and Palmer, D. L. Alcoholism, infection and altered host defenses: A review of clinical and experimental observations. *J. Chron. Dis.* **29,** 35–49 (1976).
36. *New York Times,* June 8, 1977.
37. Terris, M. Breaking the barriers to prevention: Legislative approaches. *Bull. N.Y. Acad. Med.* **51,** 242–257 (1975).

EDITORIAL COMMENT

As pointed out by Terris, the medical costs of excessive use of alcohol are very high, even though the available data are far from adequate. It is not easy, however, to sort out the difference between "alcohol-related" and "alcohol caused"; they are not necessarily the same, yet the usual reaction is to equate the two. The question of cost is further complicated by the difficulty in identifying the etiological role of alcohol use in various disease syndromes, as illustrated by the papers of Dreyfuss in respect to central nervous system disorders (Chapter 22), Klatsky with reference to cardiovascular disease (Chapter 21), and Oeullette concerning effects on the fetus (Chapter 28) or the differing nutriture of alcoholics (pages 392–426). Data presented by Klatsky suggesting certain beneficial effects of alcohol use should probably be entered on the credit side of the ledger, when cost estimates move from the realm of "guesstimates" to a more reliable base.

Since this monograph deals primarily with fermented food beverages, it might be pertinent to indicate the relative role of the excessive use of wine and beer in contributing to the total of medical costs. It is known that cirrhosis can arise from the excessive use of wine or beer, but the question is, Do these beverages in fact play a major role? The data are meager and convey no clear picture. Terris in an earlier paper on the epidemiology of cirrhosis of the liver (1) wrote as follows: "Cirrhosis mortality in the United States, Canada, and the United Kingdom is associated with differences in the apparent consumption of alcohol from spirits and wine. No such association exists for beer." Brenner (2), studying both short- and long-term alcohol consumption in the United States in relation to changes in the national economy, found per capita consumption of alcohol and mortality from liver cirrhosis increased during periods of short-cycle recessions, but the rise in consumption was entirely accounted for by increase in spirit consumption, not in wine or beer.

Mezey and Santora in Chapter 20 of this volume present limited data on beverage choice among persons with liver disease. In their series of patients drawn mostly from the eastern seaboard, only 13% were predominantly wine drinkers and 8% were predominantly beer drinkers. Matching this statistic for beer drinkers, for example, against the fact that approximately 47% of all beverage alcohol sold in the United States is in the form of malt beverages again suggests the conclusion that beer does not play a major role in the cirrhosis picture in the United States. Much the same can probably be said for wine.

REFERENCES

1. Terris, M. Epidemiology of cirrhosis of the liver: National mortality data. *Am. J. Public Health* **57,** 2076–2088 (1967).
2. Brenner, M. H. Trends in alcohol consumption and associated illnesses. *Am. J. Public Health* **65,** 1279–1293 (1975).

31

SOCIOECONOMIC CONSIDERATIONS AND CULTURAL ATTITUDES: AN ANALYSIS OF THE INTERNATIONAL LITERATURE

M. H. Brenner

I.	Introduction	489
II.	The Supply-Demand-Stress Model	491
III.	Indicators of Key Economic, Sociocultural, and Stress Variables	492
	A. Supply	492
	B. Demand: Societal Values or Attitudes	493
	C. Social-Psychological Stress	493
IV.	Overview and Conceptual Organization of the Effects of Government Policies	494
	A. Prohibition	494
	B. Interdiction	495
	C. Restriction	495
	D. State Monopoly	495
	E. Taxation	496
	F. Licensing	496
	G. The Punitive Approach	497
V.	Discussion	498
	References	499
	Editorial Comment	504

I. INTRODUCTION

A significant proportion of the United States' and international scholarly litera-ture on the epidemiology of alcohol problems and government policies toward those problems has been reviewed. An attempt has been made to categorize and list the key factors that have been found to be associated with alcohol consumption and alcohol-related problems. The literature in this field resembles other literature in the mental and general health areas insofar as it treats the heavy or addictive use of alcohol as representing a form of illness, or at least social deviance which is strongly related to social–psychological stresses and to disturbances in the social structural environment (Baldie, 1932; Gabriel, 1933; Jellinek, 1942; Fleetwood and Diethelm, 1951; Harper and Hickson, 1951; Dent, 1955; Blane *et al.*, 1963; Bailey *et al.*, 1965). The principal academic disciplines involved in alcohol research from this standpoint are epidemiology, psychology, psychiatry, and sociology.

A second approach begins with the viewpoint that consumption of alcohol can be understood in much the same way as consumption of any other industrial product or service. In this sense, alcohol resembles other foods or beverages (tobacco, coffee) from which excessive indulgence may lead to medical or social pathologies. From this viewpoint, the standard economic considerations of supply and demand, and resulting prices, should determine the degree to which an alcoholic beverage is consumed in the society and the extent of resulting pathologies. Included in the economic view of factors affecting supply are the quantity of alcoholic beverages produced and imported (Karpio, 1945; Dérobert, 1953; Seidel, 1955; Sjöhagen, 1960; Wortis, 1963; Schmidt, 1966; Lustig, 1965; Coffey, 1966), the quantity of alternative beverages (Karpio, 1945; Coffey *et al.*, 1966), and the extent of government restriction of alcoholic beverages to various population subgroups in the society (Feeney *et al.*, 1955; Trice and Belasco, 1968; Bresard, 1972).

On the demand side, the economic view is based principally on income (Seidel, 1955; Ledermann, 1956a,b) and what is referred to in the economist's language as "tastes and preferences." These tastes and preferences represent the sociocultural attitudes of society toward the use of specific alcoholic beverages and to the behavioral conditions that result from the ingestion of varying amounts of these beverages. In other words, beyond the issue of disposable income, that of the normal or habitual mode of utilization of alcoholic beverages (and their substitutes) is of paramount concern (Jean, 1956; Aujaleu and Jean, 1957).

The economic researcher does not generally investigate the effects of sociocultural factors on consumption. Rather, he is more likely to assume that certain sociocultural factors are operative, as he is able to glean these from literature, and may or may not subsequently allow these factors to enter his general consumption model (Dérobert, 1953). The economist is more likely, on the other hand, to allow government policy-related variables to enter his model, since they affect supply directly through restrictive regulations, or indirectly through pricing policies.

The highly significant, and almost universally assumed, factor of social–psychological stress is hardly ever taken into consideration in the economic analysis of consumption patterns. As in the case of sociocultural factors, stress phenomena have been traditionally treated as lying outside the purview of economic analysis. It is for this reason that the most important research on the use of alcohol, as a mode of stress reduction, arises from psychiatry, psychology, sociology, and anthropology.

II. THE SUPPLY-DEMAND-STRESS MODEL

In organizing a conceptual framework for a review of the key factors that tend to affect consumption of alcohol and related problems, a threefold scheme has been developed. This scheme identifies supply and demand and social–psychological stress as the principal categories of factors affecting use of alcoholic beverages. The supply factors include those which bear on the absolute quantities of these beverages (Chopra *et al.*, 1942; Jellinek, 1942; Ledermann, 1956a; Müller-Dietz, 1957; Røgind, 1959; Sjöhagen, 1960; Wortis, 1963; Keller and Efron, 1974), or alternatives to them (Karpio, 1945; Nelker, 1947; Wechsler, 1972), available in the society through production or importation. They also include government regulation, either directly restricting amounts available or sold (Dent, 1942; Nelker, 1947; Immonen, 1967), or price policies indirectly affecting consumption (Karpio, 1945; Landis, 1945a; Røgind, 1959; Seeley, 1960; Wortis, 1963; New York State Moreland Commission, 1963d; Schmidt, 1966).

The second set of factors, namely, those related to demand, are of two types, and probably of equal importance. One is personal income computed either in absolute terms (Gabriel, 1933; Söderberg, 1962; New York State Moreland Commission, 1963b,d; Nielsen, 1964; Simon, 1966a; Cisin and Cahalan, 1968; Popham, 1969; Keller and Efron, 1974), controlling for the effects of inflation, or as disposable income (New York State Moreland Commission, 1963d). In addition, the prices of alcoholic beverages (Nelker, 1947; Røgind, 1949, 1959; Müller-Dietz, 1957; Seeley, 1960; Schmidt and Bronetto, 1962; Wortis, 1963; Simon, 1966a,b; Nielsen and Strömgren, 1969) or products alternative to them (Popham, 1969), are included since they clearly affect the availability of alcoholic beverages from a financial standpoint. A still more refined analysis would take into account the proportion of income expended on the ''necessities'' of life including food for basic nutritional subsistence, clothing, and shelter. In practice, however, it is fairly well established that the higher the level of income, the greater is the proportion that is used for nonnecessity expenditures (including consumption of alcoholic beverages). The second major category of demand factors involves tastes and preferences or, generally speaking, sociocultural values and attitudes with respect to the use of alcoholic beverages. These values and attitudes, in turn, tend to focus on two subjects: the normal mode of use of alcoholic beverages, e.g., tension-reduction, promotion of conviviality or intimacy, or recreation (Myerson, 1945; Jacobsen, 1951; Sariola, 1954; Tongue, 1955; Kennedy and Fish, 1959; Moore and Wood, 1963; Lustig, 1965; Bättig, 1967; Bruun, 1967; Immonen, 1967; Cisin and Cahalan, 1968) and the societal tolerance of inebriation, acute and chronic, under specific conditions (Myerson, 1945; Jacobsen, 1951; Wortis, 1963; Schuckit, 1972).

The third major group of factors concerns the area of social–psychological stress. From the literature pertaining to the impact of stress, in general, on mental and physical health, as well as from that dealing specifically with alcohol problems, a fourfold classification of major societal stresses has been identified for inclusion in the general model. These four major categories are loss, job-related stress, stress related to chronic ill health and pain, and family and marital instability. The categories of loss and job-related stress are themselves quite complex, but have been subclassified in order to identify the major distinctions found to be significant in the literature. In the case of loss, we find an important distinction between loss in socioeconomic status (Bailey *et al.*, 1965) and loss of a person to whom one is emotionally attached (Horn and Wanberg, 1970), either by death or by geographic separation. In the case of job-related stress, we have attempted to distinguish the issues of socioeconomic remuneration and prestige from those of the skill, decision-making responsibility, and variability of the work itself (Myerson, 1945; King *et al.*, 1969; Horn and Wanberg, 1970; Kielholz, 1970; Horn *et al.*, 1974).

The entire set of categories within the supply–demand–stress model are identified below.

III. INDICATORS OF KEY ECONOMIC, SOCIOCULTURAL, AND STRESS VARIABLES

A. Supply

1. Ecological factors: extent of indigenous agricultural produce and importation (Jellinek, 1942; Létinier, 1946; Dérobert, 1953; Berkowitz, 1957).

2. Supply of "alternatives" to alcohol, such as tobacco or coffee (Karpio, 1945; Nelker, 1947; Wechsler, 1972).

3. Price: absolute price and price relative to other "necessary" commodities or services and to alternatives to alcohol (Nelker, 1947; Røgind, 1949, 1959; Müller-Dietz, 1957; Seeley, 1960; Schmidt and Bronetto, 1962; Wortis, 1963; Simon, 1966a,b; Nielson and Strömgren, 1969).

4. Governmental regulation: (a) Taxation policies affecting prices (Landis, 1945b; Røgind, 1949; Bureau of Medical Economic Research, 1951; Couléon, 1964). (b) Direct regulation of supply (Dent, 1942; Nelker, 1947; Immonen, 1967).

B. Demand: Societal Values or Attitudes

1. Intoxication to alleviate psychic or physical pain (Fox, 1967).

2. Per capita personal income, or perhaps "disposable" income is the more nearly accurate measure (New York State Moreland Commission, 1963b). Also, one would like to account for expenditures for such necessities as food, clothing, and shelter according to the standard of living in the area (Landis, 1945a), i.e., real income.

3. Drinking to promote conviviality: relation to tension-release mechanism, especially used in groups of strangers or groups having internal status differentials, such as organizations (Moore and Wood, 1963).

4. Drinking to promote intimacy: related to conviviality, except occurs under more intimate circumstances, especially sex related.

5. Drinking as a recreational effort (Moore and Wood, 1963; Lustig, 1965; Horn *et al.*, 1974).

6. Attitude of society to inebriation (Myerson, 1945; Jacobsen, 1951; Wortis, 1963; Schuckit, 1972).

7. Attitude of society toward use of alcohol in relation to food (Vernon, 1928; Watson, 1955; Lustig, 1965).

8. Use of alcohol as "truth-letting" device where deviance is permitted under such intoxication or pseudointoxication.

9. Ritual use of alcohol (Jacobsen, 1951; Huxley, 1957).

10. Regulation of intoxication by criminal sanction (Landis, 1945a; Nelker, 1947; Ledermann, 1956a,b; Forbes, 1950; Müller-Dietz, 1957; Moore and Wood, 1963; New York State Moreland Commission, 1963a; Zieliński, 1969).

11. Regulation of intoxication and alcoholism by state treatment programs (Moore and Wood, 1963; Landis, 1945a; Salomaa, 1953; Knudsen, 1959; Greenland, 1960; Wortis, 1963; Lustig, 1965; Immonen, 1967; Keller and Efron, 1974). This could affect supply as well as demand.

12. Public efforts to control consumption or intoxication or alcoholism through health education (Alexander *et al.*, 1941; Cavaillon, 1947; Peterson, 1949; Steiger, 1949; Salomaa, 1953; Wortis, 1963; Keller and Efron, 1974). Again this could also affect supply.

C. Social–Psychological Stress

This factor, of course, influences the demand level for alcohol as a method of tension reduction. However, its placement in a category distinct from demand is meant to indicate that the demand category refers to cultural definitions of the appropriate use of intoxicating beverages. It is, of course, true that in a great many cultures a manifest and explicit purpose of the use of intoxicating beverages is to reduce pain, either psychic or physical. In this category, social–

psychological stress, we attempt to identify those sources of stress which increase the probability of intoxication.

1. Loss. This category is probably the largest source of severe stress in modern society. Perhaps the two most important sources of loss, as they are ordinarily imagined, are (a) loss in socioeconomic status (Bailey *et al.,* 1965); (b) loss of loved one or close friend through death or geographic mobility (Horn and Wanberg, 1970).

2. Job-related stress (Meyerson, 1945; Horn and Wanbergn, 1970; Kielholz, 1970; Schuckit and Gunderson, 1974). (a) Comparatively low socioeconomic renumeration of job; (b) low or demeaning social prestige of job; (c) low decision-making character of job; (d) comparatively unskilled work; (e) invariant, or comparatively routine, work schedules; (f) extraordinarily high levels of responsibility at high tension levels.

3. Stress related to chronic ill health and pain (Bailey *et al.,* 1965; Haberman, 1965).

4. Family and marital instability (Bailey *et al.,* 1965; Horn and Wanberg, 1970; Kielholz, 1970; Horn *et al.,* 1974).

IV. OVERVIEW AND CONCEPTUAL ORGANIZATION OF THE EFFECTS OF GOVERNMENT POLICIES

Social problems related to the heavy consumption of alcoholic beverages have been a widely recognized problem for several centuries. It appears that virtually every major civilization since the Sumerians have regulated the production, importation, and consumption of alcoholic beverages (see, e.g., Baird, 1944, 1945, 1946, 1948; Idel'chik *et al.,* 1972). During the nineteenth and twentieth centuries in Europe and North America, a great variety of state policies have governed the utilization of alcohol. Within any single country the matrices of state policies have varied through time and have been enforced with varying degrees of scrupulousness. Thus, at any moment in time the comparison among countries, or provinces within a country, with regard to state alcohol policy is a research project of significant scope. Not only is it necessary to list carefully the types of policies in use at any time in each region, but in addition, it is important to identify (ideally, quantitatively) the extent or subcategories of each regulatory mechanism.

In reviewing the great variety of statements on state policy which can be derived from the international literature on alcoholism, the following policies appear to have commanded the greatest attention over sustained periods of time.

A. Prohibition

The states have prohibited the production, importation, and consumption of all, or specific, alcoholic beverages. States which prohibit production do not necessarily always prohibit importation, or even consumption. Frequently, it is found that importation will be left unrestricted because of political pressure by a neighboring country which is a large producer of alcohol and whose economy therefore depends on the viability of sales (Sariola, 1954; Røgind, 1959). Occasionally, consumption of certain beverages is found to be especially noxious to a legislature (e.g., gin and absinthe in eighteenth and nineteenth century England, and rum in the Spanish and French colonies) (Coffey, 1966). This partial prohibition obviously allows for the continued or expanded consumption of other alcoholic beverages. In addition, the issue of enforcement is paramount, and lax, corrupt, or differential enforcement may clearly obviate the entire legal proscription (Vernon, 1928; Sariola, 1954).

B. Interdiction

Interdiction is a legal order prohibiting a described group of individuals from purchasing, possessing, or consuming beverage alcohol within a specific province (Porter, 1966). It is actually a moderate form of prohibition in that it is imposed on a specific class of individuals for a definite period of time for a stated reason. Typical interdictions relate to minimal age levels for consumption of alcohol (Wangenheim, 1964), proscription of individuals with a record of intoxication from purchasing alcoholic beverages, and proscription on sales to persons visibly intoxicated (the latter two occurring in the Scandinavian countries) (Baird, 1945; Marcus, 1946; Forsberg, 1947).

C. Restriction

This method attempts to regulate the amount of alcohol sold and consumed by restriction of liquor store licenses or, in a state monopoly system, restricting the stock of a state monopoly store, or restricting the actual number of state monopoly stores. These methods have been in use at various times for specific states in the United States and in the Scandinavian countries.

D. State Monopoly

The monopoly system may control the importation, production, or direct sales of alcoholic beverages, or all three of these (Rostow, 1940; Rostow and Hull, 1942; Sellers, 1943; Baird, 1945; Peterson, 1949; Joint Committee of the States to Study Alcoholic Beverages Laws, 1950–51; Sariola, 1954; Barker, 1957; Diesen, 1959; Røgind, 1959; New York State Moreland Commission, 1963a; Simon, 1966a). In the complete state monopoly system, the state prohibits importation or acts as the sole importer in addition to being the sole indigenous producer and providing the only legally available stores for public purchase of these beverages. In several states in the United States, monopolies have existed, and continue to exist, on the production or sales of beverage alcohol (Rostow, 1940; Rostow and Hull, 1942; Joint Committee of the States to Study Alcoholic Beverage Laws, 1950; Garrett, 1954; Barker, 1957; New York State Moreland Commission, 1963a,c; Simon, 1966a). Virtually all of the European countries have, during some time span, experimented with a total or partial state monopoly system (Fleming, 1937; Marcus, 1946; Immonen, 1952; Sariola, 1954; Elmer, 1957; Diesen, 1959; Røgind, 1959; Sjöhagen, 1960). It is important to identify the specific alcoholic beverages over which governments maintain a monopoly, as well as the extent of monopolistic coverage in a given region (Jean, 1956; Aujaleu and Jean, 1957; Barker, 1957; Sjöhagen, 1960).

E. Taxation

Again, virtually all European and Northern American countries have utilized a system of taxation to attempt to regulate the importation, production, sales, and actual level of consumption of alcoholic beverages. The outstanding rationale for the use of taxation as a means for curbing consumption is that increased prices essentially make alcohol less available financially, thus in effect limiting use (Rostow, 1940; Dent, 1942; Watson, 1955; Sjöhagen, 1960; Schmidt, 1966). Taxation on production has the effect of increasing cost and ultimately raising price levels for consumers. Import tariffs on nondomestically produced alcohol similarly serve to raise prices to consumers, but have the additionally significant political motive of protecting native industries from foreign competition.

It should not be overlooked that a most important factor in the taxation of alcoholic beverages has been the revenue accruing to governments (Henderson, 1941; Baird, 1945; Traylor, 1949; Hu, 1950; Røgind, 1959; Sjöhagen, 1960). Indeed, numerous scholarly articles published in the United States and Europe attest to the fact that prohibition measures were overturned, not only because of public sentiment, but also because of the enormous loss of revenue to government institutions (Berkshire, 1940; Karpio, 1945; Hu, 1950; Sariola, 1954; Coffey, 1966).

F. Licensing

States typically issue licenses, generally for a fee, for the sale of alcoholic beverages, both in stores and restaurants (Rostow, 1940; Dent, 1942; Peterson, 1949; Social Sweden, 1952; New York State Moreland Commission, 1963c). Licenses are also typically issued for the production, and occasionally the importation, of such beverages. The effects of licensing are very similar to those of taxation, both with respect to price increases to the consumer and increased government revenue. In addition, however, licensing also serves as a type of restriction by which the absolute number and types of sellers, producers, and importers are legally controlled by the State (Rostow, 1940; New York State Moreland Commission, 1963b).

G. The Punitive Approach

The above six types of government policy all concern methods of regulation of the supply of alcoholic beverages, either through regulation of the amounts and types of alcohol available to the public (at any price) or through some manipulation of the price structure. An entirely different approach is punitive in nature and involves the use of the criminal legal system. In terms of the criminal law, three types of activity based on involvement with alcohol are proscribed as follows.

1. Public Intoxication

There is great variation on the type of legal restriction on intoxication in public. In many countries and provinces, there is some difficulty in clearly defining the degree to which it is necessary for intoxication to be "public" before the legal sanction is applied (Forbes, 1950). One feels that much of the use of the term "public" is meant to, and does in fact, invoke the issue of the degree of *disturbance* to the public. In other words, much lies at the discretion of the individual policeman as to whether an offense is deemed sufficiently intolerable for criminal action to be taken against the offender.

2. Intoxication with Criminal Offense

In this case, the primary emphasis is on the criminal offense itself, e.g., homicide, assault, rape, robbery, or larceny. There is great variance among countries and states whether, or to what degree, perpetration of a criminal offense associated with alcohol intoxication can be excused on the grounds of intoxication. This is an extraordinarily wide-ranging problem, since it is reported that anywhere from 30 to 60% of all arrests are related to alcoholic intoxication in cities in North America and Europe (Nelker, 1947; New York State Moreland Commission, 1963a). It appears that in most countries and states, serious offenses are not excusable by virtue of intoxication (Forbes, 1950; Moore and Wood, 1963; Lustig, 1965). In those regions where the law allows such an excuse, it is usually the case that the offender either undergoes involuntary treatment or serves a sentence in a hospital for the criminally insane.

3. Intoxication While Driving an Automobile

The legal codes in different European countries and different states in the United States vary considerably as to degree of penalty legally mandated for two different types of offenses. (a) Driving while intoxicated: this offense pertains only to the act of driving under the significant influence of alcohol. Government regulations vary according to how much alcohol is necessary in the bloodstream in order that the individual be adjudged "intoxicated," (Landis, 1945a; Forbes, 1950; Lustig, 1965; Christie, 1967; Zieliński, 1969) as well as the means by which level of intoxication is measured (blood analysis, breath analysis, tests of perception or judgment, etc.) (Forbes, 1950; Moore and Wood, 1963; Lustig,

1965). (b) Another set of laws pertains solely to traffic violation offenses. These may concern negligent driving resulting in accidents, exceeding the speed limit, disregarding traffic signals. improper parking, or driving on the pedestrian walkway. Once it is established that such laws have been violated, there is a question whether the individual's behavior is deemed negligent and thus excusable to some degree on the grounds of intoxication (*Proc. Int. Conf.,* 1951) or whether violation of the law is punishable on its own (regardless of the intoxication) (Forbes, 1950; Moore and Wood, 1963), or finally, whether any violation of traffic laws is considered to be more serious—and therefore subject to more severe punishment—when associated with intoxication.

This entire area of regulation of alcohol consumption is of very great importance, since as much as 50% of automobile accidents involving mortality have been attributed to alcohol intoxication (Ledermann, 1956a,b; Europe police. . ., 1966).

4. Violations of Laws Pertaining to Regulation of Consumption, Production, Sales, or Importation

In many countries and states, violations of the laws regulating supply and consumption, including prohibition, interdiction, restriction, state monopoly, taxation, and licensing, are criminal offenses. Violation of these laws or the aiding or abetting of such violations is illegal. Punitive measures vary widely both nationally and internationally from fines to imprisonment, loss of license, or interdiction from further use.

V. DISCUSSION

It is particularly since World War II that alcohol abuse and alcoholism are being viewed as illnesses rather than as criminal offenses in Europe and North America. In the Scandinavian countries especially, laws have been set in motion to relieve the criminal justice system of its responsibilities in alcoholism and to substitute treatment methods of managing alcoholics (Landis, 1945a; Knudsen, 1959; Greenland, 1960; Immonen, 1967). In these countries there is mandatory treatment of alcoholics (Salomaa, 1953; Knudsen, 1959), usually on an ambulatory basis with psychotherapy and drugs, where prognosis is reasonably good. If repeated warnings to appear for treatment are disregarded, hospitalization is legally enforced (Salomaa, 1953; Knudsen, 1959). There remains a strong partnership, however, between the criminal justice and medical treatment systems, where alcoholics who are considered dangerous to society are turned over to custodial care. Furthermore, persons selling or giving alcohol to alcoholics can be prosecuted.

It is not possible in this brief summary to even list the many different types of legally mandated therapeutic modes or environments that are routinely used in different European countries (Moore and Wood, 1963; Knudsen, 1959; Greenland, 1960; Wortis, 1963; Lustig, 1965; Immonen, 1967; Keller and Efron, 1974). Another major source of public policy toward regulation of alcohol consumption in Europe and North America concerns public health education. Efforts in this respect have been particularly substantial in the Scandinavian countries, France, and Communist countries, but such efforts have only recently begun in North America (Alexander *et al.,* 1941; Cavaillon, 1947; Peterson, 1949; Steiger, 1949; Salomaa, 1953; Wortis, 1963; Keller and Efron, 1974). In fact, there seems to be at least a loose connection between the use of some form of state monopoly and sustained expenditures on public health education and problems of intoxication and alcoholism.

REFERENCES

Alexander, L., Moore, M. and Myerson, A. (1941). A proposal for changes in present methods of sale of alcoholic beverages to conform with the Federal Food, Drug, and Cosmetic Act. *Trans. Am. Neurol. Assoc.* **67,** 230-232.

Aujaleu, E., and Jean, P. (1957). Bilan et orientation de la lutte contre l'alcoolisme. [Evaluation and orientation of the fight against alcoholism.] *Rev. Hyg. Med. Soc.* **5,** 41-85.

Bättig, K. (1967). Alkoholismus: Epidemiologische Zusammenlänge und Folgen. [Alcoholism: Epidemiological relationships and sequels.] *Naturwiss. Rundsch.* **20,** 200-204.

Bailey, M. B., Haberman, P. W., and Alksne, H. (1965). The epidemiology of alcoholism in an urban residential area. *Q. J. Stud. Alcohol* **26,** 19-40.

Baird, E. G. (1945a). Controlled consumption of alcoholic beverages. *In* ''Alcohol, Science, and Society'' (Lecture 21). *Q. J. Stud. Alcohol,* New Haven, Connecticut.

Baird, E. G. (1945b). The alcohol problem and the law. *Q. J. Stud. Alcohol* **5,** 126-161.

Baird, E. G. (1946). The alcohol problem and the law. *Q. J. Stud. Alcohol* **6,** 335-383; **7,** 110-162, 271-296.

Baird, E. G. (1948). The alcohol problem and the law. *Q. J. Stud. Alcohol* **9,** 80-118.

Baldie, A. (1932). The causation, treatment and control of the alcohol habit. *Br. J. Inebriety* **30,** 45-68.

Barker, T. W., Jr. (1957). The states in the liquor business. Some observations on administration. *Q. J. Stud. Alcohol* **18,** 492-502.

Berkowitz, M. I. (1957). A comparison of some ecological variables with rates of alcoholism. *Q. J. Stud. Alcohol* **18,** 126-129.

Berkshire, S. (1940). Uniform control and taxation of alcoholic beverages. *Q. J. Stud. Alcohol* **1,** 558-562.

Blane, H. T., Overton, W. F., Jr., and Chafetz, M. E. (1963). Social factors in the diagnosis of alcoholism. I. Characteristics of the patient. *Q. J. Stud. Alcohol* **24,** 640-663.

Bresard, M. (1972). Alcoolisme et promotion sociale. [Alcohol and social advancement.] *Bull. Acad. Natl. Med.* **156,** 47-51.

Bruun, K. (1967). Drinking patterns in the Scandinavian countries. *Br. J. Addict.* **62,** 257-266.

Bureau of Medical Economic Research. (1951). Medical care expenditures, prices and quantity, 1930-1950. *J. Am. Med. Assoc.* **147,** 1354-1359.

Cavaillon, A. (1947). ''Report on Alcoholism,'' Interim Commission, Rep. No. 104. World Health Org., Geneva.

Chopra, R. N., Chopra, G. S., and Chopra, I. C. (1942). Alcoholic beverages in India. *Indian Med. Gaz.* **77**, 224–232, 290–296, 361–367.

Christie, N. (1967). Scandinavian experience in legislation and control. The Scandinavian hangover. *Trans-Action* **4**(3), 34–37.

Cisin, I. H., and Cahalan, D. (1968). Comparison of abstainers and heavy drinkers in a national survey. *Psychiatr. Res. Rep.* **24**, 10–21.

Coffey, T. G. (1966). Beer Street: Gin Lane; some views of 18th century drinking. *Q. J. Stud. Alcohol* **27**, 669–692.

Couléon, H. (1964). The problem of the private distillers. *Rev. Prat.* **14**, Suppl. 4, xxvii–xxxii.

Dent, J. Y. (1942). Alcohol legislation and taxation in Britain in wartime. *Q. J. Stud. Alcohol* **3**, 221–229.

Dent, J. Y. (1955). "Anxiety and Its Treatment: With Special Reference to Alcoholism," 3d Ed. Skeffington, London.

Dérobert, L. (1953). L'économie de l'alcoolisme.[The Economics of Alcoholism.] Monogr. Inst. Nat. Hyg., No. 2. Min. Sante Publique, Paris.

Diesen, H. (1959). Den alkoholpolitiska situationen i Norge. [Current alcohol policy in Norway.] *Alkoholpolitik* **22**, 93–99, 121–122.

Elmer, A. (1957). The change of temperance policy in Sweden. *Br. J. Addict.* **54**, 55–58.

Europe police crack down on the drinking driver. (1966). *Mil. Police J.* **16**(5), 7–10.

Feeney, F. E., Mindlin, D. F., Minear, V. H., and Short, E. E. (1955). The challenge of the Skid Row alcoholic. A social, psychological and psychiatric comparison of chronically jailed alcoholics and cooperative alcohol clinic patients. *Q. J. Stud. Alcohol* **16**, 645–667.

Fleetwood, M. F., and Diethelm, O. (1951). Emotions and biochemical findings in alcoholism. *Am. J. Psychiatry* **108**, 433–438.

Fleming, R. (1937). The management of chronic alcoholism in England, Scandinavia and Central Europe. *N. Engl. J. Med.* **216**, 279–289.

Forbes, G. (1950). Drunkenness and the criminal law. *Med. Pr.* **223**, 74–77.

Forsberg, J. (1947). Linturi-systemet; en presentation och kritisk bedomning [The Linturi system, a description and critical evaluation]. *Tirfing* **41**, 130–133.

Fox, R. (1967). Alcoholism and reliance upon drugs as depressive equivalents. *Am. J. Psychother.* **21**, 585–596.

Gabriel, E. (1933). Der Einfluss der Wirtschaftskrise auf den Alkoholismus. [The effect of the depression on alcoholism.] *Muench. Med. Wochenschr.* **80**, 815–818.

Garrett, E. W. (1954). Special purpose police forces. *Ann. Am. Acad. Polit. Soc. Sci.* **291**, 31–38.

Greenland, C. (1960). Habitual drunkards in Scotland, 1879–1918. A historical note. *Q. J. Stud. Alcohol* **21**, 135–139.

Haberman, P. W. (1965). Psychophysiological symptoms in alcoholics and matched comparison persons. *Community Ment. Health J.* **1**, 361–364.

Harper, J., and Hickson, B. (1951). The results of hospital treatment of chronic alcoholism. *Lancet* No. 261, 1057–1059.

Henderson, Y. (1941). The proof gallon: A federal tax strait jacket. *Q. J. Stud. Alcohol* **2**, 46–56.

Horn, J. L., and Wanberg, K. W. (1970). Dimensions of perception of background and current situation of alcoholic patients. *Q. J. Stud. Alcohol* **31**, 633–658.

Horn, J. L., Wanberg, K. W., and Adams, G. (1974). Diagnosis of alcoholism; factors of drinking, background and current conditions in alcoholics. *Q. J. Stud. Alcohol* **35**, 147–175.

Hu, T. Y. (1950). "The Liquor Tax in the United States. 1791–1947. A History of the Internal Revenue Taxes Imposed on Distilled Spirits by the Federal Government," Monogr. Public Finance Natl. Income, No. 1. Grad. Sch. Bus., Columbia Univ., New York.

Huxley, A. (1957). The history of tension. *Ann. N.Y. Acad. Sci.* **67**, 675–684. (Also in: *Sci. Mon.* **85**, 3–9.)

Idel'chik, K. I., Aruin, M. I., and Nesterenko, A. I. (1972). I Vserossiiskii s'yezd po bor'be s p'yanstvom. [The first All-Russian Congress on the Fight Against Inebriety.] *Sov. Zdravookhr.* **31**(2), 61–65.

Immonen, E. J. (1952). The alcohol problem in Finland. *Q. J. Stud. Alcohol* **13**, 685–688.

Immonen, E. J. (1967). The alcohol problem in Finland. *Br. J. Addict.* **62**, 287–293.

Jacobsen, E. (1951). Alcohol as a social problem. *In* "Rauschgifte und Genussmittel" (K. O. Møller, ed.), Ch. 8, pp. 231–268. B. Schwabe, Basel.

Jean, P. (1956). L'organisation de la lutte contre l'alcoolisme. [Organization of the fight against alcoholism.] *Rev. Hyg. Med. Soc.* **4**, 748–761.

Jellinek, E. M. (1942). The interpretation of alcohol consumption rates with special reference to statistics of wartime consumption. *Q. J. Stud. Alcohol* **3**, 267–280.

Joint Committee of the States to Study Alcoholic Beverage Laws (1950). "Alcoholic Beverage Control (ABC). An Official Study." New York.

Karpio, V. (1945). Finsk kronika för 1939–44 [Finnish chronicle for 1939–44]. *Tirfing* **39**, 1–19.

Keller, M., and Efron, V. (1974). Alcohol problems in Yugoslavia and Russia; some observations of recent activities and concerns. *Q. J. Stud. Alcohol* **35**, 260–271.

Kennedy, A., and Fish, F. J. (1959). Alcoholism and alcoholic addiction. *Recent Prog. Psychiatry* **3**, 277–285.

Kielholz, P. (1970). Alcohol and depression. *Br. J. Addict.* **65**, 187–193.

King, L. J., Murphy, G. E., Robins, L. N., and Darvish, H. (1969). Alcohol abuse: a crucial factor in the social problems of Negro men. *Am. J. Psychiatry* **125**, 1682–1690.

Knudsen, R. (1959). "Public and Private Measures in Norway for the Prevention of Alcoholism." Int. Bur. Against Alcohol., Lausanne.

Landis, B. Y. (1945a). Estimated consumer expenditures for alcoholic beverages in the United States, 1890–1943. *Q. J. Stud. Alcohol* **6**, 92–101.

Landis, B. Y. (1945b). "Some Economic Aspects of Alcohol Problems. II. Estimated Consumer Expenditures for Alcoholic Beverages in the United States, 1890–1943," Mem. Sect. Alcohol Stud., No. 4, pp. 35–44. Yale Univ., New Haven.

Lederman, S. (1956a). La production et la consommation de vin et d'alcool en France [Production and consumption of wine and alcohol in France]. *In* "Alcool, Alcoolisme, Alcoolisations Donnees Scientifiques de Caractere Physiologique Economique et Social," Ch. 1, pp. 17–80. Presses Univ. Fr., Paris.

Lederman, S. (1956b). Les accidents de la circulation [Traffic accidents]. (Ch. 7). Les accidents du travail [Industrial accidents]. (Ch. 8). *In* "Alcool, Alcoolisme, Alcoolisations. Donnees Scientifiques de Caractere Physiologique, Economique et Social," pp. 181–216. Presses Univ. Fr., Paris.

Létinier, G. (1946). Elements of the national balance-sheet on alcoholism. *Population (Paris)* **1**, 317–328.

Lustig, B. (1965). Über Alkoholismusprobleme in sowjetischer und europäischer Sicht [Problems of alcoholism in Soviet and European views]. *Wien. Med. Wochenschr.* **115**, 400–403.

Marcus, M. (1946). "The Liquor Control System in Sweden." Norstedt, Stockholm.

Moore, R. A., and Wood, J. T. (1963). Alcoholism and its treatment in Yugoslavia. *Q. J. Stud. Alcohol* **24**, 128–137.

Müller-Dietz, H. (1957). Der Alkoholismus und seine Bekampfung in der Sovjetunion [Alcoholism and the fight against it in the Soviet Union]. *Int. J. Alcohol Alcoholism* **2**, 34–39.

Myerson, A. (1945). Roads to alcoholism. *Surv. Graph.* **34**, 49–51.

Nelker, G. (1947). *Svensk Krönika* [Swedish chronicle]. *Tirfing* **41**, 49–64.

New York (State) Moreland Commission on the Alcoholic Beverage Control Law (1963a). "The Relationship of the Alcoholic. Beverage Control Law and the Problems of Alcohol," Study Pap. No. 1. New York.

New York (State) Moreland Commission on the Alcoholic Beverage Control Laws (1963b). "The Relationship Between the Number of Off-Premise Licenses and the Consumption of Alcoholic Beverages: A Statistical Analysis," Study Pap. No. 3. New York.

New York (State) Moreland Commission on the Alcoholic Beverage Control Law (1963c). "Some Economic and Regulatory Aspects of Liquor Store Licensing in New York State: A Summary of Research." Study Pap. No. 4. New York.

Nielsen, J. (1964). Nogle aspekter vedr rende alkoholisme og alkoholjorskning i Norden [Some aspects of alcoholism and alcohol research in the Scandinavian countries]. *Nord. Psychiatr. Tidsskr.* **18**, 500–520.

Nielsen, J., and Strömgren, E. (1969). Über die Abhängigkeit des Alkoholkonsums und der Alkoholkrankheiten vom Preis alkoholischer Getränke [On the dependence of alcohol consumption and alcoholic diseases on the price of alcoholic beverages]. *Bibl. Psychiatr. Neurol.* **142**, 165–170.

Peterson, V. W. (1949). Vitalizing liquor control. *J. Crim. Law Criminol.* **40**, 119–134.

Popham, R. E. (1969). Tupakan hinta ja alkoholin kulutus [Tobacco prices and alcohol consumption—test of a folk-hypothesis]. *Br. J. Addict.* **64**, 219–221.

Porter, R. L. (1966). Interdiction: Its implications for correctional practice. *Can. J. Crim. Corr.* **7**, 423–429.

Proc. Int. Conf. Alcohol Traffic (1951). 1st, Stockholm. Kugelbergs Botryckeri, Stockholm.

Røgind, S. (1949). Alcohol consumption and temperance conditions in Denmark. *Q. J. Stud. Alcohol* **10**, 471–478.

Røgind, S. (1959). Monopol och förbud—Särdag i islandsk alkohollagstiftn'ng [Monopoly and prohibition—peculiarities in Icelandic alcohol legislation]. *Alkoholpolitik* No. 1, 13–16, 38–41.

Rostow, E. V. (1940). State control of trade in alcoholic beverages. *Q. J. Stud. Alcohol* **1**, 563–581.

Rostow, E. V., and Hull, T. (1942). Self-regulation in the liquor industry. *Q. J. Stud. Alcohol* **3**, 125–137.

Rostow, E. V., and Rostow, P. L. R. (1942). Federal and interstate regulation of alcoholic beverages. *Q. J. Stud. Alcohol* **2**, 816–830.

Salomaa, N., ed. (1953). "Social Legislation and Work in Finland." Min. Soc. Aff., Helsinki.

Sariola, S. (1954). Prohibition in Finland, 1919–1932; its background and consequences. *Q. J. Stud. Alcohol* **15**, 477–490.

Schmidt, L. (1966). Soziale und wirtschaftlische Probleme des Alcoholismus in West Berlin. [Social and economic problems of alcoholism in West Berlin]. *Excerpta Criminol.* **6**, No. 1757.

Schmidt, W., and Bronetto, J. (1962). Death from liver cirrhosis and specific alcoholic beverage consumption: An ecological study. *Am. J. Public Health* **52**, 1473–1482.

Schucket, M. A. (1972). The alcoholic woman: A literature review. *Psychiatr. Med.* **3**, 37–43.

Schuckit, M. A., and Gunderson, E. K. E. (1974). The association between alcoholism and job type in the U.S. Navy. *Q. J. Stud. Alcohol* **35**, 577–585.

Seeley, J. R. (1960). Death by liver cirrhosis and the price of beverage alcohol. *Can. Med. Assoc. J.* **83**, 1361–1868.

Seidel, H. (1955). Die Entwicklung der Alkoholfrage in Deutschland seit dem Kriege. [The development of alcohol problems in Germany since the war.] *Int. J. Alcohol Alcoholism* **1**, 86–97.

Sellers, J. B. (1943). "The Prohibition Movement in Alabama 1702 to 1943." Univ. of North Carolina Press, Chapel Hill.

Simon, J. L. (1966a). The economic effects of state monopoly of packaged-liquor retailing. *J. Polit. Econ.* **74**, 188–194.

Simon, J. L. (1966b). The price elasticity of liquor in the U.S. and a simple method of determination. *Econometrica* **34**, 193–205.

Sjöhagen, A. (1960). Borde vi ha bibehållit restriktionssystemet? [Should we have kept the restriction system?] *Soc. Med. Tidskr.* **37**, 389-392.

"Social Sweden. A Government Survey" (1952). (Temperance Welfare.) Ch. 8, pp. 263-284. Social Welfare Board, Stockholm.

Steiger, V. J. (1949). Collaboration between government and private initiative in the fight against alcoholism in Switzerland and other countries. *Q. J. Stud. Alcohol* **9**, 544-55.

Tongue, A. (1955). "Alcohol Production and Consumption in Great Britain and Northern Ireland, with Notes on Legislation and Taxation." Swiss Alcohol Adm., Berne. (Mimeo.)

Traylor, O. F. (1949). Patterns of state taxation of distilled spirits, with special reference to Kentucky. *Q. J. Stud. Alcohol* **9**, 556-608.

Trice, H. M., and Belasco, J. A. (1968). Job absenteeism and drinking behavior. *Manage. Personnel Q.* **6**, 7-11.

Vernon, H. M. (1928). "The Alcohol Problem." Baillière, London.

Wangenheim, K. H. (1964). Liquor sale to minors: Yugoslav law. *Q. J. Stud. Alcohol* **25**, 153.

Watson, C. (1955). The problem of alcoholism in France. *Br. J. Addict.* **52**, 99-117.

Wechsler, H. (1972). Marihuana, alcohol and public policy. *N. Engl. J. Med.* **287**, 515-516.

Wortis, J. (1963). Alcoholism in the Soviet Union: Public health and social aspects. *Am. J. Public Health* **53**, 1644-1655.

Zieliński, J. (1969). The antialcoholism campaign in Poland. *Q. J. Stud. Alcohol* **30**, 173-177.

EDITORIAL COMMENT

Dr. Brenner's comprehensive citation of the world literature bearing on socioeconomic considerations and cultural attitudes affecting the epidemiology of alcohol abuse emphasizes the enormous complexity of the problem and the inherent difficulty in measuring the effect of these multiple factors. The review gives little cause for optimism that simplistic approaches to the prevention of alcoholism are likely to be effective. Rather, the paper, which brings together the literature pertaining to supply, demand, and various stress factors, points to the need for continuing research on the epidemiology of alcohol abuse in the broad interpretation of that phrase. Search should be made, too, for natural experiments in progress in states or other large political divisions in which it might be possible to isolate and study the effects of a few of the many factors that seem to influence the prevalance of alcoholism.

APPENDIX*

Compounds Identified in Whisky, Wine, and Beer: A Tabulation

By J. H. KAHN (Research Department, National Distillers and Chemical Corp., 1275 Section Rd., Cincinnati, Ohio 45237)

Components which have been identified in alcoholic beverages by this laboratory and by other workers are listed with the pertinent references.

In the course of the work of identifying components in whisky and determining whether any particular component had been previously reported in an alcoholic beverage, a tabulation was prepared of compounds identified by all workers. The compilation includes approximately 400 compounds reported from this laboratory and elsewhere, as well as those recently identified in this laboratory but not as yet reported. While this list was intended for internal use, a decision was made to make it available to workers in distillery, enology, and brewing research (see Table 1).

A complete literature search was not made; however, it is believed the cited bibliography contains references to most of the components which have been found in whisky, beer, wine, or related beverages produced by yeast fermentation. No effort was made to obtain the original publication in which each compound was first reported, and no claim is made that this tabulation is complete. Undoubtedly, additional compounds have been identified but not published. With a few exceptions, the table lists the compounds by their Geneva or ACS names, in alphabetical order by chemical groups, together with their empirical formulae, alternative names, and references.

REFERENCES

(1) Steinke, R. D., and Paulson, M. C., *J. Agr. Food Chem.* **12**, 381–387 (1964).

(2) Nykanen, L., Puputti, E., and Suomalainen, H., *J. Food Sci.* **33**, 88–92 (1968).

(3) Martin, G. E., Schoeneman, R. L., and Schlessinger, H. L., *This Journal* **47**, 712–713 (1964).

(4) Baldwin, S., Black, R. A., Andreasen, A. A., and Adams, S. L., *J. Agr. Food Chem.* **15**, 381–385 (1967).

(5) Galetto, W. G., Webb, A. D., and Kepner, R. E., *Amer. J. Enol. Viticult.* **17**, 11–19 (1966).

(6) Kayahara, K., Mori, S., Taguchi, T., and Miyachi, N., *Hakko Kogaku Zasshi* **42**, 512–517 (1964); **43**, 187–190 (1965).

(7) Taira, T., *Hakko Kyokaishi* **24**, 71–72 (1966).

(8) Suomalainen, H., and Ronkainen, P., *Nature* **220**, 792–793 (1968).

(9) Webb, A. D., Kepner, R. E., and Galetto, W. G., *Amer. J. Enol. Viticult.* **17**, 1–10 (1966); **15**, 1–10 (1964).

(10) Tripp, R. C., Timm, B., Iyer, M., Richardson, T., and Amundson, C. H., *Proc. Amer. Soc. Brewing Chemists* 65–74 (1968).

(11) Rosculet, G., and Rickard, M., *ibid.* 203–213 (1968).

(12) Harrison, G. A. F., and Collins, E., *ibid.* 101–105 (1968).

(13) Marinelli, L., Feil, M. F., and Schait, A., *ibid.* 113–119 (1968).

(14) Clarke, B. J., Harold, F. V., Hildebrand, R. P., and Morieson, A. S., *J. Inst. Brewing* **68**, 179–187 (1962).

(15) Strating, J., and Venema, A., *ibid.* **67**, 525–528 (1961).

(16) Kepner, R. E., Maarse, H., and Strating, J., *Anal. Chem.* **36**, 77–82 (1964).

(17) Lawrence, W. C., *Wallerstein Lab. Commun.* **27** 123–152 (1964).

(18) Kepner, R. E., Webb, A. D., and Maggiora, L., *Amer. J. Enol. Viticult.* **19**, 116–120 (1968).

(19) Bayer, E., *J. Gas Chromatog.* **4**(2), 67–73 (1966).

(20) Webb, A. D., and Kepner, R. E., *Amer. J. Enol. Viticult.* **13**, 1–14 (1962).

(21) Hirose, Y., Ogawa, M., and Kusuda, Y., *Agr. Biol. Chem. (Tokyo)* **26**, 526–531 (1962).

(22) Hashimoto, N., and Kuroiwa, Y., *J. Inst. Brewing* **72**, 151–162 (1966).

(23) Stevens, R., *ibid.* **66**, 453–471 (1960).

*Reprinted by permission from *J. Assoc. Off. Anal. Chem.* **52**, 1166–1178. Source of original article: National Distillers and Chemical Corporation.

(24) Jones, K. and, Wills, R., *ibid.* **72**, 196–201 (1966).

(25) Pisarnitskii, A. F., *Prikl. Biokhim. Mikrobiol.* **2**, 215–218 (1966).

(26) Webb, A. D., *Biotech. Bioeng.* **9**, 305–319 (1967)

(27) Suomalainen, H., and Keranen, A. J. A., *J. Inst. Brewing* **73**, 477–484 (1967).

(28) Ronkainen, P., and Suomalainen, H., *Suomen Kemistilehti* **39**, 280–281 (1966).

(29) Suomalainen, H., and Nykanen, L., *J. Inst. Brewing* **72**, 469 (1966).

(30) Wiley, H. W., *Beverages and Their Adulteration*, Blakiston and Co., Philadelphia, 1919, p. 281.

(31) Bernstein, L., Blenkinship, B. K., and Brenner, M. W., *Proc. Amer. Soc. Brewing Chemists* 150–157 (1968).

(32) Christensen, E. N., and Caputi, A., *Amer. J. Enol. Viticult.* **19**, 238–245 (1968).

(33) Martin, G. E., Schmit, J. A., and Schoeneman, R. L., *This Journal* **48**, 962–964 (1965).

(34) Nykanen, L., and Suomalainen, H., *Teknillisen Kemian Aikakausilehti* **20**, 789–795 (1963).

(35) Brunelle, R. L., Schoeneman, R. L., and Martin, G. E., *This Journal* **50**, 329–334 (1967).

(36) Webb, A. D., Kepner, R. E., and Maggiora, L., *Amer. J. Enol. Viticult.* **20**, 16–24, 25–31 (1969).

(37) Bober, A., and Haddaway, L. W., *J. Gas Chromatog.* **1**, 8–13 (1963).

(38) Suomalainen, H., and Nykanen, L., *Wallerstein Lab. Commun.* **31**, 5–13 (1968).

(39) deBecze, G. I., Smith, H. F., and Vaughn, T. E., *This Journal* **50**, 311–319 (1967).

(40) Powell, A. D. G., and Brown, I. H., *J. Inst. Brewing* **72**, 261–265 (1966).

(41) Van der Kloot, A. P., and Wilcox, F. A., *Proc. Amer. Soc. Brewing Chemists* 113–116 (1960).

(42) Dalgliesh, C. E., *Brewers J. (Philadelphia)* **136**, 31–37 (1967).

(43) Ahrenst-Larsen, B., and Hansen, H. L., *Wallerstein Lab. Commun.* **27**, 41–49 (1964).

(44) Kepner, R. E., Strating, J., and Weurman, C., *J. Inst. Brewing* **69**, 399–405 (1963).

(45) Clarke, B. J., Harold, F. V., Hildebrand, R. P., and Murray, P. J., *ibid.* **67**, 529 (1961).

(46) Diemair, W., and Schams, E., *Z. Lebensm. Untersuch.-Forsch.* **112**, 457–463 (1960).

(47) Waygood, W. A., *Analyst* **68**, 33–34 (1943).

(48) Dimotaki-Kourakou, V., *Ann. Fals. Expert. Chim.* **55**, 149–158 (1962).

(49) Carles, J., *Rev. Espan. Fisiol.* **15**, 193–200 (1959).

(50) Braus, H., and Miller, F. D., *This Journal* **41**, 141–144 (1958).

(51) Braus, H., Eck, J. W., Mueller, W. M., and Miller, F. D., *J. Agr. Food Chem.* **5**, 458–459 (1957).

(52) Kahn, J. H., LaRoe, E. G., and Conner, H. A., *J. Food Sci.* **33**, 395–399 (1968).

(53) Kahn, J. H., Shipley, P. A., LaRoe, E. G., and Conner, H. A., *J. Food Sci.*, in press.

(54) LaRoe, E. G., and Shipley, P. A., submitted to *J. Agr. Food Chem.*

Table 1. Compilation of components in alcoholic beverages reported by many workers

No.	Name	Empirical Formula	Whisky	Wine	Beer	Misc.	Alternative Name
			Alcohols				
1.	Allyl	C_3H_6O	53				
2.	Benzyl	C_7H_8O		36		21	
3.	Borneol	$C_{10}H_{18}O$				21	
4.	n-Butanol	$C_4H_{10}O$	6, 34, 52	36	11, 22		Butanol-1
5.	sec.-Butyl	$C_4H_{10}O$	34, 52	36	11		Butanol-2
6.	tert.-Butyl	$C_4H_{10}O$	30				
7.	Citronellol	$C_{10}H_{20}O$				21	2,6-Dimethyl-octen-1-ol-8
8.	d-Coriandrol (tent.)	$C_{10}H_{18}O$	N[a]				d-Linalool
9.	Decyl	$C_{10}H_{22}O$				21	
10.	2,3-Dihydroxybutane	$C_4H_{10}O_2$					2,3-Butyleneglycol
11.	Ethyl	C_2H_6O	Yes	Yes	Yes		
12.	Farnesol	$C_{15}H_{26}O$	Yes	Yes			3,7,11-Trimethyl-2,6,10-decatrien-1-ol
13.	Furfuryl	$C_5H_6O_2$	37	25	15		
14.	Geraniol	$C_{10}H_{18}O$	34				3,7-Dimethyl-2,6-octadien-1-ol
15.	Heptanol-2	$C_7H_{16}O$				21	
16.	Heptyl	$C_7H_{16}O$	53	36			
17.	Hexenol	$C_6H_{12}O$		5			
18.	Hexyl	$C_6H_{14}O$	34, 53	9, 36	43		
19.	Isobutyl	$C_4H_{10}O$	6, 34, 52	36	11		
20.	Isopentyl	$C_5H_{12}O$	6, 52	36	11		
21.	Isopropyl	C_3H_8O		9	11		
22.	l-Linalool	$C_{10}H_{18}O$	N	25			3,7-Dimethyl-1,6-octadien-3-ol
23.	Methanol	CH_4O		36			
24.	2-Methylbutanol	$C_5H_{12}O$	6, 34, 53	36	11		act.-Amyl alcohol
25.	5-Methylhexanol	$C_7H_{16}O$	34, 52	36	11		Isoheptyl alcohol
26.	3-Methylpentanol	$C_6H_{14}O$	30	36			
27.	4-Methylpentanol	$C_6H_{14}O$	53	26			Isohexyl alcohol
28.	Nonanol-1	$C_9H_{20}O$	30				
29.	Nonanol-2	$C_9H_{20}O$				21	
30.	1-Octen-3-ol	$C_8H_{16}O$				21	Matsutakeol
31.	Octyl	$C_8H_{18}O$	53				
32.	Pentanol-2	$C_5H_{12}O$	30	36			
33.	Pentyl	$C_5H_{12}O$	6	36	11		
34.	β-Phenethyl	$C_8H_{10}O$	6, 34, 38, 53	9	40		
35.	n-Propyl	C_3H_8O	6, 34, 52	36	11		
36.	Tryptophol	$C_{10}H_{11}NO$			17		3-Indole ethanol
37.	Undecyl	$C_{11}H_{24}O$				21	

(Continued)

Table 1. (Continued)

No.	Name	Empirical Formula	Whisky	Wine	Beer	Misc.	Alternative Name
			References Cited For:				
		Acids: A. Aliphatic, Monobasic					
1.	Acetic	$C_2H_4O_2$	34,53	36	11		
2.	Butenoic	$C_4H_6O_2$		26			
3.	Butyric	$C_4H_8O_2$	2,53	36	11		Butanoic
4.	Decanoic	$C_{10}H_{20}O_2$	2,34,53	9	40		Capric
5.	9-Decenoic	$C_{10}H_{18}O_2$		26			Caproleic
6.	Docosanoic	$C_{22}H_{44}O_2$			14		Behenic
7.	Formic	CH_2O_2	6	36	11		
8.	Heptadecanoic	$C_{17}H_{34}O_2$	2				
9.	Heptanoic	$C_7H_{14}O_2$	2,53	9	45		Enanthic
10.	Hept-3-enoic	$C_7H_{12}O_2$			14		
11.	Hexacosanoic	$C_{26}H_{52}O_2$			14		Cerotic
12.	Hexanoic	$C_6H_{12}O_2$	2,34,53	9	40		Caproic
13.	Hex-2-enoic	$C_6H_{10}O_2$			14		
14.	Hex-3-enoic	$C_6H_{10}O_2$			14		
15.	Isobutyric	$C_4H_8O_2$	2,53	9,36	11		
16.	Isovaleric	$C_5H_{10}O_2$	2,53	9	11		
17.	Lauric	$C_{12}H_{24}O_2$	2,53	36	14		
18.	Linoleic	$C_{18}H_{32}O_2$	2,N		14		
19.	Linolenic	$C_{18}H_{30}O_2$	34		14		
20.	2-Methylbutyric	$C_5H_{10}O_2$	53				
21.	2-Methylvaleric	$C_6H_{12}O_2$		26			
22.	Myristic	$C_{14}H_{28}O_2$	2,N		14		
23.	Nonanoic	$C_9H_{18}O_2$	2,34,53	9	14		Pelargonic
24.	Octanoic	$C_8H_{16}O_2$	2,34,53	9,18	40		Caprylic
25.	Oleic	$C_{18}H_{34}O_2$	2,N		14		
26.	Palmitic	$C_{16}H_{32}O_2$	2,N		14		
27.	Palmitoleic	$C_{16}H_{30}O_2$	2		14		Hexadecenoic
28.	Pentadecanoic	$C_{15}H_{30}O_2$	2		17		
29.	Propionic	$C_3H_6O_2$	2	36	11		
30.	Stearic	$C_{18}H_{36}O_2$	2,N		14		
31.	Tridecanoic	$C_{13}H_{26}O_2$	2		10		
32.	Undecanoic	$C_{11}H_{22}O_2$	2		10		
33.	Valeric	$C_5H_{10}O_2$	2,53	36	11		
		Acids: B. Aliphatic, Dibasic, etc.					
1.	Aspartic	$C_4H_7NO_4$	N		17		Aminosuccinic acid
2.	3-Ethylheptanedioic	$C_9H_{16}O_4$					
3.	Citric	$C_6H_8O_7$	33		13		2-Hydroxy-1,2,3-propane-tricarboxylic acid
4.	Fumaric	$C_4H_4O_4$		35	13		Butenedioic acid

No.	Name	Formula		N			Synonym
5.	Glutamic	$C_5H_9NO_4$				17	Aminopentanedioic acid
6.	Glutaric	$C_5H_8O_4$	36			14	Pentanedioic acid
7.	Malonic	$C_3H_4O_4$				13	Propanedioic acid
8.	Mesaconic	$C_5H_6O_4$				14	trans-Methylfumaric acid
9.	Nonanedioic	$C_9H_{16}O_4$	18				Azelaic
10.	Oxalic	$C_2H_2O_4$				13	Ethanedioic acid
11.	Succinic	$C_4H_6O_4$	36	N		13	Butanedioic acid
12.	Tartaric	$C_4H_6O_6$	35			14	
13.	Tricarballylic	$C_6H_8O_6$	18				1,2,3-Propanetricarboxylic acid
	Acids: C Aromatic						
1.	Anthranilic	$C_7H_7NO_2$	19				o-Aminobenzoic
2.	Benzoic	$C_7H_6O_2$	18	N		13	
3.	Cinnamic	$C_9H_8O_2$	19				
4.	2-Hydroxy-3-phenylpropionic	$C_9H_{10}O_3$	18				
5.	β-Phenylacetic	$C_8H_8O_2$	36	N			
6.	Phenylpropionic	$C_9H_{10}O_2$		N			
7.	β-Phenylpyruvic	$C_9H_8O_3$			27		
8.	Phthalic	$C_8H_6O_4$				14	
	Acids: D. Keto and Hydroxy						
1.	Acetoacetic	$C_4H_6O_3$			27		3-Ketobutyric acid
2.	α-Acetolactic	$C_5H_8O_4$			27		
3.	Citramalic	$C_5H_8O_5$	48,49				α-Methylmalic acid
4.	Glycolic	$C_2H_4O_3$				14	Hydroxyacetic acid
5.	Glyoxylic	$C_2H_2O_3$			27		Glyoxalic acid
6.	2-Hydroxyhexanoic	$C_6H_{12}O_3$	18				
7.	2-Hydroxyisovaleric	$C_5H_{10}O_3$	18				
8.	γ-Hydroxy-α-ketobutyric	$C_4H_6O_4$			27		
9.	γ-Hydroxy-α-ketoglutaric	$C_5H_6O_6$			27		
10.	2-Hydroxy-2-methylvaleric	$C_6H_{12}O_3$	26				
11.	3-Hydroxyoctanoic	$C_8H_{16}O_3$	18				
12.	Hydroxypyruvic	$C_3H_4O_4$			27		
13.	α-Ketoadipic	$C_6H_8O_5$			27		α-Ketohexanedioic acid
14.	α-Ketobutyric	$C_4H_6O_3$				12	
15.	α-Ketoglutaric	$C_5H_6O_5$				12	
16.	α-Ketoisocaproic	$C_6H_{10}O_3$				12	
17.	α-Ketoisovaleric	$C_5H_8O_3$				12	
18.	α-Keto-γ-methiolbutyric	$C_5H_8SO_3$			27		
19.	α-Keto-β-methylvaleric	$C_6H_{10}O_3$				12	
20.	Lactic	$C_3H_6O_3$	36	N		13	α-Hydroxypropionic acid
21.	Levulinic	$C_5H_8O_3$				14	4-Ketovaleric acid

(Continued)

Table 1. *(Continued)*

No.	Name	Empirical Formula	Whisky	Wine	Beer	Misc.	Alternative Name
22.	Malic	$C_4H_6O_5$		35	13		Hydroxysuccinic acid
23.	Oxalacetic	$C_4H_4O_5$				27	α-Ketosuccinic acid
24.	Pyruvic	$C_3H_4O_3$			12		α-Ketopropionic acid

Acids: E. Miscellaneous

No.	Name	Empirical Formula	Whisky	Wine	Beer	Misc.	Alternative Name
1.	2-Furoic	$C_5H_4O_3$	30	18			
2.	Hydrocyanic	CHN				21	

Esters: A. Aliphatic, Monobasic

No.	Name	Empirical Formula	Whisky	Wine	Beer	Misc.	Alternative Name
1.	Butyl acetate	$C_6H_{12}O_2$ (2)[b]		36	17		
2.	sec.-Butyl acetate	$C_6H_{12}O_2$ (2)			17		
3.	Decyl acetate	$C_{12}H_{24}O_2$ (2)					
4.	Ethyl acetate	$C_4H_8O_2$ (2)	6,34,52	20	11		
5.	Ethyl behenate	$C_{24}H_{48}O_2$ (22)	24,N				Ethyl docosanoate
6.	Ethyl butyrate	$C_6H_{12}O_2$ (4)	52		40,44		
7.	Ethyl carbonate	$C_5H_{10}O_3$ (1)	N				Diethyl carbonate
8.	Ethyl decanoate	$C_{12}H_{24}O_2$ (10)	6,24,34,38,53	9	15		Ethyl caprate
9.	Ethyl-9-decenoate	$C_{12}H_{22}O_2$ (10:1)	N	26	15		Ethyl caproleate
10.	Ethyl eicosanoate	$C_{22}H_{44}O_2$ (20)	N				Ethyl arachidate
11.	Ethyl-3-ethoxypropionate	$C_7H_{14}O_3$ (5)	52				
12.	Ethyl formate	$C_3H_6O_2$ (1)	6,34,52	36	23		
13.	Ethyl heneicosanoate	$C_{23}H_{26}O_2$ (21)	24				
14.	Ethyl heptadecanoate	$C_{19}H_{38}O_2$ (17)	24,N				Ethyl margarate
15.	Ethyl heptanoate	$C_9H_{18}O_2$ (7)	34,53	9			Ethyl enanthate
16.	Ethyl hexanoate	$C_8H_{16}O_2$ (6)	24,34,52	9	11		Ethyl caproate
17.	Ethyl-9-hexadecenoate	$C_{18}H_{34}O_2$ (16:1)	29,38,53				Ethyl palmitoleate
18.	Ethyl-3-hydroxybutyrate	$C_6H_{12}O_3$ (4)		26,36			
19.	Ethyl-4-hydroxybutyrate	$C_6H_{12}O_3$ (4)		26,36			
20.	Ethyl-2-hydroxyisohexanoate	$C_8H_{16}O_3$ (6)		26			
21.	Ethyl-3-hydroxypropionate	$C_5H_{10}O_3$ (3)		36			
22.	Ethyl isobutyrate	$C_6H_{12}O_2$ (4)	6,53	36			
23.	Ethyl isovalerate	$C_7H_{14}O_2$ (5)	N	36			
24.	Ethyl lactate	$C_5H_{10}O_3$ (3)	39,53	36	43		
25.	Ethyl laurate	$C_{14}H_{28}O_2$ (12)	6,24,34,38,53	9			
26.	Ethyl lignocerate	$C_{26}H_{52}O_2$ (24)	24				Ethyl tetracosanoate
27.	Ethyl linoleate	$C_{20}H_{36}O_2$ (18:2)	38,53				
28.	Ethyl linolenate	$C_{20}H_{34}O_2$ (18:3)	N				
29.	Ethyl myristate	$C_{16}H_{32}O_2$ (14)	6,24,34,38,53	9			
30.	Ethyl nonadecanoate	$C_{21}H_{42}O_2$ (19)	24				
31.	Ethyl nonanoate	$C_{11}H_{22}O_2$ (9)	24,34,38,53	9			Ethyl pelargonate

No.	Name	Formula				
32.	Ethyl octanoate	$C_{10}H_{20}O_2$ (8)	6, 24, 34, 52	9	40	Ethyl caprylate
33.	Ethyl oleate	$C_{20}H_{38}O_2$ (18:1)	38, 53	9		
34.	Ethyl palmitate	$C_{18}H_{34}O_2$ (16)	6, 24, 38, 53	9		
35.	Ethyl pentadecanoate	$C_{17}H_{34}O_2$ (15)	53			
36.	Ethyl propionate	$C_5H_{10}O_2$ (3)	6, 52	36	42	
37.	Ethyl pyruvate	$C_5H_8O_3$ (3)		36		
38.	Ethyl stearate	$C_{20}H_{40}O_2$ (18)	6, 24, 34, 38			
39.	Ethyl tricosanoate	$C_{25}H_{50}O_2$ (23)	24			
40.	Ethyl undecanoate	$C_{13}H_{26}O_2$ (11)	24, 38			
41.	Ethyl valerate	$C_7H_{14}O_2$ (5)	N			
42.	Heptyl acetate	$C_9H_{18}O_2$ (2)			21	
43.	Hexyl acetate	$C_8H_{16}O_2$ (2)		20		
44.	Hexyl hexanoate	$C_{12}H_{24}O_2$ (6)		9	17	Hexyl caproate
45.	Hexyl isobutyrate	$C_{10}H_{20}O_2$ (4)		9		
46.	Hexyl octanoate	$C_{14}H_{28}O_2$ (8)		36		Hexyl caprylate
47.	Isobutyl acetate	$C_6H_{12}O_2$ (2)	34, 53	36	11	
48.	Isobutyl decanoate	$C_{14}H_{28}O_2$ (10)	24	9		Isobutyl caprate
49.	Isobutyl eicosanoate	$C_{24}H_{48}O_2$ (20)	24			Isobutyl arachidate
50.	Isobutyl hexanoate	$C_{10}H_{20}O_2$ (6)		20		Isobutyl caproate
51.	Isobutyl isobutyrate	$C_8H_{16}O_2$ (4)		20		
52.	Isobutyl lactate	$C_7H_{14}O_3$ (3)		36		
53.	Isobutyl laurate	$C_{16}H_{32}O_2$ (12)	24			
54.	Isobutyl myristate	$C_{18}H_{36}O_2$ (14)	24			
55.	Isobutyl octanoate	$C_{12}H_{24}O_2$ (8)	24	9		Isobutyl caprylate
56.	Isobutyl palmitate	$C_{20}H_{40}O_2$ (16)	24			
57.	Isobutyl stearate	$C_{22}H_{44}O_2$ (18)	24			
58.	Isobutyl tridecanoate	$C_{17}H_{34}O_2$ (13)	24			
59.	Isobutyl valerate	$C_9H_{18}O_2$ (5)				
60.	Isopentyl acetate	$C_7H_{14}O_2$ (2)	34, 52	36	11	Isoamyl acetate
61.	Isopentyl decanoate	$C_{15}H_{30}O_2$ (10)	24, 34	9		Isoamyl caprate
62.	Isopentyl hexanoate	$C_{11}H_{22}O_2$ (6)	24, 38, N	9		Isoamyl caproate
63.	Isopentyl isovalerate	$C_{10}H_{20}O_2$ (5)		20		Isoamyl isovalerate
64.	Isopentyl lactate	$C_8H_{16}O_3$ (3)		20	17	Isoamyl lactate
65.	Isopentyl laurate	$C_{17}H_{34}O_2$ (12)	24	9		Isoamyl laurate
66.	Isopentyl-2-methylbutyrate	$C_{10}H_{20}O_2$ (5)		20		Isoamyl-2-methylbutyrate
67.	Isopentyl myristate	$C_{19}H_{38}O_2$ (14)		9		Isoamyl myristate
68.	Isopentyl nonanoate	$C_{14}H_{28}O_2$ (9)				Isoamyl pelargonate
69.	Isopentyl octanoate	$C_{13}H_{26}O_2$ (8)	24, 38	20		Isoamyl caprylate
70.	Isopentyl pentadecanoate	$C_{20}H_{40}O_2$ (15)	24			Isoamyl pentadecanoate
71.	Isopentyl tridecanoate	$C_{18}H_{36}O_2$ (13)	24			Isoamyl tridecanoate
72.	Isopentyl undecanoate	$C_{16}H_{32}O_2$ (11)	24			Isoamyl undecanoate
73.	Isopropyl acetate	$C_5H_{10}O_2$ (2)			22	
74.	Methyl acetate	$C_3H_6O_2$ (2)	3	36	44	

(Continued)

Table 1. (Continued)

No.	Name	Empirical Formula	Whisky	Wine	Beer	Misc.	Alternative Name
			— References Cited For:				
75.	2-Methylbutyl acetate	$C_7H_{14}O_2$ (2)		26			act.-Amyl acetate
76.	2-Methylbutyl decanoate	$C_{15}H_{30}O_2$ (10)		9			act.-Amyl caprate
77.	2-Methylbutyl hexanoate	$C_{11}H_{22}O_2$ (6)		9			act.-Amyl caproate
78.	2-Methylbutyl isovalerate	$C_{10}H_{20}O_2$ (5)		20			act.-Amyl isovalerate
79.	2-Methylbutyl laurate	$C_{17}H_{34}O_2$ (12)		9			act.-Amyl laurate
80.	2-Methylbutyl myristate	$C_9H_{38}O_2$ (14)		9			act.-Amyl myristate
81.	2-Methylbutyl octanoate	$C_{13}H_{26}O_2$ (8)		9			act.-Amyl caprylate
82.	Methyl formate	$C_2H_4O_2$ (1)		46			
83.	Methyl hexanoate	$C_7H_{14}O_2$ (6)	24				Methyl caproate
84.	Nonyl acetate	$C_{11}H_{22}O_2$ (2)				21	
85.	2-Nonyl acetate	$C_{11}H_{22}O_2$ (2)				21	
86.	Octyl acetate	$C_{10}H_{20}O_2$ (2)				21	
87.	Pentyl acetate	$C_7H_{14}O_2$ (2)	N		41		Amyl acetate
88.	Pentyl butyrate	$C_9H_8O_2$ (4)	30				Amyl butyrate
89.	Pentyl decanoate	$C_{15}H_{30}O_2$ (10)				21	Amyl caprate
90.	Pentyl formate	$C_6H_{12}O_2$ (1)	30				Amyl formate
91.	Pentyl hexanoate	$C_{11}H_{22}O_2$ (6)					Amyl caproate
92.	Pentyl isobutyrate	$C_9H_{18}O_2$ (4)			17	21	Amyl isobutyrate
93.	Pentyl nonanoate	$C_{14}H_{28}O_2$ (9)	30				Amyl pelargonate
94.	Pentyl octanoate	$C_{13}H_{26}O_2$ (8)				21	Amyl caprylate
95.	Propyl acetate	$C_5H_{10}O_2$ (2)	53	36			
96.	Propyl octanoate	$C_{11}H_{22}O_2$ (8)		9			Propyl caprylate
97.	Triethyl orthoformate	$C_7H_{16}O_3$ (1)	52				
98.	Undecyl acetate	$C_{13}H_{26}O_2$ (2)				21	

Esters: B. Aliphatic, Dibasic, etc.

No.	Name	Empirical Formula	Whisky	Wine	Beer	Misc.	Alternative Name
1.	Diethyl malate	$C_8H_{14}O_5$ (4)	6	20			Diethyl hydroxysuccinate
2.	Diethyl nonanedioate	$C_{13}H_{24}O_4$ (9)	N				Diethyl azelate
3.	Diethyl propanedioate	$C_7H_{12}O_4$ (3)	N				Diethyl malonate
4.	Diethyl succinate	$C_8H_{14}O_4$ (4)	53	9, 36			Diethyl butanedioate
5.	Diethyl tartrate	$C_8H_{14}O_6$ (4)		36			
6.	Ethyl acid malate	$C_6H_{10}O_5$ (4)		26			
7.	Ethyl acid succinate	$C_6H_{10}O_4$ (4)		26			
8.	Ethyl acid tartrate	$C_6H_{10}O_6$ (4)		26			

Esters: C. Aromatic

No.	Name	Empirical Formula	Whisky	Wine	Beer	Misc.	Alternative Name
1.	Dibutyl phthalate	$C_{16}H_{22}O_4$	N				
2.	Diethyl phthalate	$C_{12}H_{14}O_2$	53				
3.	Dimethyl phthalate	$C_{10}H_{10}O_4$		26			
4.	Ethyl anthranilate	$C_9H_{11}NO_2$	N	19			Ethyl-2-aminobenzoate

No.	Name	Formula						Common name
5.	Ethyl benzoate	$C_9H_{10}O_2$	53	19				
6.	Ethyl cinnamate	$C_{11}H_{12}O_2$		19				
7.	Ethyl-2-hydroxy-3-phenylpropionate	$C_{11}H_{14}O_3$		36				
8.	Ethyl phenylacetate	$C_{10}H_{12}O_2$						
9.	Phenethyl acetate	$C_{10}H_{12}O_2$	6,53	9	40		21	
10.	Phenethyl formate	$C_9H_{10}O_2$	53					
11.	Phenethyl hexanoate	$C_{14}H_{20}O_2$		20				
12.	Phenethyl propionate	$C_{11}H_{14}O_2$					21	

Carbonyls: A. Aliphatic

No.	Name	Formula						Common name
1.	Acetaldehyde	C_2H_4O	6,34,52	36	11			
2.	Acetoin	$C_4H_8O_2$		47	11			3-Hydroxy-2-butanone
3.	Acetone	C_3H_6O	6,34,N	36	11			
4.	Acrolein	C_3H_4O	6,52		11			
5.	2,3-Butanedione	$C_4H_6O_2$	6,28	47	11	8,27		Diacetyl
6.	Butan-2-one	C_4H_8O	6,52		11			
7.	Butyraldehyde	C_4H_8O	N	36	11			
8.	Decan-3-one	$C_{10}H_{20}O$			10			
9.	3-Ethoxypropionaldehyde	$C_5H_{10}O_2$	52					
10.	Formaldehyde	CH_2O	6,34	36	10			
11.	Glyoxal	$C_2H_2O_2$	28		10	27		Ethanedial
12.	Heptanal	$C_7H_{14}O$			10			
13.	Heptan-3-one	$C_7H_{14}O$			16			
14.	Heptan-4-one	$C_7H_{14}O$			10			
15.	Hexanal	$C_6H_{12}O$	38	26				Caproaldehyde
16.	Isobutyraldehyde	C_4H_8O	6,52	36	11			
17.	Isovaleraldehyde	$C_5H_{10}O$	53	36	11			
18.	3-Methylbutan-2-one	$C_5H_{10}O$			11			
19.	2-Methylbutyraldehyde	$C_5H_{10}O$	26	26	10			
20.	Methylglyoxal	$C_3H_4O_2$	28,N			27		Pyruvaldehyde
21.	Nonan-2-one	$C_9H_{18}O$				21		
22.	Nonen-2-one	$C_9H_{16}O$				21		
23.	Octanal	$C_8H_{16}O$		36	10			Caprylaldehyde
24.	Octan-2-one	$C_{15}H_{30}O$				21		
25.	Pentadecan-2-one	$C_5H_8O_2$	28		42	21		
26.	2,3-Pentanedione	$C_5H_{10}O$	52		11	27		
27.	Pentan-2-one	C_3H_6O			10			
28.	Propan-2-one				11			
29.	Propionaldehyde	$C_{11}H_{22}O$	6,52	36	11			
30.	Undecan-2-one	$C_5H_{10}O$			17			
31.	Valeraldehyde		N		11			

(Continued)

Table 1. *(Continued)*

No.	Name	Empirical Formula	Whisky	Wine	Beer	Misc.	Alternative Name

No.	Name	Empirical Formula	Whisky	Wine	Beer	Misc.	Alternative Name
	Carbonyls: B. Cyclic and Aromatic						
1.	Acetophenone	C_8H_8O	N, 53			21	Phenylmethyl ketone
2.	Benzaldehyde	C_7H_6O				7	
3.	Cyclopentanone	C_5H_8O					
4.	Cinnamic aldehyde	C_9H_8O		25			
5.	2-Furaldehyde	$C_5H_4O_2$	6, 52	9	11		
6.	Hydroxymethylfurfural	$C_6H_6O_3$			17		
7.	α-Ionone	$C_{13}H_{20}O$	54			21	
8.	β-Ionone	$C_{13}H_{20}O$	54	25, 36			
9.	5-Methyl-2-furaldehyde	$C_6H_6O_2$	53				
10.	4-Phenyl-butan-2-one	$C_{10}H_{12}O$			44		Benzyl acetone
	Acetals						
1.	Acetal	$C_6H_{14}O_2$	34, 52	36	15		Acetaldehyde diethylacetal
2.	1-Acetoxy-1-ethoxyethane	$C_6H_{12}O_3$	N				1-Ethoxy ethyl acetate
3.	Diethoxymethane	$C_5H_{12}O_2$	53				Formaldehyde diethylacetal
4.	1,1-Diethoxy-2-methylpropane	$C_8H_{18}O_2$	52				Isobutyraldehyde diethylacetal
5.	1,1-Diethoxypropane	$C_7H_{16}O_2$	52				Propionaldehyde diethylacetal
6.	1,1-Diethoxy-2-propene	$C_7H_{14}O_2$	52				Acrolein diethylacetal
7.	1,1-Di-(isopentoxy)ethane	$C_{12}H_{26}O_2$					Acetaldehyde diisoamylacetal
8.	1,1-Di-(2-methylbutoxy)ethane (tent.)	$C_{12}H_{26}O_2$	N	5			Acetaldehyde di-act.-amylacetal
9.	1-Ethoxy-1-isopentoxyethane (tent.)	$C_9H_{20}O_2$	N				Acetaldehyde ethylisoamylacetal
10.	1-Ethoxy-1-isopentoxy-3-hydroxypropane	$C_{10}H_{22}O_3$	53				β-Hydroxypropionaldehyde-ethyl-isoamylacetal
11.	1-Ethoxy-1-(2-methylbutoxy)propane	$C_9H_{20}O_2$		5			Acetaldehyde ethyl-act.-amylacetal
12.	1-Ethoxy-1-phenethyloxyethane	$C_{12}H_{18}O_2$		5			Acetaldehyde ethylphenethylacetal
13.	1-Ethoxy-1-propoxyethane	$C_7H_{16}O_2$	52				Acetaldehyde ethyl propylacetal
14.	1-Isopentoxy-1-phenethyloxyethane	$C_{15}H_{24}O_2$		5			Acetaldehyde isoamylphenethylacetal
15.	1-(2-Methylbutoxy)-1-isopentoxyethane	$C_{12}H_{26}O_2$		5			Acetaldehyde act.-amyl-isoamylacetal
16.	1-(2-Methylbutoxy)-1-phenethyloxyethane	$C_{15}H_{24}O_2$		5			Acetaldehyde act.-amyl-phenethylacetal
17.	1,1,3-Triethoxypropane	$C_9H_{20}O_3$	52				Ethoxypropionaldehyde diethylacetal
	Phenolic Compounds						
1.	Butyl caffeate	$C_{13}H_{16}O_4$		19			Butyl-3,4-dihydroxycinnamate
2.	Caffeic acid	$C_9H_8O_4$		32			3,4-Dihydroxycinnamic acid
3.	Coniferaldehyde	$C_{10}H_{10}O_3$				4	4-Hydroxy-3-methoxycinnamic aldehyde
4.	p-Coumaric acid	$C_9H_8O_3$		32			4-Hydroxycinnamic acid
5.	o-Cresol	C_7H_8O	50				2-Methylphenol
6.	2,6-Dimethoxyphenol	$C_8H_{10}O_3$	N				
7.	2,5-Dimethylphenol	$C_8H_{10}O$	N				
8.	3,5-Dimethylphenol	$C_8H_{10}O$	N				

No.	Compound	Formula				Synonym
9.	p-Ethylguaiacol	$C_9H_{12}O_2$	1,50	19		
10.	Ethyl hydroxycinnamate	$C_{11}H_{12}O_3$	1,53	36		Ethyl coumarate
11.	4-Ethylphenol	$C_8H_{10}O$	N			
12.	o- or m-Ethylphenol	$C_8H_{10}O$				
13.	Ethyl salicylate	$C_9H_{10}O_3$	N	19		Ethyl-2-hydroxybenzoate
14.	Ethyl vanillate	$C_{10}H_{12}O_4$		32		Ethyl-4-hydroxy-3-methoxybenzoate
15.	Ferulic acid	$C_{10}H_{10}O_4$		32		4-Hydroxy-3-methoxycinnamic acid
16.	Gallic acid	$C_7H_6O_5$		32		3,4,5-Trihydroxybenzoic acid
17.	Gentisic acid	$C_7H_6O_4$				2,5-Dihydroxybenzoic acid
18.	Guaiacol	$C_7H_8O_2$	50			2-Methoxyphenol
19.	Hordenine	$C_{10}H_{15}NO$			31	p-(2-Dimethylaminoethyl) phenol
20.	p-Hydroxybenzaldehyde	$C_7H_6O_2$	N			
21.	o-Hydroxybenzoic acid	$C_7H_6O_3$			14	
22.	p-Hydroxybenzoic acid	$C_7H_6O_3$	N,32		17	
23.	p-Hydroxyphenylacetic acid	$C_8H_8O_3$			31	
24.	p-Hydroxyphenylethylamine	$C_8H_{11}NO$				Tyramine
25.	p-Hydroxyphenylpyruvic acid	$C_9H_8O_4$				
26.	p-Methylguaiacol	$C_8H_{10}O_2$	1,50		27	4-Methyl-2-methoxyphenol
27.	Methyl salicylate	$C_8H_8O_3$			21	Methyl-2-hydroxybenzoate
28.	Methyl tyramine	$C_9H_{13}NO$			31	p-(2-Aminopropyl) phenol
29.	Phenol	C_6H_6O	50			
30.	Protocatechuic acid	$C_7H_6O_4$		32		3,4-Dihydroxybenzoic acid
31.	Salicylic acid	$C_7H_6O_3$		32		2-Hydroxybenzoic acid
32.	Scopoletin	$C_{10}H_8O_4$	4			
33.	Sinapaldehyde	$C_{11}H_{12}O_4$	4			4-Hydroxy-3,5-dimethoxycinnamic aldehyde
34.	Syringaldehyde	$C_9H_{10}O_4$	4,33			4-Hydroxy-3,5-dimethoxybenzaldehyde
35.	Syringic acid	$C_9H_{10}O_5$		32		4-Hydroxy-3,5-dimethoxybenzoic acid
36.	Tetramethylphenol	$C_{10}H_{14}O$	53			
37.	Tyrosol	$C_8H_{10}O_2$			17	p-Hydroxyphenylethanol
38.	Vanillic acid	$C_8H_8O_4$	N,32	32,36		4-Hydroxy-3-methoxybenzoic acid
39.	Vanillin	$C_8H_8O_3$	33,34,50,53			4-Hydroxy-3-methoxybenzaldehyde
40.	4-Vinylguaiacol	$C_9H_{10}O_2$	1			4-Vinyl-2-methoxyphenol
41.	4-Vinylphenol	C_8H_8O	1			

Hydrocarbons

No.	Compound	Formula			
1.	Benzene	C_6H_6	52		
2.	1,3-Dimethylnaphthalene	$C_{12}H_{12}$	N		
3.	2,6-Dimethylnaphthalene	$C_{12}H_{12}$	N		
4.	n-Heptane	C_7H_{16}	52		
5.	Isobutane (tent.)	C_4H_{10}	N		
6.	Limonene	$C_{10}H_{16}$			21
7.	2-Methyl-2-butene	C_5H_{10}			17
8.	Naphthalene (tent.)	$C_{10}H_8$	N		21
9.	2-Pinene	$C_{10}H_{16}$	52		
10.	Styrene	C_8H_8	52		
11.	Toluene	C_7H_8	52		

(Continued)

Table 1. (Continued)

No.	Name	Empirical Formula	References Cited For:				Alternative Name
			Whisky	Wine	Beer	Misc.	
	Miscellaneous: A. Nitrogen Compounds						
1.	Ammonia	NH_3			17		
2.	Butylamine	$C_4H_{11}N$			17		
3.	Diethylpyrazine	$C_8H_{12}N_2$				23	
4.	Dimethylamine	C_2H_7N			17		
5.	2,5-Dimethylpiperazine	$C_6H_{14}N_2$	30				
6.	2,5-Dimethylpyrazine	$C_6H_8N_2$	30				
7.	N-Ethylacetamide	C_4H_9NO		9			
8.	Ethylamine	C_2H_7N			17		
9.	N-Isoamylacetamide	$C_7H_{15}NO$		9			
10.	Isoamylamine	$C_5H_{13}N$		9			
11.	Methylamine	CH_5N			17		
12.	N-(2-Phenethyl)-acetamide	$C_{10}H_{13}NO$		9			
13.	2-Phenethylamine	$C_8H_{11}N$		9			
14.	Pyridine	C_5H_5N	30			23	
15.	Tetramethylpyrazine	$C_8H_{12}N_2$	N				
16.	Triethylpyrazine	$C_{10}H_{16}N_2$				23	
17.	Trimethylamine	C_3H_9N	30				
18.	Trimethylpyrazine	$C_7H_{10}N_2$	30			23	
	Miscellaneous: B. Sulfur Compounds						
1.	Carbon bisulfide	CS_2	52		43		
2.	Dimethyl sulfide	C_2H_6S	N				
3.	Ethyl mercaptan	C_2H_6S	30		11		
4.	Ethyl sulfite (?)	—	30				
5.	Hydrogen sulfide	H_2S			17		
6.	3-Methyl-2-butene-1-thiol	$C_5H_{10}S$			17		
7.	Methyl ethyl sulfide	C_3H_8S	52				
8.	Methyl mercaptan	CH_4S			11		
9.	Pentyl sulfide (?)	—	30				
10.	Pentyl sulfite (?)	—	30				
11.	Sulfur dioxide	SO_2			17		
	Miscellaneous: C. Lactones						
1.	γ-Butyrolactone	$C_4H_6O_2$		9, 20			
2.	4-Carboethoxy-γ-butyrolactone	$C_7H_{10}O_4$		36			
3.	δ-Nonalactone	$C_9H_{16}O_2$	53				
4.	δ-Nonalactone (branched)	$C_9H_{16}O_2$	53				
5.	γ-Nonalactone	$C_9H_{16}O_2$	53				
6.	Pantoyl lactone	$C_6H_{10}O_3$		18			

Miscellaneous: D. Sugars

1.	α-D-Fructose	$C_6H_{12}O_6$	33
2.	β-D-Fructose	$C_6H_{12}O_6$	33
3.	α-D-Glucose	$C_6H_{12}O_6$	33
4.	β-D-Glucose	$C_6H_{12}O_6$	33

Miscellaneous: E. Unclassified

1.	Bromoform	$CHBr_3$	N	
2.	Camphor	$C_{10}H_{16}O$	N	21
3.	Dichlorobenzene	$C_6H_4Cl_2$	N	
4.	Ethanol lignin	—	4	
5.	Ethyl vinyl ether	C_4H_8O	N	
6.	"Geosmin"	$C_{12}H_{22}O$	N	trans-1,10-Dimethyl-trans-9-decalol
7.	Isobutylene glycol	—	30	
8.	α-Methyl-α'-(β-furyl)-tetrahydrofuran	$C_9H_{12}O_2$	N	21
9.	Sitosterol-β-d-glucoside	$C_{35}H_{80}O_6$	51	
10.	Terpene	$C_{10}H_{20}O$	30	
11.	Terpene hydrate	$C_{10}H_{20}O_2 \cdot H_2O$	30	

[a] N = National Distillers and Chemical Corp., unpublished data.
[b] Numbers in parentheses refer to number of carbon atoms in acid radical where applicable.

INDEX

A

Abarkah, 16
Abnormalities, of children of alcoholic mothers,
    442, 443, 444
Absorption
    intestinal, alcohol and, 66, 459
    of thiamine, alcoholism and, 419
Abstemience
    in Petra, 14
    pre-Islamic, 14
Abstinence syndrome, *see also* Withdrawal
    symptoms, 372
Accidents, deaths from, 482
    alcohol-related, 482–483
Acetaldehyde
    cerebellar degeneration and, 352–353
    condensation with biogenic amines, 249
    effect of, 269–270
    ethanol metabolism and, 199, 200, 222, 247,
        266, 458
    fermentation and, 123, 124, 129
    placenta and, 445, 446
    pyridoxal 5-phosphate degradation and, 313
Acetaldehyde dehydrogenase, depression of,
    269
Acetate
    citric acid fermentation and, 104–105
    ethanol metabolism and, 199–200, 202, 247
    fermentation and, 124, 125
Acetoacetate, reduction of, 264
Acetoacetyl coenzyme A, ethanol metabolism
    and, 201
Acetobacter, kaffir beer and, 47
Acetoin, fermentation and, 123, 124
Acetylcarnitine, ethanol metabolism and, 200
Acetyl coenzyme A
    ethanol metabolism and, 200, 201, 202, 222
    ketogenesis and, 269

Acidosis, ethanol metabolism and, 199–200,
    201
Acids, in alcoholic beverages
    aliphatic, 508
    aromatic, 509
    keto and hydroxy, 509
    miscellaneous, 510
Acylcarnitine transferase, chronic alcohol con-
    sumption and, 268
Adams, Samuel, beer and, 146
Addiction, in experimental animals, 363, 387,
    388
Adenosine diphosphate, chronic alcohol con-
    sumption and, 269
Adenosine triphosphatase
    Na^+-, K^+-dependent, alcoholism and, 286,
        288
    uncoupling of, alcohol and, 202, 207
Adenosine triphosphate
    ethanol metabolism and, 200, 202, 222
    hydrocarbon dissimilation and, 109
Adenyl cyclase, alcohols and, 286
Aflatoxin, liver disease and, 311
Africa, mead production in, 37
Agave
    mezcal and, 41
    pulque and, 39
Age, use of wine and, 62–65, 182
Aging
    constructive approach to, 236–237
    withdrawal symptoms and, 388
Agriculture, Andean, importance of maize in, 27
Aguamiel
    composition of, 40
    sugars of, 39
Aguardiente, production of, 33
Air, expired, alcohol content of, 199
Aka, 22
Alcohol, *see also* Ethanol
    accidents and, 482–483

Alcohol (*cont.*)
  administration procedures in withdrawal studies
    experimental assessments, 375–376, 386
    programmed drinking, 374, 386
    spontaneous drinking, 374, 386
    work-contingent drinking, 375, 386
      delayed reinforcement, 375, 386
      immediate reinforcement, 374–375, 386
  assault and, 484
  beverage choice and
    comments, 307–311
    data on, 304–305
    laboratory abnormalities, 305–307
    liver histology, 307
    patients studied, 304
  blood level, withdrawal symptoms and, 381–385
  calcium carbamide and, 249
  caloric intake and, 66, 403–404
  cancer and, 484–485
  carcinogenesis and
    alcohol as solvent for carcinogens, 428
    ethanol and other alcohols, 428
  carcinoid tumor and, 255
  cardiovascular disease and
    arsenic-beer drinkers' disease, 323–324
    cardiovascular beriberi, 325–326
    cobalt-beer drinkers' disease, 324–325
  cirrhosis and, 481–482
  consumption in relation to human cancer
    alcohol and tobacco, 431–434
    case of primary liver cancer, 431
    cohort studies of alcoholics, 429
    correlation studies, 428–429
    retrospective case-control studies, 429–431
  coronary circulation and, 321
  coronary disease and, 335
    angina pectoris, 332
    relation to major coronary events, 332–334
  depressants and, 250–251
  direct action on fetus, 445
  diseases associated with high blood levels
    blackouts, 343
    combativeness, 343–344
    inebriation leading to coma, 343
  disorders associated with zero or diminishing
      blood levels
    delirium tremens, 345
    hallucinosis, 345–347
    rum fits, 344–345
    tremulousness, 344
  disulfiram and, 248–249, 255

  dosage, withdrawal and, 387–389
  effect on absorption and metabolism of nutri-
      ents, 312–313
  effects on pancreas, 291–292
    acute, 292–293
    chronic, 293–294
  energy value of, 213–223
  excessive use, medical costs of, 481–486
  exchange for fat in diets, 274
  government policy and, 494–498
  higher
    fermentation and, 126–128
    intoxication and, 246
  homicide and, 484
  hypertension and, 330–331, 335
  intake, underestimation of, 401
  life span and, 237
  lymphoma and, 255–256
  mental disorders and, 485
  metabolic effects, ignorance of, 258–260
  myocardial biochemistry and, 321–322
  myocardial function and, 319–321
  myocardial structure and, 322–323
  psychoactive drugs and, 249–250
  sudden deaths and, 485
  sulfonylurea and, 260
  synergism with barbiturates, 248
  total world consumption, 153
    by country, 162
  use by diabetics, 258, 273–274
  use in gerontology
    factors precluding government subsidy, 235
    factors precluding unsubsidized use, 235–236
  utilization, 87–88
    by alcoholics, 222–223
    by normal man, 220–222
  volume consumed, onset of withdrawal symp-
      toms and, 378–386
  weight gain in undernourished and, 261–262
  word, derivation of, 4
  yield, factors affecting, 122–123
Alcohols, higher, in alcoholic beverages,
    507
Alcohol abuse, *see also* Alcoholics, Alcoholism
  duration, onset of withdrawal symptoms and,
      376–377
  gerontology and
    early life, 226
    middle age, 226–227
    old age, 227–228
  prevention

public health approach, 229–230
societal control, 228–229
Alcohol dehydrogenase
ethanol formation and, 203–205
ethanol identification and, 203
ethanol metabolism and, 199, 201, 222, 247,
266, 279
hepatic, low protein diet and, 459–460
inhibition of, 313
tissue distribution of, 279–280
Alcohol dependence, temporal development of,
386–387
Alcoholic
alcohol utilization by, 222–223
Australian, clinical studies, 411–412
average nutrient intake of, 207
a case history, 362
cohort studies, cancer and, 429
elderly, rehabilitation of, 229–230
nutritional status, patients studied, 399–400
skid row versus more common, 397–399
Alcoholic beverages, controlled consumption by
institutionalized elderly
beer, 231–233
deficiencies of studies, 233, 235
five-wine study, 233
red port study, 231
studies allowing choice, 234
Alcoholic dementia, brain lesions in, 355
Alcoholic foods
saliva as amylolytic agent, 41
chica, 42–43
related fermentations, 43
sugar as major fermentable substrate
mead, 37
Mexican *pulque,* 39–41
miscellaneous, 44
palm toddies, 38–39
Alcoholic heart disease, 335
clinical features, 329–330
diagnosis, 328–329
evidence for entity, 327–328
Alcoholics Anonymous, success of, 363
Alcoholism
in Australia, 409–411
determinants
prospective, longitudinal studies needed,
239–240
diagnosis of, 198
etiology, common concepts of, 359–361
etiology of thiamine deficiency in
impaired absorption, 419

impaired utilization, 419–421
low intake, 419
genetic factors and, 369
habituation, dependence, problems, 365–366
intestinal absorption and, 286–287
amino acids, 290–291
folate, 289–290
thiamine, 288–289
vitamin B_{12}, 290
water and sodium, 287
D-xylose, 287–288
neurological disorders of nutritional cause
and, 347
amblyopia, 350–351
nutritional neuropathy, 348
Wernicke–Korsakoff syndrome, 349–350
neurological disorders of undetermined cause
and, 351–352
alcoholic dementia, 354–355
central pontine myelinolysis, 353–354
cerebellar degeneration, 352–353
Marchiafava–Bignami disease, 354
pancreatic carcinoma and, 296
role of the pleasurable experience from alcohol, 364–365
small intestine and
absorption, 286–291
enzymes, 285–286
metabolism, 284–285
morphology, 284
suicide and, 483–484
supply–demand–stress model, 491–492
tolerance and need, 366–367
underlying psychopathology and, 361–364
vulnerability to, 367
among women, 440
Aldehyde dehydrogenase
antidiabetic drugs and, 255
disulfiram and, 248, 255
ethanol metabolism and, 199, 201, 222
Aldosterone, ethanol consumption and, 206, 346
Ale, 86
definition of, 138
Alkaline phosphatase, alcoholism and, 305, 306
Alkalosis, respiratory, rum fits and, 345
Al-mukdi, 13
Amazake, preparation of, 53
Amblyopia, alcoholism and, 350–351
Americans, as wine consumers, 191–193
Amino acid
absorption, alcoholism and, 290–291
in beer, 78

Amino acid (*cont.*)
  biosynthesis of, 115–116
  deficiency, effects on offspring, 447
  plasma, ethanol consumption and, 460
  production of higher alcohols and, 126–128
  of yeast protein, 114–115
p-Aminobenzoic acid, in wine, 69
Amphetamine, alcohol and, 246, 250
Amylase, *see also* Diastase
  mashing and, 136
Amylomyces rouxii
  tapé ketan and, 43
  tapé ketella and, 46
Andes, exchange in, 25–26
Anemia
  alcoholism and, 305, 306, 307
  folate deficiency and, 312
Angina pectoris, alcohol and, 332
Animal model, for human alcoholism, 459–460
Antabuse, *see* Disulfiram
Antidepressants, tricyclic, alcohol and, 250–251
Antidiuretic hormone, ethanol and, 205
Antihistamines, alcohol and, 251
Aqua vitae, 4
Arabia, pre-Islamic, fermented beverages, 13–14
Arabs, use of beer by, 145
Armenia, *mazun* of, 55
Arsenic-beer drinkers' disease, symptoms,
    323–324
Arthur (king of Britain), use of beer by, 145
Ascorbic acid, *see also* Vitamin C
  alcoholics and, 405, 406
  in grapes and wine, 69
Ashbyia gossypii, riboflavin synthesis by, 111–112
Aspergillus oryzae, koji and, 52
Aspirin, alcohol consumption and, 246
Assault, alcohol use and, 484
Atherosclerosis
  alcohol consumption and, 273
  fat and, 259
  hypertriglyceridemia and, 273
Australia
  alcoholism in, 409–411
  population, thiamine status of, 421–422
Automobile driving, intoxication and, 497–498
Autonomic blockade, alcohol effects on
    myocardial function and, 319–320

B

Babylon, beer in, 144–145

Bacteria, intestinal, ethanol formation by, 205
Bakhar, composition of, 46
Balkans, *kefir* of, 54
Bantu, beer of, 47
Barbiturates, synergism with alcohol, 248
Barley
  malt preparation from, 134
  mead and, 37
  talla and, 51
Basi, 41
Bata^c, 13
Beer, *see also* Maize beer
  amino acids in, 78
  antiquity of, 4
  cirrhosis and, 488
  commercial, nutrient contents of, 70
  competition with wine, 184–185
  compounds identified in, 505–517
  consumption trends
    geographical region, 149–150
    historical and national, 148–149
    imports, 151–152
    in specific states, 151
  definition of, 138
  in early Egypt, 6–9
  early man and, 86
  economic contributions, 153–154
  Egyptian myth and, 5
  fatty acids in, 78
  flavor of, 138
  history of
    ancient times, 144–145
    United States, 145–146
  institutionalized elderly and, 231–233
  intestinal disaccharidases and, 285
  isocaloric exchanges for, 260
  in Mesopotamia, 13
  price, increase in, 147
  production trends, 146
    geographical changes, 147
    packaging shifts, 147–148
    purchasing habits, 148
  prohibition of, 15
  trace elements in, 71
  world consumption of, 152–153
  world production, 152
  yeast and, 91–92
Beer drinking, onset of withdrawal symptoms
    and, 377–378, 387
Behavior
  of miniature swine, withdrawal and, 462–463
  of young of alcoholic mothers, 445

Benzodiazepines, alcohol detoxification and, 250

Beriberi heart disease, alcoholics and, 325, 441

Beverage, nature of nutritional contributions, 61–62

Beverage choice, of alcoholic patients, 304–305

Bilirubin, serum, alcoholism and, 305–306

Bios, discovery of, 95

Biotin, in wine, 69

Bisulfite, glycerol fermentation and, 103–104

Blackouts, alcohol and, 343

Blood alcohol levels, 343

Blood loss, atrophic gastritis and, 282

Blood pressure, alcohol and, 319

Bock beer, definition of, 138

Body weight, ethanol and energy consumption and, 461

Bolivia, beer making in, 43

Boorde, Andrew, on beer, 87

Borassus flabellifer, 38

Botrytis cinerea, grape infection by, 131

Bouza
  composition of, 8–9, 50–51
  preparation of, 50

Brain, lesions, Korsakoff's psychosis and, 350

Brazil, beer of, 43

Bread, enrichment of in Australia, 421–422

Breweries
  decline in numbers, 146
  at Huánuco Pampa, 28–30

Brewers' yeast, types of, 138–139

Brewing
  constituents
    cereal adjunct, 134
    hops, 134
    malt, 133–134
  steps in, 135
    beer flavor, 138
    finishing and packaging, 137
    kettle boil, 136
    lagering, 137
    lautering, 136
    mashing, 134, 136
    wort cooling and trub removal, 136–137
    yeast pitching and primary fermentation, 137
    yeast system, 137
  terminology of, 138–139

Bromides, alcohol and, 250

Bromosulfonphthalein test, alcoholism and, 305, 307

Busa, nature of, 54

C

Caffeine, alcohol and, 250

Calcium carbamide, alcohol and, 249

California
  number of wineries in, 193
  wine making by colonists, 188

Calories
  ethanol and, 207
  intake, by alcoholics, 405–406

Calorimeter, human, 213–214

Calorimetry
  energy income and outgo, 216, 218–219
  gain or loss of protein, fat and water, 216, 217

Cancer, *see also* Human cancer
  alcohol and, 484–485
  esophageal, alcoholism and, 428, 429, 430
  rectal, beer drinking and, 429

Candida lipolytica, single cell production and, 109–110

Candida utilis
  as food, 108
  gross composition, 110

Carbohydrate, intake by hospitalized alcoholics, 403

Carbonation
  alcohol absorption and, 246
  of wines, 129–130

Carbon dioxide
  ethanol metabolism and, 200
  yeast growth and, 125

Carbon tetrachloride, toxicity, alcohol and, 248

Carbonyls
  in alcoholic beverages
    acetyls, 514
    aliphatic, 513
    cyclic and aromatic, 514
    hydrocarbons, 515
    phenolic compounds, 514, 515
  as by-products of fermentation, 123–124

Carcinogenesis, alcohol and
  ethanol and other alcohols, 428
  as solvent for carcinogens, 428

Carcinoid tumor, alcohol and, 255

Carcinoma, pancreatic, alcoholism and, 296

Cardiac output, alcohol and, 319

Cardiovascular beriberi, symptoms of, 325–326

Cardiovascular disease, partially defined syndromes
  arsenic-beer drinkers' disease, 323–324
  cardiovascular beriberi, 325–326
  cobalt-beer drinkers' disease, 324–325

Cardiovascular risk factors and alcohol, 334
Cardiovascular system, effects of alcohol on
    cardiovascular disease, 323–334
    historical review, 317–318
    physiology, biochemistry, and structure,
      319–323
    summary and research needs, 335–336
Caryota urens, 38
Cassava
    beer from, 43
    fermented food from, 46
Catalase, ethanol metabolism and, 200, 209,
    247
Central nervous system, ethanol tolerance of,
    210
Central pontine myelinolysis, alcoholism and,
    353–354
Cereal, beer brewing and, 134
Cerebellar degeneration, alcoholism and,
    352–353
Cheese, 89
Chicha
    foods used in preparation, 22
    production of, 42
Children, alcohol abuse and, 226
Chillproofing, brewing and, 139
China
    esophageal cancer and, 437
    fermented rice beverage of, 46
Chiu-yueh, composition of, 46
Chlamydomucor oryzae, lao-chao and, 46
Chloral hydrate, alcohol and, 250
Chloramphenicol, alcohol and, 249
Chlordiazepoxide
    alcohol and, 250, 251
    withdrawal and, 376, 385
Chlorpropamide
    alcohol and, 260
    aldehyde dehydrogenase and, 255
Cholecystokinin, alcohol and, 292, 293
Cholesterol
    serum, diabetics and, 259
    synthesis, ethanol metabolism and, 201, 222,
      268, 273, 285
Choline, requirement, alcohol and, 311, 459
Choline oxidase, level in human, 311–312
Chromium
    in beer, 71
    in yeast, 71
Cider, 85
    distillates, nitrosamines in, 428
Cirrhosis
    age and, 227
    alcoholism and, 209
    deaths from, 481–482
    nutritional deficiency and, 398
    wine or beer and, 488
Citrate, production by fermentation, 104–106
Citrate synthase, 105
Citric acid cycle, *see* Tricarboxylic acid cycle
Cobalt, synergism with ethanol, 459
Cobalt-beer drinkers' disease, symptoms of,
    324, 325
Cocos nucifera, 38
Coenzymes, yeasts as source of, 110–111
Collagen proline hydroxylase, hepatic, al-
    coholism and, 310
Colonists, attempts at wine making by, 188
Columbus, Christopher, beer and, 145
Coma, blood alcohol levels and, 343
Combativeness, alcohol and, 343–344
Compounds, miscellaneous unidentified in
    alcoholic beverages, 517
Congeners, 505
    intoxication and, 246
Consumer, protection, wine production and, 182
Convenience containers, beer sales and, 147–
    148
Copper, in beer, 71
Corn, *see also* Maize
    beer brewing and, 134
Corn syrup, brewing and, 136
Coronary circulation, alcohol and, 321
Coronary disease, alcohol and, 335
    angina pectoris, 332
    relation to major coronary events, 332–334
Corpulence, *see also* Obesity
    drink and, 65–66.
Cortisol
    ethanol consumption and, 206, 346
    lipolysis and, 266
Corynebacterium glutamicum, lysine production
    by, 116
Cozymase, discovery of, 94
Cuzco, other cities and, 28
Cytochrome *P*-450, ethanol metabolism and,
    200–209, 222

D

Dates (fruit), early Egyptian beer and, 8
Death
    alcohol-related, 485

sudden, alcohol and, 485
Deficiency disease, civilization and, 68
Dehydration, withdrawal and, 388
Delirium tremens, withdrawal and, 345, 372
Demand, alcohol consumption and, 491–492
   societal values or attitudes, 493
Dementia, alcoholism and, 412
Deoxyribonucleic acid
   recombinant, technique of, 101–102
   synthesis of, 312
Dependence, alcoholism and, 366
Depressants, alcohol and, 250–251
Diabetes
   alcohol use and, 258, 273–274
   fetal abnormalities and, 446
Diarrhea, alcoholism and, 287
Diastase, 93, *see also* Amylase
   source, for maize beer, 22
Diet
   deficient, experimental production of liver
      disease and, 311–312
   fermented beverages, early recommendations,
      87
   incorporating alcoholic beverages
      problems
         ignorance of metabolic effects of al-
            cohol, 258–260
         inadequate dietary histories, 257–258
         sensitivity to effects of sulfonylurea, 260
   ketogenic, alcohol and, 264
   for miniature swine, 460–461, 462
Dietary allowances, recommended, alcoholics
   and, 405–407
Dietary histories, inadequate, 257–258
Digestive tract, cancer, alcohol and, 427, 429,
   431–434, 484
Digitalis, sensitivity, alcohol and, 247
Dihydroxyacetone phosphate, ethanol
   metabolism and, 201, 266
Dilantin
   alcohol withdrawal and, 251
   metabolism by alcoholics, 248
3,5-Dinitrobenzoyl chloride, ethanol identifica-
   tion and, 203
Disaccharidases, intestinal, alcoholism and,
   285–286
Disease, associated with alcoholism, 66–67
Disulfiram, alcohol and, 248–249, 255
Diuresis, ethanol and, 205–206
Dopamine β-hydroxylase, disulfiram and, 249
Draught beer, share of market, 147
Drinking, context of, 33–34

Drinking patterns, alcohol withdrawal and,
   389–390
Drug
   alcohol absorption and, 246
   metabolism, chronic alcohol ingestion and,
      202, 268
   psychoactive, alcohol and, 249–250
   required by elderly patients consuming beer,
      232
Drug abuse, effects on fetus, 448
Drunkenness, in early Egypt, 5

E

Eating, life span and, 237
Economic factors, alcohol consumption and,
   490
Education, wine consumption and, 182
Egypt
   bouza of, 50–51
   early
      beer making in, 6–9, 145
      drunkenness in, 5
      recognition of excess in, 12–13
      winemaking in, 9–12
   Islam in, 15
Elaeis guineensis, 38
Elderly
   alcoholic, rehabilitation of, 229–230
   institutionalized, controlled alcohol consump-
      tion by, 230–234
Electrocardiogram, alcoholic heart disease and,
   329
Electrolyte(s)
   imbalance, central pontine myelinolysis and,
      353
   withdrawal symptoms and, 389
Eleusine coracana, kaffir beer and, 47
Elizabeth I (queen of England), use of beer by,
   145
Elizabeth II (queen of England), use of beer by,
   145
Embden–Meyerhof–Parnas scheme, fermentation by-products and, 123
Endomycopsis
   cassava fermentation and, 46
   sugar cane wines and, 41
Endomycopsis burtonii, tapé ketan and, 43
Endoplasmic reticulum, *see also* Microsomes
   chronic alcohol ingestion and, 202, 248,
      268–269

Energy
    alcohol and, 216–220
    consumption, calculation of, 221
Enzymes, alcohol metabolism and, 221–222
Eremothecium ashbyii, riboflavin synthesis by,
    111–112
Escherichia coli, recombinant DNA and, 101–
    102
Esters, in alcoholic beverages
    aliphatic, 510–512
    aromatic, 512
Ethacrynic acid, ethanol metabolism and, 247
Ethanol, *see also* Alcohol
    absorption of, 198, 278–279
    clearance from plasma in miniature swine,
        463, 464, 471
    direct effect in ketoacidosis, 266–267
    distribution in body, 199–220
    effect of chronic ingestion, 268–269
    effect on fluid balance, 205–206
    effect on gastric mucosa
        acute, 281–282
        chronic, 282–283
    elimination from blood, 458
    endogenous biosynthesis in animals, 202–
        205, 210
    equivalents in selected amounts of various
        beverages, 75–77
    excretion of, 202
    experiments of Atwater and Benedict, 214–
        220
    experiments on physical dependence
        discussion, 471–472
        miniature swine and, 460–471
    gastric motility and, 283
    gastric secretion and, 283
    intake, nutritional status and, 206–209
    ketogenesis and
        dietary studies in humans, 264
        direct effect of ethanol, 266–267
        effect of acetaldehyde and acetate, 269–270
        effect of chronic ethanol consumption,
            268–269
        substrate availability, 266
        summary, 270
    lethal level of, 210
    metabolism of, 199–202, 220–222, 247–248,
        279–281
        research needs, 209
    plasma concentrations in miniature swine, 462
    in *tapé ketan,* 45
    tolerance by miniature swine, 463–471

Ethiopia
    fermented beverage of, 51
    mead production in, 37
Ethyl acetate, formation of, 125–126
Ethylene glycol, metabolism of, 247
European Economic Community
    wine export and import by, 167
    wine-growing areas, 160
        minimum and maximum alcohol contents
            of wines, 159
    wine production in, 166
Excess, recognition in early Egypt, 12–13
Exercise, life span and, 237

F

Family, stress and, 494
Fat
    diabetic diets and, 258–259
    dietary, obesity and, 261
    storage, ethanol and, 267
Fatty acid
    alcoholic ketogenesis and, 266
    in beer, 78
    oxidation, ethanol and, 267, 268, 269, 322
    synthesis, ethanol metabolism and, 201, 222
Fatty infiltration, of liver, alcoholism and, 307
Feed supplement, brewing and, 136, 137
Fermentation
    and aging of wine, 121
    alcoholic, a look into the future, 101–102
    by-products of, 123–126
    chronology of major events in, 84
    citric acid and, 104–106
    early chemistry of, 90–91
    food preservation and, 36
    glycerol and, 103–104
    higher alcohols and, 126–128
    indirect by-products of, 126
    involving a mold and yeast
        Chinese *lao-chao,* 46
        Indian rice beer, 46
        Indonesian *tapé ketan,* 43–45
        Indonesian *tapé ketella,* 46
    involving use of a *koji*
        Japanese rice wine, 53–54
        koji, 52
    of maize beer, 25
    malo-lactic, 130–131
    processes
        metabolic pathways, 100
        revival of interest in, 99

products of, 122–123
research needs, 131
Fermented beverages
clarification, filtration, and distillation, 69–70
direct nutrient contributions of, 67–69
early
discovery, 85
types of, 85–86
ethanol equivalents in, 75–77
excesses and nutritional deficiencies, 66–67
health value of, 86–87
early dietary recommendations, 87
fermented milks, 89–90
health hazards, 88
spruce beer, 88–89
utilization of alcohol, 87–88
native American, 21–22
organic compounds in, 71–72
social and cultural importance of, 32–34
time-honored precepts of use, 62–66
trace elements in, 71
Fetal alcohol syndrome, 485
discussion
alcohol and its metabolites, 445–446
amino acid deficiencies, 447
drug abuse, 448
hypoglycemia, 446
malnutrition, 447–448
pathology, 448
smoking, 448
trace metal deficiencies, 447
vitamin deficiencies, 446
historical review
clinical, 440–441
laboratory investigations, 441
interest in, 440
modern studies
clinical investigations, 441–443
current animal studies, 443–445
research needs, 448–449
Finishing, brewing and, 137
Fixed acids, fermentation and, 124, 125
Fluid balance, ethanol intake and, 205–206
Folacin, coenzyme forms, 111
Folate
deficiency
alcoholism and, 284, 289–290, 312
fetal abnormalities and, 446
thiamine transport and, 289
water and sodium absorption and, 287
D-xylose absorption and, 288
jejunal glycolysis and, 285

levels in alcoholics, 307
Food
dietary deficiencies and, 36
ethanol absorption and, 246
ethanol elimination from blood and, 458
intake, thiamine intake and, 419
taste, fermented beverages and, 75
weight, composition, and heat of combustion
of, 215
Food yeasts, definition of, 108
Formaldehyde, methanol metabolism and, 247
Formic acid, methanol metabolism and, 247
Free Enrichment Program, wine and, 233
Fructose, alcohol overdose and, 250
Fructose 1,6-diphosphate, fermentation and, 94
Fructose effect, ethanol elimination from blood
and, 458
Furazolidine, alcohol and, 249
Fusel oils, carcinogenesis and, 428

G

Gas, fermentation and, 90
Gastrectomy, alcoholism and, 366
Gastric motility, ethanol and, 283
Gastric mucosa
alcohol dehydrogenase of, 279–280
effect of alcohol
acute, 281–282
chronic, 282–283
Gastric secretion, ethanol and, 283
Gastrin, alcohol and, 292
Gastritis
alcohol consumption and, 281
chronic atrophic, alcoholism and, 282
Gastrointestinal symptoms, blood alcohol level
and, 381–385, 389
Gastrointestinal tract, ethanol absorption in,
278–279
Genetic factors, alcoholism and, 369
Geographic changes, in beer production, 147
Geographic region, beer consumption and,
149–151
Germany, domestic wine consumption in, 179–
180
Gerontology
alcohol abuse and
early life, 226
middle age, 226–227
old age, 227–228
alcohol in
factors precluding government subsidy, 235

Gerontology (*cont.*)
  factors precluding unsubsidized use, 235–236
  definition of, 226
  research needs
    understanding the causes, 239–240
    understanding the situation, 239
Ginger, *pachwai* and, 46
Glucagon, lipolysis and, 266
Gluconeogenesis
  alcohol and, 201, 247, 273–274
  ethanol metabolism and, 201
Glucose
  citric acid fermentation and, 104–105
  intestinal metabolism, alcoholism and, 284–285
Glucose tolerance factor, yeast and, 71
Glutamate-oxalacetate transaminase
  leakage from heart, 322
  serum, alcoholism and, 305, 306
Glutamate-pyruvate transaminase, ethanol consumption and, 463, 471
Glutethimide, alcohol and, 250
Glycerol, production by fermentation, 103–104, 124, 125
α-Glycerophosphate, ethanol metabolism and, 201, 266
Glycolysis, jejunal, ethanol and, 285
Glyoxylate, citrate synthesis and, 105
Glyoxylate cycle, fermentation and, 125
Gout, wine drinking and, 201, 247
Government policy, alcohol consumption and
  interdiction, 495
  licensing, 496
  prohibition, 494–495
  punitive approach, 497–498
  restriction, 495
  state monopoly, 495–496
  taxation, 496
Grapes, ascorbic acid in, 69
Griseofulvin, alcohol and, 249
Growth hormone, lipolysis and, 266

H

Habituation, alcoholism and, 365
Hallucinosis
  blood alcohol level and, 381–385, 388–389
  withdrawal and, 345–346
    ~vers
    .nd, 365

cause of, 128
Hansenula
  cassava fermentation and, 46
  tapé ketan and, 44
Hawthorne effect, elderly patients and, 233, 235
Health
  protection, wine production and, 182
  stress and, 494
  value of fermented beverages, 86–87
    early dietary recommendations, 87
    fermented milks, 89–90
    hazards, 88
    spruce beer, 88–89
    utilization of alcohol, 87–88
Health care, expenditures, alcoholism and, 397
Heart rate, alcohol and, 319
Heat
  alcohol metabolism and, 221
  fermentation and, 129
Hemachromatosis, alcohol and, 325
Henry VIII (king of England), use of beer by, 145
Hepatitis, alcoholic, 307
  etiology of, 207
Hepatoma, primary, alcoholism and, 431, 484
High-density lipoproteins, 334
Holy Communion, wine for, in Egypt, 15–16
Homicide, alcohol use and, 484
Hops, 86
  beer brewing and, 134, 136
  beer flavor and, 138
  mead production and, 37
Huánuco Pampa, research in, 27–32
Human cancer, in relation to alcohol consumption; epidemiological evidence
  alcohol and tobacco, 431–434
  case of primary liver cancer, 431
  cohort studies of alcoholics, 429
  correlation studies, 428–429
  retrospective case-control studies, 429–431
Humulons, beer flavor and, 134
Humulus lupulus, 134
Hydrocarbons
  polycyclic aromatic, alcohol and, 428
  yeast growth on, 109
Hydrogen peroxide, ethanol metabolism and, 200
Hydrogen sulfide, fermentation and, 126
Hydromel, 85
β-Hydroxybutyrate, alcoholic ketoacidosis and, 264, 265

5-Hydroxyindoleacetaldehyde, disulfiram and, 255
5-Hydroxytryptophol, metabolism of, 255
Hyperlipidemia, ethanol metabolism and, 201
Hypertension, alcohol and, 330–331, 335
Hyperuricemia, ethanol metabolism and, 201, 247, 266
Hypoalbuminemia, alcoholism and, 460
Hypoglycemia
  alcohol abuse and, 226, 247
  alcohol use by diabetics and, 274
  fetal alcohol syndrome and, 446
  social drinking and, 260
Hypomagnesemia, rum fits and, 345

I

Iceland, fermented milk of, 89–90
Imports, of beer, 151–152
Inca
  empire, establishment of, 21–22
  entertainment by, 30–32
  power, cities and, 28
Income, wine drinking and, 182, 185
Incontinence, in elderly patients consuming beer, 232
India, rice beer of, 46
Indonesia, fermented beverages of, 43–46
Inebriation, coma and, 343
Inebriety, stages of, 13
Infarction, alcohol and, 332–334
Infants, weight at birth, alcoholic mothers and, 441
Inositol, in wine, 69
Insomnia, blood alcohol level and, 385
Insulin
  lipolysis and, 266
  requirements, dietary fat and, 259
Interdiction, alcohol and, 495
Intestinal absorption, alcoholism and, 286–287
  amino acids, 290–291
  folate, 289–290
  thiamine, 288–289
  vitamin B_{12}, 290
  water and sodium, 287
  D-xylose, 287–288
Intestinal enzymes, alcoholism and
  disaccharidases, 285–286
  enzymes involved in transport, 286
Intestinal metabolism, alcoholism and
  glucose, 284–285

lipids, 285
Intestinal transport, alcoholism and, 286
Intestine, alcohol dehydrogenase of, 280–281
Intoxication
  estimation of, 199
  punitive approach to, 497–498
Iran, esophageal cancer in, 437
Iraq, wines of, 14
Iron
  absorption, alcohol and, 71
  in wines, 71
Islam, changing attitudes with, 14–18
Isocitrate dehydrogenase, leakage from heart, 322
Isocitrate lyase, citric acid fermentation and, 105
Isoniazide, metabolism by alcoholics, 248

J

Japan, rice wine of, 53–54
Jefferson, Thomas
  beer and, 146
  wine and, 188
Job, stress and, 494
Johnson, Lyndon, wine and, 191
Jora, maize beer making and, 22, 43

K

Kaffir beer
  composition of, 47, 49
  preparation of, 47
  unclarified and clarified nutrient contents of, 69
Kanamycin, ethanol and, 246
Kaschiri, making of, 43
Kefir, 89
  nature of, 54
Ketoacidosis
  alcohol and, 147
  alcoholic, clinical picture, 263–264, 265
  treatment of, 264
Ketogenesis
  ethanol and
    dietary studies in humans, 264
    direct effect of alcohol, 266–267
    effect of acetaldehyde and acetate, 269–270
    effect of chronic ethanol consumption, 268–269
    substrate availability, 266
    summary, 270
  fat and, 259

α-Ketoglutaric acid, fermentation and, 123
Kettle boil, brewing and, 136
Ki-moto, saké production and, 53–54
Koji
    preparation of, 52
    saké and, 43
Koumiss, 86, 89
Korsakoff syndrome, *see* Wernicke-Korsakoff
    syndrome
    nature of, 55
Krebs cycle, *see* Tricarboxylic acid cycle
Kuranga, milk and, 89
Kvass, preparation of, 52
Kwashiorkor
    liver disease and, 311
    pancreatic function and, 294

L

Lactase, alcoholism and, 285–286
Lactate
    ethanol metabolism and, 201, 266
    ketoacidosis and, 264, 265
Lactobacillus
    pulque and, 39–40
    sugar cane wines and, 41
Lactobacillus brevis, kefir and, 54
Lactobacillus delbruckii, kaffir beer and, 47
Lactobacillus plantarum, palm toddies and, 38
Lactobacillus sake, saké fermentation and, 53
Lactones
    in alcoholic beverages, 516
    fermentation and, 124
Lager beer, definition of, 138
Lagering, brewing and, 137
Lao-chao, preparation of, 46
Lautering, brewing and, 136
Lead, fermented beverages and, 88
Leuconostoc, pulque and, 40
Leuconostoc mesenteroides
    palm toddies and, 38
    saké fermentation and, 53
Leuconostoc oenos, malo-lactic fermentation
    and, 130
Licensing, alcohol and, 496
Life-span, increase of, 237
Life style, life span and, 237
Lipase, steatorrhea and, 295
Lipids
    blood, alcohol and, 259
    intestinal metabolism, alcoholism and, 285
Lipolysis, factors affecting, 266

Lipoproteins
    ethanol consumption and, 268, 334
    very low-density, ethanol metabolism and,
        201
Lithium, alcohol withdrawal and, 251
Liver
    alcohol dehydrogenase of, 280
    ethanol metabolism in, 199–200, 201
    histology, alcoholism and, 307, 308–309
Liver disease
    alcoholic, 67
        research needs, 313
    apotransketolase and, 419–421
    malnutrition and
        effect of alcohol on absorption and
            metabolism of nutrients, 312–313
        experimental production of disease, 311–
            312
    occurrence, 311
Longevity, factors related to, 227
Loss, stress and, 494
Luk-paeng, rice wine and, 44
Lungs, elimination of alcohol in, 202
Lymphoma, alcohol and, 255–256
Lysine
    biosynthesis by yeast, 116–118
    in *tapé ketan,* 45

M

Magnesium, deficiency, fetal abnormality and,
    447
Maguey plant, making *pulque* from, 68
Maize, *see also* Corn
    germinated, beer making and, 42–43
    importance in Andean agriculture, 27
    kaffir beer and, 47
    religious significance of, 42
    tesgüino from, 51–52
Maize beer, *see also* Beer
    preparation of, 22–25
    in pre-Spanish time, 26–27
    state management of brewing, 27–32
Malate dehydrogenase, leakage from heart, 322
Mallory bodies, alcoholism and, 207
Malnutrition
    alcoholism and, 66–67
    effects on fetus, 447–448
    liver disease and
        effect of alcohol on absorption and
            metabolism of nutrients, 312–313

experimental production of disease, 311–312
  occurrence, 311
  pancreatic function and, 294–295
Malt
  beer brewing and, 133–134
  beer flavor and, 138
  mashing of, 134, 136
Maltase, beer drinking and, 285
Malt beverages, nutritional properties of, 77–79
Malting
  alcoholic food production and, 46–47
    Egyptian *bouza,* 50–51
    Ethiopian *talla,* 51
    kaffir (sorghum) beer, 47–50
    Mexican *tesgüino,* 51–52
    Russian *kvass,* 52
  in early Egypt, 6–7
  nutritional value of, 56
Malt liquor, definition of, 138
Mamakuna, function of, 28
Manhattan Island, use of beer on, 145–146
Marchiafava–Bignami disease, alcoholism and, 354
Mary (queen of Scots), use of beer by, 145
Masata, making of, 43
Mashing, beer brewing and, 134–136
Mayflower, use of beer on, 146
Mazun, nature of, 55
Mead, 85
  in Arabia, 13
  production of, 37
Meals
  sharing, alcohol abuse and, 229–230
  wine consumption with, 178
Media, for yeast propagation, 108
Medications, over-the-counter, alcohol in, 256
Megalocytosis, of intestinal epithelium, alcoholism and, 284
Memory, Korsakoff syndrome and, 349, 350
Mental disorders, alcohol and, 485
Meprobamate, alcohol and, 250
Merissa, composition of, 8–9
Mesopotamia, beer in, 13, 144, 145
Methanol
  metabolism of, 247
  wines and, 126
Metheglin, 86
Methionine, yeast and, 118
Methylphenidate, alcohol and, 250
4-Methylpyrazole, alcohol dehydrogenase and, 313

Methylquaalude, alcohol and, 250
5-Methyltetrahydrofolate, ethanol and, 312
Metronidazole, alcohol and, 249
Mexico, *tesgüino* of, 51–52
Mezcal, Mexican, 41
Mezr, 15
Microcephaly, in children of alcoholic mothers, 442, 443, 444
Microsomes, ethanol metabolism and, 200–201, 202, 222–223, 247–248, 268, 445, 446
Middle age, alcohol abuse in, 226–227
Milk
  alcoholic beverages from
    busa, 54
    kefir, 54
    koumiss, 55
    mazun, 55
  fermentation of, 13–14, 89–90
Miniature swine, as models for study of withdrawal syndrome, 457–460, 472
Mitochondria
  acetaldehyde and, 269
  ethanol metabolism and, 200, 267, 268–269
  protein synthesis in, ethanol and, 460
Molasses, yeast propagation and, 108
Mormons, cancer in, 429
Moromi, saké production and, 53, 54
Morphine, alcohol withdrawal and, 249
Mortality, perinatal, alcoholic mothers and, 441
Moto, saké production and, 53–54
Mouse, teratogenesis by alcohol in, 445
Mozambique, beer of, 43
Mucor, bakhar and, 46
Muko, preparation of, 42
Mum, 88
Muslims, unorthodox, arguments of, 15
Myocardial biochemistry, alcohol and, 321–322
Myocardial function, alcohol and
  acute studies, 319–320
  chronic studies, 320–321
  general summary, 319
Myocardial structure, alcohol and, 322–323

N

Naloxone, alcohol withdrawal and, 249
Native Americans, fermented food beverages of, 21–22
Near East, pre-Islamic, fermented beverages of, 13–14
Nervous system, alcohol effects on cardiovascular system, 319

Neurological disorders
  of nutritional cause associated with al-
      coholism, 347
    amblyopia, 350–351
    nutritional neuropathy, 348
    Wernicke–Korsakoff syndrome, 349–350
  of undetermined cause associated with al-
      coholism, 351–352
    alcoholic dementia, 354–355
    central pontine myelinolysis, 353–354
    cerebellar degeneration, 352–353
    Marchiafava–Bignami disease, 354
Niacin
  alcoholics and, 405, 406
  coenzyme form, 111
Nicotinamide adenine dinucleotide
  ethanol metabolism and, 199, 201, 202, 222,
      247, 266–267, 269
  oxidized:reduced ratio, ethanol metabolism
      and, 201, 255, 266
Nicotinamide adenine dinucleotide phosphate,
      ethanol metabolism and, 200, 201, 222,
      266
Nicotine
  alcohol and, 250
  dependence on, 363
Nicotinic acid, *see also* Niacin
  in wine, 69, 70
Nigeria, palm wine consumption in, 38
Nipa fructicans, 38
Nitrogen compounds, in alcoholic beverages, 516
Nitroprusside test, for ketones, 264
Nitrosamines, in cider distillates, 428
Noludar, withdrawal and, 376
Norepinephrine, deamination product, alcohol
      and, 346
Nutrients
  absorption and metabolism, effect of alcohol
      on, 312–313
  synthesis by yeasts, 95–96
Nutrition
  alcoholic beverages and, 33
  bouza and, 8–9, 51
  kaffir beer and, 48–50
  palm wine and, 38–39
  pulque and, 41
  significance of indigenous alcoholic foods in,
      55–56
  tapé ketan and, 45
  tapé ketella and, 46
Nutritional data, methods for, 400–401
Nutritional neuropathy, alcoholism and, 348

Nutritional status
  of alcoholics, 408
    alcohol intake and, 401, 402
    blood alcohol concentration, 401–402
    ideal weight and, 402, 403
    nutrient intake before and after admission,
        402–405
    patients studied, 399–400
    recommended dietary allowances, 405–407
  ethanol intake and, 206–209
Nutrition Program for Older Americans, alcohol
    abuse and, 229–230
Nystagmus, blood alcohol level and, 381–385

O

Obesity, alcohol consumption and, 221, 260–
    261, 273, 274
Old age, alcohol abuse and, 227–228
Omar Khayyam, final toast, 18
Opiates, alcohol and, 251
Osiris, as inventor of beer, 5
Otomi Indians, diet of, 68
Ouabain, thiamine transport and, 288
Oxalacetate, citric acid fermentation and, 105
Oxalic acid, ethylene glycol metabolism and,
    247

P

Pachucho, maize beer and, 43
Pachwai, preparation of, 46
Packaging, of beer, 137
  shifts in, 147–148
Palm, sap, composition of, 38
Palm toddies
  composition of, 38
  production of, 38–39
Pancreas
  effects of alcohol on, 291–292
    acute, 292–293
    chronic, 293–294
  function, malnutrition and, 294–295
Pancreozymin, alcohol and, 293
Pantothenic acid, in wine, 69
Papain, chillproofing and, 139
Paraldehyde
  alcohol and, 250
  withdrawal and, 376, 385
Paranoid schizophrenia-like state, withdrawal
    and, 346

Pasteur, Louis
  contribution to studies of fermentation, 92–93
Pasteurization, brewing and, 139
Pathology, fetal alcohol syndrome and, 448
Peh-yueh, 46
Pellagra, yeast and, 96
Penicillin, ethanol and, 246
Pentylenetetrazol, alcohol overdose and, 249–250
Pepsin, 93
Peripheral neuropathy, alcoholics and, 411–412
Phenobarbital, synergism with alcohol, 248
Phenothiazines, alcohol withdrawal and, 250
Phentolamine, alcohol and, 249
Phenylbutazone, ethanol metabolism and, 247
Philippines, sugar cane wine of, 41
Phospholipid, synthesis, ethanol metabolism and, 201
Phylloxera, American grape vines and, 188
Phoenix sylvestris, 38
Phosphate, fermentation and, 94
Physicians, discourses on wine, 16, 18
Picrotoxin, alcohol overdose and, 249–250
Plasmids, recombinant DNA and, 101–102
Plato, precepts of beverage use, 62
Pleasure, from alcohol, alcoholism and, 365
Porter, definition of, 138
Pressing, wine making and, 10
Problems, alcoholism and, 366
Processing, of foods, dietary deficiencies and, 36
Prohibition
  as government policy, 494–495
  wine industry and, 189
Propoxyphene, alcohol and, 250
Propranolol, alcohol withdrawal and, 251
Protein
  deficiency, cobalt-beer drinkers' disease and, 325
  ethanol consumption and, 460
  intake, in alcoholics, 398–399, 404, 405, 406
  loss, gastritis and, 282–283
  synthesis, alcohol intake and, 202
  synthesis by pancreas, alcohol and, 294
  in *tapé ketan,* 45
  yeast, 96, 107
    nutritional quality, 114–115
Pseudomonas, saké fermentation and, 53
Pseudomonas lindneri, pulque and, 40
Psychopathology, alcoholism and, 361–364
Public health approach, alcohol abuse and, 229–230

Pulque
  composition of, 39, 40
  Mexican, production of, 39–41
  nutrient content of, 69
  preparation of, 68
Punitive approach, to alcohol
  intoxication with criminal offense, 497
  intoxication while driving an automobile, 497–498
  public intoxication, 497
  violation of laws controlling, 498
Purchasing habits, beer and, 148
Puromycin C, alcohol and, 247
Pyridoxal 5-phosphate, alcoholism and, 312–313
Pyridoxine
  deficiency, ethanol and, 459
  in wine, 69
Pyruvate
  ethanol metabolism and, 201, 266
  ethanol formation from, 203–204
  fermentation and, 123
Pyruvate carboxylase, ethanol metabolism and, 201
Pyruvate decarboxylase, glycerol fermentation and, 104
Pyruvate dehydrogenase, ethanol formation and, 203–204

Q

Quinacrine, alcohol and, 249

R

Ragi
  cassava fermentation and, 46
  rice wine and, 44
Raphia hookeri, 38
Raphia vinifera, 38
Reciprocity, Andean economy and, 25
Recreation, life span and, 237
Redistribution, Andean economy and, 26
Red port, controlled consumption by institutionalized elderly, 231
Renin, ethanol consumption and, 206
Repeal, recovery of wineries following, 190
Reserpine, alcohol and, 250
Respiratory tract, cancer, alcohol and, 427, 429, 431–434, 437, 484
Restriction, alcohol and, 495
Rhamnus tsaddo, mead and, 37

Rhizopus, bakhar and, 46
Rhizopus chinensis, lao-chao and, 46
Rhizopus oryzae, lao-chao and, 46
Riboflavin
  biosynthesis by yeast, 111–114
  coenzyme form, 111
  in wine, 69, 70
Rice
  beer brewing and, 134
  fermented beverages from 43–46
  protein, amino acids of, 114
  saké and, 52–54
Romans, use of beer by, 145
Rum fits, withdrawal and, 344–345
Russia
  koumiss of, 55
  kvass of, 52

S

Saccharomyces, 92
  early Egyptian beer and, 7
  sugar cane wines and, 41
Saccharomyces carbajali, pulque and, 40
Saccharomyces carlsbergensis, 138
  as food, 108
Saccharomyces cerevisiae, 138
  as food, 108
  gross composition, 110
  kaffir beer and, 47
  palm toddies and, 38
  tesgüino and, 52
Saccharomyces fragilis, as food, 108
Saccharomyces sake, saké fermentation and, 53
Saké
  preparation of, 53–54
  primitive, 43
Saliva
  alcoholic foods and, 41–43
  maize beer making and, 22, 42
Sap, palm, collection of, 38
Saxons, use of beer by, 145
Schizosaccharomyces pombe, palm wine and, 38
Scurvy, spruce beer and, 88–89
Secretin, alcohol and, 292, 293
Sedation, beer or wine and, 75
Selenium, in beer, 71
Serotonin
  deamination product, alcohol and, 346
  disulfiram and, 255
Serra, Junipero (father), wine and, 188

Seventh Day Adventists, cancer in, 429
Sex, wine consumption and, 181–182
Shaosing chu, 53
Sharab, 14
Shedeh, 12
Sherry, production of, 129
Skeletal muscle, myositis, alcohol and, 328
Skr, 13
Sleep, life span and, 237
Small intestine, alcoholism and
  absorption, 286–291
  enzymes, 285–286
  metabolism, 284–285
  morphology, 284
Smoking
  coronary events and, 332, 333
  effects on fetus, 448
Social change, alcoholism and, 366
Social competence, in elderly patients consuming beer, 232–233
Social Security recipients, alcoholism or psychiatric impairment in, 363
Society, control of alcohol use by, 228–229
Sociopolitical affairs, fermented beverages and, 21–22
Sodium
  absorption, alcoholism and, 287
  in fermented beverages, 75
Sokujo-moto, saké production and, 54
Sophocles, diet of, 145
Sorghum caffrorum, beer from, 47
Spanish, views on maize beer, 26
Spruce beer, as antiscorbutic, 88–89
Starvation, alcoholic ketogenesis and, 266
State, specific, beer consumption in, 151
State monopoly, alcohol and, 495–496
Steatorrhea, alcoholism and, 295
Stomach
  effect of alcohol on, 459
  emptying time, rate of alcohol absorption and, 198
Stout, definition of, 138
Stress
  alcoholism and, 359, 492, 493–494
  hypertension and, 331
Sucrase, beer drinking and, 285
Sugar
  in alcoholic beverages, 517
  alcoholic foods from
    mead, 37
    Mexican *pulque,* 39–41

miscellaneous, 41
palm toddies, 38–39
concentration, alcohol yield and, 122
Sugar cane
alcohol production from, 33
wines from, 41
Suicide, alcoholism and, 483–484
Sulfonylurea, alcohol and, 260
Sulfur compounds, in alcoholic beverages, 516
Sulfur dioxide, fermentation and, 123, 126
Sultan, cupbearers, crest of, 16
Supply, alcohol consumption and, 491, 492
Sweating, blood alcohol level and, 381–385
Syria
use of beer in, 145
wines of, 14

T

Talla, preparation of, 51
Tane-koji, preparation of, 52
Tannins, esophageal cancer and, 437
Tapé ketan
biochemical changes during fermentation, 44–45
making of, 43–44
Tapé ketella, preparation of, 46
Taxation, alcohol and, 496
Tej, production of, 37
Temperature, alcohol yield and, 122
Tesgüino, preparation of, 51–52
Tetrahydropapaveroline, alcohol-seeking behavior and, 249
Thiamine
absorption, alcoholism and, 288–289
alcoholics and, 405, 406
coenzyme form, 111
conversion to, 420
deficiency
biochemical studies, 412–418
cardiovascular beriberi and, 326
clinical diagnosis, 411–412
etiology in alcoholism, 419–421
Wernicke–Korsakoff syndrome and, 349, 250
excretion, deficiency and, 416, 418
intestinal transport, alcohol and, 286
status of Australian population, 421–422
in *tapé ketan,* 45
in wine, 69, 70
Thiokinase, ethanol metabolism and, 200

Thymidine kinase, alcoholism and, 284
Tobacco, *see also* Smoking
addiction to, 363
cancer and, 431–434
life span and, 237
Token Economy Program, for wine, 233
Tolbutamide, 249, 260
aldehyde dehydrogenase and, 255
metabolism by alcoholics, 248
Torula, protein and, 96
Torula curanga, 89
Torulopsis holmii, kefir and, 54
Trace elements
deficiency, fetal abnormalities and, 447
in fermented beverages, 71
Trade, in wines, 166–170
Transketolase
abnormality of, 426
liver disease and, 419–421
thiamine deficiency and, 412–416
Wernicke–Korsakoff syndrome and, 349
Traumatic experience, alcoholism and, 366
Treading, wine making and, 10
Tremor
blood alcohol level and, 381–385, 389
withdrawal and, 344
Tricarboxylic acid cycle
ethanol metabolism and, 201, 222, 267, 268, 269
fermentation and, 125
Triglyceride, synthesis, ethanol metabolism and, 201, 207, 266, 285, 322
Triticum vulgaris, bouza and, 50
Trub, removal
brewing and, 136–137
Tryptophan synthetase, recombinant DNA and, 101–103
Tubulin, ethanol and, 207
Turkey (country), *busa* of, 54
Tyramine, in wine, 246

U

Undernourishment, alcohol and, 261–262
United States
Congress, Report on Alcohol and Health, 66
malt beverage output by region, 146
types of wine consumed in, 180–181
use of beer in, 144, 145–146
Upi, maize beer and, 42
Urine, ethanol excretion into, 202

V

Vascular resistance, alcohol and, 319
Venner, Tho.
  as author of "Via Recta," 63, 87
  on use of alcohol, 64
Vikings, use of beer by, 145
Virginia (state), lost colony of, beer and, 145
Vitamin, *see also* specific vitamins
  content of kaffircorn and maize, 49
  conversion to coenzyme form, alcohol and,
    312
  deficiencies
    alcoholism and, 207
      fetal alcohol syndrome and, 446
    ethanol catabolism and, 458
    kaffir beer and, 49–50
    metabolites, excretion by alcoholics, 208, 209
    nutritional neuropathy and, 348
    palm wine and, 38–39
    in *pulque,* 40
    yeasts as source of, 110–111
    yeast growth and, 95
Vitamin A
  alcoholics and, 405, 406
  in wine, 69, 70
Vitamin B$_{12}$
  absorption, alcoholism and, 290
  amblyopia and, 351
  palm wine and, 39
Vitamin C, *see also* Ascorbic acid
  serum levels, *Pulque* intake and, 68–69
Vitis labrusca, use for winemaking, 188
Vitis vinifera, grafting to native rootstock, 190

W

Wadd, William, on corpulency and drink, 65
Warfarin, metabolism by alcoholics, 248
Washington, George, beer and, 146
Water
  absorption, alcoholism and, 287
  ethanol metabolism and, 200
  unwholesomeness of, 3–4
Wernicke–Korsakoff syndrome
  abnormality of transketolase and, 426
  alcoholism and, 349–350
  incidence in Australia, 412
Wernicke's encephalopathy
  alcoholics and, 411
  transketolase and, 412–416

Wheat
  protein, amino acids of, 114
  talla and, 51
Whiskey
  compounds identified in, 505–517
  laryngeal cancer and, 437
  nutrient content of, 70
Whiskey drinking, onset of withdrawal symp-
    toms and, 377–378, 387
Wine
  American consumers of, 191–193
  antiquity of, 4
  as blood, 6
  cirrhosis and, 488
  compounds identified in, 505–517
  consumption
    consumer and health protection, 182
    factors influencing, 181–182
    occasions for, 178–181
    per capita, 170, 174–178
    quantities, 170–173
  definition of, 158–161
  in early Egypt, 9–12
  Egyptian, difference from Greek, 11
  export and import
    by continents, 167
    by country, 168–169, 170
  gouty attacks and, 201
  health value of, 87
  institutionalized elderly and, 233
  interest in, 183
  malo-lactic fermentation and, 130–131
  place, compared to other beverages, 161–163
  of pre-Islamic Near East, 14
  present status in United States, 191
  processing, by-products of, 129–130
  production
    by continent, 163
    by members of EEC, 163
    world, by year, 164
  production and consumption, forecasts, 183
  production and distribution
    countries of wine cultivation and harvest,
      163–164
    trade (or commerce), 166–170
    types of wine and quality, 164–166
  public tastes and, 184–185
  quality, designation of, 165
  red, making of, 10
  rosé, making of, 10
  from sugar cane, 41

table, nutrient contents of, 70
virtues of, 3
Wine drinking, onset of withdrawal symptoms
    and, 377–378, 387
Wine Institute, California wineries and, 189
Wineries, during Prohibition, 189
Withdrawal, symptoms in miniature swine, 470,
    471
Withdrawal studies
  discussion
    alcohol dosage and alcohol withdrawal,
      387–389
    drinking patterns and alcohol withdrawal,
      389–390
    temporal development of alcohol depen-
      dence, 386–387
  procedures
    alcohol administration procedures, 374–
      376
    research ward facilities, 374
    sequence of, 374
    subjects, 373–374
  research needs, 390–391
  results, 376–378
    volume of alcohol and withdrawal onset,
      378–386
Women
  alcoholics, 440
  as wine consumers, 192
World Health Organization, diagnosis of
    thiamine deficiency and, 411
Wort, cooling and trub removal, 136–137

X

D-Xylose, absorption, alcoholism and, 287–288

Y

Yeast, *see also* specific yeasts
  beer flavor and, 138
  ethanol tolerance of, 209
  gross composition, 110
  as living organism
    identification of growth of, 91–92
    Pasteur's contribution, 92–93
  lysine biosynthesis by, 116–118
  mutant, pyruvate decarboxylase and, 104
  nutritional aspects
    nutrient synthesis, 95–96, 110–118
    nutritive value, 69, 96
    requirements for growth, 94–95
  pitching and primary fermentation brewing
    and, 137
  protein of, 114–115
  single cell production, 109–110
Yeast-juice, action, chemistry of, 94
Yeast system, brewing and, 137
Yogurt, 89
Yuca, beer from, 43

Z

Zinc
  deficiency, fetal abnormalities and, 447
  excretion, alcohol and, 71
Zymase, discovery of, 93–94